# NUCLEAR POWER
# TECHNOLOGY

# NUCLEAR POWER TECHNOLOGY

*Available separately*

# NUCLEAR POWER TECHNOLOGY

EDITED BY

W. MARSHALL

Volume 1
Reactor Technology

CLARENDON PRESS · OXFORD
1983

*Oxford University Press, Walton Street, Oxford OX2 6DP*

*London  Glasgow  New York  Toronto*
*Delhi  Bombay  Calcutta  Madras  Karachi*
*Kuala Lumpur  Singapore  Hong Kong  Tokyo*
*Nairobi  Dar es Salaam  Cape Town*
*Melbourne  Auckland*

*and associated companies in*
*Beirut  Berlin  Ibadan  Mexico City  Nicosia*

*Oxford is a trade mark of Oxford University Press*

*Published in the United States*
*by Oxford University Press, New York*

*British Library Cataloguing in Publication Data*

*Nuclear power technology*
*Vol. 1: Reactor technology*
*1. Atomic power*
*I. Marshall, W.*
*621.48    TK9145*
*ISBN 0–19–851948–6*

*Library of Congress Cataloging in Publication Data*

*Nuclear power technology.*
*Contents: v. 1. Reactor technology – v. 2. Fuel cycle –*
*v. 3. Nuclear radiation.    1. Atomic power.*
*I. Marshall, W. (Walter), 1932–*
*TK9145.N83 1983    621.48    83–4124*
*ISBN 0–19–851948–6 (v. 1)*

*Printed in Great Britain*
*by The Thetford Press,*
*Thetford, Norfolk*

# PREFACE

Some time ago the Oxford University Press asked if I would be interested in preparing a book on nuclear power oriented towards increasing public knowledge and for the public interest. The United Kingdom Atomic Energy Authority was approaching its twenty-fifth anniversary and it therefore seemed appropriate to gather a set of chapters reviewing the state of nuclear science. The obvious way of doing this was to list all the subjects involved and then persuade an expert in each subject to write a chapter. Having chosen authors of wide experience in their fields, the result is an authoritative source book on the major aspects of nuclear science and nuclear power.

To achieve some cohesion, I set up an editorial board which has reviewed all the chapters to make sure they are factually correct, and we believe that in all the papers the choice of language makes it clear what is fact and what is opinion.

The chapters do not represent an official UKAEA point of view. Each one is very much the personal opinion of the individual author, and I have not attempted to influence his perspective of the subject. The chapters are, therefore, presented in a variety of styles, but it has been our intention throughout that they should be free of unnecessary technical jargon and comprehensible to the non-specialist. While in a publication of this standard it has been necessary to treat the technical and scientific basis for some of the subjects in depth. I hope that the book will prove helpful and informative to the general reader who, conscious of the 'Nuclear Debate' taking place around him, wishes to have access to a reliable reference document.

Many of the chapters contain bibliographies so that the reader may delve further into the technical facts and figures if he so chooses. There is also a summary of each chapter and, at the end, a comprehensive glossary and index to complete the texts.

Most of the authors come from within the UKAEA. My special thanks are due to Dr S. H. U. Bowie, who was Chief Consultant Geologist to the UKAEA from 1955 to 1977, for completing the series by contributing an article.

I especially thank Dr R. H. Flowers for the part which he has played, as Deputy Editor, in bringing this work to completion and Mrs N. M. Hutchins, as secretary to the Editorial Board, for organizing and progressing the contributions.

*May 1981*                                    W. MARSHALL
                                              *Chairman,*
                              *United Kingdom Atomic Energy Authority*

# ACKNOWLEDGEMENTS

Acknowledgements are due to the Editorial Board set up by the Chairman of the UKAEA to review the articles during their various stages of preparation. The members of this Editorial Board were:

Dr W. Marshall, Chairman (UKAEA) and Chairman of the Editorial Board
Dr R. H. Flowers, Deputy Editor

| | |
|---|---|
| Mr J. G. Collier | Dr R. G. Sowden |
| Mr J. M. Hutcheon | Mr J. G. Tyror |
| Dr P. M. S. Jones | Mr P. N. Vey |
| Dr A. B. Lidiard | Dr B. O. Wade |

Acknowledgements are due also to Mr G. Gibbons who was responsible for the artwork, to Mr J. A. G. Heller, who commented extensively on the articles from the lay-reader's point of view, to Miss P. E. Barnes, who undertook the vast typing task and to Mrs N. M. Hutchins who, as secretary to the Editorial Board, organized the authors contributions and did a large amount of editorial work.

Special thanks are due to Mr S. U. Bowie, now retired from the Institute of Geological Sciences, who prepared the article on Uranium and Thorium Raw Materials.

R. H. FLOWERS
*Deputy Editor*

# CONTENTS

# CONTRIBUTORS

J. R. ASKEW  *Reactor Development Division, Winfrith*

C. G. CAMPBELL  *Reactor Development Division, Winfrith*

J. G. COLLIER  *Director of Technical Studies and Head of Atomic Energy Technical Branch, Harwell*

P. A. DAVENPORT  *Applied Physics and Technology Division, Culham Laboratory*

K. H. DENT  *Thermal Reactors Development Division, Risley*

R. H. FLOWERS  *Chemical Technology Division, Harwell*

A. M. JUDD  *PFR/Technical Operations, Dounreay*

J. SMITH  *Technical Assessments and Studies Division, Winfrith*

R. J. SYMES  *Winfrith Education Centre, Winfrith*

J. G. TYROR  *Reactor Development Division, Winfrith*

# SYMBOLS AND ABBREVIATIONS

| Abbreviation or Symbol | Name of unit and quantity measured | Notes, Units, Value |
|---|---|---|
| A | ampere, electric current | |
| $\alpha$ | alpha particle ($He^{++}$) | |
| amu | atomic mass unit | $1.66 \times 10^{-27}$ kg |
| bar | bar, pressure | $10^5$ Pa |
| barn | unit of cross-section | $10^{-28}$ m$^2$ |
| $\beta$ | beta particle (electron) | |
| Bq | becquerel, radioactivity | s$^{-1}$ |
| $c$ | velocity of light | $3 \times 10^8$ m s$^{-1}$ |
| C | coulomb, electric charge | A s |
| °C | celsius temperature interval | degree celsius |
| Ci | curie, radioactivity | $3.7 \times 10^{10}$ Bq |
| d | day, time | |
| EFPH | effective full power hours | |
| eV | electronvolt, energy | $1.59 \times 10^{-19}$ J |
| g | gram, mass | |
| $\gamma$ | gamma radiation (photon) | |
| Gy | gray, absorbed dose | J kg$^{-1}$ |
| h | hour, time | |
| $h$ | Planck's constant | $6.6255\ 10^{-34}$ J s |
| Hz | hertz, frequency | s$^{-1}$ |
| J | joule, energy | |
| $k$ | Boltzmann constant | $1.3805 \times 10^{-23}$ J K$^{-1}$ |
| K | kelvin, absolute temperature interval | kelvin |
| m | metre, length | |
| $M$ | concentration | $10^3$ mol m$^{-3}$ |
| min | minute, time | |
| mol | mole, amount of substance | gram-molecule |
| MW(e) | unit of power station output | megawatt, electrical |
| MW(t) | unit of power station output | megawatt, thermal |
| n | neutron | |
| N | newton, force | kg m s$^{-2}$ |
| p | proton | |
| Pa | pascal, pressure | 1 newton m$^{-2}$ |
| pH | unit of acidity | $-\log_{10} M_{H^+}$ |
| rad | absorbed dose | $10^{-2}$ Gy |
| rem | absorbed dose equivalent | $10^{-2}$ Sv |
| s | second, time | |
| s.t.p. | standard temperature and pressure | 1 bar 273.16 K |
| SWU | separative work unit | see Chapter 2 |
| Sv | sievert, absorbed dose equivalent | J kg$^{-1}$ |
| t | metric tonne | 1000 kg |
| T | tesla, magnetic flux density | V s m$^{-2}$ |

| tce | tonne coal equivalent | |
| tha | tonne heavy atoms | |
| V | volt, potential difference | |
| W | watt, power | $J\,s^{-1}$ |
| y | year, time | |

## Prefixes

| Multiple | Prefix | Abbreviation |
|---|---|---|
| $10^{-18}$ | atto | a |
| $10^{-15}$ | femto | f |
| $10^{-12}$ | pico | p |
| $10^{-9}$ | nano | n |
| $10^{-6}$ | micro | $\mu$ |
| $10^{-3}$ | milli | m |
| $10^{-2}$ | centi | c |
| $10^{3}$ | kilo | k |
| $10^{6}$ | mega | M |
| $10^{9}$ | giga | G |
| $10^{12}$ | tera | T |

Additional less commonly used symbols and units are explained and defined as they occur in the text.

## The Elements

A complete list of the elements and their symbols, is included here for reference.

| Element | Symbol | Element | Symbol | Element | Symbol |
|---|---|---|---|---|---|
| Actinium | Ac | Dysprosium | Dy | Lutetium | Lu |
| Aluminium | Al | Einsteinium | Es | Magnesium | Mg |
| Americium | Am | Erbium | Er | Manganese | Mn |
| Antimony | Sb | Europium | Eu | Mendelevium | Md |
| Argon | Ar | Fermium | Fm | Mercury | Hg |
| Arsenic | As | Fluorine | F | Molybdenum | Mo |
| Astatine | At | Francium | Fr | Neodymium | Nd |
| Barium | Ba | Gadolinium | Gd | Neon | Ne |
| Berkelium | Bk | Gallium | Ga | Neptunium | Np |
| Beryllium | Be | Germanium | Ge | Nickel | Ni |
| Bismuth | Bi | Gold | Au | Niobium | Nb |
| Boron | B | Hafnium | Hf | Nitrogen | N |
| Bromine | Br | Helium | He | Nobelium | No |
| Cadmium | Cd | Holmium | Ho | Osmium | Os |
| Caesium | Cs | Hydrogen | H | Oxygen | O |
| Calcium | Ca | Indium | In | Palladium | Pd |
| Californium | Cf | Iodine | I | Phosphorus | P |
| Carbon | C | Iridium | Ir | Platinum | Pt |
| Cerium | Ce | Iron | Fe | Plutonium | Pu |
| Chlorine | Cl | Krypton | Kr | Polonium | Po |
| Chromium | Cr | Lanthanum | La | Potassium | K |
| Cobalt | Co | Lawrencium | Lr | Praseodymium | Pr |
| Copper | Cu | Lead | Pb | Promethium | Pm |
| Curium | Cm | Lithium | Li | Protactinium | Pa |

| Element | Symbol | Element | Symbol | Element | Symbol |
|---------|--------|---------|--------|---------|--------|
| Radium | Ra | Sodium | Na | Titanium | Ti |
| Radon | Rn | Strontium | Sr | Tungsten | W |
| Rhenium | Re | Sulphur | S | Uranium | U |
| Rhodium | Rh | Tantalum | Ta | Vanadium | V |
| Rubidium | Rb | Technetium | Tc | Xenon | Xe |
| Ruthenium | Ru | Tellurium | Te | Ytterbium | Yb |
| Samarium | Sm | Terbium | Tb | Yttrium | Y |
| Scandium | Sc | Thallium | Tl | Zinc | Zn |
| Selenium | Se | Thorium | Th | Zirconium | Zr |
| Silicon | Si | Thulium | Tm | | |
| Silver | Ag | Tin | Sn | | |

# 1

## How reactors work

R. H. FLOWERS

This introductory chapter sets out in simple form the basic facts upon which the working of a nuclear reactor depends. The reader is taken through the logical steps which lead to the choice of fuels, moderators, and coolants for reactors that have reached the stage of commercial operation. Differences between reactor types are explained and, in particular, the way in which fast-neutron reactors utilize the abundant fuel $^{238}$U is described.

## Contents

# 1  Introduction

There is a widespread public misunderstanding of the way in which nuclear reactors generate power, despite the fact that in 1980 around 12 per cent of all electricity in the UK came from nuclear power stations. Perhaps the fact that nuclear energy has its origins in the physics laboratory gave it a reputation for being incomprehensible, or perhaps the public shock at the power of atomic bombs in 1945 was transferred in some way to the nuclear reactors which followed. Whatever the reasons, a wider appreciation of how reactors work would allow more informed discussion of the nuclear installations around the country which are serving us today.

It is helpful first to appreciate that nuclear reactions and nuclear radiation are no newcomers to this earth. When the solar system was formed over four thousand million years ago, the earth was a very radioactive place. Naturally occurring radioactive substances have decayed away substantially since then, but even today there are immense quantities of uranium, thorium, radium, and other radioactive elements in the earth's crust. Life has evolved in the presence of the significant radiation levels generated by these substances and of the higher radiation levels of early history. Heat from nuclear reactions in the earth is responsible for continental drift, earthquakes, volcanoes, hot springs, and the so-called geothermal energy reserves. Nuclear reactions in the sun and stars give us heat and light at the earth's surface, accompanied also by significant levels of invisible nuclear radiation which we call cosmic rays.

In understanding how nuclear reactors work to produce electricity, there is no need to study the basic physics of the forces between fundamental particles. The essential points are that nuclear reactions can be made to produce heat, just as the chemical reactions of burning oil or coal produce heat, and that this heat can be converted to electricity in power stations.

Figure 1.1 illustrates the essential features of a nuclear power station. The reactor core, which comprises fuel elements and possibly a *moderator* material, becomes hot when the nuclear reaction is started up. A coolant is pumped through the core to take the heat away to a boiler or *heat exchanger* in which steam is raised. The steam, which is produced at high temperature and pressure, then flows through a turbine, which is coupled to an electrical alternator. Electricity is generated and fed into the distribution grid.

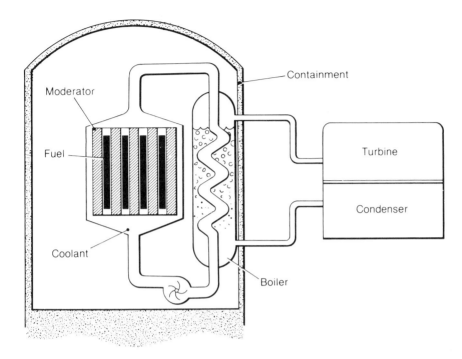

Fig. 1.1 Schematic diagram of a nuclear reactor

The waste products of the nuclear reaction are very toxic, on account of their radioactivity, for some time after their formation. In most reactors, the fuel itself is therefore contained within a close-fitting metal *can* which retains these fission products: the assembly of fuel and can forms the *fuel element*. It follows that there is no exhaust gas equivalent to the flue gases discharged from chimneys at coal- and oil-fired stations. As a precaution against any leakage from faulty fuel elements reaching the environment, the nuclear reactor is placed within a high-quality building, the *containment*, which encloses the entire reactor core. In modern plant the boilers too are inside this containment.

There is another important difference between the fossil-fired and the nuclear furnaces. The coal or oil burner is surrounded by heat-resistant material to confine the hot gases and to protect the operators from the radiation of light and heat from the combustion process. The core of the nuclear furnace must be surrounded by several metres of a dense solid, such as concrete, to protect operators from more penetrating types of radiation called *gamma radiation* and *neutrons*, which are released during the nuclear reactions.

## 2  Nuclear fission

### 2.1  **The source of nuclear energy**

The present inquiry into the source of nuclear energy can best be started by picturing the structure of atoms, those exceedingly small building bricks from which all substances are made. The atom has a nucleus, which carries a positive electrical charge characteristic of the chemical element in question, and a set of orbiting electrons, which carry a compensating negative electrical charge. The nucleus comprises particles called *nucleons*. These are of two kinds: *protons*, which carry the positive electrical charge, and *neutrons*, which are of similar mass but electrically neutral. Figure 1.2 is a diagram illustrating these basic features of the atom.

The release of the energy which is stored in fuels of all kinds involves the conversion of part of the mass of their atoms into usable energy. The equivalence of mass and energy was expressed by Einstein as:

$$E = mc^2,$$

where $c$ is the velocity of light. As this velocity is extremely large ($3 \times 10^8$ m/s), very small mass changes are equivalent to large amounts of energy. For example, a mass loss of one thousandth of a gram would provide about 3 kW of heat continuously for a year.

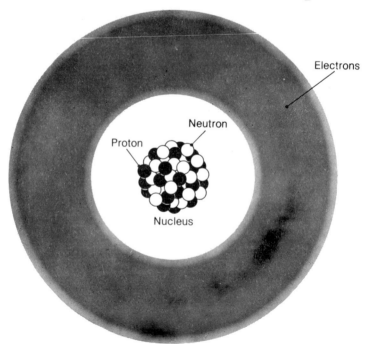

Fig. 1.2  Basic structure of the atom

The burning of wood, coal, oil, etc. is essentially a process in which carbon and hydrogen atoms in the fuel combine with oxygen to form molecules of carbon dioxide and water. The consequent rearrangement of electrons in the molecule produces changes in the forces binding the atoms together and is accompanied by a tiny reduction in the net mass. The energy so released is called chemical energy. Chemical fuels of this kind have been the predominant sources of energy for industrial and domestic use in the nineteenth and twentieth centuries; but the energy which supplies heat to the sun and to the interior of the earth is of a different kind, nuclear energy. Just as chemical fuels release energy by a rearrangement of electrons, so nuclear fuels release energy by a rearrangement of nucleons, this in both cases resulting in a loss of mass.

Let us look further into the structure of the nucleus of the atom in order to understand better how it tends to take up certain preferred configurations, releasing energy—and losing mass—in the process. Nucleons are held together in the nucleus by very strong *nuclear forces*, which bind together the mixture of protons and neutrons despite the electrical repulsion that exists between the positively charged protons. The stability of the nucleus depends upon the precise ratio of neutrons to protons, and also upon the overall number of nucleons.

To take the first point, the number of protons in the nucleus determines the chemical species to which the atom belongs and is referred to as the *atomic number* of the element. One proton in the nucleus makes a hydrogen atom, two a helium atom, and so on up to the heaviest element commonly occurring in nature, uranium with 92 protons in its nucleus. For each chemical element it is possible to vary the number of neutrons in the nucleus over a small range, producing *isotopes* of differing mass but identical chemical properties. Thus nuclei of the element gold, for example, are known with a neutron content between 98 and 125 giving *mass numbers* (after adding the 79 protons) between 177 and 204. The most stable mixture of nucleons for gold is 79 protons and 118 neutrons, giving the naturally occurring isotope of mass 197 atomic mass units (written as $^{197}$Au); if there are more or fewer neutrons than this, the nucleus will be less stable than $^{197}$Au and will spontaneously change its proton to neutron ratio. We shall see that these spontaneous changes are actually the phenomenon known as *radioactivity*.

A nucleus can *increase* its proton to neutron ratio by converting a neutron into a proton plus an electron. This results in the formation of an atom of a new element (i.e. one having different chemical properties) and also the ejection of the negatively charged beta particle (an *electron* or *negatron*). The process is known as *beta* ($\beta$) *radioactivity*. Alternatively, the nucleus may emit a neutron directly, resulting this time in an atom of the same chemical species having a nucleus lighter by one mass unit. As will be seen later, this process is important in controlling the power in a nuclear reactor.

A nucleus can *decrease* its proton to neutron ratio by the following processes.

1. It can emit an *alpha particle* ($\alpha$), which is a stable group of two protons plus two neutrons (4 atomic mass units). This results in the formation of an atom of the element with atomic number less by two and mass less by four units. The process is known as $\alpha$ -radioactivity and takes place amongst the heavier elements, where an additional requirement for stability is the reduction of the total nuclear mass.
2. It can convert a proton into a neutron. This is another form of beta radioactivity, in which a positively charged beta particle (a *positron*) is emitted, leaving an atom with atomic number less by one but having approximately the same mass as before.

All of these *radioactive decay* processes are accompanied by a small loss of mass and therefore a release of energy. The energy, usually in the range $10^{-14}$–$10^{-12}$ joules per nucleus, appears in the form of kinetic energy of the ejected particles and sometimes partly as *gamma* rays ($\gamma$), which are very short wavelength electromagnetic radiations.

The second factor affecting stability of a nucleus is its overall size. Assuming that the neutron/proton ratio has been adjusted to give maximum stability for every chemical element, there are still differences in stability between the elements which depend on mass number.

In order to describe this size effect a quantity called *binding energy* must be defined. If a nucleus were separated into its component nucleons, it would be found that the total mass of the separated nucleons was slightly greater than the mass of the assembled nucleus. This mass difference is the consequence of the energy release which occurs as the strong nuclear forces pull the individual protons and neutrons together into a stable nucleus. It happens that nuclei of the atoms with medium atomic number are more strongly bound together by nuclear forces than those of light and heavy atoms. This is illustrated in Fig. 1.3, where the relatively greater mass loss per nucleon of these medium-weight atoms is shown. The curve illustrates the two ways in which binding energy might be released for our use. If two light nuclei are joined together to make a heavier nucleus, it can be seen from the steeply falling left-hand side of the curve that there is a net loss of mass in the process. This loss is balanced by a release of energy, via the Einstein relation. (The mass differences per nucleon in Fig. 1.3 have been expressed in energy units to emphasize this equivalence.) This is the basis of nuclear fusion, which occurs in the interior of the sun where hydrogen under extremely high pressure and temperature conditions fuses to form the heavier element helium. Alternatively, as the right-hand portion of Fig. 1.3 illustrates, energy is also released if a nucleus of high mass number can be made to divide or *fission* into two parts. This is the source of the energy which is produced in all present nuclear reactors, the heavy nuclei being those of uranium or plutonium and the products of fission being isotopes of elements, such as barium and krypton, which fall in the central region of the curve.

It is one thing to know that energy will be released in the process of

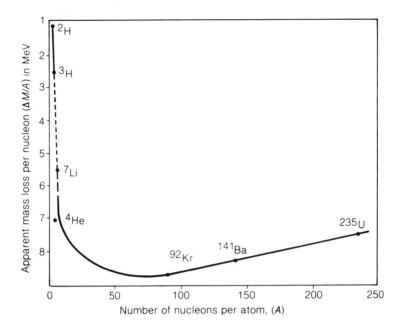

Fig. 1.3 Nuclear binding energy

radioactive decay of an unstable isotope, or fission of a heavy atomic nucleus, but quite another to cause these processes to occur as required. Radioactive decay remains to this day a process that occurs at a rate determined by the inherent instability of the nucleus in question: we cannot yet control it to provide a major power source even supposing that suitable isotopes were widely available as a fuel. Fortunately, the situation is quite different in the case of nuclear fission—and may also prove to be so for fusion.

Hahn and Strassmann[1] showed in 1939 that the nuclei of the heaviest atoms can be caused to fission by bombarding them with neutrons. All heavy nuclei can undergo fission when bombarded by neutrons that are sufficiently fast, but in 1939 Bohr and Wheeler[2] predicted that only very heavy nuclei containing an odd number of neutrons would be *fissile* to neutrons of all velocities down essentially to zero. The only naturally occurring nucleus which satisfies this condition is uranium-235, containing 92 protons and 143 neutrons. An example of the fission of $^{235}$U is shown in Fig. 1.4.

The uranium nucleus absorbs a neutron to form momentarily the new uranium isotope $^{236}$U, which then splits into isotopes of barium and krypton releasing three neutrons in the process. As will be seen later, there are many other pairs of isotopes which this fission process can generate, but they all release two or three neutrons.

In accounting for the total mass on each side of the reaction in Fig. 1.4, it is found that 0.215 atomic mass units (the mass of $^{12}$C is exactly 12 atomic mass units; 1 amu $= 1.66 \times 10^{-27}$ kg) have been lost in the fission process. Using Einstein's formula it can be calculated that the energy ($E$) in joules (J) released by the fission of one $^{235}$U nucleus is

$$E = (0.215 \times 1.66 \times 10^{-27} \text{ kg}) \times (3 \times 10^8 \text{ m/s})^2 = 3.2 \times 10^{-11} \text{ J}.$$

Compare this with the energy released by burning one carbon atom in oxygen

$$C + O_2 \longrightarrow CO_2 (+ 7 \times 10^{-19} \text{ J}),$$

and it is clear that, as an energy source, one uranium atom is equivalent to abuot fifty million carbon atoms. Expressed in another way, the fission of one tonne of $^{235}$U produces as much energy as the combustion of 2.7 million tonnes of coal.

## 2.2  The actinide elements

Although $^{235}$U is the only naturally occurring atom which will fission readily when bombarded by neutrons of any velocity (therefore called a fissile isotope), there are other heavy metals and other isotopes of uranium itself which play a part in nuclear reactors.

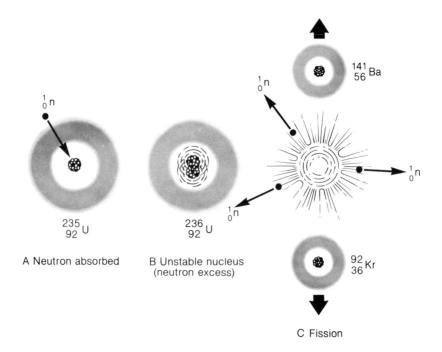

$^{141}_{56}$Ba

$^{1}_{0}$n

$^{1}_{0}$n

$^{1}_{0}$n

$^{1}_{0}$n

$^{1}_{0}$n

$^{235}_{92}$U

$^{236}_{92}$U

$^{92}_{36}$Kr

A Neutron absorbed

B Unstable nucleus (neutron excess)

C Fission

Fig. 1.4  An example of uranium fission

Chemists refer to all the metals which follow actinium in the *Periodic Table of the Elements* as actinides; these are the heaviest atoms known, and all are in the process of radioactive decay.

Uranium and thorium are naturally occurring actinide elements. They are widespread in the earth's crust at average concentrations of about two and seven parts per million respectively. The uranium is always a mixture of one part of $^{235}$U (the fissile isotope) to 138 parts of $^{238}$U—i.e. the fissile uranium is only 0.71 per cent by weight; thorium occurs almost entirely as $^{232}$Th, which is not fissile.

Although naturally occurring fissile material is thus rather rare, other fissile atoms can be made in nuclear reactors. Plutonium–239 ($^{239}$Pu), which is a heavier atom than uranium and occurs in uranium ores only in trace quantities, can be made from $^{238}$U by the process of neutron capture followed by beta decay. Likewise $^{233}$U can be made from $^{232}$Th. By these conversion processes, which will be described later, the natural resources of uranium and thorium can be turned into fuel for nuclear reactors.

## 2.3  The chain reaction

Early workers observing the fission of $^{235}$U realized that if one of the neutrons emitted during fission (Fig. 1.4) were to strike another $^{235}$U nucleus and cause it to fission, the process would repeat itself over and over again. Such a sequence of fission events is called a *chain reaction*. If more than one neutron per fission were to cause further fissions, the reaction rate would build up rapidly, while if less than one neutron per fission were available to cause further fissions, the reaction rate would slow down and stop.

Using natural uranium, which as we have seen is 99.3 per cent $^{238}$U and 0.7 per cent $^{235}$U, there are considerable problems in achieving a chain reaction of fissions. Several fates, other than fissioning a further $^{235}$U nucleus, lie in wait for the neutrons emitted during fission. They may be

(a) absorbed in $^{238}$U to form, as will be seen later, the new heavy metal isotope $^{239}$U (and subsequently $^{239}$Pu),
(b) absorbed in $^{235}$U to form $^{236}$U without subsequent fission,
(c) absorbed in materials such as structural metals and impurities, which play no further part in the nuclear reactions, or
(d) lost from the reactor core altogether.

To understand the basic factors which determine the behaviour of neutrons in a piece of uranium it is useful to introduce the concept of *nuclear cross-sections* and also to look in more detail at the number and energies of neutrons emitted during the fission process.

Cross-sections are a measure of the target area which a nucleus appears to present to an incident neutron when considering a particular reaction between the neutron and the nucleus. Put another way, the cross-section of a nucleus is a measure of the likelihood of a particular nuclear reaction occurring. Cross-sections are expressed in units of area called *barns* (1 barn = $10^{-28}$ m$^2$).

Reactions which they can relate to include fission, neutron capture, and neutron *scattering* or bouncing. Figure 1.5 shows the cross-sections[9] for the two isotopes of natural uranium ($^{235}$U and $^{238}$U) for collision with a neutron resulting in

(a) fission (denoted by $\sigma_f$), and

(b) neutron capture (denoted by $\sigma_c$) to produce a new uranium isotope.

In this figure the cross-sections are shown to vary according to the energy of the incident-neutron and some important features are apparent even at a quick glance. The unit of energy used here is the electronvolt (eV) as used by nuclear physicists (1 eV = $1.6 \times 10^{-19}$ J).

$^{235}$U shows a very high cross-section for undergoing fission with neutrons of very low velocity—in fact when the neutrons have lost all their initial velocity and are merely in thermal equilibrium with the surrounding atoms the cross-section for fission is over 500 barns. With such neutron energies the fission cross-section of $^{238}$U is zero. At higher neutron energies both isotopes of uranium have appreciable fission cross-sections, although in the case of $^{238}$U, the energy must be above 1 million electronvolts (1 MeV).

The picture is quite different for neutron capture. Both $^{235}$U and $^{238}$U capture neutrons over the whole energy range, but $^{238}$U in particular has some very large cross-section peaks for neutron energies between 5 and 500 eV.

Looking now at the neutrons released in the process of fissioning the uranium isotopes, two factors are very relevant to this problem of achieving a chain reaction. First, the number of neutrons released per fission affects the chances of causing an average of *one* neutron to trigger off another $^{235}$U fission. In Fig. 1.6 this number ( $\nu$ ) is shown plotted against the energy of the neutron which caused the fission.[9] The important point to notice is that, on average, about 250 neutrons are released per 100 fissions, there being a gradual increase towards the higher energies. Secondly, the energy of the neutrons released at the instant of fission determines the cross-section values and $\nu$ –values which apply. Figure 1.7 shows the type of energy spectrum displayed by *fission* neutrons.

The problem of sustaining a chain fission reaction in natural uranium can be appreciated by following the progress of neutrons released in that material. Under favourable conditions there is no shortage of fission neutrons to initiate a chain reaction, because $^{238}$U has the property of spontaneous fission at a very low rate—in fact 1 kg of $^{238}$U produces about 20 neutrons per second by this mechanism. However these neutrons, as they slow down, are captured so efficiently by $^{238}$U, in the cross-section peaks between about 5 and 500 eV, that there is no chance of the necessary one neutron per fission reaching a $^{235}$U atom and triggering of another fission. In fact, even if all impurity atoms could be eliminated from natural uranium metal, and if we had an infinitely large piece of it to eliminate neutron losses from the surfaces, a chain fission reaction would still be impossible.

This might have been the end of the prospects for extracting energy from

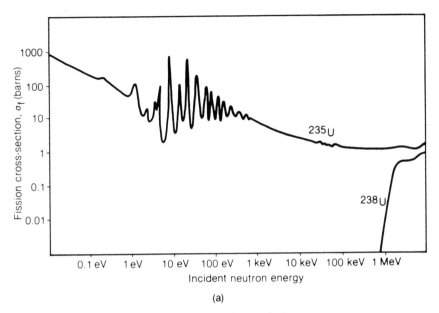

Fig. 1.5 (a) Neutron cross-sections for uranium to fission

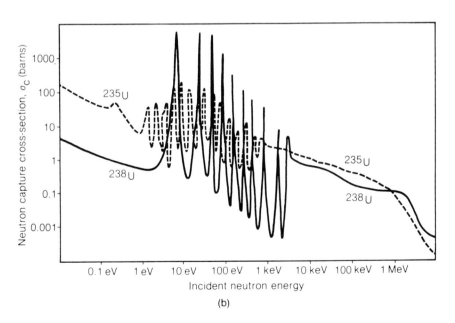

Fig. 1.5 (b) Neutron cross-sections for uranium to capture a neutron

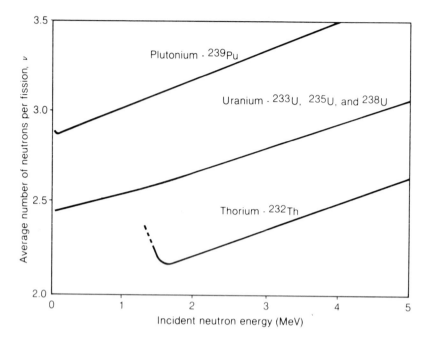

Fig. 1.6  Average number of neutrons released from fission

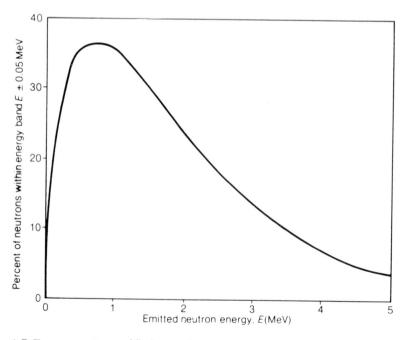

Fig. 1.7  Energy spectrum of fission neutrons

fission of heavy elements were it not for two technical 'tricks' which increase the chances of the neutrons triggering off further fissions rather than being captured, without fission, in $^{238}$U. The technical tricks, both of which are used in nuclear reactors around the world today, are the use of *moderators* and the *enrichment* of natural uranium.

## 2.4 Moderators

It can be seen from Fig. 1.5 that neutrons with energies in the range of 0.1—10 MeV, as released in the fission event, can have no chance of sustaining a chain fission reaction in natural uranium. The capture cross-sections always exceed the fission cross-sections to such an extent that of the two or three neutrons released in a fission event less than one will succeed on average in causing a fission in $^{235}$U and $^{238}$U.

The situation is quite reversed with neutrons in the energy range 0.01–0.1 eV, which are often called thermal neutrons because they have been slowed down to the point where their kinetic energy is in equilibrium with that of atoms at ordinary temperatures.

The fission cross-section of $^{235}$U is now much larger than the total capture cross-section in $^{235}$U and $^{238}$U, and a chain fission reaction is feasible.

The function of a moderator is to slow the fission neutrons down from their initial velocity to a velocity corresponding to an energy within thermal range, keeping them out of contact with uranium nuclei as it does so. It achieves this by allowing the fission neutrons effectively to bounce off the nuclei of the moderator atoms until their energy is dissipated. A good moderating material has three important qualities to fit it for this role.

1. Its atoms are small in mass so that at each collision between the neutron and the nucleus of a moderator atom there is a large transfer of energy from the former to the latter. This is analogous to the way in which a billiard ball is slowed down more through collision with another billiard ball than through collision with the floor, which is much more massive. Those readers who care to experiment will observe that a billiard ball dropped on to the floor rebounds with almost the same speed at which it struck.
2. Its nuclei have a large cross-section for scattering neutrons, so that the probability of these *elastic* collisions is high.
3. Its nuclei have a small cross-section for capturing neutrons, so that the probability of neutrons being lost in that way is small.

A good moderator, like pure carbon, reduces the energy of a fission neutron from 1 MeV to 0.03 eV within the space of about a tenth of a millisecond, the neutron undergoing about 100 collisions in the process.

Designers of the early nuclear reactors realized that by spacing rods of uranium metal throughout a heavy water or graphite moderator, neutrons would escape from the uranium metal into the moderator where they would be thermalized by collisions with a succession of nuclei (Fig. 1.8). Those

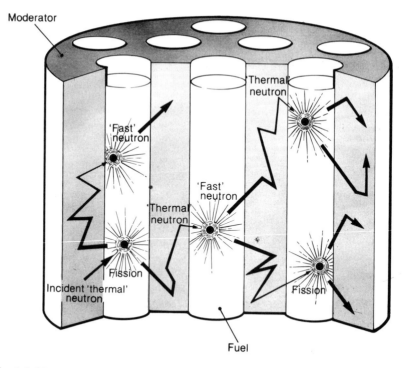

Fig. 1.8 Use of a moderator to slow down fission neutrons

neutrons scattered back into a uranium fuel element would have a very high probability of causing fission in other $^{235}$U nuclei, after which there would continue a chain reaction of $^{235}$U fission events. In 1942 a chain fission reaction was achieved in Chicago using natural uranium in the form of metal rods and oxide-packed tubes in a graphite block. Of the 250 neutrons released for every 100 atoms of $^{235}$U fissioned with thermal neutrons, 150 were being lost in captures by $^{235}$U, $^{238}$U or carbon or by escaping from the reactor, and exactly 100 were going on to cause fission in other $^{235}$U nuclei. It was necessary to assemble a graphite block (or pile) of side about 6 m in order to minimize the proportion of neutrons lost from the outside, and about 36 tonnes of natural uranium were needed.

## 2.5 Uranium enrichment

Just as neutron moderation facilitates a chain fission reaction in natural uranium by increasing the probability of $^{235}$U fission relative to neutron capture by $^{238}$U, so an artificial increase in the ratio of $^{235}$U to $^{238}$U atoms in the uranium achieves a similar result by increasing the number of fissile nuclei in a given volume of the metal. This process is called *enrichment* of the uranium.

There are many techniques by which the isotopic composition of an element may be changed from its naturally occurring state, but in the case of uranium enrichment it is done on an industrial scale either by causing uranium hexafluoride gas to diffuse through membranes or by centrifuging the gas at very high velocity (see Chapter 12).

If uranium containing, for example, 3 per cent of [235]U is available instead of naturally occurring uranium (0.7 per cent [235]U), a self-sustaining fission reaction can be achieved more easily; ordinary water will suffice as the moderator, despite the fact that it captures a significant number of neutrons in its hydrogen nuclei, and structural metals can be employed in the apparatus without so much concern about their tendency to capture neutrons. If uranium is enriched beyond about 10 per cent [235]U, then it is even possible to sustain the chain fission reaction without any moderator. Enrichment of uranium is a very expensive process but, as will be seen later, the advantages which it offers to a nuclear reactor designer have proved decisive and the majority of the world's reactors now use enriched fuel.

It is interesting to note that earlier in the earth's history the natural abundance of the [235]U isotope was higher. Figure 1.9 illustrates how the percentage of the 235 isotope must have varied during the earth's history. It follows that natural uranium reactors used to be much easier to construct than is the case now, a point dramatically illustrated by the discovery of the remains of a natural 'nuclear reactor' in the Gabon.[3] This reactor, and probably others like it, must have been moderated by ordinary water; the

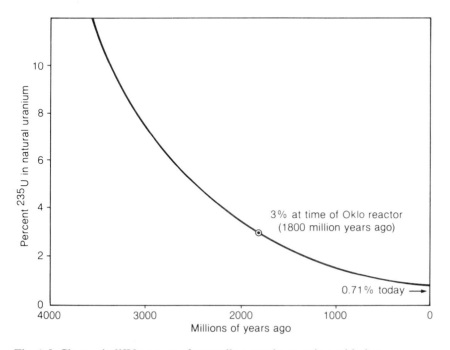

Fig. 1.9 Change in [235]U content of naturally occurring uranium with time

$^{235}$U content of naturally occurring uranium at that time is estimated to have been about 3 per cent.

## 2.6  Fast fission

If uranium is sufficiently enriched in $^{235}$U, the fission neutrons, at their energy of about 1 MeV, can sustain a chain fission reaction. This is because the $^{238}$U nuclei are no longer competing effectively for capturing neutrons. No moderator is then required. The neutrons in this situation exist for only about one microsecond before reacting with another uranium nucleus. The fission events are called *fast fissions* because the neutrons are moving faster than in a moderated system.

It has been shown earlier (Fig. 1.5) that even $^{238}$U has a sizable neutron cross-section for fission ($\sigma_f$) when the neutron is more energetic than 1 MeV. In an unmoderated system, with a fraction of the neutrons actually possessing energies well over 1 MeV, fission of $^{238}$U nuclei by these fast neutrons is a useful bonus added to the $^{235}$U fission. It can contribute around 10 per cent of the total fission events.

The feasibility of sustaining a chain reaction of fast fission in a simple lump of uranium metal, provided that the $^{235}$U isotope has been artificially increased in concentration by an enrichment process, gives rise to the concept of a *bare sphere critical mass*. For any particular $^{235}$U concentration there is a critical size of metal sphere where, after allowing for losses of neutrons from its surface, exactly one neutron per fission event will on average succeed in triggering a further fission. Figure 1.10 shows how the mass of such a sphere—the critical mass—increases as the concentration of $^{235}$U decreases. Any attempt to cast or forge a uranium sphere slightly larger than the critical mass results in a rapidly accelerating rate of fission events, and the heat produced disrupts the critical conditions by destroying the spherical shape. There can be no explosion in such an experiment, but the resultant emission of neutrons and $\gamma$ radiation is very dangerous to anyone nearby.

The critical mass is not a unique number of kilograms for a given $^{235}$U concentration. Fig. 1.10 shows the numbers relevant to a perfect sphere[10] of pure uranium metal at room temperature, but higher or lower results can be obtained by for example:

(a)  using other geometrical shapes,
(b)  using a uranium compound such as uranium oxide,
(c)  placing neutron reflectors around the uranium,
(d)  leaving neutron-absorbing impurities in the uranium, or
(e)  changing the density of the uranium by metallurgical treatments or by altering its temperature or pressure.

## 2.7  Fission products

About 166 MeV of the energy released when a nucleus fissions is in the form

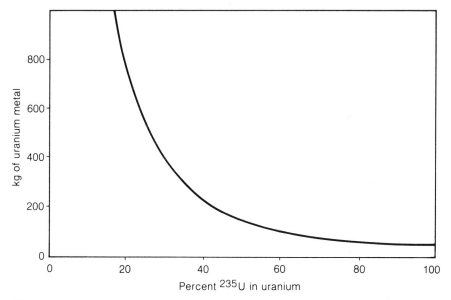

Fig. 1.10 Critical mass of a bare sphere of uranium

of the velocity of the two fragments, or *fission products*. The remainder, totalling about 37 MeV, is distributed between the neutrons and $\gamma$-radiation emitted at the instant of fission (the so-called prompt neutrons and prompt $\gamma$-rays), the inherent radioactive decay energy of the fission products, and neutrinos. Neutrinos are exceedingly small particles, carrying no electric charge, which are emitted during the beta decay of a nucleus. Table 1.1 shows this breakdown of the fission energy.

The most important role of the fission products is to collide with other atoms, dissipating their very high velocity as heat in the surrounding materials. Thereafter, interest in the fission products centres upon their physical and chemical properties.

A uranium nucleus can fission in many different ways, producing two fission products which may be about equal in mass or may differ by up to a factor of two. Observation of the fragments from the fission of large numbers of nuclei of, say, $^{235}$U has shown that there is a well-defined probability of occurrence for each pair of fragments. Figure 1.11(a) shows these probabilities plotted in a mass yield curve to indicate the way that a $^{235}$U nucleus fissions when it absorbs a thermal neutron. A typical example of a single fission event from that curve is:

$$^{235}U + \text{neutron} \longrightarrow {}^{141}Ba + {}^{92}Kr + 3\,\text{neutrons}.$$

As the incident neutron energy increases, the shape of the mass yield curve changes. Fission of $^{235}$U by neutrons of fission energies (Fig. 1.7) gives the results shown in Fig. 1.11(b). It can readily be seen that the principal

Table 1.1 Energy released when a $^{235}$U atom is fissioned by a thermal neutron

| | |
|---|---|
| Fission products kinetic energy | 166 |
| Fission products $\beta$-decay | 7 |
| Fission products $\gamma$-decay | 7 |
| Prompt neutrons | 5 |
| Prompt gamma radiation | 8 |
| Neutrinos | 10 |
| Total | 203 MeV |

difference between (a) and (b) is the higher probability of formation of a pair of fragments of nearly equal mass from the higher-energy fission process.

Most fission product atoms, as formed, have nuclei with too many neutrons for stability. They are all chemical elements familiar to us in everyday life, but their nuclei contain a larger number of neutrons than found in the well-known stable isotopes. Neutron-rich isotopes break down, as has already been shown, by emitting an electron ( $\beta$-particle), thereby converting one neutron to a proton and forming an isotope of a new element. This isotope is almost invariably formed in an energetic state, which rapidly emits gamma ( $\gamma$ ) radiation as it changes to its normal state. It is frequently found that a fission product produces several ( $\beta + \gamma$ ) emissions before arriving at a stable isotope. The probabilities of these transitions vary widely from one isotope to another. If the *half-life* is defined as the time required for half of any assembly of atoms of a given isotope to change or *decay* to the next step in its sequence of changes, then fission products with half-lives as short as one second and as long as ten million years can be found.

One consequence of these radioactive decay sequences is that an assembly in which a chain fission reaction has been occurring continues to generate heat from beta-decay reactions long after the fissions themselves have ceased. Figure 1.12 shows an example of the thermal power generated from the fission of $^{235}$U in an assembly which has been in operation for one year. Immediately after shutdown, the heat generated by the beta radioactivity of the fission products is as much as 6 per cent of the fission power, a fact which has obvious implications for the design of cooling systems of nuclear reactors.

Fission products also have an impact on reactor control, because they capture neutrons. As they accumulate in the system they constitute an additional neutron loss which must be overcome by provision of sufficient fissile material to maintain the chain reaction. It follows that as the consumption of the fissile material proceeds, the margin of excess *reactivity* over and above that needed to sustain the chain fission reaction gradually falls, until eventually the reaction stops.

## 2.8 Conversion and breeding

It has been shown that $^{238}$U exhibits a strong tendency to capture neutrons (Fig. 1.5) but that it is not fissile in the sense defined earlier. The product of

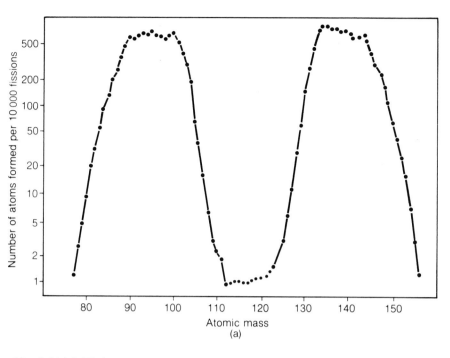

Fig. 1.11 (a)  Fission product yields from $^{235}$U in thermal neutron spectrum

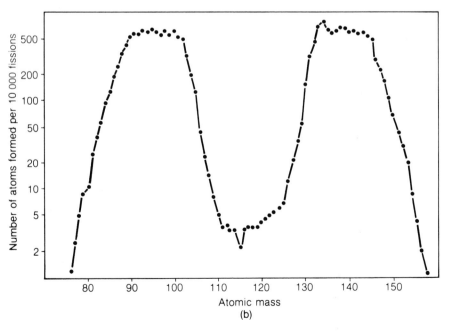

Fig. 1.11 (b)  Fission product yields from $^{235}$U in fast fission spectrum

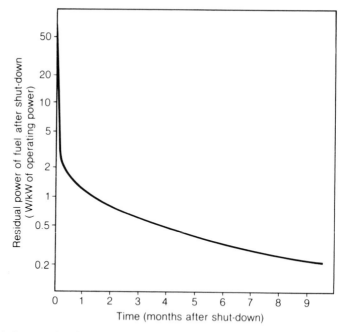

Fig. 1.12  Decay of self-heating in fuel discharged after 1 year irradiation

neutron capture is $^{239}$U, which has too many neutrons to be stable and emits two beta particles in quick succession to produce first neptunium and then plutonium:

$$^{238}U + \text{neutron} \longrightarrow \quad ^{239}U \quad \text{(half-life 24 minutes)}$$

$$^{239}U - \beta \longrightarrow \quad ^{239}Np \quad \text{(half-life 2.4 days)}$$

$$^{239}Np - \beta \longrightarrow \quad ^{239}Pu.$$

$^{239}$Pu is a radioactive metal which emits alpha particles and has a half-life of 24 000 years, but the really important point is that it is fissile like $^{235}$U. The $^{239}$Pu cross-sections for fission and neutron capture are shown in Fig. 1.13.

Neutron capture by $^{239}$Pu produces $^{240}$Pu, which is not fissile, and neutron capture by $^{240}$Pu produces $^{241}$Pu which is fissile. Moreover $^{241}$Pu decays by $\beta$-emission to produce americium, which itself can capture neutrons to produce isotopes of the heavier element curium.

Now although all these reactions occur in any nuclear reactor in which $^{238}$U is present, they are relatively unimportant compared to the first neutron capture—producing $^{239}$Pu. This step, which *converts* $^{238}$U to a useful fissile fuel, is the key to the efficient utilization of natural uranium in nuclear reactors. It is the reason why $^{238}$U is often referred to as a *fertile* material.

In an analogous way, naturally occurring thorium, which is almost entirely $^{232}$Th, is a fertile material. Like $^{238}$U, it can capture neutrons over a wide energy range. It will not however fission unles the energy is greater than a

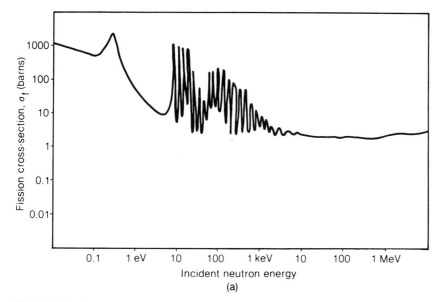

Fig. 1.13 (a)  Neutron cross-section for ²³⁹Pu to fission

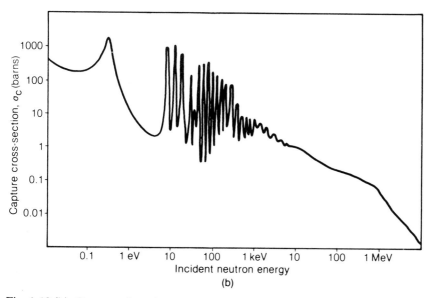

Fig. 1.13 (b)  Cross-section of neutron capture by ²³⁹Pu

million electron volts. The product of neutron capture is $^{233}$Th, which decays by two successive $\beta$-particle emissions, first to protactinium and then to the new fissile isotope of uranium, $^{233}$U:

$$^{232}\text{Th} + \text{neutron} \rightarrow {}^{233}\text{Th} \text{ (half-life 22 minutes)}$$
$$^{233}\text{Th} - \beta \rightarrow {}^{233}\text{Pa} \quad \text{(half-life 27 days)}$$
$$^{233}\text{Pa} - \beta \rightarrow {}^{233}\text{U}.$$

In order to understand better the fertile behaviour of $^{238}$U in a nuclear reactor, whether based on thermal or fast fissions, it is helpful to consider the simple arithmetic of neutron population. First a convenient quantity ( $\eta$ ) representing the number of fission neutrons released by a fissile nucleus for every neutron absorbed must be defined. This takes account of neutrons lost through capture in the fissile nucleus and for present purposes is a more illustrative number than $\nu$ alone.

$$\eta = \frac{\nu \times \sigma_f}{\sigma_f + \sigma_c}.$$

Figure 1.14 shows values[9] of $\eta$, as a function of the energy of the triggering neutron, for all of the actinide isotopes which are of any interest as fissile or fertile reactor fuels. If it is now supposed that of the $\eta$ neutrons released, $B$ are captured by the fertile material, $L$ are lost by escape from the system or captured in an impurity nucleus and exactly one neutron goes on to react with a new fissile nucleus, then

$$\eta = B + L + 1.$$

From Fig. 1.14 it is clear that at low neutron energies all the fissile nuclei yield about two fission neutrons per neutron absorbed, with $^{233}$U showing slightly higher values than $^{235}$U and $^{239}$Pu. At neutron energies above 100 keV, which is the range found in *fast* fission systems, the neutron yield rises steadily above two, with $^{239}$Pu being in the lead.

The value of $B$ represents the number of new fissile nuclei formed for every one destroyed. This number is called the *breeding ratio*, for obvious reasons. If $B$ is greater than one, fissile material is being replaced faster than it is consumed by fission, so there is the possibility of building a *breeder* reactor. If $B$ is less than one, the fissile material is being replaced at a slower rate than the fission rate, and a reactor operating in that manner is called a *converter*.

In moderated systems, that is to say in thermal reactors, $B$ is never quite able to reach one. The losses ($L$), by neutron capture in moderator and fission products, and by escape to the outside, usually total around 0.6, so that $\eta$ would have to exceed 2.6 to make a breeder. Even using $^{233}$U as the fissile material and heavy water as the moderator, on account of its low neutron capture cross-section, it is difficult to raise $B$ above 0.8; more

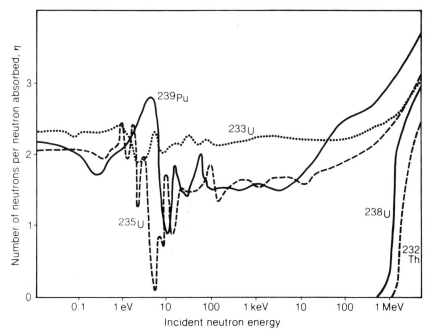

Fig. 1.14 Neutron yield per neutron absorbed

usually it has a value around 0.6. These reactors are therefore converters: Fig. 1.15(a) shows the fate of neutrons in a typical thermal reactor.

In unmoderated systems the chances of breeding are better because:

(1) the value of $\eta$ rises well above two (see Fig. 1.14) when the neutron energy is around 1 MeV, and

(2) the neutron losses by capture in fission products and structural materials are somewhat lower at the high energies, and there is no neutron loss in moderator nuclei.

Plutonium is the best fast reactor fuel, with an $\eta$ ·value around 2.6. If $L$ is kept to 0.4, then $B$ can be as high as 1.2 in a reactor containing $^{239}$Pu and $^{238}$U. Other combinations of fissile and fertile material can also be made to breed, but with breeding ratios closer to one. Figure 1.15(b) shows the fate of neutrons in a typical fast reactor.

## 3 Controlling the chain reaction

So far it has been shown that if *exactly* one fission neutron goes on to trigger a new fission event, a steady chain reaction of nuclear fissions will take place. However, any slight deviation from this delicate balance would cause the chain reaction either to die away or to accelerate continuously were it not for certain effects which tend to stabilize the reaction. In this section these

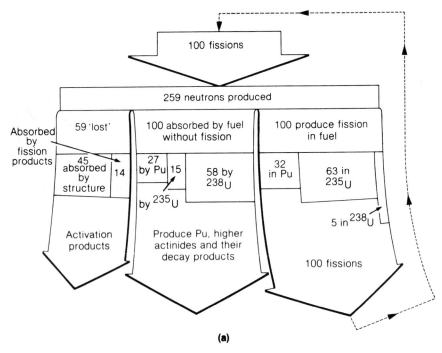

Fig. 1.15 (a)  Neutron utilization in a thermal reactor

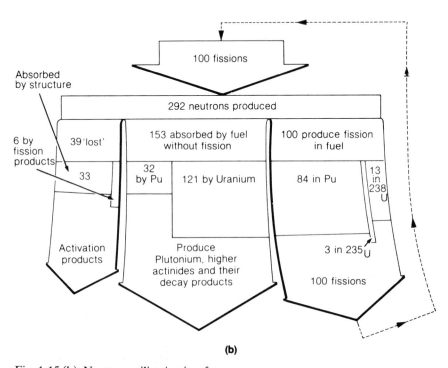

Fig. 1.15 (b)  Neutron utilization in a fast reactor

stabilizing mechanisms which, in practice, make it very easy to achieve that delicate balance will be examined.

## 3.1 Delayed neutrons

The first control mechanism depends upon a natural property of the fission products, without which it might have been impossible to develop nuclear reactors. To understand it the behaviour of the fission neutrons must again be considered.

Figure 1.6 indicates the number of neutrons emitted in the process of fissioning the various actinide isotopes. A small fraction—about 1 per cent—of these neutrons is not emitted at the instant of fission. Instead they are emitted by particular fission products which choose this process, rather than $\beta$-decay, to lower their neutron to proton ratio. The neutrons appear from a fraction of a second to a few minutes after the fission event itself, since their emission has to await the formation by $\beta$-decay of the particular fission products having this property. Table 1.2 shows the half-lives and yields of such neutrons from thermal and fast fission of $^{235}U$ and $^{239}Pu$ respectively. They are grouped according to half-life of the fission product precursor. The energies of these *delayed neutrons* are lower than those of the prompt neutrons.

Delayed neutrons play a vital role in the control of both thermal and fast reactors. Without them, changes in the rate at which uranium atoms are fissioned would occur at great speed whenever the neutron balance was not set exactly to produce one new fission from the neutrons released in the previous fission. Such exactness is impossible to set up, and even if it were achieved it would soon be upset by changing concentrations of fissile materials and fission products. The fact that about 1 per cent of the neutrons produced in fission is delayed by an averge of 10–20 seconds means that changes in fission rate do not occur rapidly and there is time to make adjustments to keep it at any desired level.

Table 1.3 again shows the number of delayed neutrons emitted when the

Table 1.2 Delayed neutron yields from $^{235}U$ (thermal) and $^{239}Pu$ (fast) fission

| Group | Effective half-life (s) | Yield (delayed neutrons per 10 000 fissions) $^{235}U$ (thermal)[4] | $^{239}Pu$ (fast)[5] | Principal fission products contributing to neutron emission in the group |
|---|---|---|---|---|
| 1 | 55 | 5 | 2 | $^{87}Br$ |
| 2 | 22 | 35 | 18 | $^{88}Br$ $^{137}I$ |
| 3 | 6 | 31 | 11 | $^{89}Br$ $^{93}Rb$ $^{138}I$ |
| 4 | 2.3 | 62 | 22 | $^{85}As$ $^{90}Br$ $^{94}Rb$ $^{139}I$ |
| 5 | 0.6 | 18 | 6 | $^{91}Rb$ $^{99}Y$ $^{140}I$ $^{145}Cs$ |
| 6 | 0.2 | 7 | 1 | $^{95}Rb$ $^{96}Rb$ $^{97}Rb$ |
| Total | | 158 | 60 | |

Table 1.3 Delayed neutrons as a percentage of total neutrons released in fission

| Fissioning actinide | Neutrons delayed (%) |
|---|---|
| Thorium-232 | 2.3 |
| Uranium-233 | 0.28 |
| Uranium-235 | 0.64 |
| Uranium-238 | 1.5 |
| Plutonium-239 | 0.21 |

common nuclear fuel isotopes are fissioned, this time expressed as a percentage of the total number of neutrons emitted by these fission events. Expressed in this way, the values are not very sensitive to the energy of the neutrons causing the fissions.

### 3.2 Control rods

In a practical nuclear reactor it is necessary not only to be able to add and remove fissile material, but also to trim the neutron balance quickly and simply. This is the function of control rods.

Control rods are remotely operated mechanical devices which can insert materials of high neutron capture cross-section into the reactor core. Their movement, being very much more rapid than the maximum rate of power change determined by the delayed nuetrons, is used to control the *neutron multiplication factor* of the assembly and therefore allow the reactor to be started up, held at the desired power, or shut down.

The neutron multiplication factor, $k_{eff}$, is the average number of fission neutrons resulting from the various interactions of one fission neutron; $k_{eff} = 1$ when the reactor is just critical and running at a steady power.

Provided that $k_{eff}$ is kept below $(1 + f)$, where $f$ is the fraction of fission neutrons which appear as delayed neutrons, the reactor power cannot increase very rapidly; the delay in the production of that small fraction of the fission neutrons means that it takes at least a few seconds to double the rate at which the fuel is fissioning. In starting up a nuclear reactor the control rods are used to keep $k_{eff}$ within that regime.

Control rods, whether for thermal or fast reactors, usually contain boron, but cadmium, europium, and gadolinium are also used occasionally. Table 1.4 shows the approximate neutron capture cross-sections of these elements in thermal and fast neutron spectra respectively.

Table 1.4  Absorption cross-section for control rod materials[19] (barns)

| | Boron | Cadmium | Europium | Gadolinium |
|---|---|---|---|---|
| Thermal neutrons | 767 | 3 269 | 4 239 | 49 000 |
| Fast neutrons | 0.105 | 0.048 | 0.178 | 0.1 |

## 3.3  The effect of temperature on neutron multiplication factor ($k_{eff}$)

Both thermal and fast reactors can be designed so that $k_{eff}$ decreases as the core temperature rises. This property makes a reactor very stable, as deviations from the steady power level are automatically self-correcting without the need for control rod movements.

There are three principal causes of these temperature effects. Firstly, in thermal reactors, increases in moderator and fuel temperatures result in moderated neutrons of slightly higher energy. If there is $^{239}$Pu in the fuel, this tends to increase the fission cross-section and $k_{eff}$, owing to the peak in $\sigma_f$ at just about 0.1 eV (see Fig. 1.13(a) ). Secondly, in both thermal and fast reactors, an increase in fuel temperature results in a reduction of $k_{eff}$, owing to so-called *Doppler broadening*. This is an effect, particularly evident in fuels containing $^{238}$U, in which the random thermal motions of neutron-capturing nuclei effectively spread out the very sharp peaks in capture cross-section above 5 eV (see Fig. 1.5(b)). The result of increasing this thermal motion is that the fuel captures a larger fraction of the neutrons and $k_{eff}$ is reduced. Thirdly, thermal expansion of a reactor core provides a larger surface area from which neutron leakage can occur, again causing a reduction in $k_{eff}$.

Particular reactor designs often exhibit other temperature effects, usually caused by changes in the geometry of the fuel or coolant. Some of these are likely to be the less desirable *positive reactivity* effects, which tend to increase $k_{eff}$ as the temperature rises.

# 4  Nuclear power station design

Having discussed how the fission of uranium and plutonium can be made to occur in a controlled manner the fundamental aspects of commercial nuclear reactor design will now be considered.

Exploitation of nuclear fission is almost exclusively for the generation of electricity in large power stations: the nuclear reactor is merely a source of heat used to raise steam, just as in a coal- or oil-fired power station the burning of fuel in a furnace fulfils that function (Fig. 1.16). Most of the hardware in a nuclear power station is therefore exactly the same as that found at a conventional power station.

Instead of allowing the hot combustion products to pass directly over pipes carrying water for steam raising, nuclear fuels are sealed into metal cans, and coolant transfers the heat from the cans to the water pipes by means of a pumped circuit.

In this chapter it is not intended to describe either the steam-raising circuits or the turbo-alternator machinery of nuclear power stations. These components have much in common with their counterparts in conventional power stations and there exist already some excellent books on the theory and practice of cycles for converting heat to work.[11,12] The source of heat, however, will be considered in some detail.

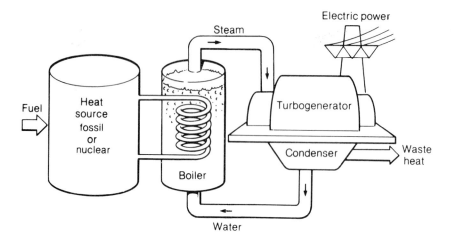

Fig. 1.16  Schematic diagram of a power station

The basic elements of a nuclear reactor are (Fig. 1.1)
    (a)  the fuel and fuel can,
    (b)  the moderator,
    (c)  the coolant, and
    (d)  the containment,
the various types of reactor in use around the world being characterized by
the particular materials chosen for these elements.

### 4.1  Fuel and fuel can

Nuclear fuel is sealed in impervious metal or ceramic cans, known as the
*cladding*, to prevent the escape of fission products. Thermal reactors must
be fuelled with actinide mixtures which contain a least 0.5 per cent of a fissile
isotope, that is to say, an isotope having a very large cross-section for fission
by neutrons at thermal energies. In practice this means $^{235}$U, $^{233}$U or $^{239}$Pu,
with the bulk of the fuel made up from $^{238}$U or $^{232}$Th. The use of these fertile
isotopes as the diluent has two advantages: some neutrons, which would
otherwise be captured uselessly by the diluent, are captured in $^{238}$U or $^{232}$Th
and thereby produce $^{239}$Pu or $^{233}$U to supplement the fissile content, and, in
the case of $^{238}$U, naturally occurring uranium is available as a ready-made
mixture of 0.7 per cent $^{235}$U with 99.3 per cent $^{238}$U.

    The actual choice of fissile content is determined by the amount of
neutron-capturing material in the particular core arrangement and by the
amount of energy it is desired to extract before renewing the fuel. Thus, by

the use of metal fuel, and moderators, cladding, and coolant having very low neutron capture cross-sections, a Magnox nuclear reactor can produce useful power from natural uranium until its fissile content has fallen from the original 0.7 per cent to 0.3 per cent. On the other hand the advanced gas-cooled reactor (AGR) uses uranium oxide ($UO_2$) as fuel and stainless steel as cladding in order to operate at a higher temperature. It also extracts five times as much energy from a given weight of fuel before refuelling. To meet those conditions the AGR must use uranium oxide which has been enriched to a $^{235}U$ content of about 2.5 per cent.

The water-cooled reactors, i.e. the pressurized-water reactor (PWR), the boiling-water reactor (BWR), the steam-generating heavy water reactor (SGHWR) and the CANDU also use uranium oxide fuel. In these reactors the cladding temperature is no greater than for a Magnox reactor, but designers, taking advantage of the compact nature of water-cooled (and moderated) cores, have used very high fission rates per weight of fuel (Table 1.5), which cause the fuel itself to reach temperatures of around 2000 °C in its centre. Consequently, uranium metal, which swells uncontrollably at 650 °C, is unsuitable, and $UO_2$, which is stable even above 2000 °C, is used. Unlike the AGR, the water reactors do not require cladding which will withstand high temperatures. The requirements for low neutron absorption, compatibility with water at 320 °C, and good mechanical strength at up to 400 °C led to the use of zirconium alloys to clad water reactor fuel; stainless steel, which has inferior neutron-absorption properties, was used to clad some early PWR fuels in the US.

The high-temperature gas-cooled reactor (referred to as HTR), which is fuelled with uranium and the THTR, which is fuelled with uranium and thorium, must again use fuels in the form of oxides or carbides. There is no metal of low neutron absorption to fill the role of cladding at temperatures up to 1250 °C and consequently a fuel element has been developed in the form of compacted small spheres of fissile material, individually coated with impervious layers of silicon carbide and carbon. Fuel of about 10 per cent fissile content is used to allow very high *burn-up* of the fuel before replacement.

Table 1.5 shows the fuel choice for the important thermal reactor variants. We can readily see that, since 1 megawatt-day ($8.6 \times 10^{10}$ J) corresponds to the fission of about 1 gram of metal, the terminal burn-up of all these fuels corresponds roughly to the fission of the original $^{235}U$ or $^{239}Pu$ content. This is possible only because some new fissile material has been generated from $^{238}U$ or $^{232}Th$. In fact, about 25 per cent of fissions in the PWR (and rather less in the lower burn-up systems) are of $^{239}Pu$.

The fast reactor must be fuelled with an actinide mixture which is around 20 per cent fissile (Table 1.6). Both $^{235}U$ and $^{239}Pu$ are used, with the balance made up of $^{238}U$. Since these reactors, like the AGR, operate at high temperature, an oxide fuel clad in stainless steel is used. If a high breeding ratio is required, it is advantageous to keep the fission neutrons at as high an

Table 1.5 Fuel and cladding materials in commercial reactors

| Reactor type | Fuel | Clad | Peak coolant temp. (°C) | Average burn-up (MW d/t) | Average fuel rating (W/g of U or Th) |
|---|---|---|---|---|---|
| Magnox | natural uranium metal | magnesium alloy | 400 | 4 500 | 3 |
| AGR | uranium oxide: 2.5% $^{235}$U in $^{238}$U | stainless steel | 650 | 18 000 | 15 |
| PWR | uranium oxide: 3% $^{235}$U in $^{238}$U | zirconium alloy | 320 | 30 000 | 38 |
| BWR | uranium oxide: 2.2% $^{235}$U in $^{238}$U | zirconium alloy | 285 | 28 000 | 20 |
| SGHWR | uranium oxide: 2.3% $^{235}$U in $^{238}$U | zirconium alloy | 280 | 20 000 | 14 |
| CANDU | natural uranium oxide | zirconium alloy | 305 | 7 000 | 19 |
| HTR | uranium oxide: 10% $^{235}$U in $^{238}$U | carbon and silicon carbide | 800 | 100 000 | 112 |
| THTR | U/Th oxide: 10% $^{233}$U in $^{232}$Th | carbon | 800 | 100 000 | 100 |

Table 1.6 Fuel and cladding for the liquid-metal-cooled fast reactor

| Reactor type | Fuel | Clad | Peak coolant temperature (° C) | Peak burn-up (MW d/t) | Breeding ratio | Rating (W/g of U+Pu) |
|---|---|---|---|---|---|---|
| LMFR | 20% $PuO_2$ (or $^{235}UO_2$) in $^{238}UO_2$ | stainless steel | 590 | 100 000 | 1.24 | 120 |

energy as possible, so moderating materials such as oxygen atoms are deleterious. Future fast reactors may therefore use uranium and plutonium carbides or nitrides which, still being stable at the high temperature, have only half as many moderating atoms per actinide atom.

## 4.2 **Moderators**

One way of comparing the effectiveness of moderator atoms is by means of *moderating ratios*. If $L$ is defined as a coefficient expressing neutron energy loss per collision[6], $\sigma_s$ as the scattering cross-section for epithermal neutrons (those in energy range $1\,eV - 1\,MeV$), $\sigma_a$ as the absorption cross-section for thermal neutrons, then

$$\text{moderating ratio} = L \times \frac{\sigma_s}{\sigma_a}$$

Table 1.7 shows the relative values of moderating ratio for some materials containing atoms from the first row of the Periodic Table, which are the lightest elements of all. A moderator *quality factor*, which is moderating ratio multiplied by a factor to take account of the effect of differing numbers of atoms in a given volume of moderator, is also shown in Table 1.7.

Water and heavy water are included simply because they are convenient forms of hydrogen and deuterium respectively. Hydrogen and deuterium are not so suitable for use in the elemental form because they are gases and because they are highly inflammable.

If the data in Table 1.7 are considered, it is clear that heavy water is a very good moderator indeed. Although it is a very expensive material and has

Table 1.7 Moderating ratios for light elements

| Moderator | Moderating ratio | Quality factor (relative to $H_2O$) |
|---|---|---|
| Deuterium, D (as $D_2O$) | 12 000 | 160 |
| Carbon, C (as graphite) | 218 | 5 |
| Beryllium, Be | 159 | 4 |
| Hydrogen, H (as $H_2O$) | 72 | 1 |
| Oxygen, $O_2$ | 1 490 | 0.016 (at atmospheric pressure) |
| Helium, He | 83 | 0.000 5 (at atmospheric pressure) |
| Lithium, Li | 0.0038 | 0.000 4 |
| Boron, B | 0.001 | 0.000 3 |
| Nitrogen, $N_2$ | 0.79 | 0.000 01 (at atmospheric pressure) |

some undesirable properties (e.g. it produces the radioactive isotope of hydrogen, tritium), it is used in the CANDU reactors, where natural uranium oxide fuel requires a first-class moderator to achieve criticality.

Carbon is another good moderator but, being inferior to heavy water, is used in conjunction with natural uranium metal or with enriched uranium oxide to achieve criticality.

Beryllium is a good moderator, but it is not a commonly used metal and so the cost of production and fabrication is high. It is also a very toxic material and consequently, although BeO has been used in research reactors, this element is not a candidate as a power reactor moderator.

Ordinary water (*light water*) is quite a good moderator. Its use in practicable only with enriched fuel, but in that combination it represents the most successful commercial reactor systems—the PWR and BWR.

Oxygen gas is a good moderator but must be used at high pressure to obtain a good 'slowing-down power', and that is impossible owing to its high chemical reactivity.

Helium gas comes next in order of merit, but as with oxygen it would have to be used at very high pressure to achieve a good moderation. So, despite its chemical inertness, helium is unattractive as a moderator.

Finally, lithium, boron, and nitrogen must be rejected on grounds of poor moderating power.

The moderators in common use are therefore $D_2O$, $H_2O$ and graphite.

There are many kinds of water-moderated reactor. They always take the form of a metal tank containing the water, in which the fuel elements are immersed. Water is quite reactive towards metals and thus gives rise to some chemical corrosion problems.

Graphite, being a strong material, is used differently. In this case, reactors are essentially piles of graphite bricks into which holes are drilled to accept fuel elements and cooling gas. There are again some chemical corrosion problems, this time by reaction of graphite with the coolant gas.

Table 1.8 shows the moderators employed in the commercial thermal reactor types.

Table 1.8 Reactor types

| Reactor type | Moderator | Fuel required |
|---|---|---|
| Thermal: | | |
| Magnox | graphite | natural uranium metal |
| AGR | graphite | 2.5% enriched $UO_2$ |
| PWR | $H_2O$ | 3% enriched $UO_2$ |
| BWR | $H_2O$ | 2.2% enriched $UO_2$ |
| SGHWR | $D_2O$ | 2.3 enriched $UO_2$ |
| CANDU | $D_2O$ | natural $UO_2$ |
| HTR | graphite | 10% enriched $UO_2$ |
| THTR | graphite | 10% enriched $UO_2$ |
| Fast: | | |
| LMFR | none | 20% enriched $UO_2$ |

## 4.3  Coolants

A nuclear reactor coolant has the job of carrying heat from the fuel elements to boilers, where high-pressure steam is generated. As it is pumped through the reactor core the coolant must pick up heat from the fuel element cladding. It must then travel to the boilers and flow over the pipes carrying water in a manner which ensures that its heat is transferred efficiently to the water. The effectiveness of these heat transfer processes depends upon the flowrate of the coolant, the shape of the coolant passages in the core, the nature of the cladding material, the physical properties of the coolant, and the magnitude of the temperature differences between cladding and coolant. The system is usually designed so that the fuel and its cladding operate at the maximum temperature which they can satisfactorily withstand, because that improves the heat transfer and also allows the coolant to be heated to a high temperature. Coolants are heated to the highest temperature which is acceptable on grounds of their physical and chemical properties, since the efficiency of the steam turbines increases as the steam inlet temperature is raised. In gas-cooled reactors, the coolant itself can usually tolerate very high temperatures indeed; the upper limit is therefore set by the undesirable chemical reactions occuring between the coolant and the reactor circuit materials.

In choosing a coolant for a given design of reactor core the following criteria are applied.
1. The coolant must be chemically compatible with the core and boiler materials over the desired temperature range.
2. The coolant must be commercially available in large quantities at an acceptable price.
3. The pumping power required to move heat at a given rate from core to boilers must be a minimum.
4. The coolant pressure must be acceptable in terms of reactor safety and structural material costs.
5. The coolant must be stable under intense $\gamma$-ray and neutron irradiation in the reactor core.

The third criterion relates to the physical properties of the coolant, in particular the specific heat, density, and viscosity. Obviously a coolant capable of taking up a lot of heat per unit volume requires a lower flowrate to do its job. If it has a low viscosity it requires less pumping power to achieve that flowrate. An approximate figure of merit to express these factors is[7]:

$$\frac{C_p^{2.8} \rho^2}{\mu^{0.2}}$$

where

$C_p$ = specific heat at constant pressure (J kg$^{-1}$ K$^{-1}$)
$\rho$ = density (kg m$^{-3}$)
$\mu$ = viscosity (kg m$^{-1}$ s$^{-1}$).

The other four criteria are less suitable for such systematic analysis but must be examined for each candidate coolant material.

Actually the choice of cheap, chemically stable, radiation-resistant fluids, capable of tolerating the temperatures needed to raise high-pressure steam, is fairly restricted. In terms of the figure of merit given for physical properties, liquids are superior to gases and high-pressure gases are superior to low-pressure gases. Table 1.9 lists possible reactor coolants and their figures of merit.

Organic liquids, such as Dowtherm, have relatively poor radiation resistance, whilst fused salts are chemically corrosive and awkward to start up from cold. Liquid metals are quite good coolants but have one disadvantage—a high chemical reactivity with water and oxygen. Although high pressure gases are relatively poor in terms of pumping power requirements they can be chosen to have excellent radiation and chemical stability at high temperatures.

Consequently, the important reactor coolants are water, heavy water, carbon dioxide, helium, and liquid sodium, as shown in Table 1.10.

Water and heavy water are excellent coolants; they cool the bulk of existing and planned nuclear reactors. However, because water boils at a fairly low temperature, it is not possible in these reactors to heat the coolant to temperatures in the region of 500–600 °C, which is necessary to operate boilers and steam generators at high efficiencies. Consequently, water-cooled reactors are relatively inefficient for electricity generation. Typically 30 per cent of the fission heat is converted to electricity. Note that even to heat the

Table 1.9 Figure of merit for physical properties of coolants

| Material | Figure of merit (Relative to Na) |
|---|---|
| *Liquids* | |
| Water, $H_2O$ or $D_2O$ (300 °C) | 60 |
| Sodium hydroxide, NaOH (350 °C) | 13 |
| Dowtherm oil (300 °C) | 7 |
| Potassium and lithium chlorides, KCl, LiCl (400 °C) | 2.5 |
| Sodium, Na (300 °C) | 1 |
| Tin, Sn (300 °C) | 0.45 |
| Mercury, Hg (300 °C) | 0.31 |
| Lead, Pb (500 °C) | 0.27 |
| Bismuth, Bi (300 °C) | 0.21 |
| Potassium, K (300 °C) | 0.21 |
| *Gases* | |
| Carbon Dioxide, $CO_2$(300 °C, 5 MPa)† | $2.8 \times 10^{-3}$ |
| Helium, He (300 °C, 5 MPa)† | $1.9 \times 10^{-3}$ |
| Hydrogen, $H_2$ (300 °C, 1 MPa)† | $4.0 \times 10^{-4}$ |
| Carbon dioxide, $CO_2$ (300 °C, 1 MPa)† | $1.1 \times 10^{-4}$ |
| Helium, He (300 °C, 1 MPa)† | $7.6 \times 10^{-5}$ |
| Air (300 °C, 1 MPa)† | $4.5 \times 10^{-5}$ |

† 1 MPa is approximately equal to 10 times atmospheric pressure

water to 320 °C in a PWR requires the circuit to be operated at a pressure of 15 MPa (150 times atmospheric pressure) to prevent boiling.

Carbon dioxide is the best, in terms of physical properties, of the available gaseous coolants; accordingly it was chosen for the large UK installation of Magnox and AGR power stations. High pressure improves the properties of gaseous coolants and there is a trend towards progressively higher pressures from the prototype reactors through to the optimized commercial design; in Table 1.10 typical $CO_2$ pressures are indicated. In these reactors the coolant is able to withstand temperatures up to 650 °C; beyond that point its chemical reactivity begins to introduce corrosion problems. However, the AGR stations are able to supply steam from their boilers at 170 × atmospheric pressure and 540 °C, conditions which match the best capabilities of modern steam turbines and allow a 40 per cent conversion of heat to electricity. This is in line with the most efficient coal- or oil-fired stations.

The liquid metals, whilst being excellent coolants, present a whole range of novel handling problems on account of their chemical reactivity. Consequently, they were not considered for commercial reactors until the fast reactor added an extra criterion to the coolant selection considerations. In a fast reactor where a high breeding ratio is required, the neutron energy must be maintained as high as possible. Therefore moderating materials, which means elements of low atomic weight, must be kept out of the core. This eliminates water from the contenders and leaves a choice between liquid metals and gases, the latter being acceptable on account of their low density of moderating atoms. In current fast reactors liquid sodium has been chosen because its higher figure of merit (Table 1.9) allows the use of higher *power density* reactors with less pumping power. (Power density is the fission rate per weight of fuel.) It has the advantage of requiring no circuit pressurization, provided the boiling point of sodium at normal pressure (about 900 °C) is not exceeded.

## 4.4  Layout of commercial nuclear reactors

The circulation of the coolant between reactor core and water boilers requires that an enclosed system of vessels and pipes be constructed. Moreover, this system must be capable of withstanding the pressure (Table 1.10) under which the particular coolant is operated. In some reactors the provision of a strong pressure vessel around the reactor core ensures that the nuclear radiation from the fission process and from the subsequent $\beta \gamma$ -decay of the fission products is absorbed before it can reach the outside. When this is not so, a *biological shield*, consisting of about 3 m thickness of special concrete to absorb $\gamma$ -rays and neutrons, is built outside the pressure vessel for that purpose.

The particular containment arrangements which have been used in commercial practice are best described by means of examples.

Table 1.10 Coolants used in commercial nuclear reactors

| Reactor type | Coolant | Maximum temperature (°C) | Pressure (MPa) | Max. efficiency, electricity/heat (%) |
|---|---|---|---|---|
| Magnox | $CO_2$ | 410 | 1.5 | 35 |
| AGR | $CO_2$ | 650 | 4.0 | 44 |
| PWR | $H_2O$ | 320 | 15.0 | 33 |
| BWR | $H_2O$ | 280 | 7.0 | 33 |
| SGHWR | $H_2O$ | 280 | 7.0 | 33 |
| CANDU (PHWR) | $D_2O$ | 320 | 8.0 | 32 |
| HTR | He | 800 | 4.5 | 45 |
| THTR | He | 800 | 4.5 | 45 |
| LMFR | Na | 580 | 0.1 | 42 |

(a) *The Magnox reactors*. Figure 1.17 shows a simplified picture of Oldbury power station. The $CO_2$ pressure is contained around the core, and four boilers by a large concrete pressure vessel, which also serves as the biological shield. Steam to drive the turbine generators leaves the pressure vessel through steel ducts and is similarly returned as boiler feedwater after leaving the condensers. All radioactive materials and the entire $CO_2$ circuit are inside the concrete vessel.

Concrete pressure vessels are much easier to guarantee against major gas leakage but they have to be cooled and insulated from the hot gas. Earlier Magnox reactors employed steel pressure vessels inside a concrete biological shield; the boilers were outside in their own pressure containments.

(b) *The advanced gas-cooled reactors*. The AGR inherits from the later Magnox stations the concrete pressure vessel enclosing reactor and boilers. The Hartlepool and Heysham reactors (Fig. 1.18) moreover have their eight boilers in *pods* within the thick wall of the pressure vessel. This arrangement allows a smaller diameter cavity to be employed for the reactor core, which is an advantage in these large 660 MW(e) installations.

(c) *The pressurized-water reactors*. These reactors, which are the most widely used commercial design for a thermal reactor, employ a steel pressure vessel which is contained in a concrete biological shield. The boilers, in which the hot water from the reactor core heats water for the steam turbines, are separately shielded. Modern PWRs have an additional steel-lined concrete pressure vessel surrounding the reactor – boiler system, so that even in the event of a primary circuit leak this secondary containment will safely contain the pressure of steam produced. Figure 1.19 shows the general arrangement.

The pressure in the primary coolant circuit of a PWR is very much higher than in any other reactor type (Table 1.10). The pressure vessels and boilers of these reactors are consequently massive constructions, the integrity of which is the subject of a large investment in metallurgical development and

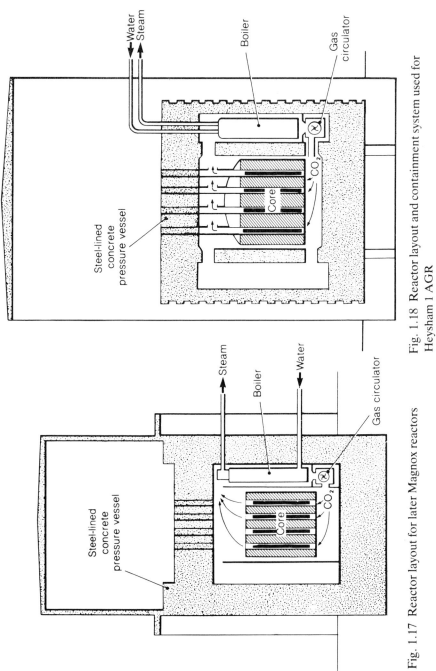

Fig. 1.18  Reactor layout and containment system used for Heysham 1 AGR

Fig. 1.17  Reactor layout for later Magnox reactors

manufacturing control. A 1100 MW(e) reactor vessel is about 14 m long, 6 m in diameter, and of wall thickness up to 280 mm. Its weight can be up to 600 tonnes. The boilers themselves are of a similar size.

(d) *Boiling-water reactors*. The BWR differs from the PWR in generating steam for the turbines directly in the reactor core. Boilers are therefore eliminated. Pressure is lower at 6 MPa so that the steel pressure vessel wall thickness is somewhat less, but a 1100 MW(e) reactor vessel is longer than a PWR vessel owing to the need for steam separation equipment above the core itself. The secondary containment is again a steel-lined concrete building able to contain the total steam resulting from a primary circuit rupture.

Figure 1.20 shows the general layout of a BWR.

(e) *CANDU reactors and SGHWR*. Heavy-water-moderated reactors for commercial use are of two main types, sometimes called the pressurized-heavy-water reactor (PHWR) and the boiling-light-water reactor (BLWR). The Canadian CANDU reactors include both types and the SGHWR is the UK version of the BLWR.

These reactors dispense with the massive steel pressure vessel for primary containment of the coolant and instead employ pressure tubes of zirconium alloy, which contain the fuel elements and which carry the high-pressure water coolant through an unpressurized tank of heavy water moderator. The relationship between thickness of a cylinderical pressure vessel, $t$, its diameter, $D$ and the pressure it is required to retain, $P$, is given approximately by the expression,

$$t = \frac{PD}{2s},$$

where $s$ is the strength of the material from which it is made. Thus, the thickness of the vessel has to be increased in direct proportion to both the pressure and the diameter. These considerations have influenced the design of pressure-tube reactors such as CANDU and SGHWR, where a tube-wall thickness of only 5 mm is sufficient to contain the high-pressure coolant, compared with the 280 mm thickness of a PWR pressure vessel. In the PHWR, $D_2O$ is used not only as the coolant in the horizontal pressure tubes but also to raise steam in separate heat exchanges (boilers) as in the PWR. The coolant operates at a pressure of 8 MPa, about half that of the PWR, but, to guard against the release of this heavy water containing some radioactive tritium to the outside atmosphere, a large secondary building surrounds the reactor and the boilers. Figure 1.21 shows the general layout of the PHWR.

The BLWR again employs zirconium pressure tubes in a tank of heavy water, but this time the tubes are vertical and ordinary water ($H_2O$) is used as coolant. Obviously this involves some neutron penalty but natural uranium oxide can still be used as fuel (e.g. Gentilly reactor in Canada). Like the

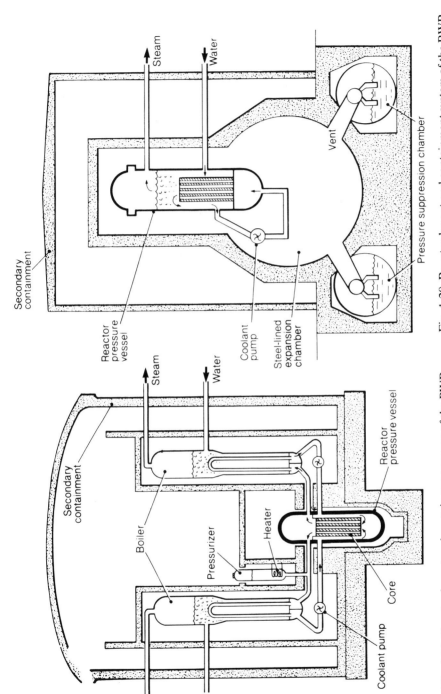

Fig. 1.19  Reactor layout and containment system of the PWR

Fig. 1.20  Reactor layout and containment system of the BWR

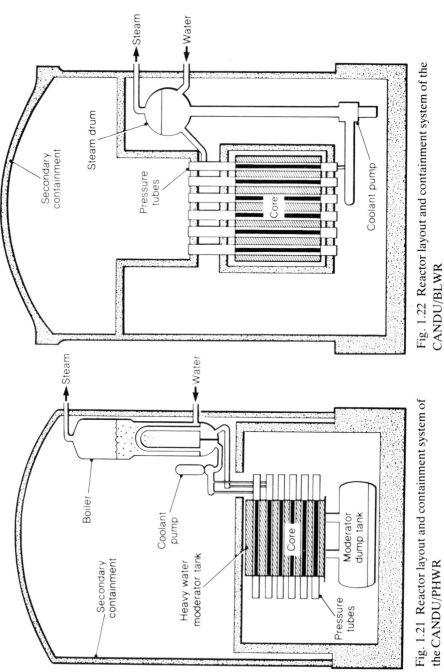

Fig. 1.21 Reactor layout and containment system of the CANDU/PHWR

Fig. 1.22 Reactor layout and containment system of the CANDU/BLWR

BWR, these reactors dispense with boilers by allowing boiling of the $H_2O$ in the core and separating steam for the turbines by means of steam drums. Circuit pressure is low (5 MPa) compared to other water reactors, but again the reactor and its steam drums are enclosed in a secondary leak-tight building (Fig. 1.22) to eliminate the chance of radioactive discharges to the atmosphere.

(f) *The sodium-cooled fast reactor*. Two basic layouts of LMFR (liquid-metal fast reactor) are in use by designers—the so-called *pool* and *loop* variants. The pool type, as favoured by UK designers, is described here.

The core of the LMFR is contained, together with the heat exchangers which transfer heat from the coolant to a secondary sodium circuit, in an unpressurized tank of sodium. Neutron shielding is provided between core and heat exchangers and the tank has a concrete shield to stop $\gamma$-radiation from reaching the outside, but massive structural strength is not required in any of the primary containment. Hot secondary sodium is used to raise steam at high pressure in boilers outside the reactor tank. They require no radiation shielding. As a precaution against accidental release of radioactive fission products from the reactor, it is enclosed in a leak-tight building together with its boilers and the turbines. Figure 1.23 shows the main features of the LMFR layout.

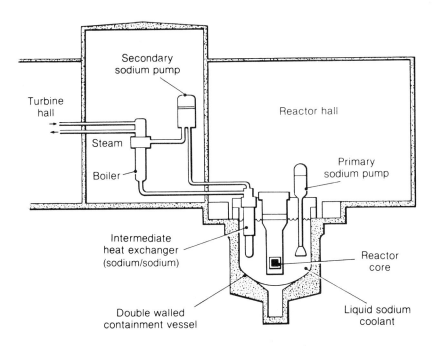

Fig. 1.23 Reactor layout and containment system of the Sodium-Cooled Fast Reactor

# 5  Operating a nuclear power station

## 5.1  **Thermal reactors**

It has been shown that thermal reactors are fuelled almost exclusively with $^{235}U$–$^{238}U$ mixtures, ranging from the 0.7 per cent of naturally occurring uranium in Magnox to 10 per cent $^{235}U$ in HTR fuel. These fuels achieve burn-ups approximately equivalent to their initial $^{235}U$ content, although 20–30 per cent of the fissions actually occur in $^{239}Pu$, so that when removed from the reactor the used fuel contains $^{235}U$, $^{238}U$, and $^{239}Pu$.

The initial fissile content of the fuel does not have a large effect on the amount of energy which can be extracted from a given quantity of uranium ore; it is chosen principally to match the parasitic neutron absorptions of the core, and to limit the burn-up to a value which can be accommodated by the metallurgical properties of the fuel and cladding. High burn-up reduces fuel throughput costs but leads to swelling of fuels and to failure of cladding.

In order to achieve continuous operation of reactors without sudden large changes in fissile loading, it is usual to replace fuel elements batch-wise rather than all at once. By the time that they are removed, the fuel elements have a fissile content ($^{235}U$ + $^{239}Pu$) which is getting towards the lower limit for sustaining criticality. When a reactor has been started up with all new fuel, it is necessary to remove the first few batches before they have reached the normal burn-up; subsequent batches can all be operated to their normal value.

When a thermal reactor is loaded with a new batch of fuel it has much more fissile material in its core than is needed to sustain the chain fission reaction. The control rods are inserted to compensate for this. As burn-up of fissile material proceeds the control rods are gradually withdrawn to keep the chain reaction going. Fuel may be added whilst the reactor is at full power (CANDU, Magnox, AGR) or alternatively it may be shut down for the operation; in the latter case, the control rods are used to hold the reactor in the shutdown condition whilst fuel is changed.

The spent fuel from thermal reactors contains both fissile uranium and plutonium. It is removed when the fissile content has fallen too low for efficient operation of the reactor. Absorption of neutrons by fission products and alteration of the metallurgical properties of the fuel and its cladding are additional factors which affect efficiency. Obviously, by removing the fission products, raising the fissile content, and providing a new clad for the fuel, it could be used again in a similar reactor or in a reactor of different type. This is the reason for reprocessing thermal reactor fuels.

The chemical plant for separating uranium, plutonium, and fission products from each other is based upon solvent extraction from a nitric acid solution containing the irradiated fuel. The process must be carried out in a well-contained and heavily shielded plant because the fission products are still emitting $\gamma$-rays and because many of them are harmful if inhaled or ingested.

Commercial reprocessing plants have long been established in the UK: the plutonium produced is being stored for use in fast reactors, the uranium will be used as fast reactor blanket material or returned to an enrichment plant, and the fission products are stored as waste.

Figure 1.24 illustrates the complete cycle of operations which is involved in operating thermal reactors.

It is worth noting that, although the above description is applicable to almost all operating thermal reactors, it is, in principle, possible to operate an analogous cycle based on the alternative fertile fuel thorium. The main difference is that thorium contains no fissile material at all, so it would be necessary to separate some $^{235}$U, $^{239}$Pu or $^{233}$U from another source to mix with the thorium in order to make the fuel.

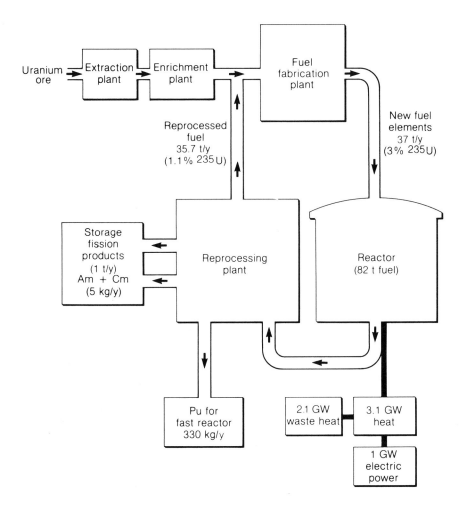

Fig. 1.24 Fuel cycle for thermal reactor (1 GW (e) )

## 5.2 **Fast reactors**

Fast reactors can be designed with a breeding ratio greater than one. This means that having started up the reactor with a core of about 20 per cent fissile material (80 per cent $^{238}$U) and a surrounding blanket of $^{238}$U, the total inventory of fissile material slowly increases. The $^{238}$U inventory, both in the core and the blanket, falls correspondingly, and of course fission products gradually build up. The flow chart in Fig. 1.25 illustrates these changes for the operation of a 1 GW(e) sodium-cooled fast reactor for one year.

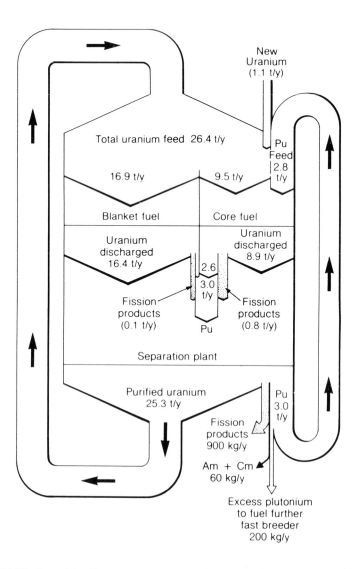

Fig. 1.25  Fissile and fertile material flow diagram for the fast reactor (1 GW (e) )

In a fast reactor of this type, one of the reasons for removing spent fuel is eliminated; the inventory of fissile material does not as in a thermal reactor fall until criticality can no longer be maintained. However, there remain three reasons for removing fuel from this reactor for reprocessing when it has reached a burn-up of about 10 per cent.

1. The fissile loading of the core fuel falls, despite the fact (Fig. 1.25) that the total reactor inventory of fissile material rises.
2. The fission products accumulate in the fuel, and by absorbing neutrons, gradually spoil the breeding ratio of the reactor.
3. The cladding, stainless steel in this case, and other structural components of the fuel elements are unable to withstand greater neutron bombardment than is received in attaining that level of burn-up.

In consequence, operation of the liquid-metal fast reactor (LMFR) entails removal of the core fuel, after about a year, for reprocessing. Fission products are removed as waste, more plutonium is added to restore the fissile fraction to 20 per cent, and the $UO_2$–$PuO_2$ mixture is then clad in new stainless steel to form new core fuel. At longer intervals the blanket elements are also reprocessed to remove fission products and to separate out the plutonium for use in fabricating the core fuel. New blanket elements are made from the purified $^{238}U$ and from the new $^{238}U$ feed material. $^{238}U$ is abundantly available as a waste product from uranium enrichment plants.

Since the reactor can produce overall slightly more plutonium than it consumes, this excess can be accumulated until it is sufficient to fuel a second fast reactor. Alternatively, if no system expansion is required, the $^{238}U$ blanket can be reduced until only enough plutonium is generated to replace that which is consumed in the core.

When system expansion is the requirement, it is obviously advantageous to minimize the quantity of plutonium involved in the whole reactor and reprocessing system; in that way it takes less time to accumulate enough plutonium to start up a second one. Consequently, designers of fast reactors have specified high fuel ratings (high heat output per kilogram of fuel) and rapid reprocessing of spent fuel.

If $B$ represents the breeding ratio of the reactor, and $T$ represents the time in years required to generate enough plutonium to install a second fast reactor together with its fuel cycle, (sometimes called the *doubling time*), then[8]

$$T = \frac{2.7R(1 + F)}{P(B - 1)(1 + \alpha)},$$

where:

$R =$ ratio of plutonium in the whole fuel cycle to plutonium actually in the reactor

$F =$ number of $^{238}U$ atoms fissioned per fission of fissile atoms

$P =$ rating of plutonium in the whole reactor (MW/kg)

$\alpha =$ ratio of neutron captures to fissions resulting from neutron interactions with fissile atoms.

Fission of 1 kg of fuel generates 2.7 megawatt-years of heat. The complete cycle of operations involved in the fast reactor system is depicted in Fig. 1.26.

The so-called *uranium-plutonium cycle* described above is the most attractive way to operate a fast reactor if the aim is to achieve a good breeding ratio. In principle, an analogous cycle using thorium as the fertile fuel could be operated as already discussed, breeding $^{233}$U instead of $^{239}$Pu. There is little incentive to develop such a fast reactor because the fertile fuel $^{238}$U is abundant enough to fuel the uranium-plutonium cycle for many hundreds of years. The value of $B$ in the doubling time equation is smaller for the *thorium-uranium cycle*, principally because a lower number of neutrons are released in fast fission of $^{233}$U rather than $^{239}$Pu, and the fact that direct fast fissions of $^{232}$Th are fewer than for $^{238}$U. A longer doubling time is the result.

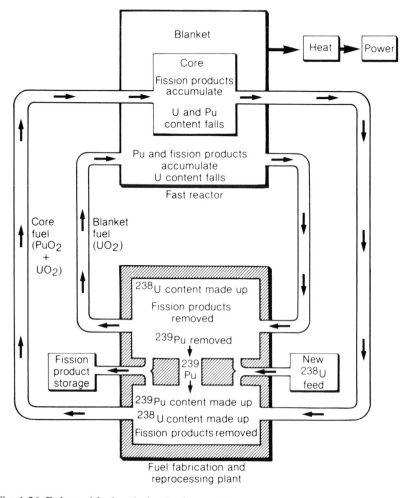

Fig. 1.26  Balanced fuel cycle for the fast reactor.

# 6 Some of the problems

Nuclear reactors are built to generate electricity for sale. Like any other machine, a nuclear reactor is designed in such a way that the cost of building, operating, and maintaining it over its useful life is minimized, whilst agreed standards of safety are met.

In practice, most problems in nuclear power station construction and operation are of the conventional type met in any large engineering project. Likewise the injuries to people which occur are of the type which is met at any major construction site. However, the nuclear reactor itself is one part of the plant which is unconventional by past standards so a few comments on its design problems are needed.

Broadly speaking, there are two aspects to reactor safety. First, during normal operation there must be acceptably low risk of injury from all causes to the staff who operate the reactor, and there must be suitable control over the amounts of potentially toxic effluents released to rivers, the sea, and the atmosphere. Second, the probabilities of accidents to the reactor must be kept below levels where risk to the public is comparable with other commonly accepted risks.

## 6.1 Normal operation

Nuclear reactors do not contain quantities of volatile toxic chemicals but they do of course contain radioactive fission products, which emit $\beta$- and $\gamma$-radiation, and actinide materials, which emit $\alpha$-radiation. Maintenance of safe conditions during normal operation involves making sure that these substances are effectively contained inside the primary circuit of the reactor. When maintenance has to be carried out on primary circuit components, it must be possible to gain access to them without exposing personnel to harmful amounts of radiation.

## 6.2 Fault conditions

A nuclear power station carries some potential risks in common with many other types of large-scale industrial plant in that it contains fluids at high temperature and high pressure and it involves the movement of heavy objects by crane in buildings of considerable height. Prevention of accidents to operating staff in such circumstances is a matter of adherence to safe working practices and compliance with established standards of engineering.

A greater interest however centres on the types of accident which could breach one or more of the three barriers which prevent fuel and fission product materials from escaping to the open air. The containment of large quantities of radioactive substances in an industrial plant is a relatively novel task, and so it is not surprising either that it attracts public attention or that it has been so thoroughly examined by engineers in the nuclear industry.

The range of faults which can occur in a complex power station is very wide. Designers have approached this problem by ensuring that those faults which could lead to the breach of even the first barrier, the fuel cladding, to a limited extent are very improbable. Such a fault might for example be expected once during the life of a fraction of the power stations in use. Less severe faults, such as those requiring operators to shut down briefly for repairs, may be expected more frequently than this, whilst more severe faults, leading to release of some radioactive substances from the containment building, are not likely to occur more than once in about a thousand years at any one power station. Even in this last category the design of the barriers is such that the release of radioactive substances is within acceptable limits specified by national licensing authorities.

The key event in an accident sequence is damage to the nuclear fuel cladding, which gives rise to radioactive materials in the primary circuit of a reactor. It follows that control of the cladding temperature under fault conditions is the prime objective of designers. Since cladding temperature is determined by the balance between the heat generation rate in the fuel and the heat removal rate by the coolant, nuclear reactors are equipped with extremely reliable multiple arrangements for stopping the fission reaction and for cooling the fuel elements. As described previously and shown in Fig. 1.12, nuclear reactor fuel continues to generate heat long after the fission reaction has been stopped. The cooling arrangements must therefore be designed to operate reliably after shutdown, even though at a much reduced rate, for a period of time which varies from one reactor type to another.

## 7 Conclusions

The working of nuclear reactors, no less than that of fossil-fuelled power plant, can be understood by the non-specialist reader. A wider understanding of the basic principles of nuclear fission, and of the fission products produced, would do much to allow the public to appreciate the points made by supporters and opponents of the industry in the 'nuclear debate'.

All nuclear reactors use uranium as their basic fuel. Uranium is a radioactive metal which is widespread in the earth's crust. Unlike coal- or oil-fired power plant the nuclear reactor has its fuel sealed inside metal or ceramic cans so that the waste products, instead of being discharged into the air and on to the land, remain encapsulated when the fuel is finally discharged. These fission products are themselves radioactive and toxic but the quantity produced in generating a given amount of energy is so very small, by comparison with other fuels, that it is entirely practicable to engineer their confinement until such time as their radioactivity has decayed to a level where dispersal is safe.

The fast reactor does not differ in basic principle from the thermal reactor; it does however have the important capability of utilizing nearly the whole of the uranium fuel available, rather than just 1 per cent or less which can be used in current thermal reactors. The reprocessing of fuel after it is discharged is an essential feature of fast reactors as we know them.

The potential hazards in operating nuclear reactors are very well understood, by comparison with other hazards which we face in everyday life, and can be controlled to any reasonable level of risk by application of appropriate standards of engineering design and manufacture.

## Acknowledgement

I would like to thank Dr. F. A. Johnson (Chemical Technology Division, Harwell) for devising the diagrams used in this chapter.

## References

1. HAHN, O. AND STRASSMAN, F. *Naturwiss*, **27**, 11 (1939).
2. BOHR, N. AND WHEELER, J. A. *Phys. Rev.*, **56**, 426 (1939).
3. BODU, R., BOUZIGUES, H., MORIN, N., AND PFIFFELMANN, J. P. *C.R. Acad. Sci. Paris*, **275D**, 1731 (1972).
4. Reactor Physics Constants, Argonne National Laboratory, *ANL* 5800 (2nd edn 1963).
5. BESANT, C. B., CHALLEN, P. J., McTAGGART, M. H., TAVOULARIDIS, P., AND WILLIAMS, J. G., *J. Brit. Nucl. Energy Soc.* **16**, 161, (1977).
6. GLASSTONE, S., AND EDLUND, M. C. *The elements of nuclear reactor theory*, Van Nostrand New York (1952).
7. ETHERINGTON, H., *Nuclear engineering handbook*, section 9–3, pp. 9–91. McGraw Hill New York (1958).
8. STORRER, F., *Introduction to the physics of fast power reactors*, IAEA Technical Report No. 143, p. 247. International Atomic Energy Agency, Vienna (1973).
9. POPE, A. L., *Current edition of the main tape NDL–1 of UK Nuclear Data Library*. *AEEW–M* 1208. UK Nuclear Data Library, Harwell (1973). (See also other items in Library.)
10. HILDENBRAND, G., *Nuclear energy, nuclear exports and the non-proliferation of nuclear weapons*. In *Proc. AIF Conf. on International Commerce and Safeguards for Civil Nuclear Power, March 1977*, Atomic Industrial Forum, New York (1977).
11. WOOTTON, W. R. *Steam cycles for nuclear power plant*. Nuclear Engineering Monographs. Temple Press, London (1958).
12. HAYWOOD, R. W. *Analysis of engineering cycles*, (3rd edn). Pergamon Press, Oxford (1980).

# 2

# Reactor physics

## J. R. ASKEW, C. G. CAMPBELL, J. G. TYROR

Reactor physics is concerned with understanding in a detailed, quantitative manner the interactions which take place between the free neutrons and the various materials and structures which make up a nuclear power reactor. It is a very fundamental and basic aspect of reactor technology and determines many of the key operational and safety features of a nuclear reactor. It embraces the physics of the interactions of neutrons with many materials, the mathematical modelling of the flow and characteristics of the neutron field, and the sophisticated computer strategy required to handle the mathematical solutions and the vast amounts of data involved.

This article attempts to summarize the key aspects and current state of the art and to do so in a descriptive manner. The mathematical content has been kept to a minimum throughout in order to render the material digestible for the non-expert reader. The first three sections are indeed rather general and may provide a sufficient discussion of the topic for many purposes. For those with rather deeper interests, however, §4 and §5 provide a more detailed description of current techniques. The article is a free-standing one with no references and it is hoped that it will provide the non-specialist with a broad understanding of what reactor physics is about.

## Contents

# 1  Reactor physics and its objectives

Perhaps the outstanding characteristic of an operating nuclear power reactor is that it contains a vast number of free neutrons—about a thousand million per cubic centimetre. These neutrons are born in the reactor with a velocity of about 20 000 km/s and have a lifetime of perhaps a millisecond before they are swallowed into the nucleus of an atom of the reactor's material. Neutrons are continually being born and removed, and their density is so high that they can be pictured as forming a continuous regenerating sea, enveloping and permeating the entire reactor structure and flowing freely between its atoms. They are the life blood of the reactor. Reactor physics is concerned with the study and understanding of the interactions between this sea and the reactor material through which it flows. These interactions are most conveniently described in mathematical terms, but in the present chapter the subject is presented in a descriptive manner.

The scientific issues involved are, however, rather complex, and the lay reader may find the going rather difficult. Nevertheless, the topic is a basic one, underpinning much of the development of nuclear power and it is hoped that the determined reader will find rewarding the glimpses obtained of the issues faced and the approach used in the field of reactor physics.

Neutrons react with the materials present in a reactor in a wide variety of ways. The detailed study of how and why such reactions take place between the neutron and the nucleus of a particular atom is the province of nuclear physics and is not described in this chapter. It is sufficient to say that these reactions may be characterized by nuclear physics in the form of nuclear cross-sections. The particular cross-section $\sigma_r^N(E)$ is, for example, a measure of the probability that a neutron of energy $E$, will have a reaction $r$ with the nucleus of an atom of material $N$.

The reactor physicist is concerned with evaluating the rate at which reactions taken place and their consequences, so that he has a thorough understanding of the life cycle of the neutron sea. This is of vital importance because the number of fissions, the amount of neutron capture in structural material, the capture in control mechanisms, the number of neutrons leaking from the reactor, and other important reactions determine all the performance and safety characteristics of all fission reactors.

In the early 1940s the questions were largely those of criticality and critical mass; was it possible to obtain a self-sustained chain reaction in which neutrons born in fission just balanced those absorbed in the arrays of uranium and other materials? Knowledge of the basic cross-section was extremely sketchy and experiments involving sizeable quantities of the materials were required to observe whether and in what circumstances criticality could be achieved. Early work of this nature soon demonstrated that critical arrays could be formed using natural uranium rods with graphite, heavy water, or beryllium, but that otherwise enrichment in the fissile $^{235}U$ isotope was necessary. The results of large numbers of these integral experiments were correlated to very simple theoretical models of the competing neutron processes in order to determine the preferred design points of the early UK, US, Canadian, and French natural-uranium reactor systems.

In the thirty years of its history since then, the questions posed to reactor physicists have increased enormously in complexity and range but are still based on the evaluation of competing reaction rates, albeit to increasing accuracy, and often in very localized regions of the reactor. The technical ability to deal with these questions has, however, largely kept pace as a result of immense international programmes to measure the relevant cross-section data, the development of theoretical models of sufficient detail and complexity and the associated dramatic improvements in power and performance of the large, main frame computers. This last point cannot be overemphasized. It is difficult to imagine that today's detailed understanding of what happens with a nuclear reactor could have been achieved without the enormous and parallel developments in computer technology.

Questions of criticality are of course still relevant. It is still important to establish the enrichment required of a fuel loading and attention is also directed to the safe sub-criticality of storage arrays of fuel, of irradiated fuel in transport flasks and in ponds, and of reprocessing plants. But in terms of actual reactor design and performance the reactor physicist is increasingly concerned with fine detail and the demands for improvements in accuracy that arise from economic pressures. He is now also greatly concerned with the implications of possible faults or accidents, where the accuracy required may be less than for normal operations but where the abnormal conditions provide a challenging framework outside the territory well explored by experiment.

### 1.1 Steady-state conditions

In terms of normal, steady operating performance—the design point—most attention focuses on the heat source distribution in the reactor because this determines material temperatures. The source of heat is the fission reaction so that a knowledge of the rate at which this reaction takes place and its spatial variation throughout the reactor is necessary. Other questions of importance include the ability of control or shutdown mechanisms to remove the required number of neutrons from the cycle, the irradiation damage due to neutron absorption experienced by structural materials, the ability of shielding to limit neutron and radiation escape from the reactor core and hence the irradiation field inside the reactor vessel which surrounds it.

The long-term operation of reactors involves changes in the isotopic composition of the fuel through the various neutron interactions which can take place. These changes require careful evaluation because they affect such things as criticality, power distribution, and control effectiveness. A fuel element will normally remain in a reactor for a few years. It is then replaced by a fresh piece of fuel, but the timing of the change and the sequence in which all the elements are changed require careful optimization to ensure efficient fuel utilization, efficient use of discharge equipment, and minimum loss of power in the process. This is the fuel management problem and it requires for its solution an ability to model criticality, power distribution, and control and irradiation effects over several cycles of fuel life.

For these and many other requirements, the central need is the ability to evaluate the neutron distributions throughout the reactor on a day-to-day basis as a detailed function of neutron energy (because reaction rates are very sensitive to energy) and of geometry (both local—within an individual piece of fuel—and global—from piece to piece). The physics and mathematics of the processes are now reasonably well understood, and the basic cross-section data are generally of adequate accuracy for normal design and operational requirements. Precise solutions are, however, not possible with even the most powerful of currently available computers. The need is thus for approximation and simplification to bring the problem within range, and it is here that the reactor physicist is required to demonstrate his skill and judgement.

Developments in experimental reactor physics since those early lattice measurements have been less dramatic but equally important. Today, in general, experiments are performed to validate or quantify the uncertainty in theoretical models. They are selected to throw light on particularly sensitive or uncertain areas, and the demand for accuracy is high. Such accuracy is most readily obtainable in zero-power reactor systems where the neutron levels are sufficiently high to be measurable to the required accuracy but low enough not to give rise to irradiation or heat-production problems. Zero-power experiments are thus invaluable in checking our ability to predict such quantities as criticality, power (i.e. neutron) distribution, control rod effects, and shielding efficiencies. They can however provide only limited information on irradiation effects, and for these and ultimate confidence in theoretical models, measurements from operating power reactors are required.

## 1.2 **Kinetics and safety**

So far steady-state or very slowly varying (irradiation) effects have been considered, and these are appropriate to the economic power-producing abilities of the reactor. It is equally important, however, to ensure that the reactor can be comfortably controlled, that it can be shut down and started up safely and effectively, and that it responds benignly to faults or accidents which might just conceivably occur. These considerations require an understanding of how the neutron population might change with time following some disturbance to the normal criticality-balanced situation. It is necessary to add the time variable to the original theoretical models, and this in turn introduces the need for alternative approximations in their solution.

Experimental data are of course less readily obtainable for these time-transient events, because effects which are dependent on reactor power are often of major importance (and thus not available in zero-power reactors) and there is naturally a marked reluctance to introduce accident transients into operating power reactors. Fortunately, however, in safety analyses, conservative evaluations are often sufficient and the demand for high accuracy, best-estimate calculations is not great.

This chapter represents an attempt to summarize and survey the current state of the art in reactor physics. In §2 some of the key processes and concepts are introduced and illustrated quantitatively through tables and figures. In §3 the progress and evolution in reactor physics techniques are described, from the simple models fitted to large numbers of experiments to the large, sophisticated—and computer-bound—theoretical models of today. These two sections are rather general and provide in themselves a picture of the sort of issues with which reactor physics is concerned. For many readers this will be sufficient. Readers with a more specialized interest however, will find in §4 and §5 a detailed appreciation of current techniques. Here it is necessary to distinguish between the approaches used for thermal reactors and those appropriate to fast reactors. The differences stem from differences in the relative importance of some of the basic reactor physics processes but also from differences in the state of knowledge of the relevant cross-section data and the availability of measurements from operating power reactors. Finally, in §6, current outstanding problems are considered and an attempt is made to identify future trends and requirements.

# 2 Processes of reactor physics

## 2.1 **The fission process**

With nuclear power reactors, attention is centred on the interaction of neutrons with the atomic nuclei of a range of materials. Many such reactions are possible—capture, scattering, etc.—but of paramount importance is the fission reaction. Fission is theoretically possible for materials with a mass number of 100 or more, but it is only with higher mass numbers (above about 230) that the critical energy required for fission of the compound nucleus formed on absorption of the neutron is comparable with the excitation energy supplied by the neutron. Indeed for odd/even nuclei with an odd number of neutrons and an even number of protons (such as $^{233}U$, $^{235}U$, $^{239}Pu$, and $^{241}Pu$) the excitation energy of the compound nucleus (about 6.8 MeV) is greater than the critical energy for fission (about 6.5 MeV) so that no additional kinetic energy from the interacting neutron is required. Fission in these materials can thus take place with slow-moving (thermal) neutrons. For even/even nuclei ($^{232}Th$, $^{238}U$ and $^{240}Pu$) however, the excitation energy is smaller (about 5.5 MeV), and the neutron must contribute kinetic energy for the fission process to be possible. Thus with these materials fission is only possible with *fast* neutrons—neutrons with kinetic energies in excess of about 1 MeV.

The fission process leads to the production of:
  (a) two light elements—fission products;
  (b) a few neutrons; and
  (c) beta particles, neutrinos, and gamma radiation,
and is associated with a net loss of mass relative to the initial excited compound nucleus. The equivalent energy appears largely as the kinetic energy of the fission products. The total energy account associated with a fission of a number of fissile nuclei is shown in Table 2.1 and includes an allowance for the small amount of energy released when neutron capture rather than fission occurs. All except neutrino energy is converted to heat and most of it close to the fission event. Note, however, that the fission products (and in reactors some capture products) are unstable and result in decay chains which release energy by beta-particle and gamma-ray emission over a considerable period. Thus even after a reactor is shut down and the fission processes terminated, heat continues to be generated. The characteristics of

Table 2.1 Energy produced in fission

| Product | Mean emitted energy/fission (MeV) | | | |
|---|---|---|---|---|
| | $^{235}$U | $^{239}$Pu | $^{233}$U | $^{238}$U |
| Kinetic energy of fission products | 166 | 173 | 166 | 167 |
| Kinetic energy of fission neutrons | 5 | 6 | 5 | 5 |
| Prompt gamma radiation | 8 | 8 | 8 | 8 |
| Fission product decay | | | | |
|   Beta radiation | 7 | 6 | 6 | 9 |
|   Gamma radiation | 7 | 6 | 6 | 8 |
|   Neutrinos | 10 | 9 | 8 | 12 |
| Associated capture gamma radiation | 9 | 12 | 9 | 11 |
| Total | 212 | 220 | 208 | 220 |

the large number of possible decay chains are now sufficiently well known to allow this decay heating effect to be predicted to an accuracy of about 10 per cent. The magnitude of the effect is shown in Fig. 2.1. The total heat available from a fission event is about 200 MeV, and complete fissioning of 1 g of $^{235}$U yields a total amount of energy equivalent to a rate of 1 MW for 1 day, i.e. 1 MW–day of actual energy. This equals the heat produced by burning approximately 2.5 tonnes of coal and illustrates one of the advantages of nuclear power.

In detail there is a large number of possible fissile reactions, producing a range of possible pairs of fission products. The fission mass yield (i.e. the percentage of fissions which result in fission products of a given mass number) is shown in Fig. 2.2 and illustrates the typical double-hump distribution. The fission products are, of course, unstable and the resultant

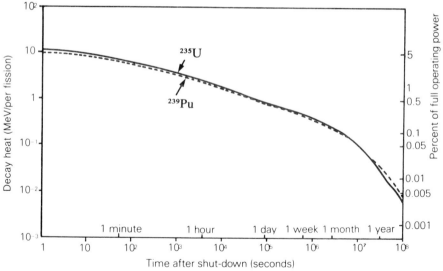

Fig. 2.1  Heat produced after reactor shutdown

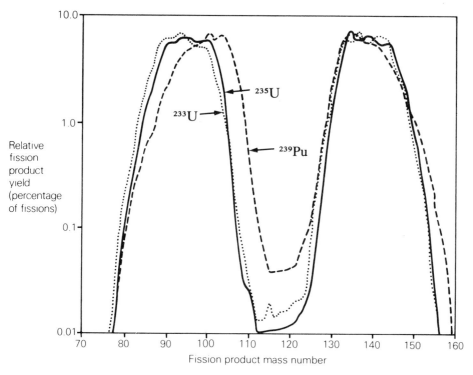

Fig. 2.2 Fission product mass yield

emissions of radiation, which might take place over long periods of time, are the cause of potential health hazard in handling them.

The number of neutrons produced on average in the fission process ($\nu$) depends significantly on the nucleus fissioned and the energy of the neutron causing the event. The values are shown in Table 2.2 for fission by a slow-moving neutron of energy 0.025 eV and also for fissions arising in a typical fast reactor with neutron energies around 1 MeV. Some absorptions in fissile materials, however, do not result in fission, the incident neutron being captured to produce an isotope of the original nucleus which is normally stable and non-fissile (e.g. $^{236}U$, $^{240}Pu$). Of more value than $\nu$ is

Table 2.2 Neutrons produced in fission ($\nu$)

| Fissile nuclide | Mean number of neutrons emitted per fission | |
| --- | --- | --- |
| | Fission by thermal neutrons | Fission by fast neutrons |
| $^{235}U$ | 2.42 | 2.46 |
| $^{239}Pu$ | 2.86 | 2.94 |
| $^{233}U$ | 2.48 | 2.51 |
| $^{232}Th$ | | 2.12 |
| $^{238}U$ | | 2.76 |

thus the number of neutrons produced per neutron absorbed ($\eta$). Values of $\eta$ are shown in Fig. 2.3 for the common fissile materials.

A self-sustaining chain reaction occurs when for each of the neutrons produced in the fission process, one further fission is ultimately achieved. A necessary condition is clearly that $\eta > 1$. It is, however, not a sufficient condition because in practice neutrons may escape (leak) from the reactor or be captured parasitically in other (e.g. structural) materials. Indeed it may be beneficial to arrange for the capture of neutrons in fertile material (such as $^{238}$U, $^{232}$Th), because by so doing new fissile material is produced. Thus for example, $^{239}$Pu is produced by neutron capture in $^{238}$U, as follows:

$$^{238}\text{U} + \text{n} \rightarrow {}^{239}\text{U}$$
$$\downarrow^{-\beta}$$
$$^{239}\text{Np}$$
$$\downarrow^{-\beta}$$
$$^{239}\text{Pu.}$$

For every neutron absorbed in fissile material there are $\eta\epsilon - 1$ available for other fates, where $\epsilon$ is a factor near unity included to allow for the small amount of fast neutron fission. In particular, the number available for capture in fertile material is:

$$C = \eta\epsilon - 1 - \text{other losses.}$$

This quantity $C$ is thus the number of new fissile atoms produced per fissile atom used and is known as the conversion ratio. Reactors with $C$ greater than unity produce more fissile material than they consume and are thus referred to as *breeders*. Clearly a requirement for breeding is that $\eta$ should be greater than 2 by an amount necessary to compensate for other losses, and this represents a major incentive to keep such losses small. Figure 2.3 shows that the best prospects for breeding lie with high-energy fission of $^{239}$Pu (Pu-fuelled fast reactors), or possibly with $^{233}$U—even with low-/intermediate-energy neutrons ($^{233}$U-fuelled thermal reactors).

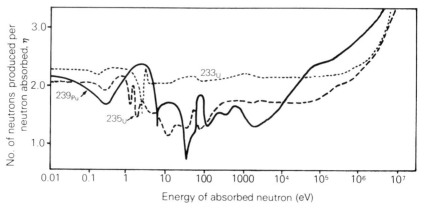

Fig. 2.3 Neutrons produced by absorption in fissile isotopes

## 2.2  **Neutron balance**

It has been noted in §2.1 that in order to sustain a critical chain reaction, one of the neutrons produced in the fission process must cause another fission before it is captured by non-fissile material or leaks from the reactor. In this way the rate of producing neutrons just balances the rate of removing them, and by appropriately adjusting the absolute level of the neutron population the desired heat production can be obtained.

Neutrons are produced in the fission process with the energy distribution spectrum shown in Fig. 2.4 and have a mean energy of approximately 2 MeV. The absorption cross-sections for the commonest fissile materials, however, increase significantly with decreasing neutron energy (Fig. 2.5 shows the effect for $^{235}U$), and at a rate greater than that for most parasitic or

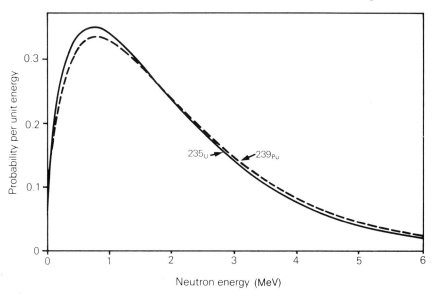

Fig. 2.4  Energy distribution of neutrons at birth

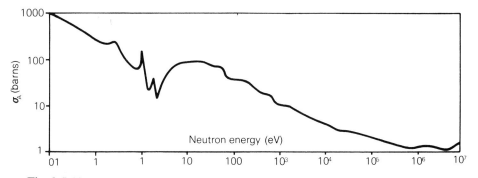

Fig. 2.5  Neutron absorption cross-section for $^{235}U$

fertile materials. In thermal reactors therefore advantage is taken of this increase by reducing the energy of the neutrons as much as possible, i.e. to that appropriate to thermal equilibrium with the material present. The loss of neutron energy is achieved by introducing a moderator—a material which slows neutrons through elastic collision without significant capture and in as few collisions as possible. Hydrogen, with the same mass as a neutron, is the most effective at slowing down neutrons, but unfortunately it does also capture neutrons. Heavy water ($D_2O$) has a lower neutron capture rate and, despite its lower slowing-down power, is by far the most effective moderator, but ordinary water ($H_2O$), graphite, and beryllium are also acceptable. In this way a significant number of neutrons in the energy range below 0.1 eV can be produced. A typical neutron energy spectrum in a thermal reactor covering eight decades of energy is shown in Fig. 2.6.

At any stage during the slowing-down and thermalization process, neutrons may be captured by non-fissile materials or leak from the reactor. Leakage may be reduced by introducing neutron reflectors around the reactor core. These reflectors consist of non-capturing, efficient neutron-scattering materials (i.e. moderators), which scatter back into the core neutrons that would otherwise completely escape.

Neutron capture may take place in a number of materials—fuel element cladding, structural members, moderators. The fission products referred to in §2.1 also capture neutrons, and to an increasing extent as they accumulate during fuel irradiation. These losses all represent wasteful and unavoidable inefficiencies in the overall neutron economy as well as leading to damage

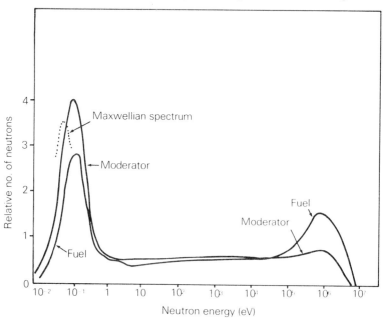

Fig. 2.6 Neutron spectra in a thermal reactor

and activation problems. Additionally, of course, capture takes place in fertile material and this may well be beneficial. Finally, the neutron balance is maintained by introducing particular neutron-capturing materials whose concentration can be readily adjusted. These control absorbers frequently take the form of movable rods containing boron or some other suitable material of high capture cross-section.

In a critical reactor these source, leakage, and removal components of the neutron economy are in balance at all parts of the reactor and the net neutron multiplication factor ($k$) is equal to unity. When $k$ is greater than 1, the reactor is supercritical and said to possess reactivity of amount $(k-1)/k$. Typical neutron balance sheets for a thermal and fast reactor are shown in Table 2.3.

## 2.3  Fast fission

It was shown in §2.1 that for the odd/even fissile isotopes the fission and absorption cross-sections increased markedly with decreasing neutron energy. It was also noted that the even/even (fertile) isotopes ($^{238}$U, $^{232}$Th,

Table 2.3  Neutron balance for typical critical thermal and fast reactors

|  | Thermal reactors | | Fast reactors | |
|---|---|---|---|---|
|  | Start of life | After several years | Start of life | After several years |
| Neutrons produced by fission in | | | | |
| $^{235}$U | 95 | 56 | 3 | 3 |
| $^{238}$U | 5 | 5 | 12 | 12 |
| $^{239}$Pu | — | 35 | 77 | 77 |
| $^{240}$Pu | — | — | 4 | 4 |
| $^{241}$Pu | — | 4 | 4 | 4 |
| Total | 100 | 100 | 100 | 100 |
| Neutrons removed by absorption in | | | | |
| $^{235}$U | 48 | 28 | 2 | 2 |
| $^{238}$U | 32 | 24 | 46 | 44 |
| $^{239}$Pu | — | 19 | 34 | 35 |
| $^{240}$Pu | — | 3 | 3 | 3 |
| $^{241}$Pu | — | 2 | 2 | 2 |
| $^{242}$Pu | — | 2 | | |
| Fission products | 0 | 5 | — | 2 |
| Structural materials | 8 | 7 | 8 | 8 |
| Moderator/coolant | 6 | 5 | | |
| Control absorbers | 2 | 1 | 3 | 1 |
| Leakage | 4 | 4 | 2 | 3 |
| Total | 100 | 100 | 100 | 100 |

$^{240}$Pu ...) are also fissile but only for high-energy (MeV) neutrons. The fission cross-sections for these fertile isotopes are shown in Fig. 2.7 and exhibit a threshold energy characteristic. We have seen that all neutrons begin life as fast neutrons even in thermal reactors, so that high-energy (fast) fission is a useful supplement to the predominantly low-energy fissions in the fissile isotopes. Typical values of the fast-fission factor (the number of neutrons appearing below 1.4 MeV for every neutron born in thermal fission) are given in Table 2.4.

Table 2.4  Values of the fast-fission factor ( $\epsilon$ ) for a simple graphite/natural uranium lattice

| Volume of graphite | Radius of natural uranium metal rod (cm) | | |
|---|---|---|---|
| Volume of uranium | 0.635 | 1.460 | 2.286 |
| 25 | | 1.045 | 1.061 |
| 50 | 1.025 | 1.041 | 1.052 |
| 100 | 1.018 | 1.039 | |

The value of this factor is highest when the fuel is in large lumps, and the neutrons emitted in the fission process have an opportunity to cause fissions in the fertile material before escaping into the moderator and being slowed down below the fast-fission threshold.

In fast reactors every effort is made not to degrade the neutron energy within the limitations posed by structural and cooling materials. Thus the contribution from fission in the fertile isotopes is significant. Typically 10 per cent of the fissions are from $^{238}$U in the core of a large Pu-fuelled fast reactor.

## 2.4  Resonance capture

The fission cross-sections exhibit rapid changes with energy in the 10–1000 eV region. For $^{238}$U the capture cross-section is shown in Fig. 2.8 and exhibits an even more widely varying form. These are resonance effects and represent the very high neutron absorption probability when the energy of the incident neutron, together with its binding energy in the compound nucleus, coincide with an energy level of the compound nucleus.

In thermal reactors, the resonance capture which takes place during the neutron slowing-down phase is very important. Although it leads to fissile $^{239}$Pu production, it is a marked debit in the neutron balance and consideration is thus frequently given to arrangements of fuel which lead to a high value for the probability of a neutron escaping resonance capture. This is especially true in the design of natural uranium systems where this effect is of major importance. This can be done by 'lumping' the fuel into rods, for then atoms of the resonance capture material within the fuel are shielded from neutrons of resonance energy owing to the filtering absorption by atoms at the

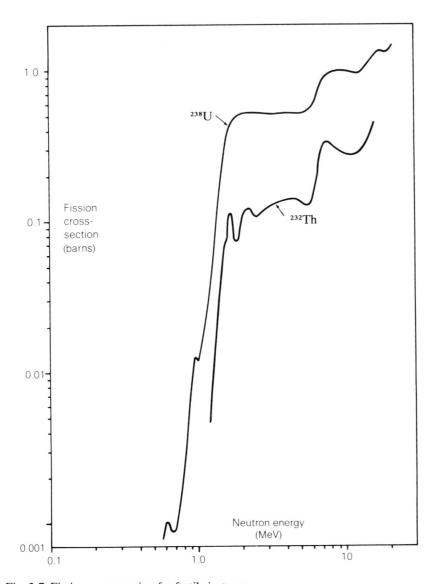

Fig. 2.7  Fission cross-section for fertile isotopes

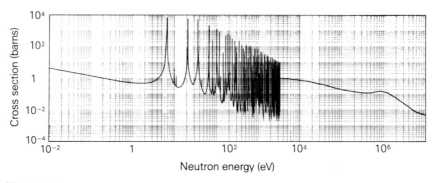

Fig. 2.8 Capture cross-section for $^{238}$U

surface. The actual amount of resonance capture depends on the effectiveness of the moderator in slowing neutrons down past the resonance-capture region, the amount of moderator, and the degree of shielding in the fuel. The evaluation of resonance capture is difficult on account of the large number of resonances to be considered (see Fig. 2.8), the rapidly varying energy dependence and high magnitude of the capture cross-section, and the shielding due to fuel heterogeneity.

It is not only the fertile materials $^{238}$U and $^{232}$Th which exhibit resonance structure in their neutron-absorbing characteristics, although it is most marked and significant for these. Fissile and structural materials also have resonances which introduce complications into the evaluation of neutron balance. Sodium too has a marked resonance at 3 keV (see Fig. 2.9), which is of considerable importance in fast reactor considerations. As an experimental technique, the capture of neutrons at a well-defined resonance energy is of value because it enables the neutron energy distribution in a reactor to be characterized by measuring the amount of activation in suitable detector materials.

## 2.5 Thermalization

During the slowing-down process, the neutron interaction with the moderator can be pictured as a series of billiard-ball elastic collisions between the energetic neutrons and atomic nuclei at rest. This results in the neutron losing energy by an amount inversely proportional to the mass of the moderating atom. Hydrogen is thus the most effective slowing-down material, although its advantage is offset somewhat by a small but significant capture process.

In fact, of course, the moderator atoms are not at rest, but exhibit motions dependent on the temperature and molecular forces involved. The effect is only important for neutrons with low energy, but in these circumstances neutron interactions with the moderator are much more complex. This process of neutron thermalization results in an energy distribution of

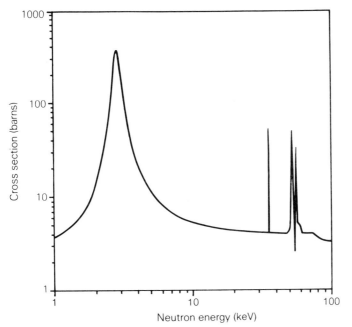

Fig. 2.9 Neutron cross-section for sodium

neutrons (energy spectrum) which depends upon the details of the moderator-scattering processes and the amount of absorption. For an idealized moderator consisting of a non-absorbing gas, the neutron energy spectrum takes a Maxwellian form determined by the temperature and is familiar from the kinetic theory of gases. In practice, the spectrum is harder than this (i.e. biased towards higher energies) (see Fig. 2.6) because of absorptions, but still peaks at about 0.1 eV.

An accurate determination of the thermal neutron energy spectrum is important because most of the reactions in a thermal reactor take place at these low energies in materials which may vary greatly in the energy dependence of their absorption cross-sections. It is a particularly difficult problem when materials are present with resonances at low energies—as is the case, for example, with $^{238}$Pu which has a strong resonance at 0.3 eV.

## 2.6 Geometric effects

It has been shown that resonance capture considerations may influence the choice of fuel element geometry. This is particularly true for natural uranium reactors where neutron economy is of major concern. With enriched-uranium fuels, however, other considerations are of more importance. Thus the heat-removal capabilities of the coolant will determine the amount of fuel surface, and the temperature at which the fuel is to operate or is allowed to reach in possible accident situations will determine the 'thickness' of the fuel.

Issues associated with the manufacture of fuel, the loading and unloading of fuel into and out of the reactor core, and the approach to the eventual reprocessing route may also have an impact on fuel element geometry. In the event, most fuel elements consist of bundles of cylindrical fuel pins of a size determined by overall optimization considerations—small diameter for sodium-cooled fast reactors and progressively larger for water- and gas-cooled reactors (see Table 2.5). For thermal reacators using efficient moderators (heavy water, graphite) the amount of such moderator present in the reactor core is large and sufficient to ensure that the fuel bundles are physically well separated in their individual cylindrical channels. For light water thermal reactors, however, the optimum amount of water required is quite small and similarly in sodium-cooled fast reactors the amount of coolant is kept to a minimum. For these systems the fuel elements are thus essentially in uniform and tightly packed lattices which are separated into bundles only for handling purposes.

The influence of the geometrical arrangements of the fuel upon reactor physics aspects is considerable, particularly in the case of thermal reactors. Heterogeneous fuel distributions lead to marked spatial variations in neutron density—neutron absorption rates are high in the fuel and the thermal neutron fluxes correspondingly low. These considerations have a marked influence on neutron balance and criticality and also on the heat source or power rating distributions. The evaluation of power distributions (i.e. fission rates) is a key aspect of reactor design since temperature in the fuel bundle invariably provides a limit on design and there is a need to optimize performance in some sense within this limit. Power distribution may be identified at two levels:

(1)  the pin-to-pin variations within a single fuel bundle or assembly,
(2)  the channel-to-channel (or assembly-to-assembly) distribution within the reactor as a whole.

In (1) the outer pins are in a higher neutron flux and with a softer spectrum than are pins near the centre of the bundle, but this variation may be compensated for by variations in enrichment to give a flatter power distribution and a more uniform approach to the limiting temperature. In (2), channels near the centre of the reactor will generally experience a higher neutron flux than those towards the core edge where neutrons are lost by leakage. Again this trend may be compensated for by using a lower-

Table 2.5 Fuel pin diameters

|  | Fast reactor | PWR | Advanced gas-cooled reactor | Magnox |
|---|---|---|---|---|
| Coolant | Sodium | Water | $CO_2$ Gas | $CO_2$ Gas |
| Fuel pin dia. (mm) | 5–8 | 8–10 | 14.5 | 28–29 |
| No. of pins per unit | 300–400 | 200–300 | 36 | 1 |

enrichment fuel in the centre of the reactor core to produce a flatter power distribution and reduce peak fuel-rating conditions.


## 2.7 Burn-up

The production of heat by 'burning-up' fissile material is the primary object-ive of nuclear power, and the modelling of this process is a vital aspect of reactor physics. Our concern is with the effect of neutron irradiation on reactor core materials—not from the mechanical or metallurgical aspects (although these are very important)—but because neutron irradiation leads to neutron absorption, changed isotopic composition, and hence changed reactor physics characteristics.

Most of the neutron absorption takes place in the fuel, and it is here that isotopic changes are most significant. The change in concentration of fissile species is, for example, given by

$$\frac{dN_f}{dt} = S_f - N_f \int_E (\sigma_f(E) + \sigma_c(E))\phi(E)\,dE, \tag{1}$$

where $S_f$ represents any source term for the fissile material, and in the removal term the energy dependence of cross-sections and neutron spectrum have been explicitly introduced. Of course, the neutron spectrum depends upon the isotopic composition and is thus itself burn-up dependent (see Fig. 2.10). In the low-enriched fuel cycle of thermal reactors there is no source term for $^{235}U(S_5 = 0)$ so that the amount of this isotope falls steadily as the burn-up proceeds. For $^{239}Pu$, however, there is a source term from neutron capture in the fertile $^{238}U$, i.e.

$$S_9 = N_8 \int_E \sigma_c(E)\phi(E)\,dE. \tag{2}$$

The variation in fuel isotopic composition as burn-up proceeds is illustrated in Figs. 2.11 and 2.12 for typical thermal and fast reactor systems.

It has already been shown that in the fission process fission products are born with a wide range of mass, energy, and degree of instability. They decay more or less rapidly by beta and gamma emission to more stable forms and in so doing present a substantial potential health hazard. The safe containment of these fission products is a major design and operational requirement. From the physics point of view, however, their main im-portance is that they can capture neutrons.

The build-up of fission products is normally cumulative so that the capture is most pronounced in highly irradiated fuel. Typical values for the proportion of neutrons captured in this way are shown in Table 2.3 and the distribution of captures amongst the most important fission products is shown in Table 2.6.

These neutron losses are the result of captures in something of the order of 100 important nuclides. The most significant fission product of all for thermal reactors is $^{135}Xe$ which has a high yield and also a very large capture

Table 2.6 Distribution of neutron captures in the fission products of irradiated fuel

| Thermal reactors | | Fast reactors | |
|---|---|---|---|
| Isotope | % of total f.p. captures | Isotope | % of total f.p. captures |
| $^{135}$Xe | 25 | $^{101}$Ru | 8 |
| $^{149}$Sm | 10 | $^{103}$Rh | 8 |
| $^{143}$Na | 8 | $^{105}$Pd | 7 |
| $^{103}$Rh | 7 | $^{133}$Cs | 7 |
| $^{147}$Pm | 6 | $^{99}$Te | 6 |
| $^{131}$Xe | 5 | $^{102}$Ru | 5 |
| $^{133}$Cs | 5 | $^{147}$Pm | 4 |
| etc. | | etc. | |

cross-section. Typically up to 2 per cent of neutron absorptions may take place in $^{135}$Xe but of more importance is the fact that the precise amount depends upon the local fission rate and this has a marked influence on variable power operation and stability considerations (see Appendix A).

Other new materials are produced as a result of neutron capture reactions, and of particular importance in reprocessing and long-term storage are the actinides with long half-lives. Again the build-up is broadly a linear function of irradiation and typical values of the amount and associated activity for some of the most important higher actinides are shown in Table 2.7.

The effect of these isotopic changes on reactor physics characteristics is very marked in thermal reactors. The neutron balance is clearly changed (see Table 2.3) and, in general, neutron-absorbing material has to be adjusted within the reactor to maintain criticality. In PWR systems, for example, there is a steady fall in the fuel fissile content as burn-up proceeds (see Fig. 2.11) and significant quantities of absorbing material must be removed from the core to balance this. In fast reactors this adjustment is modest (see Fig. 2.12), because the breeding characteristics ensure adequate production of new fissile material as the older material is burned.

Table 2.7 Production rate and associated activity of most important actinides

| Isotope | Higher actinide production per 1000 MW reactor | | | |
|---|---|---|---|---|
| | kg/year | | Ci/year | |
| | PWR | FR | PWR | FR |
| $^{241}$Am | 2 | 11 | $6.10^3$ | $4.10^4$ |
| $^{242}$Cm | 0.1 | 0.3 | $3.10^5$ | $10^6$ |
| $^{244}$Cm | 0.8 | 0.3 | $7.10^4$ | $3.10^4$ |

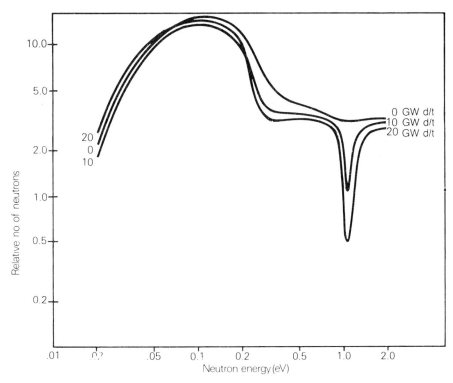

Fig. 2.10 Effect of irradiation on thermal neutron spectrum in a thermal reactor

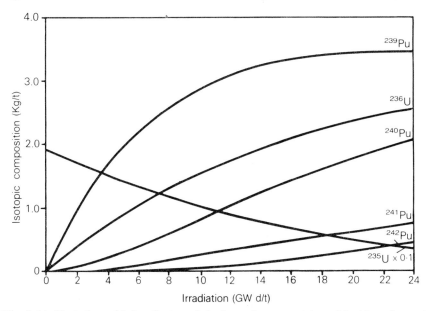

Fig. 2.11 Variation with irradiation of the isotopic composition of fuel in a thermal reactor

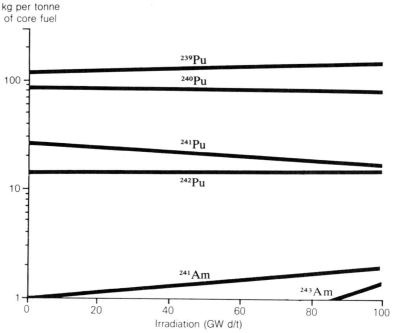

Fig. 2.12 Variation with irradiation of the isotopic composition of fuel in a fast reactor

## 2.8 Fuel management

Of course, burn-up is not uniform throughout a reactor; it proceeds most rapidly where the neutron-flux levels are highest. As a consequence there are relative changes in the balance of events and in particular the spatial distribution of the heat source determined by the fission rate will change as burn-up proceeds.

These changes in reactivity and relative power distribution are of great importance in reactor design. In general it is desirable to minimize them so that as much fuel as possible can be operated at optimum performance conditions. Fuel management systems are aimed at doing just this and many strategies have been devised.

For advanced gas-cooled reactors for example the wide separation of fuel channels permits on-load refuelling, one fuel bundle at a time, as required to maintain criticality. The fuel management problem is to determine the order in which this should be done so as to maintain full power output without excessive demand on individual fuel elements. For other reactor systems an off-load, periodic batch fuelling strategy may be necessary and this requires the introduction of techniques for changing the effectiveness of neutron absorbers with time. In pressurized-water reactors this is done by including in the fuel bundle special pins of burnable poisons, whose effectiveness falls as they capture neutrons, and by then varying the concentration of boron dissolved in the water. Other strategies may involve shuffling fuel bundles from place to place within the reactor core and the use of movable nuetron absorbers to shape the power distribution to a desirable form. The computer models developed to study and evaluate alternative fuel management stra- tegies have been amongst the most complex and challenging within the reactor physics field.

## 2.9 Mathematical representations

Some of the processes vital to the physics considerations of neutron processes within a nuclear power reactor have been discussed and, their mathematical representation will now be considered.

The equations which determine the neutron population are essentially those of neutron balance in which the rate of change of the number of neutrons in a particular class at any point is determined by the difference between the number produced and the number removed. Consider the class of neutrons with a particular energy $E$, moving in the direction represented by the vector $\Omega$ at a point $\mathbf{r}$ in the reactor. (Strictly neutrons in an energy band of width d$E$ about $E$ are considered, but this complication may be disregarded for present purposes.) The sources of neutrons into this class are several:

(a) directly, as a result of the fission process;
(b) scattering at the point $\mathbf{r}$ into $(E, \Omega)$ from some other values $(E', \Omega)$; and
(c) the flow of neutrons of $(E, \Omega)$ to $\mathbf{r}$ from elsewhere in the reactor.

Neutron losses from the class are given by the converse procedures:

(i) the absorption of neutrons in the class by any of the nuclei of the atoms of the materials present at $\mathbf{r}$;
(ii) scattering at the point $\mathbf{r}$ of a member of the class $(E, \Omega)$ which results in a new energy of the neutron $E'$ and a new direction of motion $\Omega'$; and
(iii) the outflow from $\mathbf{r}$ of neutrons of energy $E$ in direction $\Omega$.

The mathematical expression of the neutron balance in these terms is the Boltzmann equation of neutron transport and is an equation familiar from thermodynamic and other transport processes. For the present it should be noted that there are several independent variables—$\mathbf{r}$, $E$, $\Omega$, and time $t$—and that the geometry within which the solution is sought may be extremely complicated and heterogeneous.

Even using the most advanced computers, it is practically impossible to obtain solutions in anything approaching full generality. Suitable approxi- mations or simplifications are inevitably required, the choice being a matter of skill and judgement in relation to the particular question under con- sideration. Thus for steady-power operation or even slow burn-up processes, the time variable may be eliminated. For many problems the energy variable may be represented as a series of discrete energy bands or *groups* within which the cross-sections are assumed to be constant. The number of such groups depends upon the application but typically ranges from a few to a few thousand. Of course, the treatment brings with it the problem of deriving appropriate group cross-sections in such a way that the product of the group neutron flux and the group cross-section yields a correct reaction rate.

Another type of approximation frequently employed is to replace the integro-differential equation of neutron transport by the partial differential equation of the diffusion process. This diffusion approximation will be good where the material composition of the reactor is uniform and the neutron flux is slowly varying in space and direction. The advantage of simply working with sets of second order partial differential equations is so great, however, that even when these conditions are not strictly valid ingenuity is rewarded in deriving equivalent diffusion parameters to force a good diffusion solution.

Thus using the group and diffusion approximations the neutron balance equations obtained are of the form:

$$\frac{1}{v_j}\frac{\partial \phi_j}{\partial t} = \nabla \cdot D_j \nabla \phi_j - \Sigma_{aj}\phi_j - \Sigma_{sj}\phi_j + \sum_{j'} \Sigma_{sjj'}\phi_{j'} + \lambda \chi_j \sum_{j'} \nu \Sigma_{fj'}\phi_{j'}. \qquad (3)$$

This is the balance equation for the neutron flux $\phi_j$ in a unit volume and the $j$th energy group, the terms on the right hand side representing (in order):

(1) net leakage of neutrons of energy $j$
(2) absorption removal of neutrons of energy $j$
(3) scattering of neutrons out of energy group $j$
(4) source into energy group $j$ of scattered neutrons of initial energy $j'$
(5) fraction of fission neutrons bred in group $j$ as a result of the fission process in any group $j'$

while on the left hand side $v_j$ represents the velocity of neutrons with energy $j$. This type of approximation is not uncommon and has led to the extensive development of numerical techniques and computer strategies for obtaining numerical solutions.

For a reactor operating steadily at constant conditions the left-hand side of eqn (3) is zero and the remaining set of equations (one for each energy group) determines the relative neutron distribution for the steady-state reactor. The absolute level of the neutron population can be whatever is required—low for a zero-energy reactor, high for a power reactor. In the latter case the neutron levels are set by reference to the heat output, (i.e. fission rate) required as fol' jws:

$$\sum_i \Sigma_{fj}\phi_j = F. \qquad (4)$$

## 2.10 Feedback effects

Some of the coefficients of eqn (3) may be quite strong functions of other, dependent variables such as material temperature or coolant quality. These in turn will be functions of the neutron flux through the heat source term and thus there may be significant naturally occurring feedback terms associated with sets of auxiliary equations.

Temperature effects are perhaps the most important and several may be identified. One effect of increasing fuel temperature is to increase the thermal motion of the $^{238}$U atoms, broaden the relative energy of neutron interaction (a Doppler effect), and thus reduce the effectiveness of the surface absorption filter for neutrons of resonance energies. The increase in neutron flux caused by a reduction of surface absorption allows more $^{238}$U captures to take place and reduces the number of neutrons available for fission. The role of $^{238}$U is thus to oppose any tendency for the fuel temperature to change.

The effect of moderator temperature is quite different. Here it is the thermalization process which is affected, a high moderator temperature yielding a spectrum of neutrons biased towards higher energy. The effect of this spectrum change depends upon the isotopes present in the fuel, but it could be marked if significant amounts of $^{239}$Pu were present because of a large resonance at 0.3 eV. In general, the hotter the moderator the more events occur in the resonance, and since $^{239}$Pu is fissile this leads to increased heating and neutron source especially in low-enriched reactor designs.

Feedback effects from a two-phase coolant may also be important. In a boiling water reactor in which steam is generated in the reactor core, the variation in coolant density along a channel will affect local moderation and local absorption (since water is both an effective moderator and non-trivial neutron capturer). The effects are complex but it is sometimes possible to design for a net zero effect. In a fast reactor, however, voidage in the coolant sodium would have a marked and quite different effect. Local loss of sodium would reduce what little effective moderation there is in a fast reactor and harden the spectrum so that an increased number of neutrons remained at high energy to be available for the fission process there. With plutonium fuel this leads to an increase in neutron multiplication. On the other hand, loss of sodium also leads to increased neutron leakage from the core and this loss of neutrons more than compensates the increase from the hardening spectrum in small reactors and at the edges of large reactors.

One other feedback effect of particular importance in thermal reactors is that associated with the fission products. In general a high neutron flux leads to a high level of fission products and their parasitic absorption. Of particular importance in thermal reactors is $^{135}$Xe which has an enormous cross-section for thermal neutrons, but also has a natural decay leading to a loss rate comparable in magnitude with that from neutron capture. The total capture in $^{135}$Xe is thus high and sensitive to flux level and has a significant effect in 'flattening' power distributions. The effects of $^{135}$Xe are discussed more fully in Appendix A. There are many other natural feedback mechanisms and all contribute an interacting complexity to the neutron balance problem.

## 2.11  **Reactor kinetics and safety**

So far steady-state, time-independent issues have been implicitly discussed, and in so far as reactors are designed and required to operate for long periods at constant power, this is entirely appropriate. There are occasions, however, at startup and shutdown when kinetic effects such as the variation of $^{135}$Xe level with time may be important. Of possibly even more importance than these operational transients are transients resulting from postulated accident situations such as loss of coolant or inadvertent control rod withdrawal. For these situations explicit consideration of the time variable is necessary.

If, for illustrative purposes, a single energy group representation for a very large reactor in which leakage can be neglected is considered, then eqn (3) simplifies to

$$\frac{1}{v}\frac{d\phi}{dt} = (k-1)\Sigma_a\phi, \tag{5}$$

where $v$ is the velocity of the neutron and $k$ is the neutron multiplication constant of the reactor; when $k = 1$ the reactor is just critical and the neutron flux, $\phi$, remains constant. The time constant for changes in $\phi$ is $(1/(k-1))$ where $1 = (v\Sigma_a)^{-1}$ is the lifetime of a neutron. This is typically in the range $10^{-3}$ to $10^{-5}$ seconds in a thermal reactor and implies a very rapid response to changes in the value of $k$. Thus for example if $k = 1.01$, the reactor power in a graphite-moderated reactor would increase by a factor of 20 000 in 1 s.

If this were the whole story reactors would be very difficult to control. In fact, the position is eased to a very considerable extent because not all neutrons are produced directly in the fission process; a small fraction is produced after a significant delay. Under normal, steady-operating conditions, these delayed neutrons are indistinguishable from their prompt brethren. In non-steady conditions, however, they diminish the rate at which the total neutron flux can change, since for this small fraction the rate of change is limited by the delay, which may be of many seconds.

The prompt neutrons are produced directly and immediately the fission process takes place, together with the pair of fission fragments into which the fissile nucleus has been split. These fragments are basically unstable and must undergo a series of $\beta$-disintegrations before they become stable. Very occasionally such a $\beta$-decay leads to an isotope in a sufficiently excited state that it may immediately emit a neutron. Thus the delayed neutrons are only indirectly the product of the fission process and are delayed by a time which is that associated with the $\beta$-decay characteristics of the precursor fission products. Altogether it appears to be possible to identify several such precursors, but in practice consideration is limited to six groups of delayed neutrons, each characterized by a particular precursor decay constant. To a certain extent this is the result of fitting expressions of the appropriate form to the experimentally observed time behaviour of the total delayed neutron population. For some at least of the groups, however, an identification has been made of the particular decay chain responsible for the delayed neutron. Data relevant to the six delayed groups arising from fission in $^{235}U$, $^{238}U$ and $^{239}Pu$ are given in Table 2.8.

The effect of the delayed neutrons is discussed further in Appendix B and leads to a considerable increase in effective neutron lifetime and the ease with which reactors may be controlled.

Table 2.8 Delayed neutron data

| Delayed group | Decay constant, $\lambda$, $(s^{-1})$ | Fractional yield, $\beta_i$ | | | | |
|---|---|---|---|---|---|---|
| | | $^{235}U$ | $^{238}U$ | $^{239}Pu$ | $^{240}Pu$ | $^{241}Pu$ |
| 1 | 0.0127 | 0.00025 | 0.00021 | 0.00008 | 0.00008 | 0.00006 |
| 2 | 0.0320 | 0.00154 | 0.00245 | 0.00068 | 0.00081 | 0.00139 |
| 3 | 0.128 | 0.00134 | 0.00191 | 0.00033 | 0.00039 | 0.00051 |
| 4 | 0.304 | 0.00259 | 0.00713 | 0.00081 | 0.00110 | 0.00221 |
| 5 | 1.35 | 0.00089 | 0.00447 | 0.00024 | 0.00038 | 0.00104 |
| 6 | 3.63 | 0.00016 | 0.00123 | 0.00005 | 0.00008 | 0.00017 |
| Total | | 0.00679 | 0.01741 | 0.00219 | 0.00284 | 0.00539 |

The feedback effects discussed earlier in this Section will have an effect on the kinetic characteristics of reactors and hence on the systems designed to control them. The $^{238}U$ fuel temperature effect provides negative feedback since increase in flux gives increase in temperature, which leads to increased $^{238}U$ capture and hence a decrease in neutron flux. This is a very important self-stabilizing effect present in both thermal and fast reactor systems, and one which acts very rapidly.

Other feedback effects, however, may be positive and destabilizing as, for example, the effect of moderator temperature in the presence of $^{239}Pu$. The $^{135}Xe$ feedback is particularly interesting because it is present in all thermal reactors and, in large reactors, can lead to instabilities not only in the total reactor power but in the spatial power distribution also. Thus the combination of feedback effects in the large natural uranium reactors gives rise, in the absence of control systems, to a number of different modes of spatial instability in the neutron flux distribution. It would, for example, be possible for the power to rise in one half of the reactor and to fall in the other. These natural instabilties in higher modes are associated with fairly long time constants but still require special detection and control systems to suppress them.

# 3 The evolution of reactor physics methods

## 3.1 Early natural uranium systems

The reactor chosen in Britain for the first stage of the nuclear power programme was the $CO_2$-cooled graphite-moderated system using fuel rods of natural uranium metal clad in Magnox. The elimination of $^{235}U$ enrichment as a potential design variable placed emphasis on the need to optimize the reactivity achievable in natural uranium/graphite systems through appropriate selection of rod diameter, lattice pitch and coolant channel size.

In the early 1950s it was quite out of the question to calculate the performance of such reactors to anything approaching the accuracy needed. Two main reasons accounted for this. Knowledge of the magnitude and energy dependence of the nuclear cross-sections of the materials of the reactor core was poor and, even had this information been available, the arithmetical operations required to handle such information, and to solve the problems of neutron transport, were beyond the computing capability then available. Indeed all that could be handled was a simple theoretical model of the reactor which took into account the main feature of the behaviour of the neutron population. The simple model chosen was based on the four-factor formula.

$$k_\infty = \epsilon \cdot p \cdot f \cdot \eta, \tag{6}$$

which expresses the neutron multiplication, $k_\infty$, of an infinite clean lattice of the fuel/moderator mixture in terms of the fast fission factor, $\epsilon$, the resonance escape probability, $p$, the fuel-absorption probability, $f$, and the number of neutrons produced per absorption in the fuel, $\eta$. Highly simplified models were used to evaluate $^{238}U$ resonance capture (to obtain $p$) and fast fission events; to represent the thermalization process, the thermal neutron spectrum was assumed to take the idealized Maxwellian shape with a high-energy tail. In a real, finite array, the neutron multiplication is reduced by neutron leakage to

$$k = k_\infty \cdot L \tag{7}$$

where $L$ is the non-leakage probability, and the evaluation of neutron leakage and distribution were based on one- or two-energy group representations (see §2). Of course, for a practical reactor the clean, unirradiated value of $k$ given by eqn (7) must be greater than unity and by an amount sufficient to offset losses in reactivity due to such effects as burn-up.

These various lattice factors were expressed plausibly but simply in terms of such physical properties as density, volume, and temperature of the materials present and some idealized nuclear cross-sections. These expressions were then fitted to a large series of experiments so that the formulae (6) and (7) could be used to interpolate values of $k$ to any rod diameter, lattice pitch, or coolant channel size.

The experiments employed were thus of prime importance and various types were required. Perhaps their most important aspect is that they need not necessarily be made in critical systems, with $k = 1$. It is possible (and indeed desirable for operational reasons) to operate with a small, sub-critical system having a neutron source away from which the neutron population decays in a manner characteristic of the $k$-value of the lattice. Measurement of the thermal neutron flux at a number of points by the activation of appropriate detector foils enables an appropriate exponential flux distribution to be fitted, and the size at which a reactor with the same lattice would go critical can then be obtained. A photograph of an early exponential experiment is shown as Fig. 2.13.

Although these exponential experiments yielded the key information on criticality, supplementary experiments played a major role. Thus the detailed spatial distribution of the thermal neutron flux within an individual lattice cell was measured in so-called fine-structure experiments and was useful in adjusting estimates of the fuel thermal absorption probability, $f$. For these experiments again a sub-critical mass is adequate although for good accuracy a high neutron intensity source is required and is generally provided by an adjacent critical reactor.

Measurements were also made specifically to provide evidence of the accuracy of the evaluation of $\epsilon$. The quantity measured was the ratio of fissions in $^{238}U$ relative to fissions in $^{235}U$. The technique involved the irradition of uranium foils positioned to sample the whole cross-section of the fuel in a lattice of fuel pins, and the measurement of the fissions by counting the fission-product gamma activity. The use of foils of different $^{235}U$ enrichment enabled the distinction between $^{238}U$ and $^{235}U$ fissions to be made.

Yet another complementary measurement was aimed at providing information on the value of resonance escape probability, $p$. Here the quantity measured was the neutron capture in $^{238}U$ (which is largely at resonance energies) relative to fissions in $^{235}U$. The number of $^{238}U$ captures in a uranium foil can be obtained by deriving the amount of $^{239}Np$ produced through measurement of its radiation emission as it decays to form $^{239}Pu$.

In the late 1950s and early 1960s these highly simplified conceptual models of reactor physics processes were correlated and fitted to a very large and comprehensive series of experiments. The models thus effectively provided a method of interpolating within the experimental data range and for modest extrapolation outside it. They were then directly used as the basis for the selection of the lattice characteristics of the Magnox and equivalent French natural uranium systems and were entirely adequate for this limited purpose.

Prediction of the actual critical loading of the reactor and the relative power distribution from fuel channel to fuel channel requires, of course, a model for the whole reactor core and reflector. This was achieved by representing the reactor as a homogeneous paste with neutron-absorbing, scattering, production, and diffusion properties derived from the lattice

Fig. 2.13 An exponential experiment showing a simple lattice arrangement of fuel rods and moderator

constants discussed above. One- or two-energy group approximations (see §2) were used and the resulting equations solved either analytically (for say, uniform cylindrical reactor) or by some direct numerical approach. This macroscopic treatment of the overall reactor was satisfactory because the Magnox reactors are large, with slowly varying lattice properties and with low neutron absorption characteristics.

There were, of course, a number of problems of application that greatly concerned reactor physicists at this time, of which two are perhaps worthy of particular mention. The first arose from the attempt to incorporate within a homogeneous diffusion reactor model the actual neutron migration by streaming along the large gas coolant channels. Clearly neutrons are able, on average, to move more readily in this direction than in directions normal to fuel channels. This streaming effect was represented by allowing asymmetric diffusion parameters within the theoretical models and by special experiments aimed at measuring the equivalent diffusion asymmetry.

The second example related to the problem of designing appropriate control systems in the light of the destabilizing temperature and xenon effects referred to in §2. There was a need to contol and suppress a number of spatial modes of potential instability, and so a three-dimensional reactor model was required with explicit solution of the time variable and the feedback phenomena. Such a computer model was developed in 1959 but the available computer power limited it to one-neutron-group form and a discrete spatial representation of a few thousand mesh points. Nevertheless, it was proved to be adequate by comparison with spatial transient experiments performed at Calder Hall, and was much used in the design and evaluation of the automatic sector control systems characteristic of Magnox and AGR reactors.

## 3.2  Enriched thermal reactors

Elsewhere in the world the focussing of attention on enriched-uranium fuel, cooled and moderated by water, made the reactor physics problem more difficult. The degree of enrichment itself introduces an important additional parameter, but consequential effects associated with the economic desire to obtain high rating and high burn-up also introduce geometrical and other complications into the conceptual fuel element and the neutron-balance story. In the UK attempts were made to carry over to the enriched AGR the simple modelling approach so successfully employed for the natural uranium Magnox reactors. But here too the additional complexity of the bundle geometry, the enriched and hence highly absorbing fuel, and the shifting emphasis away from simple initial criticality considerations to burn-up and other effects soon proved too much.

The basic principle of separating the problem into two parts—first studying the lattice cell in detail and then synthesizing an array of such cells into a representation of the whole reactor—remains broadly acceptable. However, the fine structure neutron flux variations within the multipin fuel element and the characteristics of the neutron energy spectra as a result of the higher neutron absorption and plutonium build-up are now rather complex and impossible to represent within a simple, few-parameter model.

The response, of course, is to adopt a more detailed, basic model of the important reactor physics processes, and to change the basis of the supporting experiments from the sweeping study of parametric variables to ones rather specifically designed to investigate particular uncertainties in the basic model. This change in the UK from a correlation-based method to basic methods was a fundamental one which paralleled similar developments for water-moderated reactors, and placed the approach to enriched thermal reactor physics on a broadly consistent plane.

By the mid-1960s, the reactor physics processes and their mathematical descriptions were well understood, and detailed cross-section data were becoming available for the important isotopes. Computer power, however, was limited and so in the early basic models gross approximations and simplifications were still necessary. Particular attention was devoted to the development of approximate neutron thermalization treatments and the representation of resonance $^{238}$U capture in close arrays of fuel pins. Some form of smearing of the fuel bundle was inevitably involved and treatment of the burn-up process included very limited isotopic representation. Such lattice models of the 1960s—MUFT/SOFOCATE for PWR in the US, METHUSELAH and ARGOSY in the UK—were by today's standards certainly simple, but they indicated the basic determination to model the neutron process as accurately as possible within the current limits of basic nuclear data and computer power.

The overall reactor treatment needed to evolve also: enriched reactors are smaller than Magnox reactors, are capable of high burn-up, and thus exhibit more marked variations in composition between lattice cells. Thus whilst some form of homogenization and few-group representation of the reactor was effectively necessary, care was needed in the correct choice of the appropriate parameters and numerical solutions of the overall diffusion equations were required with sufficient detail to represent the quite marked bundle-to-bundle neutron flux variations. As computer power increased more elaborate two- and three-dimensional finite-difference codes were developed with million mesh point capabilities and considerable effort was devoted to devising numerical algorithms which would enable a convergent flux solution to be obtained rapidly and effectively.

The basic accuracy and confidence in these early fundamental models was not high and considerable reliance on experimental data was still necessary for both model validation and improvement. The emphasis turned, however, away from the simple sub-critical lattice experiments except in special circumstances such as those involving plutonium-enriched fuels. Instead, the emphasis swung towards critical systems which, because of the use of enriched fuel, could be made small enough for sensible economic and experimental purposes and yet large enough to contain typical power-reactor features. It was still required to measure reaction rate distributions and reaction rate ratios throughout these so-called integral experiments so that for access purposes the neutron flux level in these critical experiments could not be too high. These zero-energy reactor experiments did, however, of necessity involve sophisticated and well-engineered plant requiring close attention to operational practices for successful and economic programmes (see Fig. 2.14).

The target accuracies demanded of the theoretical models in order to minimize design margins and operational problems received much attention through the 1960s. In 1967, for example, an IAEA Panel agreed on economic grounds that target accuracies were to include $\pm 5$ per cent on relative power generation throughout a reactor and $\pm 2.5$ per cent on the number of fissions experienced by a fuel element at discharge. Targets such as these were formidable and called for more sophisticated models with less approximate treatments of the key reactor physics processes and also for integral experiments involving still more of the important features characteristic of operating power reactors. Experiments were performed with plutonium-bearing fuels to simulate the physics effects of fuel irradiation and in some cases actual irradiated fuel was used. Reactor heating was introduced to study temperature effects and various combinations of absorber and control rod arrays were studied. Effects not accessible on zero-energy reactors were measured, albeit with less precision, in operating power reactors. In this way, for example, the isotopic analysis of fuel irradiated in a well characterized environment provided checks on the theoretical burn-up model, and attempts to reproduce the observed behaviour of early power reactors under transient conditions and over long periods of operation provided ultimate checks on the accuracy obtainable by the theoretical models.

The required experimental techniques increased in organizational complexity (with the size and complexity of the experiment) and also in precision to meet the demanded accuracy targets. The gamma and beta counting of irradiated fuels was automated (see Fig. 2.15) and became more accurate with the introduction of advanced detectors. Increasing care was taken in the matching of detectors to the actual geometry of fuel element structure so that, for example, in experiments involving coated-particle fuels for high-temperature reactor development, the particles themselves were dissolved to measure the $^{238}$U capture by $^{239}$Np counting (Fig. 2.16).

The struggle for improved accuracy through the 1970s on an increasing number of topics continued to drive the development of theoretical methods to increasingly detailed representation of the reactor materials and geometry, and of the relevant neutron processes. The Boltzmann equation of neutron transport (rather than the simpler diffusion approximation) is increasingly solved for the fine-structure, lattice-cell part of the model. The particular technique used depends somewhat on the geometry, accuracy and neutron process of particular concern.

Fig. 2.14  A zero power experimental reactor

Two of the techniques are of particular interest. In the first a succession of individual neutrons is tracked through a detailed representation of the reactor geometry. The events experienced by the neutron are sampled from the range of possibilities on a random number basis. The technique is thus effectively one of computer simulation of what actually happens, and because it depends upon a probabilstic argument was christened the *Monte Carlo method*. The drawback is that computer power is too limited in practice, to allow more than a relatively few neutrons to be tracked in this way and the statistical basis of the results may be poor.

In the second technique less accurate but deterministic calculations are performed by simplifying the problem and evaluating events averaged over spatial regions. This is done by solving the transport equation for a mesh of lines between each pair of regions in order to determine the average probability that a neutron born in one will have its next collision in the other. These collision probabilities can then be used to link all regions together in a matrix formulation of the problem which may be readily solved by iterative or matrix-inversion techniques.

The separation between local and overall reactor representations is still a necessary but acceptable approximation. Techniques have been developed,

Fig. 2.15 A counting laboratory

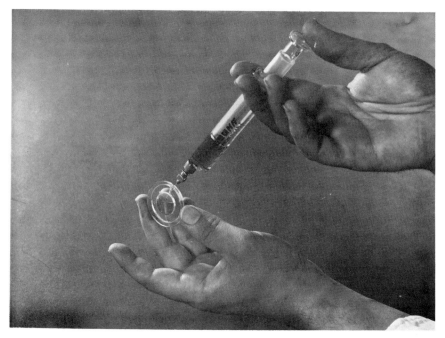

Fig. 2.16 Injection of a counting sample for measuring $^{238}$U capture in particulate fuel

however, of minimizing the errors introduced by treating groups of lattice cells or by iterating between local and reactor calculations in such a way that the environment of the individual cell as given by the reactor computation is incorporated in the lattice-cell calculation through appropriate boundry conditions.

Today the advanced theoretical methods available tend to be enshrined in inter-connected computer code modules from which the user can select his particular choice of representation or approximation to suit his needs and his purse. Such computational schemes require associated nuclear data files and an organizational capability to enable computed results to be transferred readily from one code module to another or to be printed or displayed. This aspect of the code system provides the designer, operator, or safety assessor with a simple interface to enable him to make effective use of the computational edifice and plays an increasingly important role in today's practical reactor physics. The capabilities, flexibilities, and organizational aspects of such modern systems are reviewed in more detail in the following sections.

## 3.3   Fast reactors

Fast reactors have been of concern and of interest for as long as thermal reactors and methods have correspondingly been developed for evaluating the reactor physics characteristics. The line of evolution, however, has been markedly different from that for thermal reactors for the following reasons.

1. In thermal reactors the important effects are associated with fairly clear regions of the neutron spectrum, starting with $^{238}$U fission at high energies, followed by capture in the resonances of $^{238}$U during the slowing-down regime, leading to absorption in $^{235}$U at thermal energies. No such clear partitioning with energy occurs in fast reactors. Scattering, capture, and fission events compete with one another significantly at all energies of relevance to fast reactors.

2. Because the neutron spectrum in a fast reactor peaks at about 250 keV, the thermal reactor physics problems of neutron thermalization do not exist, and $^{238}$U resonance capture is only important at energies around a few keV where the spectrum is falling in intensity. Inelastic scattering of neutrons in fuel and structural materials is the main mechanism leading to loss of neutron energy rather than elastic collisions in a moderating material.

3. Important competing events occur over a wide range of neutron energies spanning six decades. To measure the cross-sections required with sufficient accuracy especially at the higher energies poses formidable problems, and much attention has therefore been given to deriving information on nuclear data from measurements made on experimental critical fast reactor assemblies and, where necessary, 'adjusting' basic cross-sections accordingly.

Furthermore, in general terms, the cross-sections of materials for fast neutrons are considerably smaller than for thermal neutrons. The mean free path of a neutron in a fast reactor is correspondingly larger. On the other hand, fuel pin sizes are much smaller than in thermal reactors and in general the heterogeneous structure within a fast reactor is on a scale small compared with the neutron mean free path. This means that the fast reactor core can be effectively treated as homogeneous mixture and that there is little need for the lattice-cell type of evaluation which has so dominated thermal reactor physics. However, because of the relatively large neutron mean free paths, doubt is cast on the validity of the diffusion approximation, particularly in modelling events at interfaces such as the boundaries between the core region and that of the surrounding uranium blanket, in which capture of neutrons leads to breeding of plutonium. Direct solutions of the Boltzmann transport equation have been employed, albeit in simple one- or at most two-dimensional models, in order to establish the corrections required to diffusion solutions.

The evolution of fast reactor methods has thus centred on the construction of improved sets of group-averaged data for the isotopes of interest. In the late 1950s five-, six- and seven-group sets were common whilst during the 1960s sets containing of the order of 20 groups for perhaps 12 nuclides were introduced. These sets were ostensibly based on the most recent nuclear data measurements interpolated and extrapolated on the basis of nuclear model theories. During the period, many experimental fast reactor assemblies were built, particularly at the Argonne National Laboratory in the US, in which accurate information was obtained on the critical size of the reactor for a wide range of different compositions covering $^{235}$U and $^{239}$Pu fuel with different diluents simulating coolant and structural materials. Because of the relatively small heterogeneity in fast reactor cores, such a wide range of compositions could be studied by building assemblies of stacks of square coupons, each containing fissile material, fertile material, or a diluent. The arrangement of these coupons throughout the assembly could be changed from experiment to experiment without the need to refabricate material. The value of a particular data set was judged largely on the ability to reproduce the measured critical size of such fast reactor cores. The results were frequently disappointing.

The reason was not hard to find. Sensitivity analysis showed that if realistic uncertainties were assigned to nuclear cross-sections, then evaluations of reactivities of fast reactor assemblies would have a consequential uncertainty of ± 5 per cent. This value was unacceptably higher than the target uncertainly of 1–2 per cent. A very lengthy programme of differential cross-section measuremens was required to new and formidable levels of accuracy. Alternatively, and possibly as a short-term measure, the basic cross-section could be adjusted within these uncertainties to yield improved agreement with integral data.

Thus towards the end of the 1960s, uncertainties in the nuclear data dominated the uncertainties in prediction of fast reactor performance. It was realized that, with the accurate knowledge of the compositions of fast reactor assemblies available, data on the critical size provided rigid constraints on the group of parameters ($\eta-1$), and that this same group appeared in performance-prediction calculations. Studies showed that it would be feasible to measure other groups of parameters to accuracies of about 2 per cent or better, such as the capture rate in $^{238}$U and the fission rates in $^{238}$U and $^{235}$U, all relative to the fission rate in $^{239}$Pu. These parameters establish the balance between neutron events in the core, and a programme measuring them was conducted, for example, in the UK ZEBRA reactor (Fig. 2.17) with a view to using the data, together with the US critical size data, in a systematic fashion, to select cross-sections within the errors assigned to them which gave a best fit to the integral measurements and to the evaluated differential cross-sections.

The underlying philosophy was that the integral data had little energy resolution but high precision, whereas the differential cross-sections had good energy resolution and the correct shape over narrow ranges of neutron energies, but were subject to systematic error in the relative values of cross-sections separated by several decades in energy. If data adjusted within the known error limits could be selected by fitting the integral experiments, then similar integral properties on untested lattice compositions could be predicted with improved precision.

In the adjusted data set finally chosen, measurements of the neutron spectra measured by time-of-flight techniques and by hydrogen-filled proportional chambers in several ZEBRA assemblies were included amongst the integral measurements. Although this process did not make the use of the energy resolution available from time-of-flight spectrometry, the detailed shape of the spectrum in the vicinity of the 3 eV resonance in sodium, for example, added a useful commentary to the differential data for this resonance. An example of the agreement obtained between measured and calculated spectra using adjusted data is shown in Fig. 2.18. It will be noted that there still remains a tendency for more neutrons to be observed at energies below 1 keV than are predicted by calculation. The adjusted 37-group nuclear data set produced in the UK in the late 1960s has proved to be very successful in predicting fast reactor properties studies in ZEBRA in later experiments, showing that the quantities used to adjust data were correctly selected as being those of greatest importance in obtaining the correct neutron balance.

Much attention through the 1970s has been given to the validation of the methods used to predict properties having direct bearing on the safety performance of fast reactors. Experiments in ZEBRA have studied power distributions, the reactivities controlled by arrays of control rods of different compositions, and reactivity coefficients such as those associated with loss of sodium from different parts of the reactor. Much of this work has been conducted in assemblies of conventional design, in which the lower plutonium enrichment of a central cylindrical region is used to flatten the power distribution and in which the fertile breeder zones surround the fissile core.

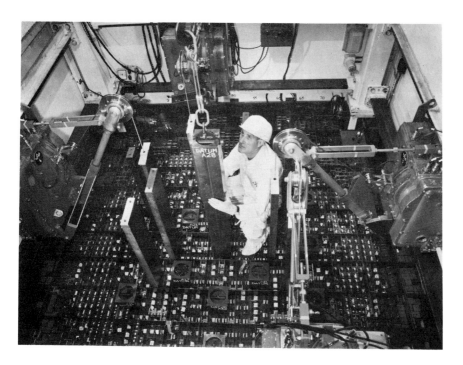

Fig. 2.17 The ZEBRA zero energy fast reactor

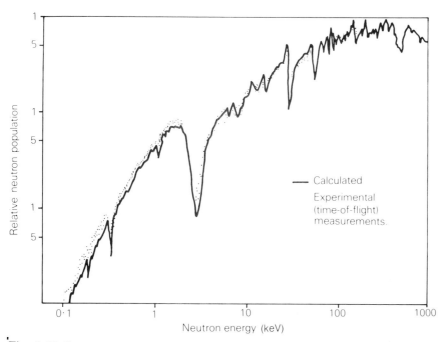

Fig. 2.18 Comparison between measured and computed neutron spectrum in a typical fast reactor

More recently initial studies have been made of fast reactors in which fertile material is introduced into the core in discrete regions. To a first approximation the amount of fissile and fertile material is unchanged, compared with a conventional core, but heterogeneity in the fertile material distribution is deliberately introduced in order to improve the sodium-voiding feedback effect.

Today's fast reactor methods still differ in several ways from thermal reactor methods on account of the differing relative importance of the various processes to be modelled. Some of the mathematical techniques involved of course are the same—Monte Carlo methods, transport solutions—and there is some indication that not only basic nuclear data but derived group data sets may come into common use. What the two approaches do have in common is a dependence on advanced computer technology and the need for organizational software to enable the user to utilize the computational system in an effective manner. This point will be discussed in a later section.

# 4  Current status: thermal reactors

This section briefly reviews some of the techniques and approaches currently used in applying reactor physics studies to thermal reactors. As previously noted the theoretical models involved are required for a wide range of applications covering design, operation and safety assessment and the accuracy and detailed capability called for is correspondingly varied. The underlying requirement is thus for an approach of considerable flexibility, simple and approximate where acceptable, but capable of refinement to the most demanding levels of accuracy.

The accuracy requirements depend upon the particular application for which the data and models are needed, and must be evaluated for each case. The considerations involved will be illustrated by reference to three such studies, the first of which relates to choice of fuel enrichment.

Many of the most widely adopted reactor types, AGR, PWR and BWR, employ fuel slightly enriched in the fissile isotope $^{235}$U. The amount of enrichment has to be specified when ordering the fuel. The designer will be aiming to have a system with a multiplication sufficient to permit the fuel to be kept in the reactor until it has either produced a target amount of heat, or achieved its target burn-up of so many MW-days/tonne. He will calculate the multiplication of the reactor, including the effects of structural materials and leakage of neutrons; he will also model the time-dependent effects of build-up of plutonium and of higher actinides, of the effect of the fission product build-up, the effect of temperature upon resonance capture, and so on. The appropriate balance of these effects over the desired fuel cycle will enable him to fix an appropriate enrichment. The calculational models currently in use will permit the reactivity of a given assembly of known composition to be predicted to better than $\pm 0.5$ per cent even where no closely parallel system is available for comparison. (The AGR reactors of 660 MW (e) were 10 times bigger than the prototype, and had significantly different fuel element designs, but their design reactivity was correct to 0.25 per cent on start-up.) The changes in isotopic composition of fuel over life add to the uncertainty, and it is expected that for a new reactor the irradiation obtainable within reactivity limitations will be predicted to $\pm 10$ per cent. The uncertainly in the enrichment required to meet certain performance targets may thus be evaluated.

The economic implications of these uncertainties however may be rather small. On operation of the reactor any error will be detected and the enrichment may be changed for subsequent fuel orders. Thus the economic penalty of a small shortfall in reactivity will be a slightly higher rate of fuel discharge than was anticipated. For excessive enrichment the penalty may be having to discharge fuel containing more useful fissile material than was necessary. Larger errors in enrichment would, of course, begin to impinge upon other aspects of the design; if the fuel were so reactive that the reactor control system was fully employed before loading was complete, or, at the other extreme, the fuel when loaded did not take the reactor critical, the fuel would have to be refabricated and the operation of the reactor delayed, with a much more significant cost penalty.

A second type of assessment study concerns the reactivity worth of control absorbers. For heavily absorbing rods ('black' to thermal neutrons) it is now usually possible to predict the reactivity effect of an array of rods to $\pm 5$ per cent of this total worth. The designer has usually to allow for a range of situations in which it is postulated that a number of rods fail to operate when called upon, in order to give a large margin of safety. Having to make additional allowance for predictive uncertainty adds to the margins needed, and results in the installation of more rod mechanisms than are really necessary. This excess provision is a capital cost which cannot be recouped once measurements on the reactor have proved it to be unnecessary. At the extreme, because it may have influenced the choice of design, for example the pitch at which the mechanisms are located, it may continue to be an expense on a whole family of reactors. It is often worthwhile, as in this case, to conduct small-scale experiments to reduce uncertainties so as to minimize margins. In this situation the model is being used to extrapolate from the small to the large scale, with a concomitant reduction in possible sources of error.

In both these types of application, because they apply to the normal operation of the reactor, the true position will be revealed by observation when it is built and commissioned. A different type of calculation is involved when attempting to assess what would happen to the reactor in a hypothetical accident situation. Here, however, a 'best estimate' evaluation is not necessarily required; a calculation which is demonstrably pessimistic at various stages may be sufficient to establish a safety argument, and the accuracy issue thus becomes one of sign and certainty rather than magnitude. An example of such a calculation is the so-called 'anticipated transient without scram' in a PWR. Here it is postulated that a situation may arise in which, following some normal operating (and hence anticipated) transient such as the trip of the turbine, the control system fails to shut down the reactor as would normally be the case. Even in this extreme case the reactor is calculated to reach an acceptable steady-state condition, albeit at a higher temperature and pressure than that of normal operation. It is not sensible to test an occurrence on a power station—indeed to do so would arguably entail more risk than that from the error itself—so that estimates rely more heavily on theory and its validation from small-scale experiments of each of the variables involved. In this particular case a controlling factor is the negative reactivity effect of rising water temperature (temperature coefficient), and this can in fact be studied in the normal ranges of temperature and pressures. Where reliance is being placed upon calculations without the prospect of ultimate direct check with observation, the required standards of validation are of course correspondingly higher.

## 4.1  The WIMS scheme

In order to illustrate the current level of modelling capability, it is necessary to discuss the methods used for the lattice-cell and for the whole-reactor type of calculation. In order to focus discussions on the lattice-cell methods, the capability of the most widely used WIMS (Winfrith Improved Multi-Group Scheme) models will be given particular consideration. The development of WIMS has been a major effort within the UK over the last decade, but its current capability is quite typical of the type of approach used world-wide in studying reactor physics problems of thermal reactors.

The WIMS family of codes models the fuel assembly—cluster or box—and thus involves two levels of heterogeneity: the pin and the array of pins. Boundary conditions of different degrees of complexity are used to allow for the rest of the reactor, and the depletion, or burn-up, of the fuel is computed at this stage of the calculation. It is evident that the real operating reactor must be critical, and this is an important constraint.

The lattice code produces simplified data—often homogenized and reduced to two energy groups—for use in whole-reactor calculations. Because the burn-up is solved at the lattice-cell stage the reactor simulation requires a library of such data depending upon the fuel design and enrichment, irradiation, and, for some types of reactor, other variables such as operating temperatures or coolant densities. The reactor calculation serves in effect to produce the appropriate normalization of the lattice solutions from one fuel channel to another and axially along the channel, including the effects of control absorbers partially inserted. A full three-dimensional model is thus needed, with the order of 5000 zones even for the most simplified model. Some feedback of the effect of the heat rating of the fuel on temperatures or coolant flows is used at this stage of the calculation.

The WIMS computing scheme for lattice cells was one of a number of similar schemes developed around 1965. Many of the individual techniques used in the code had been in use for some time prior to the assembly of this scheme, but at this time it became possible to make a major step forward in modelling capability. A number of factors combined to permit this:

(1) a sufficient number of measurements of basic nuclear data had been completed, covering all the most important nuclides and reactions, to make it worthwhile attempting detailed modelling;

(2) a sufficient basis of understanding of the most difficult modelling problems and a number of techniques for solving the Boltzmann equation had been accumulated in earlier model developments; and

(3) the development of computers had proceeded to a point where adequate calculations could be performed at a reasonable cost.

Two particular innovative asepcts of the WIMS development are outlined in the following paragraphs.

1. The use of models applicable to all types of thermal reactor, in particular to reactors having different moderators (graphite, heavy or light water) and different fuel element designs. Apart from the obvious gain of wider utilization of the scheme, this enabled a wider range of experiments to be employed in validating the methods and data used, and provided a way of eliminating some kinds of cancelling errors in nuclear data. For example, if the resonance capture were wrongly predicted, as proved to be the case, it was possible to determine whether the moderator or fuel data were at fault.

2. The provision of a general, deterministic resonance capture treatment. This is the most technically difficult model problem, and the basis for dealing with it is discussed in more detail in Appendix C as an illustration of the general class of modelling problems. Especially difficult to represent were the cluster designs of reactor, such as CANDU and SGHWR, in which a double heterogeneity of pin and cluster were present. Previous models had been unable to treat these systems in a general way from basic data, and recourse was necessary to Monte Carlo calculations when 'best' estimates were needed. Such calculations however had the twin drawbacks of high cost and a degree of random variation on predictions which made it difficult to use them for systematic design purposes.

### 4.1.1 *Characteristics of the underlying physics process*

From the modelling point of view three basic regimes of neutron energy to which different considerations apply can be distinguished.

1. *The fast region, E > 100 keV.* Neutrons from fission are born in this energy range. It is characterized by a wide diversity of possible neutron reactions, including sharp energy variations of reaction probability at thresholds, for example of fast fission in uranium-238. Scattering is often highly anisotropic, even in the centre-of-mass frame of reference. Cross-sections are usually fairly small—in other words neutrons travel longer

distances between collisions than is the case at lower energies—and the representation of geometric details is correspondingly less important.

2. *The resonance region, 100 keV > E > 4 eV.* In this range the light nuclides tend to have constant cross-sections dominated by potential scattering. Intermediate and heavy nuclides exhibit resonances in which capture and fission cross-sections reach high values over small energy ranges and are temperature dependent owing to the relative motion of neutron and nuclide (Doppler broadening). Given these rapid changes, or discontinuities, of cross-section, both in energy and space, the modelling problem is at its most difficult. Because of the high cross-sections many of the effects are of short range.

3. *The thermal region, 4 eV > E.* Here thermal motion of the light nuclides has a considerable effect upon neutron behaviour and the thermalization process. The width of resonances is no longer always small compared with the possible energy change per collision, and in this sense cross-sections can be considered to be more slowly varying.

### 4.1.2  *Energy groups*

As has been noted in §2, the most effective way of treating the energy variable is to represent it as a number of discrete energy groups. The choice of groups is conditioned by the nature of the processes involved, but is inevitably a compromise between the large number required for accuracy of representation and the small number suggested by considerations of computer capability. The basic group system in the event selected for WIMS is shown in Table 2.9 and consists of 69 groups. (Given present-day computers an increase to around 100 groups would represent a more appropriate choice.) The associated 69-group data library was assembled by appropriately averaging basic, differential data sets over these energies. The most important nuclides represented are shown in Table 2.10 together with two group reductions of the library cross-sections. Only a limited number of higher actinides are currently modelled, as shown in Fig. 2.19. These actinides cover the most significant processes from the point of view of absorption and fission events in the reactor. Separate codes and data libraries are used to give fuller representation for determining compositions of nuclides important in reprocessing or storage of fuel. The most important fission products—again from the point of view of neutron absorption rather than of health hazard—are represented explicitly and are shown in Fig. 2.20. All other relatively unimportant products are lumped together in a single pseudo-fission product.

The use of such a modest group structure presupposes that a subsidiary model is used to represent the resonance range, as here neutron cross-sections vary so rapidly that more than 100 000 groups are needed to represent them directly. Two types of subsidiary resonance model are used

Table 2.9 69-Group energy boundaries for WIMS

| Group | Energy range | | Group | Energy range | |
|---|---|---|---|---|---|
| | Upper limit | Lower limit | | Upper limit | Lower limit |
| | (MeV) | | 33 | 1.30  (eV) | 1.15 |
| 1 | 10.0 | 6.0655 | 34 | 1.15 | 1.123 |
| 2 | 6.0655 | 3.679 | 35 | 1.123 | 1.097 |
| 3 | 3.679 | 2.231 | 36 | 1.097 | 1.071 |
| 4 | 2.231 | 1.353 | 37 | 1.071 | 1.045 |
| 5 | 1.353 | 0.821 | 38 | 1.045 | 1.020 |
| 6 | 0.821 | 0.500 | 39 | 1.020 | 0.996 |
| 7 | 0.500 | 0.3025 | 40 | 0.996 | 0.972 |
| 8 | 0.3025 | 0.183 | 41 | 0.972 | 0.950 |
| 9 | 0.183 | 0.111 | 42 | 0.950 | 0.910 |
| 10 | 0.111 | 0.06734 | 43 | 0.910 | 0.850 |
| 11 | 0.06734 | 0.04085 | 44 | 0.850 | 0.780 |
| 12 | 0.04085 | 0.02478 | 45 | 0.780 | 0.625 |
| 13 | 0.02478 | 0.01503 | 46 | 0.625 | 0.500 |
| 14 | 0.01503 | 0.009118 | 47 | 0.500 | 0.400 |
| | | | 48 | 0.400 | 0.350 |
| | | | 49 | 0.350 | 0.320 |
| | | | 50 | 0.320 | 0.300 |
| | (eV) | | 51 | 0.300 | 0.280 |
| 15 | 9118.0 | 5530.0 | 52 | 0.280 | 0.250 |
| 16 | 5530.0 | 3519.1 | 53 | 0.250 | 0.220 |
| 17 | 3519.1 | 2239.45 | 54 | 0.220 | 0.180 |
| 18 | 2239.45 | 1425.1 | 55 | 0.180 | 0.140 |
| 19 | 1425.1 | 906.898 | 56 | 0.140 | 0.100 |
| 20 | 906.898 | 367.262 | 57 | 0.100 | 0.080 |
| 21 | 367.262 | 148.728 | 58 | 0.080 | 0.067 |
| 22 | 148.728 | 75.5014 | 59 | 0.067 | 0.058 |
| 23 | 75.5014 | 48.052 | 60 | 0.058 | 0.050 |
| 24 | 48.052 | 27.700 | 61 | 0.050 | 0.042 |
| 25 | 27.700 | 15.968 | 62 | 0.042 | 0.035 |
| 26 | 15.968 | 9.877 | 63 | 0.035 | 0.030 |
| 27 | 9.877 | 4.00 | 64 | 0.030 | 0.025 |
| 28 | 4.00 | 3.30 | 65 | 0.025 | 0.020 |
| 29 | 3.30 | 2.60 | 66 | 0.020 | 0.015 |
| 30 | 2.60 | 2.10 | 67 | 0.015 | 0.010 |
| 31 | 2.10 | 1.50 | 68 | 0.010 | 0.005 |
| 32 | 1.50 | 1.30 | 69 | 0.005 | 0.000 |

in WIMS. The original, and most widely used, is based upon an equivalence between the solution in the real geometry and that in a simpler problem in which the resonance absorber is mixed with an appropriate amount of hydrogen in an infinite homogeneous system. The matching process of identifying the amount of hydrogen is described briefly in Appendix C, but is aimed at getting similar flux dips through the resonances and preserving some integral of captures. The equivalent homogenous problem is solved exactly for various hydrogen concentrations and fuel temperatures, and the results stored in a library, which the code can look up during its solution of the real problem and deduce what group constants to use in the resonance range. The procedure has been extensively tested, and by great attention to

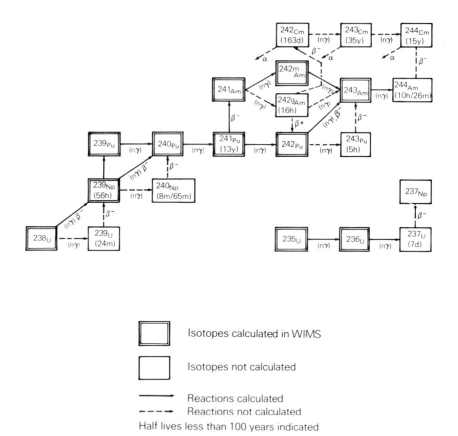

Fig. 2.19  Principal links in the heavy atom burn-up chains

detail it has proved possible to reproduce capture probabilities in uranium-238 to within 1 per cent for a wide range of geometries. The results of some of these proving tests are shown in Table 2.11.

The equivalence model has proved to work well for slab, pin and cluster geometries, but gives only average pin events. There are some applications in which more detail is needed, for example, in studying the spatial variation of the build-up of plutonium within a single pin. Here a sub-group method is used. The resonance is described as a histogram of a small number of 'standard' cross-section. The width of each step in the histogram is chosen to preserve the known homogeneous resonance integrals in hydrogen. Then the spatial problem is solved by substituting each of these cross-sections in turn as the 'fuel' cross-section, and the fluxes obtained are used to integrate and produce an average cross-section for the final calculation. Fig. 2.21 shows how such an approximation was tested against an experimentally determined distribution of plutonium build-up in a $UO_2$ fuel pin typical of a PWR. Of course, not all calculations need be performed with the full

Table 2.10 Thermal absorption and fission cross-section and resonance absorption, and fission integrals of isotopes held in the WIMS 69-Group library

| Isotope | Identifier | 2200 m/s cross-section (barns) | | Resonance integral 0.055 eV–2 MeV (barns) | |
|---|---|---|---|---|---|
| | | absorption | fission | absorption | fission |
| Hydrogen | 2001 | 0.322 | — | 0.140 | — |
| Deuterium | 8002 | 0.00056 | — | 0.00056 | — |
| Boron | 10/1010 | 3809.0 | — | 1667 | |
| Carbon | 12 | 0.00340 | — | 0.00143 | — |
| Oxygen | 16 | 0.00009 | — | 0.00001 | — |
| Aluminium | 27 | 0.229 | — | 0.181 | — |
| Silicon | 29 | 0.160 | — | 0.0669 | — |
| Chromium | 52 | 3.10 | — | 1.33 | — |
| Manganese | 55 | 13.8 | — | 13.4 | — |
| Iron | 56 | 2.53 | — | 1.21 | — |
| 'Iron' | 1056 | 2.53 | — | 2.61 | — |
| Stainless steel 304 | 9056 | 2.90 | — | 1.69 | — |
| Nickel | 58 | 4.60 | — | 2.04 | — |
| Inconel 750X | 8058 | 4.14 | — | 1.92 | — |
| Inconel 718 | 9058 | 3.85 | — | 1.80 | — |
| Zirconium | 91 | 0.183 | — | 0.891 | — |
| Zircaloy | 9091 | 0.195 | — | 0.883 | — |
| Silver | 2109 | 64.8 | — | 334 | — |
| Cadmium | 2113 | 2400 | — | 37 | — |
| Indium | 2115 | 202 | — | 3215 | — |
| Gadolinium-155 | 2155 | 60050 | — | 1316 | |
| Gadolinium-157 | 2157 | 254000 | — | 493 | — |
| Uranium-235 | 235 | 680 | 580 | 412 | 361 |
| Uranium-236 | 236 | 6.08 | 0.0 | 313 | 0.517 |
| Uranium-238 | 2238 | 2.72 | 0.0 | 272 | 0.180 |
| Uranium-238 | 3238 | 2.72 | 0.0 | 273 | 0.181 |
| Neptunium-239 | 939 | 65.0 | — | 17.6 | — |
| Plutonium-239 | 2239 | 1029 | 742 | 446 | 262 |
| Plutonium-240 | 1240 | 283 | 0.046 | 7625/7735 | 3.44 |
| Plutonium-241 | 241 | 1387 | 1031 | 749 | 560 |
| Plutonium-242 | 242 | 17.3 | — | 1043 | — |
| Americium-241 | 941 | 625 | — | 1640 | — |
| Americium-242m | 942 | 8500 | 6800 | 2300 | 1840 |
| Americium-243 | 943 | 94 | — | 1483 | — |
| 1/v Absorber | 1000 | 1.0 | — | 0.429 | — |

69-group library. Methods are available within the WIMS scheme for condensing the data to fewer groups and this may be entirely appropriate for scoping or approximate studies.

### 4.1.3 *Representation of geometry*
It has already been noted in §2 that the motion of neutrons within a reactor assembly is represented by the Boltzmann transport equation. This equation may be solved within WIMS by a variety of methods, the particular choice being largely determined by the degree of geometrical complexity involved.

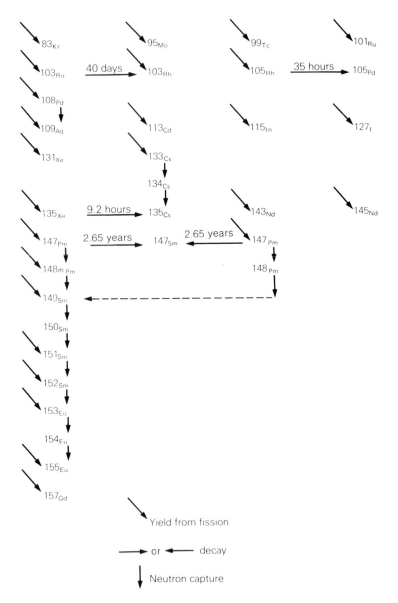

Fig. 2.20  Fission product chains in WIMS library

Many general and complex geometrical representations are treated by the collision probability method referred to in §3. The degree of geometrical detail which can be represented using such a method is considerable and is well illustrated in Fig. 2.22. This shows the spatial regions selected to represent an HTR fuel assembly in an experimental zero-energy reactor in which the effect of control absorbers was being studied.

Table 2.11 Comparison of WIMS resonance capture model for 238-U with exact Monte Carlo calculation

| Lattice description | Capture probability 5.53 keV to 4 eV | |
|---|---|---|
| | WIMS | Monte Carlo |
| Light water and 3% UO₂ regular rod array; volume ratio 1:1 | 0.2121 | 0.2105 ± 0.0018 |
| Light water and 3% UO₂ regular rod array; volume ratio 3:1 | 0.09027 | 0.0878 ± 0.0013 |
| Light water and 3% UO₂ regular rod array; volume ratio 4:1 | 0.07240 | 0.0685 ± 0.0014 |
| Heavy water and 3% UO₂ regular rod array; volume ratio 4:1 | 0.3350 | 0.3345 ± 0.0007 |
| 37-rod cluster of 0.9% UO₂ rods; light water cooled, heavy water moderated | 0.1302 | 0.1255 ± 0.0006 |
| 37-rod cluster of 0.9% UO₂ rods; air cooled, heavy water moderated | 0.1727 | 0.1733 ± 0.0013 |
| Light water and 1.3% metal regular rod array; volume ratio 1:1 | 0.2885 | 0.2843 ± 0.0021 |
| Light water and 1.3% metal regular rod array; volume ratio 1.5:1 | 0.1999 | 0.1948 ± 0.0025 |
| Graphite and 0.7% metal regular rod array; volume ratio 30:1 | 0.2017 | 0.2000 ± 0.0026 |
| 11.48 cm diameter U-238/H rod (0.00905 atoms U-238, 0.02469 atoms H/barn-cm) in a sea of heavy water | 0.1736 | 0.1747 ± 0.0009 |

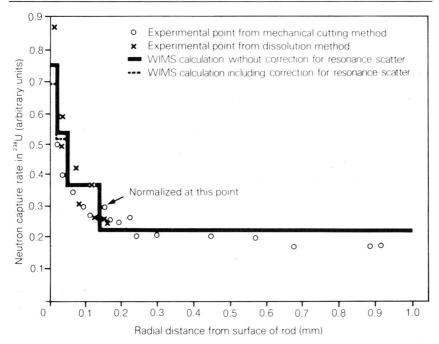

Fig. 2.21 Comparison of calculated and measured plutonium production near the surface of a oxide fuel pin

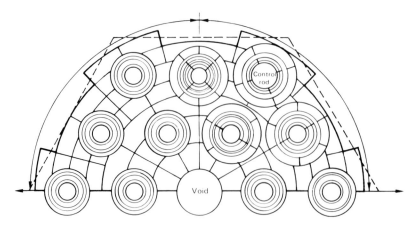

Fig. 2.22  HTR fuel block sub-divided into 139 regions for calculations using collision probability method within WIMS

Of course, in many applications much simpler geometries are involved. Thus in regular arrays of pins, a basic unit of a cylinder within a square boundary is adequate, and it is often possible to approximate the boundary conditions further to give a wholly cylindrical (and hence one-dimensional) model. Several solution techniques are available in such simple geometry involving either use of the diffusion approximation or direct integration of the transport equation. At the other extreme the Monte Carlo routines available within WIMS provide an even more general geometrial capability than that offered by the collision probability approach.

### 4.2  **Whole-reactor calculations**

It has been the UK, and largely the European, practice to condense the results of the detailed lattice-cell calculations (such as those described above) into few-group form when it comes to studying the whole reactor. These few-group data are produced as functions of irradiation and other important variables.

The subsequent representation of the reactor itself may follow either of two distinct and different routes. For rectors with fuel elements or assemblies well-separated by moderator (Magnox, AGR, SGHWR, CANDU), a heterogeneous approach is possible in which the capture and birth of neutrons in individual fuel channels is explicitly represented, and the motion of neutrons through the intervening moderator is solved using the diffusion approximation. In the alternative approach the reactor is divided into a number of homogeneous volume zones each with its own data obtained by appropriate homogenizing (averaging) of the lattice-cell results. The reactor is thus modelled by a patchwork of homogeneous zones and the model solved in the few-group diffusion approximation. This is effectively the only approach available for light water reactor systems.

### 4.2.1 *Heterogenous methods*

In its simplest form a fuel rod is represented by a line source and sink of neutrons in a sea of moderating, diffusing material. The neutron flux at any point of the reactor is then made up from contributions from all the individual source/sinks. These contributions are represented by functions (called 'kernels') which are essentially geometric in form. The actual birth and capture effects are related to the neutron flux in the moderator by boundary conditions at the line source/sinks and in this way a complete system of equations is obtained.

In order to represent the real geometric details of a reactor, progressive refinements have been made to the simple model. A cylindrical boundary to a reactor may be represented by expanding the kernels in modified functions all of which go to zero at the boundaries, whilst the finite size of fuel channels and axial variations of properties may be introduced as series expansions. For large systems the resultant integral equation (in which the events at each point in the reactor are expressed in terms of the sources at all other points) becomes laborious to solve, and this has limited the practical application of the technique.

Two possible developments of the method remain to be fully explored; one is to recast the equation in a differential form, so that the solution at a point is obtained in terms of the representation of nearest-neighbour fuel channels. A more radical prospect is to calculate kernels directly by Monte Carlo simulation in full space/energy detail for a regular array of fuel elements, and to express the solution in terms of these kernels which are much more localized in space and are guaranteed to give the exact solution for regular systems.

### 4.2.2 *Homogeneous methods*

The heterogeneous models referred to in §4.2.1 are at least superficially attractive in that they retain some of the detailed structure of a reactor assembly. They have however achieved only limited application because of the practical difficulties of developing efficient and accurate versions. The alternative homogeneous methods are much more commonly used. They are based on reducing the pin or cluster lattice cell to an 'equivalent' homogeneous paste, and then assuming that the diffusion approximation is good enough to model the flow of neutrons between homogenized regions.

The homogenization of what is a very heterogeneous structure, although a delightfully simple concept, presents considerable practical difficulty, and it has proved necessary to develop highly ingenious routes for obtaining homogeneous paste characteristics which can satisfactorily represent neutron behaviour in the real lattice cell. Many prescriptions have been developed for the purpose, often specific to different reactor types. It is not too difficult, for a fairly regular geometry, to deduce from a lattice-cell calculation some constants giving adequate representation of total events in the cell. The problem is to describe the flow of neutrons across and axially along the cell—that is, to determine appropriate average diffusion coefficients. The considerations involved in such a process can be illustrated by supposing that a heterogeneous reactor cell, consisting of a square zone of moderator with a fuel element at its centre, can be replaced by a homogenized equivalent. Imagine a neutron entering the cell across its boundary. It is desirable that the probability of fission, capture, and leakage across any other face is reproduced by our simpler homogeneous model and, at the same time, that the probability that a neutron born in the cell by fission will escape from it is preserved, even though in the homogenized model the source is spread over the whole mesh zone and not confined near the centre. The two requirements conflict and some empirical compromise is made in practice in order to obtain a solution. The differential diffusion equations are in practice solved numerically by covering the reactor model with a mesh of points and approximating differentials by finite differences between values at neighbouring points. Thus, a differential term such as $df/dx$ is represented by:

$$\frac{f_n - f_{n-1}}{\Delta},$$

where $f_n$ is the value taken by the variable $f$ at the mesh point $n$ and $\Delta$ is the $x$-distance between these mesh points. In this way the solution of the differential equations is reduced to the solution of a (large) number of inter-connected algebraic equations. The development of efficient techniques for obtaining these solutions has been an important research activity in its own right. In the case of a full three-dimensional reactor representation, values of the neutron flux at a mesh point are linked to values at six neighbouring points. The finite-difference analogue of the differential equation (3) thus becomes (in the steady state):

$$R_{jn}\phi_{jn}\Delta^2 = \sum_m D_{nm}(\phi_{jm} - \phi_{jn}) + \sum_{j'} \Sigma_{jj'}\phi_{jn'}\Delta^2. \tag{8}$$

Here the left hand side represents the removal (by absorption or scatter) of neutrons of energy group $j$ at the mesh point $n$. The first term on the right hand side represents the net flow of neutrons of energy $j$ into or out of the mesh volume around mesh point $n$. The second term represents the source of neutrons of group $j$ at mesh point $n$ by scatter or a fission process involving neutrons initially of energy $j'$.

   To solve the resultant algebraic matric equations some constraints to force the system to be just critical or balanced is needed; otherwise as the equation is progressively satisfied mesh by mesh the flux might continue to rise or fall. This constraint is used to try to represent the real way in which the reactor is held critical. The following three options are used for this purpose.

1. *Power shaping*. This simulates the automatic control system of the reactor. An AGR has sector control rods which respond to signals collected from gas outlet thermocouples and forces them to target values. The code used for

design studies of fuel management strategies simulates this operational constraint whilst the total rod insertion keeps the reactor model just critical.

2. *Power conditioning*. This represents the core-follow situation during operation. Measurements are available on the reactor, and the theoretical model is constrained to fit these. The actual technique used for the AGR is to allow the absorption cross-section and thus the multiplication to vary for each channel, as there is a measurement of gas outlet temperature on each channel. It is possible to incorporate other information in the fitting process. A further degree of sophistication is given by the use of correlated methods to give a most probable solution where the discrepancies between theory and observation are difficult to reconcile or where measuring uncertainties are significant.

3. *Eigenvalue*. In this option a multiplier is introduced on the yield of neutrons from fission so as to maintain the reactor mode just critical. Physically this may be considered as an assumption of a spatially uniform discrepancy in nuclear data to explain the difference between the observed critical reactor and the super- or sub-critical model. This is the most widely used model option, largely for historical reasons. As reactors have become larger quite a small local perturbation can give rise to a big flux tilt across the system, and the spatially uniform multiplicative eigenvalue adjustment may not represent this. This physical property is reflected mathematically in a small separation of eigenvalues and a resultant poor convergence of iterative methods of solution. This has stimulated the move to more constrained solutions of the type referred to earlier in this section.

## 4.3 The modelling of burn-up

Brief consideration has already been given to the method used for detailed solution of a lattice cell, with boundary conditions simplified to some degree to approximate its environment in the reactor, and the ways in which information extracted from this calculation are used to build up a picture of the power distribution in the reactor as a whole. So far it has been assumed that the composition of the lattice cell is known.

In fact the neutron reactions lead to major changes in composition over the lifetime of fuel in a reactor. Typically, power reactor fuels are exposed to about one fission per initial fissile atom loaded, although the production of plutonium and its consumption over life means that not all the uranium-235 loaded is used up.

The determination of the balance between the production and destruction of materials requires careful modelling, as does the effect of the fission products and of the higher actinides produced by capture in uranium and plutonium isotopes. It is thus usual to model the whole burn-up process within the detailed lattice-cell calculation. There is, however, some variation in practice of the way in which this information is carried over to the whole reactor model.

### 4.3.1 *The macroscopic method*

The simplest route is to use the cross-sections generated in the lattice burn-up as functions of irradiation and the type of fuel (enrichment, etc.) loaded. Thus the data produced from the lattice calculation, including the reduction of channel boundary conditions or homogenization, are collected and tabulated against irradiation. Burn-up effects are then simulated in the whole reactor by integrating calculated power production over time to determine irradiation and then obtaining new values of the equivalent cell parameters from the pre-stored tables.

A number of refinements may be used to deal with modest perturbations from the state calculated in the lattice model. In particular, such effects as soluble-poison and xenon concentrations should be noted.

*Soluble-poison concentration.* Although this will be simulated in the lattice calculation, the concentration at particular irradiations will not match exactly that for the reactor as a whole. Small deviations are accommodated by storing microscopic cross-sections and the number densities used in the lattice calculation, and correcting the macroscopic cross-section as the calculation proceeds.

*Xenon concentration.* As xenon-135 is produced partly by decay of iodine-135 its production follows the fission rate with some time lag. For steady operation of the reactor this can be well-simulated in the lattice calculation, although it must be modelled separately for transient studies. Even in the steady state, however, the absorption in xenon-135 is dependent upon the rating, or flux level, because the concentration is determined by competition between capture and decay. Thus it is often useful to store the rating at which the lattice calculation was performed, the microscopic xenon cross-section, and the concentration so that any mis-match between lattice calculation and reactor calculation may be corrected.

### 4.3.2 *The microscopic method*

In this case the type of correction already referred to in the treatment of soluble poison and of xenon-135 is extended to the principal heavy elements present. It is still necessary to perform lattice/burn-up calculations in order to obtain satisfactory energy spectra and spatial averaging, but number densities and microscopic cross-sections are extracted for use in the reactor calculation. A minimum of 5000 such compositions will be needed to represent a power reactor, and at least 10 nuclides should be modelled at each point. In principle it will be possible to use fewer lattice calculations, as long as care is taken to see that the microscopic cross-sections are not used outside the range of conditions to which they are applicable. It is possible to extend this range by suitable interpolation or extrapolation procedures. For example, the cross-section of plutonium-240 has a very large resonance at 1.06 eV and its effective cross-section is very dependent upon concentration due to shielding effects. Tabulation in a suitable parametric form will permit the data to be used at concentrations other than those for which they are generated.

### 4.3.3 *Spatial modelling*

For reactors having an obvious cluster configuration, as for example in AGR and CANDU systems, it is natural to associate composition with the fuel element. For a PWR or BWR box, in which pins are distributed fairly uniformly, the unit of modelling is open to choice. It can be either the pin or the assembly. In general, US practice has tended to the former, European practice to the latter. It is clear that the whole-reactor spatial solution in three dimensions is much more easily accomplished if each box is represented as one or $2 \times 2$ mesh intervals rather than by one mesh or each of the array of $17 \times 17$ pins, but, equally, a more elaborate lattice-cell calculation must be performed to give appropriate smeared data. As fuel elements are the units used for refuelling, there is no loss of modelling accuracy in this respect, and the mean distance travelled by neutrons from birth to absorption (around 10 cm in a PWR) suggests that there can only be limited variation of conditions between neighbouring pins. There is little comparative evidence of the performance of the two types of model, although *core-follow* studies of operational reactors suggest that the coarse-mesh method is at least adequate for monitoring performance and will permit the calculation of rating distributions with a standard deviation of about $\pm 3$ per cent compared with experiment.

In a homogeneous model of an AGR there will be typically four mesh points per channel in plan and perhaps 20 to represent the axial variations. Of course, the finite-difference equations are solved more accurately the finer the mesh. This is, however, increasingly time-consuming and it is not obvious that overall benefit is obtained by solving an approximate model to high accuracy.

## 4.4 **Validation**

It will be clear from the foregoing description of lattice and reactor calculational methods that the prediction of reactor behaviour depends upon very extensive measurements of data on basic nuclear reactions and upon a number of simplifications aimed at reducing the problem of describing neutron transport to a computationally tractable form.

The refinement and validation of a calculational system of this kind is perhaps the most time-consuming aspect of its development. To some extent it is possible to separate out model and data effects; in some cases, as in the calculation of resonance capture previously referred to, it is possible to perform more detailed calculations, in this case using Monte Carlo methods which introduce no significant modelling assumptions, using identical data to those in the simpler model. Calculational schemes often provide solution modules of different degrees of complexity so that inter-comparisons of this kind are facilitated. This will demonstrate the effect of the simplification. To test data it is possible to use experimental information for simple systems, such as homogeneous spheres of solutions of uranium or plutonium in water, in which even the simple calculational models would give a good representation of the geometry.

Beyond these basic checks the procedure is more complicated. The testing progresses to regular rod array experiments, the next simplest geometric option. In addition to reactivity estimates it is important to have detailed reaction rate data on individual nuclides so as to preclude the possibility of good agreement due to cancelling errors. Typically it is possible to measure the uranium-238 fast fission, the uranium-238 capture (or rate of plutonium production), and the fission rate of plutonium-239 relative to fission rate in uranium-235. These are the principal components of the multiplication calculation. Combined with measurements of fine structure (the relative neutron density in the moderator, can, and fuel) all the steps in the calculation are checked reasonably closely except leakage, which is obtained from measurement of critical size.

It is important that validation studies embrace a suitably wide range of materials—metal and oxide fuel, uranium and plutonium fissile materials, graphite, and heavy and light water moderators. In this way even experiments which do not have reaction rate data can be used to discriminate between data and model effects. Some 300 lattices of this kind were studied as part of the process of validation of the WIMS code scheme. One of the early deductions from this comparison was that the resonance data for uranium-238 then in use was in error, despite previously reported good agreement on lattice calculations. The effect had previously been obscured by combinations of model and moderator data errors. The code has used adjusted data for a number of years, until recent new measurements have identified the source of the discrepancy as the widths of the low-lying resonances which were overestimated in earlier measurements.

Having established that simple lattices can be well predicted the validation proceeds to realistic cluster or assembly geometries in which spatial representation is the major problem, and in which other variables such as control rods and poisons are introduced. In such experiments it is usual to attempt to simulate the real reactor geometry as closely as possible, so that extrapolations may be made with minimum error.

Despite the generally high degree of agreement now obtainable with the best models, there are still discrepancies in prediction which are greater than would be expected from the assessed uncertainties in basic data. Moderator temperature coefficients, involving the reactivity effects due to change of thermal neutron spectrum, are not well predicted, especially in light water reactors, and there is some evidence that the reactivity effects of boron dissolved in the moderator are also poorly predicted. The experimental base for these deductions is, however, limited, and the comparisons are complicated because changes in boron concentration or temperature are usually balanced against each other or against leakage in a critical experiment, and absolute comparisons are difficult to obtain.

A broad-based validation provides a sound basis for the final test of comparison which is that against actual power reactor performance. Measurements on operating plant are, however, notoriously difficult, and the conditions under which they are taken are often not precisely defined. It is often the case that the assessment of predictive accuracy rests as much on data from zero-energy reactors as on that from the operating power reactor.

## 4.5  The adequacy of current methods

The combined capabilities of theoretical prediction and operational measurement are clearly sufficient to permit the design and safe operation of current types of nuclear power plant, as long as sufficient margins are left to cover the various uncertainties remaining. The incentive to improve methods is therefore primarily economic, and is usually related to a desire to reduce these margins.

Typical situations are those which arise with operating plant. Suppose, for example, that it is decided that an increase in the irradiation exposure of fuel is practicable, and that this would lead to improved fuel cycle costs through a reduction in the amount of fabrication and reprocessing work required. One of the consequences would be that the reactivity difference between new and old fuel would be increased, and the peak rating of the new fuel at loading would be increased relative to the average.

It may be that a substantial margin between the operational rating and the limit set by safety considerations (typically the rating at which even a complete loss of coolant or coolant pressure could be safely accommodated) was left in the original design. Better calculations and extensive comparison between these and the observations on the reactor might lead to a reduction in uncertainties sufficient to accommodate a higher rating and thus an improved fuel cycle.

In this, as in most areas involving theoretical modelling, the feedback of measurements from specially constructed experiments or from operating plant is a source of progressive refinement of predictive capability. By way of illustration of the current situation, some of the more obvious types of physics considerations are discussed, together with the accuracy that might be expected in application to an AGR. These are typical of a larger number of predictions required for design and for the assessment of safety margins.

### 4.5.1  *Reactivity*
A substantial effort has gone into checking the application of basic nuclear

data to the prediction of the neutron multiplication cycle in low-enriched uranium-fuelled reactors. By dint of analysing some hundreds of experiments using rather exact modelling techniques it has been possible to make small refinements to the basic nuclear data and to improve predictions over a wide range of systems. The use of the same models and data to predict very different reactor types—natural or low enrichment, metal or oxide fuel, graphite, light or heavy water moderator—generates confidence that these are real-data effects and enables them to be used to extrapolate to other systems.

Reactor lattices of known composition can be predicted using these methods to a standard deviation of about ± 0.5 per cent in reactivity even where no direct experimental evidence is available. Where data are available for similar geometric arrangements, this figure will reduce to around ± 0.3 per cent for a power reactor. In a normal reactor, great care is taken with quality control of materials in the core, including testing batches of steel and graphite for their neutron-absorbing properties. The effect of tolerances allowed on geometry is small, and thus the uncertainty in reactivity introduced by these effects is small. In small systems the amount of moderator present or the leakage of neutrons from the core may be sensitive to the building tolerances, and the associated uncertainties become significant. This is reflected in differences of reactivity in different 'builds' of the same design. Fig. 2.23 shows the departure of the calculated reactivity from the observed critical state for the early operation of the Hunterston R3 reactor. In addition to the basic reactivity effects the period shown includes large changes in control rod insertion and in core irradiation. A similar accuracy is obtained for PWR fuel cycles where sequences covering a number of years of operation are available.

### 4.5.2 *Isotopic composition*

The high accuracy obtained in the prediction of reactivity is possible only if the balance of events between fission and neutron capture in fertile material (conversion) is well predicted. In fact an error of 1 per cent in the conversion ratio would itself lead to ± 0.3 per cent in reactivity in a typical thermal reactor. Table 2.12 shows comparisons between theoretical predictions of isotopic compositions and experimental measurements based upon chemical and isotopic analysis of fuel discharged from a number of reactors. The interpretation of such experiments in the operating reactor is subject to a number of uncertainties; it is, for example, difficult to determine the total heat output of a reactor to better than ± 3 per cent. The temperature distribution is also inferred from theoretical calculations.

The agreement is generally extremely good. In this case the accuracy of the isotope depletion part of the calculation is studied by relating the burn-up to the depletion of uranium-235, the principal component of heat production. The data used in these comparisons had been adjusted to fit clean critical assemblies, but had not taken into account any evidence from

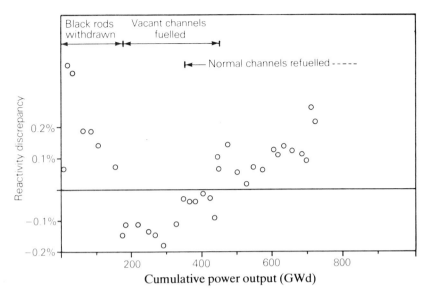

Fig. 2.23 Difference between measured and predicted reactivity for a typical AGR burn-up. The high accuracy achieved is thus good confirmation that the basic data and modelling are accurate.

### 4.5.3 *Rating distributions*
It is difficult to measure temperatures within a fuel assembly in an operating reactor. Even when this is achieved, it is often at the expense of some perturbation of conditions by the measuring equipment. The pin-to-pin variation of induced gamma activity from a discharged fuel element or assembly is thus often the most precise way to check predictions of relative pin power during operation where the irradiation is relatively short; for longer irradiations the pin-to-pin depletion is a good indication of lifetime average conditions.

Table 2.12 Values of theory/experiment for isotopic content of irradiated fuel

| Reactor | Calder Hall | | WAGR | | NPD (Canada) | | SGHWR | | Yankee (USA) | |
|---|---|---|---|---|---|---|---|---|---|---|
| Type | Nat U graphite moderated | | 2.5% UO$_2$ graphite moderated | | Nat U D$_2$O moderated | | 2% UO$_2$ D$_2$O moderated /H$_2$O cooled | | 3.4% UO$_2$ H$_2$O moderated | |
| Irradiation (MWd/t) | 2236 | 4492 | 9850 | 14000 | 5600 | 9200 | 1600 | 7080 | 7000 | 13000 |
| Pu/U | 1.00 | 1.00 | 0.98 | 0.98 | 1.03 | 1.00 | 1.01 | 1.00 | 0.99 | 0.92 |
| $^{239}$Pu/Pu | 1.00 | 1.01 | 1.00 | 0.99 | 1.00 | 1.00 | 0.99 | 0.99 | 1.01 | 1.04 |
| $^{240}$Pu/Pu | 0.96 | 0.98 | 0.98 | 0.99 | 1.00 | 1.00 | 0.99 | 1.00 | 1.00 | 1.00 |
| $^{241}$Pu/Pu | 1.01 | 1.04 | 0.98 | 1.00 | 1.01 | 1.04 | 0.9 | 0.96 | 0.97 | 0.91 |
| $^{242}$Pu/Pu | 1.1 | 1.1 | 1.23 | 1.16 | 1.03 | 1.03 | — | 1.03 | 1.01 | 0.97 |
| $^{236}$U/$^{238}$U | 1.0 | 1.0 | — | — | — | 1.01 | — | 1.02 | 0.97 | 0.98 |

On the small scale of an assembly it is possible to represent the geometry very accurately in the theoretical model, albeit at some expense in computation, but the nuclear data still give rise to uncertainties in prediction. Figure 2.24 shows the comparison between the prediction of pin power and measurement for an irradiated AGR fuel bundle. For most cluster geometries it is possible to predict relative rating in the clean assembly to about 1 per cent, but the position may be somewhat worse in a relatively large assembly such as that in a PWR, especially if there are irregularities due to the inclusion of burnable poisons or control rods. Here the uncertainty may rise to as high as ± 3 per cent.

### 4.5.4  *Control rod worth*
Most of the reactors in commercial use now use cylindrical rods loaded with boron for shutdown purposes, although other geometries (cruciform rods) and materials are sometimes used. The very high neutron capture cross-section of the boron is not in significant doubt, as it has a rather simple energy-dependence and is, indeed, used as a standard for making other measurements.

For heavily loaded ('black') rods the problem of prediction is largely a

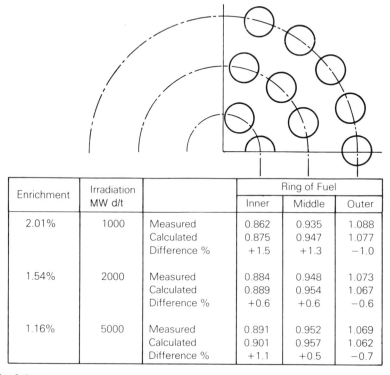

| Enrichment | Irradiation MW d/t | | Ring of Fuel | | |
|---|---|---|---|---|---|
| | | | Inner | Middle | Outer |
| 2.01% | 1000 | Measured | 0.862 | 0.935 | 1.088 |
| | | Calculated | 0.875 | 0.947 | 1.077 |
| | | Difference % | +1.5 | +1.3 | −1.0 |
| 1.54% | 2000 | Measured | 0.884 | 0.948 | 1.073 |
| | | Calculated | 0.889 | 0.954 | 1.067 |
| | | Difference % | +0.6 | +0.6 | −0.6 |
| 1.16% | 5000 | Measured | 0.891 | 0.952 | 1.069 |
| | | Calculated | 0.901 | 0.957 | 1.062 |
| | | Difference % | +1.1 | +0.5 | −0.7 |

Fig. 2.24  Relative ring-to-ring rating distribution in an AGR fuel cluster

geometric one, for any neutron entering the rod is almost certain to be absorbed. In this case it has been shown to be possible to predict the reactivity worth of rods—i.e. the fraction of total neutrons absorbed in them to ± 5 per cent. Where the rods are less absorbing the data and neutron spectrum uncertainties are greater, and the uncertainty may rise to ±10 per cent.

### 4.5.5 Macroscopic power distribution

It has already been noted that the power distribution in a reactor is calculated by synthesizing the results of a lattice calculation with those of a whole core calculation, normally involving a homogenized representation of the reactor. It appears at first sight that the different reactors vary so much in their size and geometry that the required methods would be correspondingly diverse, but in fact the important measure of size from the point of view of neutron modelling is the mean distance travelled by a neutron in its lifetime, and measured on this scale the pitch on which fuel elements are located in AGR, SGHWR, CANDU and PWR is broadly similar.

Figure 2.25 shows a typical plot of the difference between a completely theoretical estimate of channel powers in an operating AGR and values deduced from measurements of channel gas outlet temperature. Figure 2.26

| | 6 | 8 | 10 | 12 | 14 | 16 | 18 | 20 | 22 | 24 | 26 | 28 | 30 | 32 | 34 | 36 | 38 | 40 | 42 | 44 |
|---|---|---|---|---|---|---|---|---|---|---|---|---|---|---|---|---|---|---|---|---|
| 94 | | | | | | | | 3.5 | 2.9 | 3.4 | 1.7 | 3.5 | 1.1 | | | | | | | |
| 92 | | | | | | 0.3 | 2.9 | 2.1 | 3.5 | 3.0 | 4.1 | 2.6 | 1.6 | -0.8 | 1.8 | | | | | |
| 90 | | | | | -0.7 | 1.6 | 0.4 | 3.0 | 0.7 | 0.8 | 2.0 | 0.9 | 1.1 | 1.5 | -0.1 | -1.0 | | | | |
| 88 | | | | 0.1 | 2.8 | 0.5 | 2.5 | 3.2 | 2.3 | -1.7 | 0.2 | 0.6 | 2.2 | 0.6 | 3.2 | 0.3 | -1.2 | | | |
| 86 | | | 2.2 | 0.5 | 1.4 | 1.7 | 1.1 | 0.8 | 1.2 | 2.2 | 0.7 | 0.3 | 0.9 | 0 | 0.1 | -0.4 | -2.2 | -2.0 | | |
| 84 | | -0.8 | 0.5 | 2.3 | 1.6 | 2.3 | 2.9 | -1.1 | 2.7 | 0.1 | 1.2 | 1.4 | -0.5 | 0.9 | 0.4 | -0.5 | 1.5 | -2.7 | -1.3 | |
| 82 | | | 1.3 | 3.0 | 2.2 | 0.4 | 1.0 | -1.1 | 1.8 | 0.5 | -0.7 | -2.2 | -0.7 | 4.3 | -2.6 | 0.7 | -0.8 | 0.2 | 0.4 | 0.1 |
| 80 | 2.4 | 1.5 | 2.0 | 1.4 | 1.1 | -2.5 | 1.7 | 1.5 | 1.3 | -1.2 | -1.5 | 2.8 | -0.4 | -0.6 | -1.8 | -1.0 | 0.6 | 0.6 | 1.0 | 3.3 |
| 78 | 1.5 | 2.4 | 2.7 | 0.2 | -0.6 | 0.9 | -0.6 | 1.1 | -0.3 | -0.6 | -2.4 | 0.5 | 0.6 | -2.9 | -0.2 | -1.5 | -1.0 | 0.7 | 0.9 | 2.0 |
| 76 | 3.7 | 1.2 | -1.1 | 0.3 | -1.7 | -1.1 | -1.0 | -2.8 | 1.5 | 0.9 | 2.0 | -2.6 | -1.4 | 1.0 | 0.3 | 0.3 | 0.1 | 0.8 | 3.4 | 2.3 |
| 74 | 4.6 | 3.3 | 1.2 | -2.5 | -1.2 | -1.8 | -1.7 | -3.3 | -0.9 | 0 | -2.1 | -1.1 | -2.0 | -0.7 | -0.4 | -1.8 | -0.4 | 0.4 | 1.8 | 2.3 |
| 72 | 0.4 | 5.0 | 2.0 | -3.1 | -2.2 | -0.5 | -1.2 | 0.5 | 0.6 | -1.5 | 1.1 | -0.5 | -0.9 | -2.4 | -0.2 | 0 | -0.2 | 1.8 | 1.9 | 3.9 |
| 70 | 1.9 | 0 | 1.4 | 0 | -2.4 | -1.9 | -1.8 | -2.2 | -3.4 | -3.1 | -2.1 | -0.5 | -2.5 | -2.1 | -2.6 | 1.4 | 0.3 | 1.1 | 1.4 | 3.0 |
| 68 | | 0 | -1.1 | -2.0 | -2.6 | -4.8 | -5.4 | 0.7 | -4.1 | -2.1 | -3.8 | -1.1 | -1.0 | -2.5 | -0.7 | 0.6 | 3.7 | 2.9 | 1.7 | |
| 66 | | | -0.2 | -0.7 | -2.7 | -2.8 | -4.9 | -4.6 | -5.2 | -1.8 | -2.8 | -3.4 | -2.1 | -0.8 | -2.0 | -0.9 | -0.4 | 1.5 | 0.8 | -1.8 |
| 64 | | | | -4.3 | -4.3 | -2.0 | -4.1 | -3.9 | -3.5 | -1.9 | -3.0 | -1.3 | -0.9 | -2.7 | 0.2 | 3.0 | 1.9 | -0.2 | 1.1 | |
| 62 | | | | | -4.8 | -3.0 | -3.1 | 0.6 | -2.0 | -1.2 | -3.4 | -1.9 | -1.5 | 0.4 | -1.1 | 1.5 | -1.0 | -0.9 | | |
| 60 | | | | | | -0.6 | -4.5 | -1.3 | -0.2 | -1.4 | -1.8 | -0.7 | 0.1 | 2.9 | 2.2 | 0.9 | 2.7 | | | |
| 58 | | | | | | | -1.3 | -0.9 | -0.8 | 0.3 | 1.2 | 0.4 | 0.4 | 2.1 | 1.4 | 1.8 | | | | |
| 56 | | | | | | | | 2.6 | 3.0 | 2.7 | 1.1 | 2.7 | 4.1 | | | | | | | |

RMS value 2.04 %

Fig. 2.25 Typical comparison between theory and experiment for AGR channel powers difference between measured and predicted powers expressed as percentage of measured values

| | 6 | 8 | 10 | 12 | 14 | 16 | 18 | 20 | 22 | 24 | 26 | 28 | 30 | 32 | 34 | 36 | 38 | 40 | 42 | 44 |
|---|---|---|---|---|---|---|---|---|---|---|---|---|---|---|---|---|---|---|---|---|
| 94 | | | | | | | | 3.7 | 3.7 | 4.1 | 3.9 | 3.6 | 3.7 | | | | | | | |
| 92 | | | | | | | 3.9 | 4.0 | 4.2 | 4.3 | 4.2 | 4.3 | 4.3 | 4.3 | 4.0 | 4.0 | | | | |
| 90 | | | | 4.1 | 4.3 | 4.5 | 5.1 | 3.9 | 3.5 | 3.6 | 4.1 | 5.3 | 4.8 | 4.3 | 4.3 | | | | | |
| 88 | | | 3.6 | 5.1 | 4.1 | 5.4 | 4.1 | 3.5 | 3.9 | 4.1 | 5.0 | 4.3 | 5.6 | 4.4 | 4.3 | 3.8 | | | | |
| 86 | | 4.2 | 4.1 | 4.0 | 5.1 | 4.1 | 3.5 | 3.9 | 5.1 | 5.0 | 4.2 | 3.8 | 4.3 | 5.3 | 4.2 | 5.2 | 4.3 | | | |
| 84 | 3.8 | 4.2 | 4.2 | 5.1 | 3.9 | 3.5 | 3.9 | 4.8 | 4.0 | 4.2 | 5.2 | 4.1 | 3.8 | 4.2 | 5.3 | 4.4 | 4.3 | 4.0 | | |
| 82 | | 4.0 | 4.8 | 5.5 | 4.1 | 3.4 | 3.7 | 5.0 | 4.1 | 3.5 | 4.6 | 4.1 | 5.0 | 4.0 | 3.7 | 4.3 | 5.6 | 4.7 | 4.0 | |
| 80 | 3.8 | 4.3 | 5.3 | 4.3 | 3.7 | 3.9 | 4.7 | 4.1 | 4.9 | 3.8 | 3.8 | 3.8 | 4.2 | 5.0 | 4.0 | 3.7 | 4.2 | 5.1 | 4.2 | 3.7 |
| 78 | 3.5 | 4.3 | 4.2 | 5.0 | 4.2 | 5.2 | 4.2 | 3.8 | 4.0 | 3.6 | 3.5 | 4.1 | 4.9 | 4.0 | 4.7 | 4.0 | 3.6 | 3.9 | 4.1 | 3.4 |
| 76 | 4.0 | 4.2 | 3.5 | 4.2 | 5.0 | 4.3 | 4.9 | 4.0 | 5.1 | 4.0 | 4.7 | 4.8 | 4.0 | 3.6 | 4.0 | 5.0 | 4.0 | 3.4 | 4.1 | 4.0 |
| 74 | 4.2 | 4.2 | 3.5 | 4.0 | 5.2 | 4.2 | 3.7 | 4.0 | 3.9 | 4.9 | 4.1 | 4.1 | 4.0 | 4.5 | 4.0 | 4.8 | 4.1 | 3.5 | 4.1 | 3.8 |
| 72 | 3.5 | 4.7 | 3.9 | 3.6 | 4.1 | 5.0 | 4.3 | 5.1 | 4.3 | 5.1 | 5.5 | 4.2 | 3.6 | 4.0 | 5.0 | 4.1 | 4.9 | 4.1 | 4.2 | 3.5 |
| 70 | 3.7 | 4.2 | 5.2 | 4.3 | 3.8 | 4.2 | 5.2 | 4.3 | 3.8 | 4.2 | 4.3 | 5.0 | 4.0 | 4.7 | 4.0 | 3.7 | 4.2 | 5.2 | 4.2 | 3.8 |
| 68 | | 3.9 | 4.6 | 5.5 | 4.3 | 3.7 | 4.0 | 5.0 | 4.2 | 4.9 | 3.7 | 4.2 | 5.0 | 3.8 | 3.5 | 4.2 | 5.7 | 4.8 | 4.0 | |
| 66 | | 4.0 | 4.3 | 4.3 | 5.7 | 4.1 | 3.6 | 4.0 | 5.2 | 4.2 | 4.0 | 4.7 | 4.0 | 3.5 | 4.0 | 5.2 | 4.3 | 4.3 | 3.8 | |
| 64 | | | 4.2 | 5.3 | 4.3 | 5.2 | 4.2 | 3.6 | 4.0 | 4.9 | 5.1 | 4.0 | 3.6 | 4.2 | 5.5 | 4.3 | 4.4 | 4.4 | | |
| 62 | | | | 3.8 | 4.4 | 4.2 | 5.6 | 4.1 | 4.8 | 4.0 | 4.0 | 3.6 | 4.2 | 5.5 | 4.4 | 5.4 | 3.9 | | | |
| 60 | | | | 4.4 | 4.2 | 4.7 | 5.2 | 4.0 | 3.5 | 3.5 | 3.9 | 5.2 | 4.8 | 4.5 | 4.5 | | | | | |
| 58 | | | | | | 3.9 | 4.0 | 4.2 | 4.2 | 4.1 | 4.1 | 4.2 | 4.3 | 4.1 | 4.1 | | | | | |
| 56 | | | | | | | 3.8 | 3.5 | 3.9 | 4.0 | 3.5 | 3.8 | | | | | | | | |

Form factor 1.33

Fig. 2.26 Channel power distribution measurements in an AGR (MW/channel)

shows the individual channels vary in their power output between 3.4 and 5.7 MW in this case, but the peak discrepancy on prediction is 5.4 per cent and the standard deviation 2 per cent. No adjustment has been made to any of the data in the light of operating experience in this comparison.

Standard deviations of 3 per cent have been observed for a range of PWR assembly power calculations using similar nuclear data. As reactors become bigger, quite small local reactivity discrepancies are capable of introducing gross power tilts of a few per cent across the reactor, and it becomes more difficult to maintain the same accuracy unless a spatial control system is used to keep the power distribution flat.

### 4.5.6 *Temperature coefficients of reactivity*

The effect of changing temperatures in the reactor upon neutron multiplication is an important factor in control. In particular, the temperature coefficient associated with the fuel material is important as a safety factor; if, as is usually the case, a rise in temperature reduces neutron multiplication, it provides a mechanism to limit any transient upset to the state of the reactor.

Two mechanisms are involved: firstly, the Doppler effect apparently broadens the capturing resonances in uranium-238 and reduces reactivity as a result; secondly, the scattering in $UO_2$ leads to a hardening of the thermal neutron spectrum. For an AGR in which the fuel is not intimately mixed with moderator this effect is significant. At the start of life when only uranium is present, changes in spectrum have little effect, but as plutonium builds up temperature changes shift the balance between plutonium and uranium events, producing a positive reactivity change. Comparison between prediction and measurements made on an operating reactor are shown in Fig. 2.27.

### 4.5.7 *Data banking*

A reactor typically contains a few hundred fuel assemblies. Originally on loading in the reactor they will be of a few basic designs or enrichments, but they experience different histories of rating and environment and may be shuffled in position as irradiation proceeds. Controllers move in and out depending upon operating power level, and levels of dissolved poison in the core may change. The estimates of uncertainty used in predicting the content of fuel assemblies or of individual pins, and the resulting rating levels, have assumed that the operating history is known and modelled. To collect the basic information about the dimensions and enrichments of the fuel loaded and the reactor configuration, and to present it in a form usable by the reactor model is a major task. Most reactor operators now have some form of computerized system to assist with the handling of these data.

Figures 2.28 to 2.30 show some types of graphical output from the data bank on the Winfrith SGHWR to illustrate the kinds of data needed even for

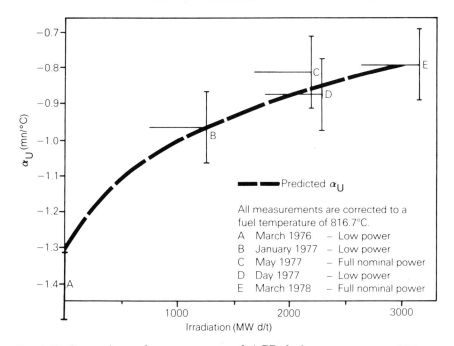

Fig. 2.27 Comparison of measurements of AGR fuel temperature coefficients ($^\alpha$U) with theoretical predictions

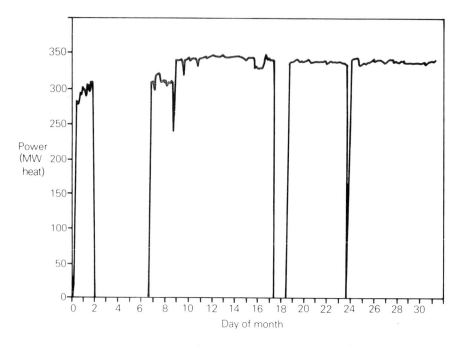

Fig. 2.28 SGHWR power history

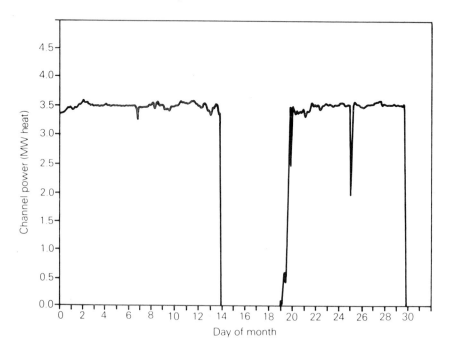

Fig. 2.29 Typical individual SGHWR channel power

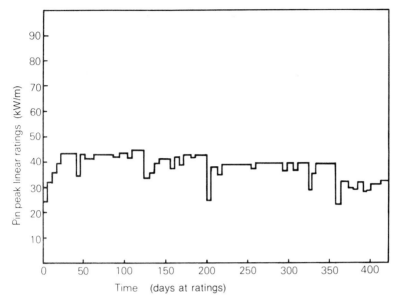

Fig. 2.30 Typical output of an SGHWR pin peak linear rating history calculation

a rather small reactor (of 104 channels). These range from the overall operating history of the reactor as shown in Fig. 2.28 for August 1977, through that for individual fuel clusters in Fig. 2.29, to deductions, combining theory and measurement of peak linear rating in individual pins in the reactor (Fig. 2.30). Fuel elements (and thus pins) must be followed from manufacture, through loading—and possibly shuffling—in the reactor until reprocessing, and the calculational scheme to do this is an increasingly important tool of the reactor operator.

# 5 Current status: fast reactors

## 5.1 **Target accuracies**

In common with thermal reactors, the main economic penalties associated with uncertainties in prediction of reactor properties relate to provisions which have to be made to guarantee the power output of the reactor. Uncertainties in reactivity prediction reflect directly on uncertainties in the required fuel enrichment. If the fuel enrichment is, in the event, too low, then one consequence may be that the burn-up which the fuel can achieve is smaller than fuel endurance would allow. This state of affairs could be corrected after the first few fuel cycles, so that the penalty lies in the higher fuel costs incurred for perhaps about a year. Other solutions may be operationally more attractive. For example, the outer core zone of higher enrichment fuel could be increased in size, either by diminishing the size of the inner core zone or by reducing the breeder thickness. There may be a penalty in loss of power output because the power profile is changed, or in loss of breeding in the blanket until corrective action on the enrichment of feed fuel can be taken. A fast reactor however allows considerable operational versatility which avoids severe penalties. Taking account of the cost of replacement electricity arising from reduced output, the accuracy needed for reactivity predictions is judged to be ± 0.5 per cent at the one standard deviation level for fresh fuel and between 0.5 per cent and 1.0 per cent for irradiated fuel.

Turning to the question of the performance targets for control rods; typically there are about ten fuel elements for each control element in a fast reactor, and this sets the design layout of the core. If there is an uncertainty in the reactivity invested in the control rods of ± 5 per cent, then the designer would plan for a contingency of at least 10 per cent. This would alter the core layout from one in which there was a rod for every ten lattice positions to one in which a rod occupied one in every nine positions. This in turn increases the core size to maintain power output as well as requiring more control rods and driving mechanisms. Alternatively, a reactor designed for natural boron rods could use boron enriched in the high neutron capture isotope $^{10}$B to provide a margin for uncertainties. If enriched material proved to be necessary, then a considerable cost penalty associated with providing enriched material remains for the life of the station. Any shortfall in operating control rod reactivity shortens the fuel cycle time and is to be avoided. This not only sets accuracies for the reactivity of the rods used for operational control, but brings in target accuracies for the reactivity effects of fission products as well as for the prediction of the breeding gain of the core.

Frequently the rods used for ensuring that the reactor can be shut down at all times are separate in function from those for operational control. These shutdown rods must cater for the reactivity changes on going from power to the shutdown condition, as well as for any possible mis-loading of the core at refuelling periods, including the removal of control rods for servicing. The power changes of reactivity are caused by changes in temperature, due to the Doppler fuel coefficient and to reductions in coolant density, and to core expansion effects such as fuel element bowing, and relative movements of the core with respect to the control rods. For both shutdown and operational rods an accuracy of ± 5 per cent in the reactivity controlled is sought.

The maximum power output which can be achieved is frequently set by the temperature of the hottest fuel element. Design and operation aim to minimize the ratio of peak to average power, and accurate predictions of power distributions are needed in cores in which the elements have different compositions arising from their irradiation histories. A typical target for peak to average power is set at ± 1 per cent.

The temperature of the coolant leaving individual fuel elements must be as nearly uniform as possible, not only to maximize efficiency but also to minimize thermal cycling and fatigue of the core structure. In assessing the heat sources throughout the core, the energy deposited in control rods and in breeder elements caused by gamma rays migrating from the fuel is significant and calls for data on gamma spectra and interaction cross-sections.

In an operating power reactor, fast neutrons cause many displacements of the atoms in the reactor components, and the consequent swelling and distortion are in general more severe than in thermal reactors, although these are offset at least partially by temperature-induced creep. These complex material changes call for predictions of the neutron doses and dose gradients leading to distortion, together with data on atomic displacement rates, on helium formation rates and material temperatures. These considerations lead to target accuracies on neutron dosimetry, calling for prediction of total neutron fluxes and damage doses to ± 5 per cent and gradients to ± 10 per cent.

The design of the shutdown cooling requirements and of the route for movements of irradiated fuel requires estimates of the heat arising from fission product decay in fuel elements. Target accuracies are set in the range 2 per cent to 5 per cent in the prediction of fission product heating for decay times exceeding 10 seconds.

Some aspects of the activities induced in irradiated fuel and structural materials assume particular importance. For example, it is necessary to know the strength and distribution of neutron sources in order to monitor the shutdown reactivity of a core. The curium isotopes have a high spontaneous fission rate and provide such a source, as do ($\alpha$, n) reactions in light isotopes such as carbon, oxygen and fluorine present in fuel elements. Data are required on the cross-section involved to achieve a target accuracy for such sources of 10 per cent.

Of prime importance in the consideration of uncertainties in performance assessment are the coefficients governing the behaviour of the reactor under conditions of malfunctioning and accidents. The most important of these coefficients in a fast reactor is that associated with loss of reactivity as the fuel temperature rises, because of the Doppler effect in $^{238}$U. This is a strongly stabilizing coefficient in determining the kinetic response of the reactor. Another safety parameter is the coefficient associated with loss of sodium coolant which for plutonium-fuelled fast reactors of conventional design is positive (leading to a gain in reactivity) in the central regions of the core but otherwise negative in regions where the flux gradient induces leakage of neutrons from the core. Typical accuracies sought for the prediction of the Doppler effect and maximum positive sodium void effect are ± 10 per cent to ± 15 per cent.

Typical target accuracies for the evaluation of reactor physics characteristics of fast reactors may be summarized as follows:

| | |
|---|---|
| reactivity | 0.5% to 1.0% |
| breeding gain | ± 0.05 |
| doppler effect | 10% to 15% |

| | |
|---|---|
| maximum sodium void effect | 10% to 15% |
| control requirements | 5% |
| control rod reactivities | 5% |
| power distributions | 1% |
| decay heat | 2% to 5% |
| activity of components | 10% |
| neutron flux | 5% |
| radiation damage dose | 5% |
| dose gradients | 10%. |

## 5.2 Nuclear data

In discussing the evolution of fast reactor physics in §3, the problems of acquiring nuclear data over the wide energy range needed for fast reactor performances prediction have been discussed, together with the technique for selection of data within the measured uncertainties, based on evidence from integral experiments. In general most fast reactor physics calculations are performed with 20 or 30 enegy groups but with the capability of determining the group data on the basis of more detailed spectrum evaluations. The UK approach to determining the group data will now be reviewed.

The nuclear data libraries currently used are in two forms: the fine-group library known as FGL5 and the broad-group library FD5. The fine-group library FGL5 contains 2240 fine groups covering the energy range 0.414 eV to 1.65 MeV together with two thermal energy groups to embrace the quite small number of neutrons below 0.414 eV. The fine-group data are supplemented by sub-group data within the fine groups to represent the distributions of resonance cross-section values for resonant materials. Up to 50 sub-groups within a fine group are possible and the sub-group data are tabulated at a number of temperatures. Data for the most important fast reactor substances are stored in the fine-group library. These include H, $^{10}$B, $^{11}$B, C, O, Na, Cr, Mn, Fe, Co, Ni, Mo, Ta, $^{235}$U, $^{238}$U, $^{239}$Pu, $^{240}$Pu, and $^{241}$Pu.

The basic source data from which this fine-group data library is formed are held in the UK Nuclear Data Library which contains evaluated data based on measurements made throughout the world. The data are held as cross-sections tabulated as a function of neutron energy, together with tabulations of parameters characterizing the resonance data, and are supplemented by resonance-parameter data derived from studies of the statistical distribution of resonances. It is necessary to select or adjust these statistical parameter data to be consistent with the smooth cross-sections derived from evaluation of measurements made at energies beyond those at which resolution of individual resonances is possible.

The process of creating an unadjusted fine-group library is outlined in the following summary.

1. The primary cross-sections in the UK Nuclear Data Library are averaged over the energy bands corresponding to the individual groups of the set.

The group structure is so fine that a $1/E$ weighting spectrum is sufficient for this averaging process.

2. Tabulations of resonance cross-sections are produced from the statistical resonance data using the multi-level Breit Wigner formula. These are checked for consistency with the smooth cross-sections of the unresolved resonance region above $\approx 25$ keV.

3. Inelastic scattering data are obtained assuming scattering is isotropic, and are represented as cross-sections to identified discrete levels or to pseudo-levels approximating the continuum distributions.

4. Secondary neutron energy distributions arising from elastic scattering are calculated for selected primary groups; distributions for the remaining primary groups are obtained by interpolation.

The standard broad-group set for design and operational calculation is the FD5 set which contains data for many additional substances. This set has 37 energy groups, 35 being condensed from the 2240 fine groups, and with the two thermal groups again added. Resonance shielding is represented by shielding factor tables, in which the data are tabulated against background cross-section and temperature. The main FD5 library contains the data normally used for core neutronics calculations. In addition there are libraries of fission product cross-sections (for about 200 fission products), activation cross-sections and atomic displacement cross-sections together with energy release data. Delayed neutron data and decay heat data form separate libraries.

The data for the principal substances in the FGL5 library have been adjusted to give a best fit to the basic cross-section measurements and simple integral or reactor measurements, taking into account the estimated uncertainties in these data. Covariance data for the adjusted cross-sections are obtained which enable the uncertainty arising from nuclear data in calculations made using FGL5 to be estimated. The effect of these adjustments is not small. Thus the combined use of the differential cross-section data and the results of measurements carried out on zero-power fast reactors such as ZEBRA lead to significant changes in the prediction of reactor properties. Illustrations of such changes in the case of PFR are shown in Table 2.13.

## 5.3 Current physics methods

Again, as in the section on thermal reactors, discussion is focused on the methods and approaches currently adopted in the UK. These are not typical of methods in use elsewhere.

### 5.3.1 *Homogenization methods and few-group data preparation*
For standard design and operational calculations, the 37-group FD5 group constant set is quite sufficient—indeed fewer groups than 37 are often adequate. For more broadly based scoping studies, sub-assembly heterogeneity effects of fast reactors can be neglected, and it is sufficient to use the FD5 broad-group data set weighted with simple average material number

Table 2.13 Changes made to PFR properties arising from the adjustment of nuclear data

| Reactor property | Change made (per cent) |
|---|---|
| $k_{eff}$ | + 3.4 |
| $^{238}$U fission/$^{239}$Pu fission | +13.7 |
| $^{238}$U capture/$^{239}$Pu fission | −11.8 |
| $^{239}$Pu capture/$^{239}$Pu fission | + 7.7 |

densities appropriate to each different sub-assembly region, such as the inner core region, the outer core region and the blankets.

The FD5 data set to be used is obtained from the FGL5 library by using the code MURAL. This has many fast reactor applications but in generating FD5 data, simplified representations of a typical fast reactor are calculated in fine groups. These calculations are used to produce group data condensed down to the structure of the FD5 library for each region of the reactor. Resonance shielding factors are obtained from a range of MURAL calculations spanning typical compositions.

Whilst homogenization is often adequate, there will be occasions on which it will be necessary to assess the error made in neglecting the detailed heterogeneity in the neutron flux distributions associated, for example, with fuel sub-assembly structures or control mechanisms. Such heterogeneities may be included in the MURAL calculation in cylindrical, slab, or spherical form and the FD5 data condensed for each region from the resulting space/energy solution.

The overall reactor geometry may also have an influence on the weighting spectrum to be used in generating the FD5 data. Thus, for example, the spectrum will vary continuously in the neighbourhood of the boundary between the core and breeder regions. These spectral transients may be evaluated by a MURAL calculation which represents the whole reactor smeared into homogeneous regions. Heterogeneity can be introduced into such studies by using energy-dependent leakages derived from these whole-reactor calculations as a correction to MURAL sub-assembly calculations.

### 5.3.2 *Whole-reactor neutron flux calculations*

The standard method for performing whole-reactor flux calculations is to use the neutron diffusion approximation. A full range of geometrical and dimensional options is used including the triangular and hexagonal mesh representations associated with the shape of fast reactor sub-assemblies. The finite-difference techniques used are similar to those referred to in the previous section. In general, however, more neutron groups (up to the full FD5 set) are used in fast reactor calculations than in those for thermal reactors and, since the scattering is predominantly downwards in energy, rather different solution procedures are followed.

As has been commented earlier, the use of diffusion solutions may be suspect at the interface between regions of markedly differing composition. Solutions of the whole-reactor equations in transport form are thus undertaken to investigate the validity of the diffusion approximation although with present computing power these investigations are limited to one- or two-dimensional geometry representations of the reactor. The method used in solving the multi-group transport equation is normally one involving direct numerical integration of the equations resulting from a representation of the angular dependence of the neutron flux by a limited number of discrete directions. Other methods referred to earlier—Monte Carlo and collision probability—may also be used when appropriate.

Because of the limitations on transport theory capability, it is usual to use a simplified model of the reactor for comparative transport and diffusion theory calculations, and to apply the derived correction factors to the three-dimensional diffusion theory flux solutions.

Illustrations of the size of the corrections are given in Fig. 2.31, which shows the ratio of the transport theory predictions of the $^{239}$Pu and $^{238}$U fission rates relative to those obtained from the diffusion theory as a function of radial position in a large ZEBRA assembly. This reactor used an inner core of lower enrichment than the outer core in order to flatten the radial power distribution, and the errors arising from the diffusion approximation at the core zone interface and at the core breeder interface are clearly illustrated.

The improvement in the ratio of the calculated to experimentally measured $^{238}$U fission rate resulting from applying these corrections is shown in Table 2.14. In this Table C/E refers to the calculated to experimental value, which should ideally be unity. The remaining discrepancies in the C/E values are

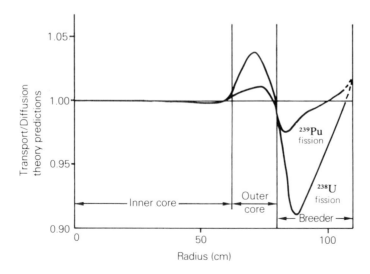

Fig. 2.31 The effects of transport theory solutions on diffusion theory calculations of $^{239}$Pu and $^{238}$U fission rate distributions

Table 2.14 The effect of applying transport theory corrections to the C/E values of the radial $^{238}U$ fission rate distribution

| Region | Distance from centre (cm) | C/E (Diffusion theory C-values) | Uncertainty on experimental value E (%) | Transport correction to C (%) | Corrected C/E |
|--------|------|--------|--------|--------|--------|
| Radial | −103.1 | 0.944 | 1.9 | +1.6 | 0.959 |
| blanket | − 92.1 | 1.055 | 1.4 | −6.9 | 0.982 |
|  | − 81.4 | 1.021 | 1.2 | −4.0 | 0.980 |
| Outer | − 76.0 | 0.941 | 0.8 | +1.5 | 0.955 |
| core | − 70.6 | 0.947 | 0.7 | +3.8 | 0.983 |
|  | − 65.0 | 0.969 | 0.6 | +2.0 | 0.988 |
| Inner core | − 59.6 | 0.994 | 0.6 | +0.5 | 0.999 |
|  | − 54.3 | 1.004 | 0.6 | −0.8 | 0.996 |
|  | − 43.5 | 1.004 | 0.5 | −0.4 | 1.000 |
|  | − 21.8 | 1.005 | 0.5 | 0 | 1.005 |
|  | 0.0 | 1.000 | 0.5 | 0 | 1.000 |
|  | 21.7 | 1.003 | 0.5 | 0 | 1.003 |
|  | 43.5 | 1.002 | 0.6 | −0.4 | 0.998 |
|  | 54.2 | 0.995 | 0.6 | −0.8 | 0.986 |
|  | 59.6 | 0.999 | 0.6 | +0.5 | 1.004 |
| Outer | 65.0 | 0.966 | 0.6 | +2.0 | 0.985 |
| core | 70.6 | 0.949 | 0.7 | +3.8 | 0.985 |
|  | 76.0 | 0.944 | 0.8 | +1.5 | 0.958 |
| Radial | 81.4 | 1.022 | 1.4 | −4.0 | 0.81 |
| blanket | 92.1 | 1.043 | 1.4 | −6.9 | 0.971 |
|  | 103.1 | 0.921 | 1.8 | +1.6 | 0.936 |

unexplained but are thought largely to arise from residual inaccuracies in the $^{238}U$ nuclear data. No allowance is made for data uncertainties in the experimental uncertainties shown. Experience has also shown that these discrepancies are greater for thin regions of the reactor and are further accentuated by markedly different region compositions. In the illustration used, the core had an unusually thin region of outer zone composition, with a correspondingly large difference between inner and outer zone enrichments.

Reference was made in §3 to the problem of neutron streaming along the the channels of gas-cooled reactors. A similar issue arises in fast reactors associated with sodium and other relatively low-density regions. The problem is studied using two-dimensional transport theory and employing the discrete ordinate method referred to above in order to obtain equivalent diffusion solutions.

The absence in fast reactor methods of the separation into lattice-cell and whole-reactor calculations typical of thermal reactor methods simplifies the treatment of burn-up. Detailed burn-up histories of fast reactors are obtained by solving the isotopic depletion or burn-up equations between successive many-group, whole-reactor diffusion solutions. The burn-up of the reactor is normally evaluated for a specified number of days at a specified total reactor power or maximum channel power. Alternatively, given reactor zones can be burnt for different numbers of days or to a specified target heavy-atom burn-up. In addition, it is possible to follow the burn-up of specific nuclides which may be of particular importance such as the destruction of boron in control rods by neutron capture.

The temperature of structural core components, such as the hexagonal wrappers enclosing fuel sub-assemblies, is obtained from the appropriate heat transfer equations, the whole-reactor fission power distribution, and the heating contribution from gamma radiation. The gamma power maps are calculated from sources based on the neutron flux solutions together with a library of gamma production data. It is frequently necessary to use Monte Carlo methods to study the transport and deposition of gamma energy.

## 5.4 Calculational scheme

Reference has already been made to the importance of harnessing computer power to ensure that reactor physics calculations are performed effectively and efficiently. This is particularly true when detailed solutions are required involving not only vast amounts of numerical manipulation but also the input, processing, and production of very large quantities of data. Given the long time-scale of fast reactor development and the involvement of scientists and engineers at widely separated sites around the country, the need for an effective computer-based scheme available to all has been particularly important. Thus, for the last decade a vitally important aspect of the development of fast reactor physics methods has been the use of the growing number of computers, not only to do the calculations but also to organize them and store the results for future access. No review of current reactor physics methods would be complete without some reference to such calculational schemes, and the COSMOS system, developed in the UK for fast reactor applications, is amongst the most advanced.

The COSMOS scheme has been designed to have the following four fundamental features.

1. *A modular database.* A logical set of physics data was identified as a generalization of the well known FORTRAN programming language array and called a *datablock*. Fast reactor source data (manufacturing details, core layout, material compositions, etc.) are now cast in this form and constitute what is called the *Fast Reactor Database*. It is one of the most important innovations of COSMOS and contributes directly to the desired consistency in the use of primary data.

2. *A comprehensive databank.* The databank is the general repository for the datablocks of the Fast Reactor Database, and was formulated after a theoretical study of filing systems. Datablocks are stored individually and

each is tagged with a label on entering the databank. The label identifies the origin of the datablock. A complete record of the subsequent usage of each datablock in COSMOS calculations is maintained. A datablock may be retrieved directly from its label or indirectly by reference to some calculation in which it was used.

The databank is serviced by some 30 utilities, but its day-to-day management is largely automated. There is a high integrity back-up system for guarding the contents against accidental loss, which is integrated with a comprehensive archiving system.

3. *Modular coding.* The use of a databank with modular data systematizes the processing of data required in the solution of the physics and engineering equations. COSMOS takes advantage of this by providing programmers with standard sub-routines which perform the burden of all such data processing. Consequently the physics and engineering are clearly separated from the rest of the program. All COSMOS programs have completely compatible input and output subroutines.

4. *Automatic job preparation.* It is possible within COSMOS to carry out long sequences of interlocking calculations requiring very large amounts of input data. As the possibilities for making calculations grow, so do the possibilities for making errors. To take full advantage of the power of the COSMOS scheme, the preparation of jobs by hand had to be eliminated by providing automatic means. For this purpose a new language has been created which combines FORTRAN facilities with a very powerful string editor and a repertoire of command verbs for interrogating the COSMOS databank filing system catalogues. In this way users can prepare sequences of jobs using simple, manual instructions.

### 5.5 Use of the COSMOS scheme

Apart from COSMOS systems programs and associated utilities, the scheme consists of programs operating in the so-called 'engineering' and 'calculational' frameworks. Programs in the former category are available for modelling the sub-assembly loading and reloading of operational reactors, such as the prototype fast reactor at Dounreay, and then making up a calculational model of the loaded reactor for physics calculations in the calculational framework. The physics programs can also be used directly for conceptual and design studies.

Two schemes exist for sequencing the physics programs in the COSMOS scheme. One scheme is designed to allow the available programs to be sequenced in a variety of ways under direct user control, whilst the other is designed to sequence the programs in a fixed well-defined way. The former scheme is used for reactor conceptual studies, whilst the latter is used to monitor the existing UK prototype fast power reactor (PFR), where the calculational routes, for monitoring fuel burn-up for example, are well established together with their associated uncertainties. Each scheme uses the COSMOS random access databank and the physics methods already described.

The core performance of various reactor designs in usually initially studied in two-dimensional calculational geometry, with perhaps some preliminary calculations in even simpler one-dimensional models. Calculations in a few reactor zones are performed often in the full energy group structure of the FD5 broad-group library. The control rod arrays are smeared into cylindrical annuli and the absorber number densities modified to allow for the increased absorber worth produced by such a smearing process. Fuel and breeder sub-assembly homogenized group constants are prepared from the FD5 library, or in the case of special studies from the FGL5 library using MURAL. Group constants for control rods are normally prepared by a MURAL cylindrical super-cell calculation. The reactivity coefficients arising from changes in fuel temperature—the Doppler effect—are examined by preparing zone-dependent group constants at a number of temperatures, temperature-dependent data for $^{238}$U, $^{235}$U, $^{239}$Pu, $^{240}$Pu, Mo, Ta and Fe being available in the fine-group library; sodium void effects are examined by preparing group constants for zones with and without sodium. The reactivity effects arising from postulated accidents, such as the slumping of the core, are examined by preparing group constants for zones of standard and high material density. Flux calculations are carried out using diffusion theory, but transport theory calculations are also required to provide corrections to the diffusion theory solutions. Large reactivity perturbations are normally calculated from the difference in reactivity between calculations performed in each geometry. Exact perturbation theory is used for distributed effects; small reactivity perturbations are calculated by first-order perturbation theory.

More detailed reactor design studies are carried out in three-dimensional models of the whole reactor using hexagonal or triangular mesh representations; otherwise broadly the same methods are employed as those already described. Control rods are then of course represented exactly as single hexagons or groups of six triangles; any control rod insertions can be modelled using the $z$ dimension. In these more detailed spatial calculations fewer energy groups are used: 28 groups for Doppler studies, 16 for sodium void studies, 9 for control rod studies, etc., the group structure chosen being the result of special supporting investigations applicable to the particular type of reactor design. Where possible, sector calculations are carried out rather than whole-plan studies in order to reduce the problem size. The burn-up of the PFR is followed by calculations carried out in whole-plan geometry with six neutron groups, whilst PFR forward-planning fuel management calculations are carried out in the same energy group structure but in triangular geometry with axial bucklings reproducing axial leakage.

### 5.6  **Validation of methods and data**

#### 5.6.1  *Comparison with more exact methods*
The recommended standard methods of calculation for the design and operation of sodium-cooled fast reactors use diffusion theory with broad-group data prepared from the FD5 library by number density smearing, together with specially derived cross-sections both for control rods, in which the homogenization takes account of the flux fine structure, and for control rod channels. There are recommended energy group condensation procedures for all performance predictions such as fuel management studies, and power distribution calculations. Two dimensional plan models are used for some calculations, with routines for calculating axial average (or mid-plane) bucklings and compositions.

The approximations involved in the use of FD5 and homogenized compositions in the calculation of various properties such as Doppler effects, sodium temperature and voiding reactivity effects, and power distributions have been assessed by comparisons with calculations using MURAL-FGL5 and cell models representing the pin and sub-assembly heterogeneity effects. Streaming effects have been separately investigated using transport theory codes. The effects of spectral transients near core-breeder boundaries have also been evaluated using MURAL-FGL5. The resonance shielding method used with FGL5 has itself been validated by comparison with calculations using the detailed cross-section data at $10^5$ energy points.

The approximations involved in the use of diffusion theory have been investigated by comparisons with two-dimensional transport theory methods. The choice of suitable finite-difference mesh structures has been studied by calculations made in fine-mesh representations. International intercomparisons of calculations of benchmark problems have provided further validation of the methods and data used.

#### 5.6.2  *Comparision with measurements on experimental facilities*

1. *ZEBRA*. Comparison of predicted performance with specially designed experiments is a key component of the validation process. Within the UK the ZEBRA zero-power critical facility is used for both simple integral measurements and for full-size simulations of power reactors. The former measurements are used mainly for data testing and adjustment, and also as a direct source of reactor spectrum-averaged data to be used in fast reactor calculations. The power reactor simulation experiments are used for studying various reactor properties and assessing the accuracy with which these can be calculated. Such experiments have been carried out in ZEBRA for the UK Prototype Fast Reactor, and for the Japanese fast reactor, MONJU, in the MOZART collaborative programme with Japan. Studies of larger fast-reactor and heterogeneous cores have been made in the BIZET collaborative programme on ZEBRA. The reactor physics aspects studied on ZEBRA include:

(i) Assembly reactivity
(ii) Power distributions
(iii) Control rod reactivity worths and reaction rates
(vi) Doppler coefficients associated with the heating of small samples of $UO_2$
(v) Sodium voiding reactivity effects (radial and axial scans of single element void worth, and large region voiding)
(vi) Gamma-ray energy deposition.

The measurements on reactor mock-ups provide a test of both methods and data. By analysing a range of measurements, the accuracy achievable by the calculations can be studied in some detail. Thus, for example, by analysing distributed sodium void reactivity effects measured in different-sized cores, the accuracy of separate components of the effect may be estimated. In the same way studies of control rods having different boron enrichments and different diameters, and also varying arrangements of rods, make possible the determination of trends in the relationship between calculation and measurement of rod worths, and reaction rates both near the rods and over the whole reactor.

In the analysis of ZEBRA experiments, the philosophy is to use in the first place the methods and data recommended for design and operational purposes. Only when these methods fail to reproduce the experimental observations, is resort made to more elaborate calculations. However, in order to achieve versatility in the range of material compositions which can be studied in ZEBRA, a large part of the material inventory is held in the form of plates or coupons 50 mm × 3.2 mm thick. These plates are stacked into cells which are repeated throughout zones to achieve the desired material compositions. The flux heterogeneity associated with this structure is more severe than that in a multi-pin rod bundle typical of a fast power reactor, and requires special treatment. The code MURAL is used for this purpose in its one-dimensional slab geometry, collision-probability option for solving the multi-group neutron transport equation.

Firstly, a converged solution is obtained for the multi-region cell in broad groups using the broad-group library, so producing a source for use in a fine-group source calculation. The converged fine-group flux is used to condense the fine-group cross-sections down to the broad-group library structure; those broad-group constants are then used to solve the broad-group equations. The multiplication produced by this second broad-group treatment is then compared with that produced by the fine-group pass, and if the difference is unacceptably large the fine-group pass is repeated with more accurate values of the spatial source derived from the second broad-group solution. Since the FD5/FGL5 libraries are well matched for most fast reactor problems, more than one fine-group pass is not normally required.

The final cell-averaged broad-group cross-sections are used as the group constant data for whole-reactor calculations which then form the basis of the ZEBRA analysis.

A comparison of measured and calculated reaction rates for ZEBRA is shown in Fig. 2.32. The figure illustrates the stack of plates which constitutes a ZEBRA plate cell of typical fast reactor composition, and compares the $^{238}$U fission rates averaged over the $UO_2$ plates with the results of MURAL calculations.

Figure 2.33 gives a comparison with greater spatial resolution. In this case, the detailed variation of the $^{238}$U capture rate is shown with increasing penetration into the thicker uranium metal plate of a breeder region cell in ZEBRA. The MURAL calculation follows the measured capture rate satisfactorily.

2. *Doppler experiments.* The Doppler temperature coefficient is an important negative feedback effect in fast reactors and makes an important contribution to limiting the magnitude of any temperature transient in an accident situation. It is thus necessary to obtain reliable confirmation of the accuracy of predicted values of this coefficient. Probably the most useful measurements of the Doppler coefficient for a whole reactor are those obtained in the US SEFOR program. The experiments, which involved reactor transients in which the oxide fuel reached temperatures of about 1800 °C were simulated using the data and methods described above and the theoretical results found to be in close agreement with the measurements. Such comparisons enable conclusions to be drawn regarding the accuracy of the nuclear data and the calculational methods. These conclusions are used in Doppler calculations for the PFR and possible fast reactor designs and form the basis of the current view that Doppler coefficient can be predicted to ± 20 per cent.

3. *PFR core-following and special PFR experiments.* Whilst zero-energy experiments provide well-designed, high accuracy data, the measurements made on power reactors provide the ultimate opportunity for the demonstration of the integral validity of the methods and data used. Accuracy and detail may be rather lower than in a zero-energy reactor but the direct relevance of the experiment/theory comparison can hardly be in doubt.

Special efforts have thus been made in the UK to validate nuclear data and calculational methods by following the critical fuel cycle operation of the 250 MW(e) Prototype Fast Reactor at Dounreay and also by analysing a number of special experiments performed in PFR. The analysis of power reactor experiments requires accurate representation of many reactor details and is greatly helped by the availability of the COSMOS scheme.

Comparisons of prediction and measurements have been made for reaction rates, such as $^{239}$Pu and $^{235}$U fission, sub-assembly coolant outlet temperatures, control and rod worths, the isothermal temperature coefficient, the power coefficient, the reactivity loss with burn-up, and reactivity noise. The predictions have made use of the experience gained in the ZEBRA PFR mock-up. The general conclusion is that there are some detailed differences between calculation and observation, but that none of these has any safety implications.

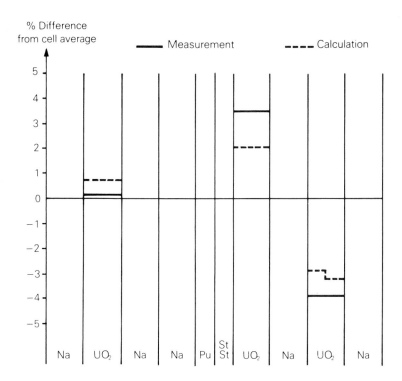

Fig. 2.32 $^{238}$U fission rate distribution in atypical ZEBRA core cell

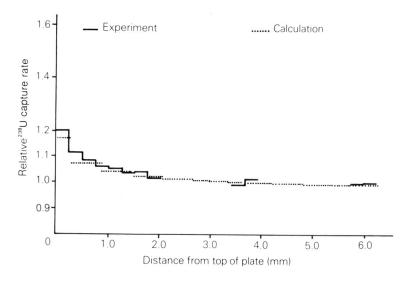

Fig. 2.33 Variation of $^{238}$U capture rate through uranium metal plate in breeder region cell.

The reactivity loss with burn-up was well predicted except during the initial period of power operation. The reaction rate measurements in the core compared well with predictions, but in the upper axial breeder discrepancies of 10-15 per cent were found, which may be due to neutron streaming effects in the central sodium-filled channel in which the measurements were made.

Various special experiments are being carried out in PFR which will contribute to the validation process, such as an irradiation of a well-characterized $^{238}U$ sample in support of Doppler validation, and integral measurements of the ratio of $^{239}Pu$ capture to fission reactions, which has a direct influence on the prediction of breeding performance. In addition core following will continue to be pursued, and the calculated fuel burn-up and sub-assembly irradiation induced swelling and distortion compared with the results of post-irradiation examination.

### 5.6.3 *Accuracy of prediction of fast reactor core neutronics properties*

The evaluation of uncertainties in both the methods of calculation and the associated nuclear data libraries rests on the following studies:

    (i) calculations of the sensitivities of reactor performance parameters to uncertainties in the adjusted nuclear data, and prediction of the errors in these parameters by making use of the variance-covariance files of the data library;

    (ii) detailed studies of the approximation made in the calculation methods, involving comparisons with more accurate methods; and

    (iii) comparisons of the predictions made using these methods and data with a wide range of experimental observations made principally in zero-power critical reactors, but also in experimental and power reactors.

The accuracies of prediction of fast reactor core neutronics properties are determined by a combination of these three approaches. They depend on the type of fast reactor design and its similarity to the mock-up assemblies which have been studied, and are illustrated by the following values which are for a typical sodium-cooled fast reactor of conventional design:

    (i) effective multiplication $\pm 0.5\%$
       (equivalent to fuel feed enrichment $\pm 1.0\%$);

    (ii) breeding gain $\pm 0.05$ (which is equivalent to an uncertainty of 20%);

    (iii) control rod reactivity worths $\pm 10\%$;

    (iv) sodium voiding reactivity effects:
       central term $\pm 13\%$, leakage term $\pm 10\%$;

    (v) Doppler reactivity effects:
       normal core $\pm 18\%$, voided core $\pm 25\%$;

    (vi) fission product reactivity effects $\pm 10\%$;

(vii) power distributions: relative rates in different core zones $\pm 5\%$; distributions within a core zone $\pm 2\%$, gradients near to control rods $\pm 10\%$.

The figures are for one standard deviation. In general, it should be noted that the accuracies currently achievable approach, but do not yet in all cases reach, the targets identified at the beginning of this section.

# 6 Future trends in reactor physics

In earlier sections an outline has been given of the way in which attempts to understand and predict the neutronics behaviour of reactor cores have progressed. Reliance upon extensive series of integral experiments has gradually been overtaken by the use of computer based-models, using the enormous fund of information on cross-sections and other nuclear properties provided by world-wide programmes of measurements.

It has been indicated that predictions of many basic quantities needed for the design and operation of present day reactors can now be made to an accuracy which is very high compared to that attainable in most engineering fields, and that, as a consequence, large investments—perhaps the introduction of new fuel designs, for example—can be made on the basis of theoretical studies alone. What then of the future? Is the need only to refine and codify the procedures now in use so that they can be easily used in new applications? There are a number of reasons that lead to the supposition that this will not be adequate.

Two factors may be identified that support this conclusion: firstly, the increased scale of the reactor industry, both in numbers of reactors in use and their individual size, and secondly, the continuing development of computing machines.

The scale of the industry means that the economic benefit of even modest improvements, say in fuel cycle performance or the required rating margin on a reactor, can be very large indeed, especially if the same results or design modifications can be applied to a number of reactors. Thus there will be an incentive to devise and demonstrate more accurate models than have hitherto been needed. These will be required to support not only new design variants and concepts (such as the 'heterogeneous' fast reactor cores) but also improvements in the operating regime of existing reactors.

The most likely route towards this last end is the development of adaptive or learning models using observations from operating plant. It is inherently more difficult to take measurements on operating power stations than on laboratory-scale experiments, and this is often reflected in lower precision on any individual measurement. Similarly, it is rarely possible to exercise close control over all the variables of plant state so as to get 'clean' measurements of single effects. To counterbalance this, however, much greater quantities of data can be accumulated over a long period of time, and it is clearly directly relevant; there is no need to extrapolate from small to large scale.

A small number of workers are currently experimenting with the problems involved in this type of study, which range from the mechanics of collecting and manipulating large amounts of information to the statistical and modelling developments needed to interpret it and produce best estimates from a combination of theory and observation. It is anticipated that such activities will receive more attention in the future.

Of course, operating plant will only provide information about quantities observable in normal operation, such as reactivity changes with irradiation, power distributions, and control rod worths. Laboratory or rig experiments will still be needed to confirm predictions of key issues which involve more extreme conditions appropriate, for example, to studies of the consequences of abnormal or accident sequences, such as high temperatures or distorted geometries. Such studies will not usually be performed in operating power stations for reasons of economy or even of safety. This is a continuation of the trend for experiments to become more specialized and less like mock-ups of reactors.

The greater knowledge and predictive capability resulting from such observations can only be harnessed by improvements in the models themselves and in their availability to reactor designers and operators. Here the immense advances in computing capability and associated reductions in unit cost will make it economic to provide more advanced plant simulators, and to have them more immediately available to operators. The application of such models is at present largely confined to off-line applications, such as the choice of refuelling strategies, because they are laborious and time-consuming to use; it is to be expected that there will be a growing tendency to incorporate them directly in plant control as they become more convenient and rapid in execution.

Similarly, the designer and assessor of the future will demand an increasing ease of use of the sophisticated models through an interface which interprets his requirements in direct physical or engineering terms. The COSMOS scheme already addresses the issue, but significant improvements based upon improved computer system software are still required.

The more widespread use of reactors equally widens the scope of associated activities such as fuel fabrication, storage, transport, and reprocessing. Such processes involve problems with criticality and radiation transport, and the elaborate models which have hitherto been applied mainly to problems of the reactor core and shields are now in demand to permit more efficient solutions to the many out-of-reactor problems. These typically involve much more complicated geometries than those of the regular lattices of the reactor core, and much still remains to be done to simulate the behaviour of these and to demonstrate that their properties can be reliably predicted with minimal recourse to experiment. The most powerful tool for studies of this kind is often the Monte Carlo technique, but the problems involved in reducing the statistical uncertainties inherent in its use are formidable and are likely to continue to receive attention.

Neutron physics problems are not, of course, confined to fission reactor applications. The energy generated in the fusion of light elements appears mainly as high-energy neutrons, and the design and development of fusion reactors poses a number of difficult problems in neutron and radiation transport.

Over the past 20 years the emphasis of reactor physics work has changed and much has been achieved; the large experimental teams have disappeared and been replaced by much smaller numbers of specialists, mainly with a theoretical and computing background. However, it does not seem necessary for reactor physics expertise to diminish with increasing exploitation of nuclear power; rather the reverse. The benefits of improved understanding are still increasing.

## Acknowledgements

The authors would like to thank the South of Scotland Electricity Board and the Central Electricity Generating Board for permission to reproduce illustrations.

## Appendix A. Xenon–135

The fission product poison $^{135}$Xe is of considerable importance in thermal reactors because it has a high yield and also an enormous cross-section for the capture of neutrons; it has a resonance at about 0.1 eV and a cross-section of $2.7 \times 10^6$ barns at 0.025 eV. In fact only a small percentage (about 5 per cent in the case of $^{235}$U fissions) of the $^{135}$Xe produced is as a direct result of the fission process. Most of the $^{135}$Xe formed arises from the radioactive decay of the direct fission product $^{135}$I. The $^{135}$Xe is removed from fuel partly through its radioactive decay (half-life 9.2 hours) and partly through its capture of neutrons. The equations which determine the concentration of $^{135}$Xe are thus

$$\frac{dI}{dt} = \gamma_I \Sigma_f \phi - \lambda_I I, \tag{9}$$

and

$$\frac{dX}{dt} = \gamma_x \Sigma_f \phi + \lambda_I I - \sigma_x X \phi - \lambda_x X, \tag{10}$$

where,

> $I$    is the concentrations of $^{135}$I
> $\gamma_I$   is the fission yield of $^{135}$I $(=0.061)$
> $\lambda_I$   is the decay constant of $^{135}$I $(=6.7\ h^{-1})$
> $X$    is the concentrations of $^{135}$Xe
> $\gamma_X$   is the fission yield of $^{135}$Xe $(=0.003)$
> $\sigma_X$   is the neutron capture cross-section for $^{135}$Xe
> $\lambda_X$   is the decay constant of $^{135}$Xe $(=9.2\ h^{-1})$
> $\Sigma_f$   fission cross-section of the fuel
> $\phi$    is the neutron flux.

### Steady-state level

Under steady reactor operating conditions the left-hand sides of eqns (9) and (10) are zero, giving

$$X = \frac{(\gamma_I + \gamma_x)\Sigma_f \phi}{\lambda_x + \sigma_x \phi} \tag{11}$$

as the level at which the $^{135}$Xe concentration saturates. The fraction of all neutrons captured in $^{135}$Xe is given by

$$F_x = \frac{\sigma_x X}{\Sigma_a}$$
$$= (\gamma_I + \gamma_x) \cdot \frac{\sigma_x \phi}{\gamma_x + \sigma_x \phi} \cdot \frac{\Sigma_f}{\Sigma_a}. \tag{12}$$

Thus

$$F_x \to (\gamma_I + \gamma_x) \frac{\Sigma_f}{\Sigma_a} = 0.04 \quad \text{as} \quad \phi \to \infty,$$

but $F_x = 0.02$ at more reasonable thermal power reactor levels.

## Transient effects

This loss of about 2 per cent of all neutrons by capture is $^{135}$Xe is a significant item in the neutron balance account. Equally important is the variation in this quantity as a result of changes in power level. Thus suppose that a reactor has been operating steadily over a period of a few days. The $^{135}$Xe will have reached the level given by eqn (11). Suppose now that there is a sudden reduction in power (i.e. the value of $\phi$ drops), then eqn (9) shows that the value of $I$ will start to fall but only at a rate determined by $\gamma_I$ of a few per cent per minute. The decrease in the major source term for $^{135}$Xe in eqn (10) is thus very slight. On the other hand the removal term $\sigma_x X \phi$ reduces immediately with $\phi$. Thus $dX/dt$ is initially positive and xenon concentration rises.

This transient response in xenon level to changes in power is of importance in two major respects. In the first place the rise of xenon concentration leads to more captures in the poison and thus a further decrease in the flux level. In other words, the effect of xenon is to reinforce or amplify any change in power level and in this sense $^{135}$Xe is an important destabilizing influence in thermal-power reactor operation.

However, in the second place, if the reduced power level is maintained (e.g. the reactor is shut down), then the $^{135}$Xe level continues to rise until the drop in the $^{135}$Xe concentration causes the right-hand side of eqn (10) to become negative. The xenon level then falls to saturate at the value appropriate to the reduced power level. The time taken for the $^{135}$Xe level to reach its maximum is about 10 hours and this peak level following shutdown is typically 50 per cent greater than the starting/steady-operating value.

## Appendix B. Delayed neutrons

In §2 it was shown that delayed neutrons are those arising from the radio-active decay of a fission product. Let $N_i$ represent the density of the $i$th group of such neutron precursor products, then

$$\frac{dN_i}{dt} = \gamma_i \Sigma_f \phi - \lambda_i N_i, \tag{13}$$

where $\gamma_i$ is the probability that the fission process leads to the particular fission product concerned with the $i$th delayed neutron group and $\lambda_i$ is the decay constant for the fission product.

Introducing the prompt and delayed neutrons separately into eqn (5) gives

$$\frac{1}{v}\frac{dN}{dt} = \{k(1-\beta)-1\}\Sigma_a \phi + \Sigma_i p_i \lambda_i N_i, \tag{14}$$

where $\beta$ is the total delayed neutron fraction and $P_i$ is the probability that the decay of the $i$th group precursor isotope will in fact lead to neutron emission.

The actual time behaviour of the variables $\phi$ and $N_i$ may be considered by assuming that they vary as $e^{\omega t}$. Then the all-prompt eqn (5) yields

$$l\omega = (k - 1). \tag{15}$$

Eqns (13) and (14) yield the corresponding relationship

$$l\omega + k\omega \sum_{i=1}^{I} \frac{\beta_i}{\omega + \lambda_i} = (k - 1), \tag{16}$$

where $\beta_i = \dfrac{p_i \gamma_i}{\nu}$ is the delayed neutron fraction in the $i$th group.

Eqn (16) can now be used to obtain two useful asymptotic expressions. Thus when the excess multiplication $(k - 1)$ is very small it may be expected that the neutron flux changes slowly and that all values of $\omega$ are small—certainly smaller than those for $\lambda_i''$—and hence may be neglected in comparison with them. On this basis, eqn (16) becomes

$$l\omega + k\omega \sum_{i=1}^{I} \frac{\beta_i}{\lambda_i} = (k - 1), \tag{17}$$

or $\tau\omega = (k - 1).$ $\tag{18}$

Comparing this last equation with (15), we see that an 'effective' neutron lifetime given by

$$\tau = l + k \sum_{i=1}^{I} \frac{\beta_i}{\lambda_i} \tag{19}$$

has been defined.

The reactor response is thus given by the all-prompt expression but with an increased neutron lifetime. The increase is very great. For $^{235}U$

$$\sum_{i=1}^{I} \frac{\beta_i}{\lambda_i} = 0.08 \text{ s},$$

whereas $l$ is typically 0.001 s. Thus if $(k - 1) = 10^{-4}$ for example, the neutron flux increases with a period of about 850 s, whereas if all neutrons were prompt the period would be about 10 s.

At the other extreme, when the excess multiplication is very large and the values of $\omega$ are much greater than any of the $\lambda_i$, then

$$l\omega + k \sum_{i=1}^{I} \beta_i = (k - 1),$$

i.e. $\qquad\qquad l\omega = (k - 1) - k\beta. \tag{20}$

The reactor behaviour is thus given by the all-prompt expression but with an excess multiplication somewhat reduced below that applied. Thus for

$(k-1) = 2 \times 10^{-2}$ the reactor period of a $^{235}$U system is about 0.077 s compared with 0.05 s if all neutrons are prompt. The delayed neutrons still have a retarding effect but much reduced below their impact at small values $(k-1)$. For values of $(k-1)$ much in excess of $k\beta$ the delayed neutrons have very little effect—the excess multiplication is so large that even without the delayed neutrons the remainder multiply so that we hardly notice their absence. For $(k-1)>\beta$ the reactor is said to be 'prompt critical'.

# Appendix C. The equivalence treatment of resonances in WIMS

The resonance treatment originally developed for the WIMS code was an elaboration of the 'equivalence theorem' model which had long been used in a simple form. The basic problem to be treated is that in the resonance range the cross-section of many of the most important nuclides present in the reactor vary rapidly with energy, reaching such high values that the fuel material looks 'black' to neutrons at resonance energies. Thus at these energies the neutron flux is correspondingly low within the fuel material. The shape of the resonances is approximately given by the Breit Wigner form:

$$\sigma_\gamma(E) = \frac{\text{Constant} \cdot \Gamma_n \Gamma_\gamma}{\left\{\dfrac{\Gamma_n + \Gamma_\gamma}{2}\right\}^2 + (E - E_0)^2} , \qquad (21)$$

which is so rapidly varying that it cannot be fitted even by a 7th order polynomial. As the quantity of interest is the reaction rate (the product of flux and cross-section) a very fine representation is needed for numerical integration, in excess of $10^5$ energy points. It is inconvenient and expensive to do this on every occasion and the approximate methods have made use of the fact that the basic flux shapes are rather similar for a range of such problems. The stratagem adopted is to solve exactly a number of simple cases of homogeneous mixtures of hydrogen and resonance absorbers, and then to match the average behaviour of the heterogeneous system being solved to the most appropriate such mixture. A simplified account of the matching process to illustrate the underlying theory is given, but readers are invited to consult the literature for a detailed description of the method.

Hydrogen is particularly simple as a slowing-down medium because the number of neutrons $q(u)$ slowing down across a lethargy bound $u$ (lethargy is defined as the logarithmic energy decrement from some initial energy) is directly related to the flux $\phi\ (u)$ and hence to the collission at $u$, $\Sigma_s\ (u)$ $\phi\ (u)$. For most of the energy range of interest, the cross-section of hydrogen is constant. Considering a purely absorbing resonance material the following flux is obtained:

$$(\Sigma_a(u) + \Sigma_s)\phi(u) = q(u),$$

hence

$$\frac{dq}{du} = \Sigma_a(u)\phi(u) = \frac{\Sigma_a(u)q(u)}{\Sigma_a(u) + \Sigma_s}. \tag{22}$$

Integrating,

$$\int \frac{dq}{q} = \int \frac{\Sigma_a(u)q(u)\,du}{\Sigma_a(u) + \Sigma_s}. \tag{23}$$

Hence, the slowing-down density at a given lethargy is

$$q(u) = \exp\left(-\frac{I}{\Sigma_s}\right), \tag{24}$$

where the integrating factor $I$ is called the resonance integral:

$$I = \int_0^u \frac{\Sigma_a(u)\Sigma_s\,du}{\Sigma_a(u) + \Sigma_s}. \tag{25}$$

For a two-region problem in which the resonance absorber is lumped into a fuel rod and the moderator surrounds it, the events in the rod may be described in terms of collision probabilities. Let $P(m \rightarrow f)$ be the probability that a neutron born in a moderator at lethargy $u$ has its next collision in the fuel zone and $\psi(u)$ be the emission density of such neutrons. Then

$$\Sigma_a^f(u)\, V^f\, \phi^f(u) = \psi^m(u)\, V^m\, P(m \rightarrow f), \tag{26}$$

which by reciprocity

$$= \psi^m(u)\frac{\Sigma_a^f(u)}{\Sigma_s^m}\, V^f[P(f \rightarrow m)], \tag{27}$$

where $V^f$ and $V^m$ are the volumes of fuel and moderator regions. Most of the principal moderating materials have scattering cross-sections which vary little with energy in the resonance region, and the emission density $\psi^m(u)$ will vary slowly due to depletion. Thus only the term in square brackets is rapidly varying with energy. It is noted that in the limit as the cross-section tends to infinity the escape probability from a body is dependent only upon the surface to volume ratio

$$\operatorname*{Lim}_{\Sigma \rightarrow \infty} P(f \rightarrow f) = \frac{\Sigma \bar{l}}{1 + \Sigma \bar{l}} = 1 - \frac{1}{\Sigma \bar{l}}, \tag{28}$$

where $l = 4 \times$ volume/surface is the mean chord length. If this is chosen as a basis for a general approximation, then

$$P(f \rightarrow f, \Sigma) \approx \frac{\Sigma \bar{l}}{a + \Sigma \bar{l}}, \tag{29}$$

and then

$$\phi^f(u) \approx \frac{\psi^m(u)}{\Sigma_s^m} \cdot \frac{a}{a + \Sigma \bar{l}}. \tag{30}$$

This is seen to be similar in form to the result for homogeneous mixtures of hydrogen and resonance absorber, if the effective scattering is replaced by

$$\Sigma_s = \frac{a}{\bar{l}}. \tag{31}$$

In this way, for a suitably averaged 'Bell factor', $a$, it is possible to relate the energy integral of the events in the heterogeneous system rather closely to that in the homogeneous one.

# 3

# Gas-cooled reactors

K. H. DENT

For historical reasons connected with their early development, most of the world's operating reactors are water-cooled; some 20 percent however are cooled by gas. Whilst gas cooling does not necessarily give the most compact or cheapest solution, it has important technical and safety characteristics. Two gases are of particular interest and have led to parallel developments in the family of gas-cooled reactors—carbon dioxide used in the Magnox reactors and AGRs, and helium used in the high-temperature reactors.

Whilst gas-cooled reactors are usually associated with Britain, a number of widely varying designs have been built around the world. In the UK a programme of natural uranium Magnox power stations was embarked on in 1955. These plants have given low-cost power and solid performance; among them they hold nuclear world records for total electricity generated and cumulative load factor. The next development, the AGR, brought a sophisticated reactor system which matched the steam conditions of the most modern steam turbines. The programme got off however to a bad start with construction delays and rising costs, but as the first stations overcome their teething troubles there is good reason for confidence in their success.

The high-temperature reactors are characterized by higher core ratings and temperatures and flexible fuel cycles. They have come too late on the scene to be serious contenders for normal steam–electric power generation, but are of interest for direct-cycle gas turbine generation and for high-temperature process heat and are under active development for these purposes in several countries.

The problem with gas-cooled reactors is that the technology is so versatile. Hundreds of diverse concepts have been studied, various experimental plants run and some fifty power reactors built. As a result no single strong line of gas-cooled reactors has emerged world-wide. Of all the reactor systems however, it is the gas-cooled family that seems to have the greatest flexibility and long-term potential for development as either a thermal or fast reactor system.

## Contents

# 1 Introduction

Of the world's operating reactors, 20 per cent are cooled by gas. The rest, except for a few experimental plants, use water. If we look for a moment at the coolant, not as something there to cool the fuel, but in its real task of transporting heat from the reactor core to the boilers to raise steam (Fig. 3.1) or to provide process heat, we begin to see why gas is of interest.

    A gas has some inherent virtues for this task. First, because the density of a gas is variable, its operating temperature can be chosen independently of the operating pressure. Thus a high gas temperature can be used, limited

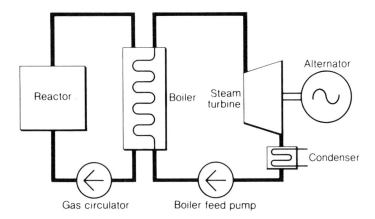

Fig. 3.1 The main elements of a gas-cooled reactor steam-electric plant

only by the core and circuit materials, to give good steam conditions from the boiler and thus good conversion of heat to electrical energy through the resulting high turbine efficiency. The optimum pressure can be selected separately on considerations of safety and of the economics of pumping power and pressure circuit costs. Gas has certain intrinsic safety advantages. It can undergo no phase change as a result of rising temperature or falling pressure, and so there cannot be any discontinuity in cooling under fault conditions, and flows and temperature can be predicted more simply and with greater confidence. Continuity of fuel cooling for on-load refuelling is more easily achieved with a gas. In addition with a gas there is no risk of a fuel – coolant interaction of the kind that in certain circumstances could result from the dispersion of melted fuel in a liquid coolant. Finally, a gas carries a relatively low burden of activated corrosion products, gives low radiation levels for maintenance round the circuit, requires small active effluent plants, and gives rise to only low radiation doses to the operators.

Offsetting these virtues is the combination of low density and low specific heat of gases. Even with fairly high pressures, this requires comparatively large temperature differences to transfer the heat between the fuel and gas, and between the gas and the boiler surface. As a result core ratings tend to be low and core and boilers tend to be large. It also requires large volume flows to transport the heat, and therefore large circulator sizes and powers.

There are many potential gases to choose from. The question of what gas to use is, however, associated with the choice of moderator to slow the neutrons to thermal velocities. The first nuclear reactor constructed in Chicago in 1942 used graphite. The good moderating properties of graphite, combined with its low neutron capture cross-section, have led to it being used almost universally for gas-cooled reactors, and commercial developments have greatly improved the qualities of graphite available. Only a handful of experimental or demonstration reactors have been built with a different moderator: these have used heavy water as moderator with pressure tubes containing the fuel and gas coolant. Indeed, graphite moderation has become almost synonymous with gas-cooled reactors.

The choice of coolant gas is influenced mainly by the thermodynamic, nuclear and chemical properties, and by its cost and supply. Table 3.1 gives properties of some of the candidates at 300 °C , a typical temperature of interest.

For good heat transfer and heat transport with low pumping power, a gas with high specific heat and high molecular weight, or density, is desirable. The coolant needs to have a low neutron absorption, to give good neutron economy and to avoid a rise in core reactivity if the reactor accidentally loses pressure. It also needs to be stable under irradiation, and preferably to have low neutron-induced radioactivity. Good chemical stability and low corrosion are obviously important. Of the many gases available the choice narrows quickly. Methane and the potentially attractive higher hydro-carbons are ruled out as they would be unstable. Hydrogen would require large circu-

Table 3.1 Properties of some possible coolant gases

| Gas | Molecular weight | Density at 1 bar (kg/m³) | Specific heat at constant pressure at 1 bar (kJ/kg°C) | Neutron absorption cross-section Maxwellian spectrum (m²/molecule) |
|---|---|---|---|---|
| Air | 29 | 0.60 | 1.0 | $1.9 \times 10^{-28}$ |
| Oxygen | 32 | 0.67 | 1.0 | 0 |
| Nitrogen | 28 | 0.58 | 1.1 | 2.4 |
| Hydrogen | 2 | 0.04 | 14.7 | 0.4 |
| Argon | 40 | 0.84 | 0.5 | 0.4 |
| Helium | 4 | 0.08 | 5.2 | 0 |
| Carbon dioxide | 44 | 0.92 | 1.1 | 0.003 |
| Steam | 18 | 0.39 | 5.8 | 0.4 |
| Methane | 16 | 0.34 | 3.2 | 0.8 |

lators and has a potential explosion risk. Air or oxygen would be too corrosive, nitrogen absorbs neutrons too readily. Two gases stand out as candidates: carbon dioxide which is dense, cheap, but not chemically fully inert, and helium which is inert, has a high specific heat, but is costly.

These two coolants have led to parallel lines of development in the family of gas-cooled reactors, sharing much common technology but with differing characteristics. The carbon dioxide reactors are characterized by relatively low specific core ratings, moderate temperatures and large size—typically the natural uranium plants of the UK and France and the advanced gas-cooled reactors (AGRs). The helium reactors aim for high ratings, small size and high temperature—the family of high-temperature reactors, or HTRs.

## 1.1 Gas-cooled reactors round the world

Gas-cooled reactors are mainly associated with Britain, and the UK programme is discussed later. There has, however, been interest in them in a number of countries. Table 3.2 lists plants outside the UK.

France was early in the field and her interests at that time closely paralleled those of Britain. Following her first air-cooled graphite-moderated reactor used for plutonium production, France built two more units, still with large horizontal-channelled cores but enclosed in prestressed concrete vessels with closed-cycle carbon dioxide cooling. She then took up commercial interest in the line, developing it dramatically through three units built successively near Chinon on the Loire. These reactors were all of vertical graphite core design with natural uranium metal fuel clad in magnesium alloy, and carbon dioxide cooled. Chinon-1, completed in 1962, was a small demonstration plant of 68 MW(e). Chinon-2 and -3 moved up to 198 and 476 MW(e) respectively and showed the commercial capability of the system. The system was taken to a further stage of development in two units at St Laurent by using highly-rated hollow fuel elements, and perhaps to the ultimate for natural uranium metal fuel at Bugey-1, the last of the line.

Table 3.2  Gas-cooled reactors outside the UK

| Plant | Country | Reactor type | Coolant | Moderator | Power MW(e) (net) |
|---|---|---|---|---|---|
| G2, G3 | | | | | 2 × 28 |
| Chinon-1† | | | | | 68 |
| Chinon-2 | | | | | 198 |
| Chinon-3 | France | natural U | $CO_2$ | graphite | 476 |
| St Laurent-1 | | | | | 487 |
| St Laurent-2 | | | | | 516 |
| Bugey-1 | | | | | 547 |
| Julich AVR | Germany | HTR | He | | 58 |
| THTR Uentrop ‡ | Germany | HTR | He | | 300 |
| Latina | Italy | Magnox | $CO_2$ | | 200 |
| Tokai Mura | Japan | Magnox | $CO_2$ | graphite | 157 |
| Vandellos | Spain | natural U | $CO_2$ | | 480 |
| Peach Bottom-1† | USA | HTR | He | | 40 |
| Fort St Vrain | USA | HTR | He | | 330 |
| EGCR § | USA | enriched U | He | | 22 |
| Bohunice Al | Czechoslovakia | | | | 110 |
| EL4 | France | | | | 70 |
| KKN† | Germany | pressure tube | $CO_2$ | heavy water | 100 |
| Lucens† | Switzerland | | | water | 8 |

† demonstration plant, now shut down    § was not started up
‡ under construction

A natural uranium reactor was built by the UK at Latina in Italy, and a further unit at Tokai Mura in Japan. The French later constructed a St Laurent type plant at Vandellos in Spain.

The Federal Republic of Germany moved straight to the development of the high-temperature reactor. The potential of this system looked good: its helium cooling and high-temperature all-ceramic fuel gave a prospect of high efficiency, of direct-cycle operation with a gas turbine, and of conserving the cheap uranium ore supplies by its good conversion ratio with the thorium–uranium fuel cycle. Whilst there was active long-term interest in the system over this period also in the UK and USA, Germany saw the HTR as a second-generation advanced system that might enable that country to overcome or even leap-frog the nuclear lead of other nations.

By 1967 the 15 MW(e) experimental AVR (Arbeitsgemeinschaft Versuchsreaktor) plant was built and generating at Jülich. The reactor was of unique design with a pebble-bed core consisting of graphite spheres containing highly enriched uranium plus thorium. The plant is still in use as a test bed for fuel. A bold plan two years later to build a 25 MW(e) direct-cycle plant at Geesthacht, with a prismatic-core reactor (that is, one built up of prismatic blocks of graphite containing the fuel) driving directly a helium turbine, fell through. By 1972 however construction started on the 300 MW(e) THTR power station at Uentrop. The THTR (thorium high-temperature reactor) is an advanced design based on the AVR concept with a pebble-bed core and conventional steam turbine plant, but with the reactor circuit integrated in a prestressed concrete pressure vessel. It is due to start up in the early 1980s.

Turning to the USA, we find the gas-cooled reactor as a small voice seeking recognition in a land dominated by light water reactors developed from the early defence and submarine plants. Curiously, the first gas-cooled power reactor there was an experimental mobile low power plant, ML-1, completed in 1961, using light-water moderator and nitrogen coolant driving a direct-cycle turbine generating 400 kW. Concern, however, that the longer-term potential of the gas-cooled reactor was being overlooked in the United States, led to the building of the 22 MW(e) demonstration plant at Oak Ridge, Tennessee, which was completed in 1965. This experimental gas-cooled reactor, the EGCR, was graphite-moderated and helium-cooled with stainless steel clad uranium dioxide fuel similar to that of the advanced gas-cooled reactor then under demonstration in the UK. By that time the experimental Peach Bottom HTR was being built; it was felt that EGCR was no longer fully representative of future gas-cooled reactor technology and the decision was taken to moth-ball the plant rather than to put it into active operation.

The 40 MW(e) HTR at Peach Bottom, Pennsylvania, went into full power operation in 1967. It was finally shut down in 1974 after providing valuable experience of the coated-particle prismatic-core concept. Only a year after Peach Bottom started up, construction began of a commercial HTR station of 330 MW(e) output at Fort St Vrain in Colorado. The reactor was of advanced design, in a prestressed concrete vessel as the German THTR but with a prismatic core. The US utilities, encouraged by the way Fort St Vrain was being built and attracted by the environmental merits of the HTR with its low active effluents and low thermal pollution, began to order large standardized plants. There was however a sharp downturn at that time in demand forecasts, and the larger plants were not proceeded with.

Interest in the HTR continues in a number of countries, and there are unconfirmed reports of a small plant being built in the Soviet Union as a precursor to larger units.

The reactors mentioned so far have almost all been graphite-moderated. Interest in the ultimate use of natural fuel cycles caused another class of gas-cooled reactor to be explored: this combined the merits of carbon dioxide cooling with the neutron economy and simple replaceability of heavy water moderation. Four plants, each quite different in style, have been built; but with the wider availability of enrichment the concept has not been exploited further.

The first of these was the Bohunice station of 110 MW(e) in Czechoslovakia, commissioned in 1965. The core comprises heavy water in a calandria pierced by vertical pressure tubes to act as fuel channels, all of which is enclosed in a steel pressure vessel linked by gas inlet and outlet pipes to three external boilers. Each pressure tube is extended up through the top head of the vessel to enable the fuel to be charged on load. The fuel is natural uranium. The second plant is the small experimental unit built in an artificial cave at Lucens in Switzerland intended for development as a possible national reactor marque. The fuel was in the form of very slightly enriched uranium metal pins clad in finned magnesium–zirconium alloy, held in a graphite support and inserted into vertical pressure tubes in the heavy water, with re-entrant gas flow from the top. Next on the scene was the French EL4 plant in Finistere. It has an output of 70 MW(e) and has been operating since 1967. The fuel is uranium dioxide (slightly enriched because of the small core size of this initial demonstration plant) clad in stainless steel; the pressure tubes forming the fuel channels are arranged horizontally through the calandria and are extended out through the concrete shield at each end to provide a convenient arrangement for on-load fuelling. Finally, the Kernkraftwerk Niederaichbach (KKN) was built in the Federal Republic of Germany in 1970. The fuel elements here were also bundles of pins of slightly enriched uranium dioxide with stainless steel cladding, but in this case the pressure tubes were arranged on a vertical lattice in the calandria with on-load refuelling from the top.

These various reactors round the world have provided a wealth of diverse experience. It is in the UK however that the gas-cooled reactor has been taken to major commercialization.

## 1.2  The UK scene

In the opening words of his contribution[1] to the Geneva conference of 1955, Sir Christopher (now Lord) Hinton recalled one of the earliest UK steps. He said:

In every engineering development there is an element of chance which may sway the balance between the choice of one design or another. When I was asked to undertake responsibility for the production of fissile material in 1946, it was the intention that Great Britain should build reactors of the original Hanford type. This type of pile is graphite-moderated and water-cooled and it cannot be made inherently stable if operated solely on a natural uranium fuel. Because of this, very large safety distances were considered essential. These safety distances are permissible in large continental areas, but in the thickly populated United Kingdom it was difficult to find suitable sites. Therefore it was thought that we would do better to choose a type of pile which was inherently stable and which did not demand isolation in the interests of safety. It was for this reason that our efforts were turned towards the graphite-moderated gas-cooled type of pile.

Thus the two production reactors built at Windscale in Cumbria, and commissioned in 1950, were graphite-moderated and made as simple as possible with once-through air cooling. The fuel was natural uranium, aluminium clad. They were preceded by two research reactors of the same type built at Harwell—the small Gleep, Britain's first reactor which started work in 1947 and still in use today, and its larger brother Bepo which was shut down in 1968.

The waste of the potentially useful heat from these reactors was a challenge to the engineers, and the design of a reactor suitable for electric power generation was studied at Harwell in 1951 and 1952. The aim was to retain the use of natural uranium fuel because no enrichment plant was then available, and without heavy water supplies the designers turned naturally to gas cooling and graphite moderation. Completion of the study[2] coincided with new defence pressures and it was decided to build the new reactor as a dual-purpose power station at Calder Hall. Such was the confidence in the concept that eight units were committed, four at Calder Hall and four more at Chapelcross in Scotland, before the first was half built. These plants were built remarkably quickly, the first Calder unit being at power only two and a half years after the site was opened up.

The Calder reactor design[3] was made as simple and reliable as could be. As far as possible components were kept within existing manufacturing technology so as not to prejudice construction time or reliabilty, the carbon dioxide coolant was kept to a pressure of 6.9 bar and a top temperature of 345 °C, refuelling was carried out off-load and at atmospheric pressure. Major features of the reactor engineering were however inevitably novel, including the vertical core arrangement, the graphite structure that had to withstand radiation-induced dimensional changes in the graphite bricks over its life, the stacked fuel elements, the top control rod and refuelling arrangements. Many of these features have been retained through all the later designs. The first reactor went on power in 1956 (Fig. 3.2) and the Calder Hall and Chapelcross plants are still in use today.

Fig. 3.2 HM The Queen opening Calder Hall, October 1956

Electric power demand in the UK at this time was still rising rapidly; all oil had to be imported, shortages were predicted of reasonable cost coal and of manpower for the mines over the coming decades. Against this background the Goverent saw it as vital to apply nuclear power commercially with all speed. A white paper[4] presented to Parliament in 1955 proposed an initial programme of 1500 to 2000 megawatts with the early stations to be of the Calder Hall type. The magnesium alloy Magnox (derived from magnesium, non-oxidizing) gave its name to this reactor system.

# 2  The Magnox reactors

Nine commercial Magnox stations were eventually built in England, Wales, and Scotland, with a total net output of nearly 5000 MW (Table 3.3).[5] Each station comprised two reactors and the associated generating plant. They were placed on fairly remote sites, away from the coal fields and not very far from the load centres so as to maximize their value (Fig. 3.3).

These reactors used natural uranium fuel. The consequent need to conserve neutrons set severe limits to the design. The fuel was in the form of metal bars, 28 mm or 29 mm diameter and between 480 mm and 1280 mm long according to the design, sealed in cans of magnesium alloyed with small additions of beryllium and aluminium, and heavily finned to improve the heat transfer. Avoidance of the alpha-beta metallurgical transition point of uranium set an upper limit to the centre temperature of the fuel. Fuel ratings were consequently restricted to around 2.5–3.6 MW(t)/t reactor mean, and the bulk gas outlet temperature to 340–410 °C. In addition the major step was taken to refuel the reactors on-load. This is an important feature: it maximizes the core reactivity by refuelling continuously, avoids the initial large excess reactivity needed with batch refuelling, enables the plant to run

Table  3.3  UK commercial Magnox stations

| Station | Commissioning date | Design output (MW(e) net) | Design net effici- ency (%) | Reactor vessel | Coolant pressure (bar) | No. of fuel channels |
|---------|-------------------|---------------------------|----------------------------|----------------|------------------------|---------------------|
| Berkeley | 1962 | 2 × 138 | 24.7 | steel cylinder | 9 | 3265 |
| Bradwell | 1962 | 2 × 150 | 28 | | 9 | 2564 |
| Hunterston A | 1964 | 2 × 160 | 28 | | 10 | 3288 |
| Hinkley A | 1965 | 2 × 250 | 26 | steel | 13 | 4500 |
| Trawsfynydd | 1965 | 2 × 250 | 28.8 | sphere | 16 | 3740 |
| Dungeness A | 1965 | 2 × 275 | 32.9 | | 18 | 3876 |
| Sizewell | 1966 | 2 × 290 | 30.6 | | 19 | 3784 |
| Oldbury | 1968 | 2 × 300 | 33.6 | concrete cylinder | 24 | 3308 |
| Wylfa | 1971 | 2 × 590 | 31.4 | concrete sphere | 27 | 6156 |

Fig. 3.3 The Trawsfynydd Magnox station in North Wales

for long periods without the need to shut down to refuel, and improves availability since an occasional cladding failure would otherwise require fairly prompt shutdown to remove the element to avoid oxidation of the metallic uranium. The fuel ran into serious problems in the development stage. The combination of higher can temperature and greater coolant mass flow in the commercial designs caused creep deformation of the fins and support features, and rattling of the elements in the channels, which took several years of intensive development to cure.

The stations were designed and built by consortia of companies set up for the purpose, of which there were up to five at one period. This arrangement fostered much inventiveness and diversity in the designs, as well as price competition. As a result however, each station differed from its predecessors, and although there was progressive overall improvement, each had its peculiar construction delays and teething problems.

The progression is shown in Table 3.3. Despite the handicap of using natural uranium, the unit outputs were successively increased from 138 to 590 MW(e), and by increasing the gas pressure the station net efficiencies were raised from the early figure of 25 per cent up to 33.6 per cent.

Hinkley A in Somerset is typical of the early Magnox stations (Fig. 3.4). The station comprises two identical reactors providing steam to six 93.5 MW main turbo-alternators, ranged on the steam and feed sides in a common turbine house, to give a net output of 500 MW(e) total. Each reactor core is

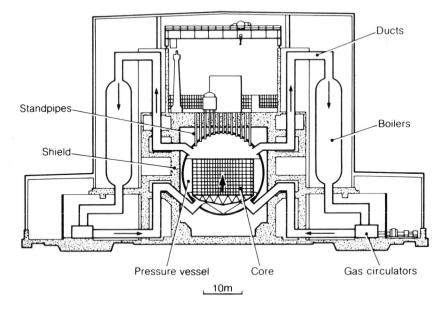

Fig. 3.4 Hinkley A Magnox station: cross-section of the reactor and gas circuit

built up of 625 t of machined graphite bricks keyed together to form a stable structure which, together with the reflector at top and bottom and sides, is 16 m diameter and 8.8 m high. The core is pierced by the vertical fuel channels 98 mm diameter, each containing a stack of eight fuel elements.

The fuel rods are 29 mm diameter by 850 mm long, and the can has a polyzonal heat transfer surface, in which gas swirl is promoted in each of four circumferential zones to give good heat transfer, with additional fins to centralize the elements in the channel. Carbon dioxide coolant circulates upwards through the channels, entering the core at 180 °C and leaving at a mixed temperature of 375 °C. The reactor is controlled by absorber rods driven by winch mechanisms above the core, some of which are operated in zones automatically to control azimuthal power instability resulting from xenon poisoning and burnout.

The core is contained in a spherical steel pressure vessel 75 mm thick and 20.4 m diameter, surrounded by a steel and concrete shield wall, and provided with standpipes at the top for refuelling and to house the control rod drives. From the vessel the coolant flows through six external circuits each with a boiler unit and gas circulator interconnected by steel ducting. The boilers, placed at a level above the core to promote natural circulation, are of the forced-circulation drum type and have serpentine tubing with welded-on studs to give an extended heat transfer surface on the gas side. Superheated steam is raised on a dual pressure cycle at 45 bar 365 °C, and

12.6 bar 354 °C, respectively; for the limited upper gas temperature, this dual-pressure steam cycle allows part of the steam to be generated at a much higher pressure and so appreciably improves the steam cycle efficiency and electrical output. The single-stage axial circulators have a variable-frequency electric drive so as to vary the mass flow to maintain correct steam conditions at part load. Refuelling is carried out continuously at full pressure and power with fairly complex equipment situated on the fuelling floor above the reactor. The required refuelling rate is three or four channels per day when operating at full load. After discharge the spent fuel is stored for some months in a cooling pond adjacent to the reactor before being taken away in shielded flasks for reprocessing.

A milestone was reached in the Oldbury station which was to set the pattern for all future gas-cooled reactors in the UK. Here an integrated reactor design was used, with the core, coolant circuit, boilers, and circulators enclosed in a prestressed concrete vessel (Fig. 3.5; cf. Dungeness B AGR station, Fig. 3.6). The Magnox reactors reached their peak of efficiency and design style at Oldbury. For the first time the boilers were of the once-through type. Whilst the design relied more on high integrity of the in-vessel components, particularly the boiler and the insulation of the inside walls of the vessels which would be difficult to repair or replace if major

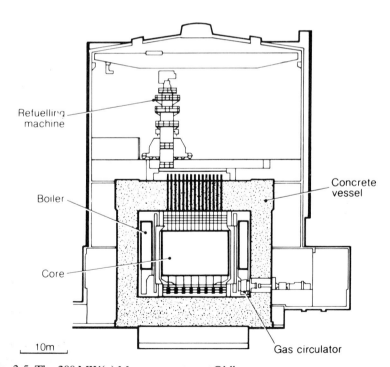

Fig. 3.5 The 300 MW(e) Magnox reactor at Oldbury

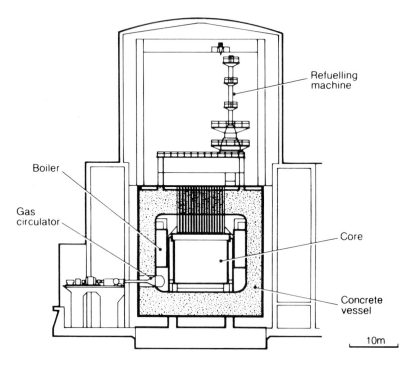

Fig. 3.6 The 600 MW(e) Dungeness 'B' AGR drawn to the same scale as Fig. 3.5

failures occurred in service, the reactor was much more compact, and the concrete vessel reduced the risk of sudden depressurization from failure of a steel circuit component.

The operation of these Magnox stations has been remarkably good, bearing in mind that there was no replication in design and that development was telescoped into the construction period. The historical capital cost of the stations was high. Fuel costs however are low and, as inflation diminished the effective capital cost, the stations have provided very cheap power relative to other plants. In 1979 the total generating costs for stations commissioned in England and Wales over the preceding twelve years were: oil-fired 1.42, coal-fired 1.23, Magnox 0.76 pence per kWh.[6]

Towards the end of the Magnox construction programme it was appreciated that the corrosion rate of mild steel in the hot carbon dioxide in the reactor outlet region was greater than expected. It was decided that, rather than continue operation at full power, it would be more economic to prolong the plant life by restricting the top gas temperature to about 360 °C. The resulting derating of the units varied between zero and 29 per cent depending upon the original top temperature and the degree to which the plant operation could be re-optimized. The earlier built stations had, however, already shown that they were capable of their planned outputs, and had actually operated at net electrical powers above their design figures.

Following the early vibration and creep problems mentioned earlier that came to light on the fuel, the differing fuel elements were much rationalized and later designs used a herringbone-finned heat transfer surface, with integral location lugs, which was aerodynamically stable and much more resistant to damage (Fig. 3.7). Fuel element failures became acceptably low—approaching one per reactor per year. Although the station outputs have mostly been restricted in the later years, the heat output from the fuel has more than exceeded expectations. The original target burn-up of 3000 MW d/t of uranium has been progressively increased by more than 60 per cent, and all stations now operate at just over 5000 MW d/t channel average burn-up.

On the whole the stations were praised by the generating boards and earned the sobriquet 'Britain's nuclear workhorses'. The availabilities have been remarkably good, and despite the derated outputs the cumulative load factors have been quite high (Table 3.4). Annual percentage load factors in the high 80s and 90s (based on the derated output where appropriate) have not been uncommon.[7]

Radiation doses to station staff have been low. In a survey[8] of doses covering the five years 1971 to 1975, the overall annual dose for all Magnox stations was 186 man-rems total per station. Significantly, for the three latest

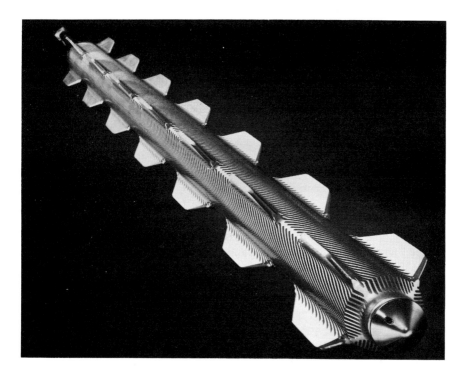

Fig. 3.7 A Magnox fuel element with 'herring bone' finning

Table 3.4 Performance of UK Magnox stations

| Station | Design output (MW(e) net) | Maximum achieved output (MW (e) net) | Restricted output† (MW(e) net) | Cumulative load factor from commissioning to late 1978 relative to: | |
|---|---|---|---|---|---|
| | | | | design output (%) | restricted output (%) |
| Berkeley | 276 | 302 | 276 | 79 | 79 |
| Bradwell | 300 | 325 | 250 | 71 | 78 |
| Hunterston | 300 | 349 | 300 | 82 | 82 |
| Hinkley | 500 | 531 | 430 | 68 | 68 |
| Trawsfynydd | 500 | 513 | 390 | 60 | 69 |
| Dungeness | 550 | 551 | 410 | 67 | 80 |
| Sizewell | 580 | 496 | 420 | 59 | 79 |
| Oldbury | 600 | 563 | 416 | 51 | 78 |
| Wylfa | 1180 | 903 | 840 | 42 | 47 |

† applies to later years of operation with the top gas temperature limited to 360° C

stations of Sizewell, Oldbury, and Wylfa this average was only 66 man-rems (an order of magnitude less than published figures for water reactors operating in the USA). Recent figures published by the Health and Safety Executive[9] show the largest and latest Magnox station, Wylfa, to have collective doses of only 39 and 62 man-rems in 1977 and 1978 respectively— figures which are among the lowest for any nuclear station in the world.

Meantime in France the Magnox type reactor was being developed boldly (Table 3.5).[5] The third of the series of reactors built at Chinon was in some ways ahead of the UK contemporary plants when completed in 1966, with its concrete vessel, higher output, and greater channel ratings. A well-conceived uniquely French marque of reactor was then developed, of which three units were built: two at St Laurent and one at Bugey. This series extended the performance and life span of the Magnox reactor beyond the point at which it was dropped in the UK.

The St Laurent reactors, completed in 1969 and 1971, were of integral design as in the earlier Oldbury station in the UK. Here however the core was arranged with downward flow and the boilers were placed immediately below (Fig. 3.8) so that the vessel diameter was minimized, the hot gas was confined to a small volume, and the vessel surface was swept with cooler gas at inlet temperature. Relatively high channel ratings were obtained by using a hollow-rod fuel element, 23 mm bore and 43 mm outside diameter, which provided a larger heat transfer surface while keeping the uranium still at an acceptable temperature. The Bugey reactor design was similar, but performance was taken to the ultimate for metallic fuel by using a thin annular fuel element, 95 mm diameter and 77 mm bore, clad and cooled both inside and out. The mean fuel rating of 5.9 MW(t)/t almost doubled previous figures.

Table 3.5 French Magnox stations

| Station | Commissioning date | Design output (MW(e) net) | Net efficiency (%) | Reactor vessel | Coolant pressure (bar) | No. of fuel channels |
|---|---|---|---|---|---|---|
| Chinon-1 | 1962 | 68 | 22.7 | steel cylinder | 24 | 1148 |
| Chinon-2 | 1964 | 198 | 25.1 | steel sphere | 24 | 2304 |
| Chinon-3 | 1966 | 476 | 31 | concrete with steel external circuit | 26 | 3264 |
| St Laurent-1 | 1969 | 487 | 29.5 | concrete integral design | 26 | 3264 |
| St Laurent-2 | 1971 | 516 | 30.5 | concrete integral design | 26 | 3264 |
| Bugey-1 | 1972 | 547 | 28.4 | concrete integral design | 41 | 852 |

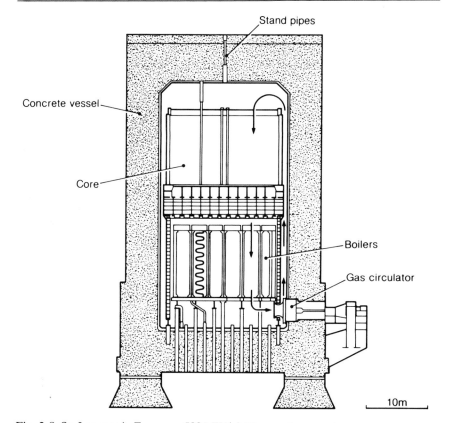

Fig. 3.8 St. Laurent in France, a 500 MW(e) Magnox type reactor

The building of gas-cooled reactors in France ended however with Bugey–1. Although there was interest in the HTR, a more advanced gas-cooled system had not been developed; the current natural uranium system was felt to be limited to a size of 600–700 MW and not competitive in cost with the fast developing light water reactors.[10,11]

## 3  The advanced gas-cooled reactor

The limitations inherent in the Magnox reactors using natural uranium metal were recognized from the start. The relatively low fuel ratings and temperatures meant that the plant was large, less efficient than it might be, and costly. Ever since Charles Parsons built his first multi-stage steam turbine in Newcastle in 1884, turbo-alternators had been continuously improved in operating conditions and output. The low steam temperature of the Magnox stations caused a step backwards in the size and efficiency of their turbines.

Studies of improved designs were therefore started in the UK in the mid-1950s. The concept that emerged was based on a fuel that was chemically inert in carbon dioxide up to very high temperatures—slightly enriched uranium dioxide arranged in the channel in clusters of small diameter pins. It was named the advanced gas-cooled reactor.

To show its capability a small demonstration plant of 30 MW(e) was built at Windscale in Cumbria[12] (Fig. 3.9). It was commissioned towards the end

Fig. 3.9 The Windscale demonstration AGR

of 1962. The reactor had some advanced engineering features in addition to the fuel: the gas circuit for instance, although of steel as its contemporaries, was of compact design with short concentric ducts connecting the vessel and boilers, with the hot gas flow in the inner pipe and the cool return gas outside.

The original plan was to clad the fuel pins with beryllium, which was suitable for the high temperature and had a low neutron absorption. It became clear, however, before the Windscale plant was completed that it was not going to be possible to develop a beryllium cladding with the required ductility; the alternative of stainless steel was therefore adopted.

The demonstration plant behaved well. Its load factor in the first year of operation was 75 per cent and its availability (allowing for outages due to experiments) was 89 per cent. The stage was set for a moderate step forward in gas-cooled reactors in the UK.

In the event the generating board, in inviting bids in 1964 for its next station Dungeness B in Kent, widened the competition to include water reactors in order to be fully informed before making a choice of system for its next nuclear programme. In this situation the advocates of the AGR understandably tightened up their designs to make them more competitive. The government was happy to endorse the generating board's choice of an AGR; the assessment of the bids was obviously[13] conditioned by the UK conditions at the time.

The trend in AGR performance as seen at the time[14] is shown in Fig. 3.10. The step improvement in efficiency does more than just reduce the reactor size and fuel construction; it also greatly reduces the amount of cooling water required for the turbine condenser and therefore the amount of thermal pollution caused. The heat rejected in the condenser for a given electrical output $E$ is $E(1-\eta)/\eta$ so that it falls rapidly with increasing efficiency, $\eta$.

## 3.1 The AGR stations

The AGR programme started with some apparent advantages. The construction industry was now experienced. There had been sufficient studies already made to know the capability of the system, so that unit size for instance was standardized at the outset at 660 MW(e) gross from a turbo-alternator based on the existing 500 MW machines. It was not until a little later however that, following a parliamentary review,[15] the number of design and construction companies was reduced to two. Nevertheless the four reactors of Hinkley B and Hunterson B (which followed the Dungeness B order) were built to a common design, and those at Hartlepool and Heysham, whilst of a different design again, were all four essentially identical.

The programme comprised five twin-unit stations. Some details of these, together with the two later stations of Heysham II and Torness, are given in Table 3.6.

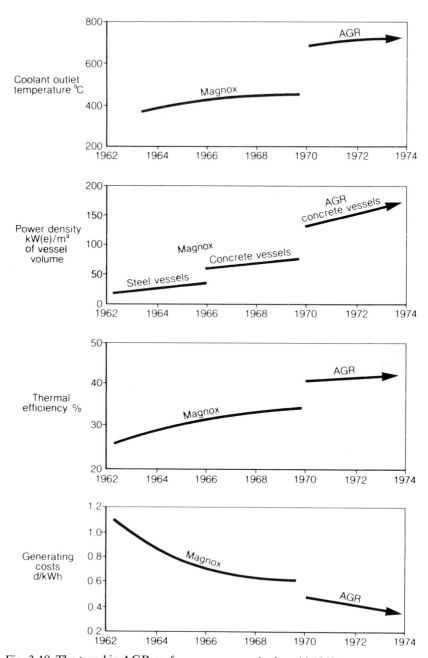

Fig. 3.10 The trend in AGR performance as seen in the mid-1960s

Table 3.6 Advanced gas-cooled reactor stations

| Station | Commissioning date | Design output (MW(e) gross) | Net efficiency (%) | Concrete vessel type | Coolant pressure (bar) | No. of fuel channels | Design steam conditions (bar) | (°C) |
|---|---|---|---|---|---|---|---|---|
| Dungeness B | 1980 | 2 × 660 | 41.0 | single cavity | 34 | 412 | 159 | 566 |
| Hinkley B | 1976 | 2 × 660 | | single cavity | 42 | 308 | 159 | 538 |
| Hunterston B | 1976 | 2 × 660 | | single cavity | 42 | 308 | 159 | 538 |
| Hartlepool | 1983 | 2 × 660 | 41.5 | multi-cavity | 41 | 324 | 159 | 538 |
| Heysham I | 1983 | 2 × 660 | | multi-cavity | 41 | 324 | 159 | 538 |
| Heysham II | 1986 | 2 × 660 | | single cavity | 43 | 332 | 159 | 538 |
| Torness | 1986 | 2 × 660 | | single cavity | 43 | 332 | 159 | 538 |

Construction of the Dungeness B station[16] began in the Autumn of 1965 (Fig. 3.6). The early civil construction work went reasonably well but a number of engineering difficulties later emerged; in addition, strikes, low productivity, and changing safety requirements took their toll and the work fell seriously behind. As a consequence, Hinkley B, ordered two years after, became the lead station. The first of its two units was commissioned in 1976, with the first of the Hunterston B units close on its heels.

The Hinkley station[17] follows the usual UK pattern of twin units. The reactors are each unitized with their own turbines and auxiliaries, but have certain common services, such as fuel handling and storage, and are housed in a single building complex (Fig. 3.11). The steam conditions of 159 bar and 538 °C (2300 lb/in², 1000 °F) with single reheat to 538 °C are the same as for the most modern fossil fuel stations; indeed the turbo-alternators of Hunterston B and the nearby later oil-fired station at Inverkip are identical and major parts have been interchanged between the two.

The reactor is of integral design (Fig. 3.12) and owes much in its engineering to the earlier Oldbury Magnox station. The carbon dioxide gas is circulated by eight blowers through the reactor core and downwards through the boilers back to the circulator inlets. Design gas temperatures are 292 °C at core inlet and 645 °C at outlet.

Fig. 3.11  Hinkley B AGR station

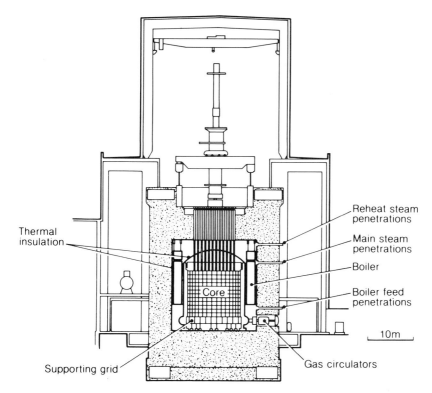

Fig. 3.12 Cross-section through a Hinkley AGR

The core is a coherent structure of bricks machined from a high-strength, low neutron absorption, isotropic graphite. It consists of twelve layers of which the upper and lower form the top and bottom reflector respectively, and the remaining ten form the moderator plus radial reflector. Each layer is made up of hollow polygonal bricks spigotted top and bottom into the adjoining ones to form the continuous vertical fuel channels in the core positions, together with smaller interstitial bricks which provide positions for control rods (Fig. 3.13). The graphite structure is mounted on a steel egg-box grid which is supported on the floor of the reactor vessel, on 16 rocking supports. Each brick is radially keyed to its neighbours. In this way a stable structure is ensured, the keys allowing thermal expansion to take place as well as the varying shrinkage that will occur in the bricks with irradiation over their life, without affecting the alignment of the channels. Above the core is a neutron shield made up of rectangular graphite blocks; around the outside is a shield wall of steel-encased calcium hydroxide. These shields stop activation of components outside the core structure and allow entry to the vessel at reactor shutdown if needed for inspection and repair.

The coolant is arranged to flow upwards through the fuel channels and

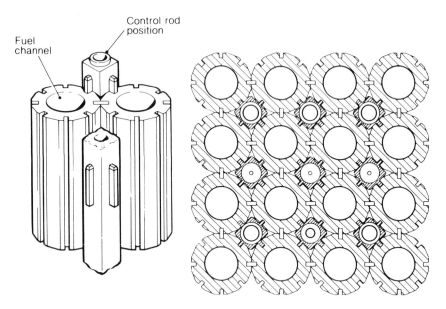

Fig. 3.13  Construction of the graphite core of the Hinkley AGR

return down through the boilers. In this way contraflow heat exchange is achieved with upward water flow in the once-through boilers, and good natural circulation occurs in the event of circulator loss.

After leaving the circulators, half the gas flows directly to the channel inlets. The rest is directed by a gas baffle to the top of the core structure where part of it flows downwards to cool the side shields while the bulk flows down through the interstitial spaces between the graphite bricks to keep them at a temperature at which thermal oxidation of the graphite is negligible (Fig. 3.14). The two streams then join to flow upward through the channels. By cooling the graphite in series in this way the total flow required is less and the core outlet temperature is not degraded.

There are 308 fuel channels. Each is continued upwards above the core by guide tubes and a standpipe in the vessel roof so as to provide individual direct access from the fuelling floor above the vessel. There are eight fuel elements stacked one above the other in the channel. Above them in the channel is a small inertial collector which collects active dust, mainly fine oxide spalling from the fuel pin surfaces, so reducing activity levels in other parts of the gas circuit to which access may be required. The elements comprise 36 fuel pins made of sintered uranium dioxide pellets, 14.5 mm diameter and 5.1 mm bore, sealed in cans a metre long of stainless steel (20 per cent Ni, 25 per cent Cr, Nb) with a roughened outside surface to improve the heat transfer. The pins are supported in three concentric rings by stainless steel grids within a graphite sleeve 190 mm bore through which the coolant flows (Fig. 3.15). The graphite sleeve has a double wall with a small

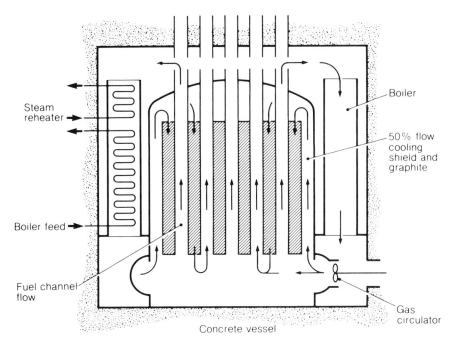

Fig. 3.14 Diagram showing gas flow in the Hinkley AGR

Fig. 3.15 AGR fuel element

gap between to reduce heat conduction between the hot gas in the channel and that on the outside flowing downward to cool the bulk moderator. A tie bar passes through the centre of the stack of elements and takes its weight during handling. This arrangement provides the degree of flexibility needed in the fuel stringer to pass freely through the charge path and channel. The tie bar is connected to a reusable plug unit which ducts the hot gas flow to outlet ports near the top of the standpipe. Each plug unit is fitted with a sampling pipe to check for pin cladding failures, a thermocouple to measure the gas outlet temperature, and a variable gag, which is controlled from the station control room to adjust the channel flow.

The reactor is designed for on-load fuelling, a feature which is an integral part of the present AGR concept. It goes with the other feature of individual channel access which enables each channel to be instrumented and flow-controlled; it gives the potential for high plant availability, enables any leaking fuel to be removed at once from the reactor so as to preserve a clean circuit and low activity levels, and minimizes both the feed enrichments and core excess reactivity. Fuel enrichment is varied in radial zones in the core so as to flatten the power distribution and improve the form factor. The initial enrichments were 1.16, 1.54, and 2.01 per cent, and feed enrichments 2.01, 2.10, and 2.55 per cent in the inner, intermediate, and outer zones respectively. Whilst the concept of on-load fuelling with single-channel access is very simple, the equipment itself is fairly massive and complex and has to be engineered to a high standard, since the fuelling machine has to be connected to, and disconnected from, the gas circuit of the operating reactor. The balance of judgement was however in its favour. The fuel is handled by a large shielded fuelling machine (Figs. 3.12 and 3.20), which will take the full-length assembly of fuel and plug unit. This is mounted on a gantry and services both reactors. It is planned that at equilibrium fuel cycle the reactor will be refuelled continuously at a rate of about one and a half channels per week for the design burn-up of 18 000 MW d/t. After discharge the fuel assembly is broken down in a shielded cell. The plug unit is then serviced for reuse, and the individual fuel elements are stored in a pond before being dispatched for reprocessing.

Core reactivity is controlled by about 80 absorber rods occupying interstitial positions between the fuel channels. The rods are formed of stainless steel tube and each is suspended individually by roller chain from a drive unit in a standpipe in the vessel top. Just over half the rods are coarse controllers used for large changes of reactivity, such as occur during startup and shutdown; these have inserts of boron–stainless steel alloy to make them wholly black to neutrons. The rest are fine controllers used for automatic spatial control and fine trim. The rods drop into the core by gravity to trip the reactor in an emergency. A completely diverse secondary method of shutdown is provided by a system of nitrogen injection into the coolant gas.

The boilers are of the once-through type. These give a more simple and compact arrangement than recirculating boilers, avoiding the need for steam drums and the associated penetrations which would be required through the concrete vessel wall. There are four independent boilers; each is made up of three units installed in a quadrant of the annulus around the core and unitized with two circulators so that an individual boiler can be shut down without affecting the rest. Each of the twelve units is built of serpentine tube platens supported in a rectangular steel casing open top and bottom. The feed and steam connections are brought out from each unit through penetrations in the side wall of the reactor vessel (Fig. 3.14). The hot gas from the core flows downwards first through the reheater section, then through the superheater, evaporator, and economizer. The construction materials are austenitic steel (18 per cent chromium, 12 per cent nickel) above 520 °C, 9 per cent chrome steel between 520 and 350 °C, which avoids stress corrosion in the evaporator section, and mild steel below 350 °C.

The eight gas circulators are installed in horizontal penetrations in the vessel wall below the boilers. Their combined power is 35 MW to give a core flow of 13 000 t/h. The machines are submerged in the carbon dioxide with a pressure closure at the outer wall of the vessel, thus avoiding the need for a shaft seal, and are designed as an integrated unit which can be removed and replaced complete from outside the vessel. The centrifugal impeller is mounted directly on the shaft of the 11 kV 50 Hz induction motor and runs at 2970 rev/min. Mass flow is varied by adjustable inlet guide vanes operated by a mechanical linkage within the motor compartment; in this way the main drive unit is kept very simple and variable speed, slip rings or commutator avoided in the motor. An external forced lubricant system is used.

The prestressed concrete pressure vessel is 18.9 m inside diameter and 19.4 m inside height, with a wall thickness of 5.03 m (Fig. 3.12). It is designed for a working pressure of 42 bar and an ultimate pressure of 115 bar. The gas-tight steel liner is 13 mm thick and is extended through the penetrations in the vessel wall required for the fuel channels, boiler connections, circulators and other services. The prestressing is applied by a system of helical tendons arranged in 16 layers in the vessel wall with alternate pairs clockwise and anti-clockwise. These provide simultaneously the axial and circumferential prestress in the vessel wall and, by means of extending the system a short distance axially at the top and bottom, provide also the required compressive stress in the top and bottom slabs. Each tendon comprises seven 17.8 mm diameter strands of high-tensile steel individually anchored at each end by a wedge and barrel anchor system.

The vessel liner is designed so as to be shop fabricated in panels, each complete with the cooling pipework welded on, and the panels are assembled on the site to form the liner shell. This is then rolled into place on to the vessel base, and the concrete walls poured in 'lifts' with the tubes for prestress tendons cast in place. On completion, the tendons are threaded through the tubes and tensioned to give the correct prestress to the concrete so that it remains in compression at the vessel's working pressure. The tendons form a highly redundant system, and each can be inspected, re-tensioned or even replaced at any time.

To keep the mild steel liner at the required temperature of 60–80 °C and avoid thermal cracking and loss of strength in the concrete, cooling is provided by means of water cooling pipes attached to the concrete side of the liner. The liner is then insulated on the gas side to assist the cooling and to reduce the heat loss and the load on the cooling system. The insulation is applied in layers of isotropic ceramic fibre, interleaved with stainless steel foil to prevent slumping and thermal convection, and retained and protected by steel cover plates attached by studs to the liner (Fig. 3.16).

A gas treatment plant is provided to maintain the correct composition of the carbon dioxide coolant, particularly the levels of carbon monoxide and methane that are added to control radiolytic oxidation of the graphite. The basic reaction in the core under irradiation can be represented by

$$CO_2 + C \longrightarrow 2CO,$$

but additional carbon monoxide and water vapour are produced by the radiolytic destruction of methane through the reaction

$$3CO_2 + CH_4 \longrightarrow 4CO + 2H_2O.$$

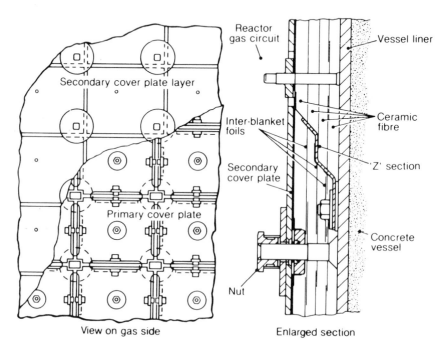

View on gas side                    Enlarged section

Fig. 3.16 Details of the insulation on the inside face of the Hinkley AGR vessels

The gas treatment circuit operates continuously, taking a small bypass flow from the delivery side of the circulators to the external treatment plant and returning it to the low pressure side of the reactor circuit. The gas passes first through a recombiner fed with oxygen, which re-oxidizes the carbon monoxide to carbon dioxide, then through a recuperative heat exchanger and a cooler, through a dryer to remove excess moisture, and back via the recuperator to the reactor. An electrolysis plant provides a stream of pure oxygen and hydrogen. The former is fed to the recombiner. The hydrogen is fed to a methanation unit together with a make-up supply of carbon dioxide to form methane through

$$CO_2 + 3H_2 \longrightarrow CH_4 + H_2O.$$

The output is injected into the treatment circuit to replace the methane destroyed in the core, the surplus moisture being removed by the dryer.

The station layout is based on a compact integrated arrangement of the two units (Fig. 3.11). The two reactors are housed side by side at 73 m centres in a steel-frame reactor building. Common fuel-handling plant serves both units, with the fuelling machine spanning above the two reactors and an irradiated fuel breakdown cell and storage ponds between them. The gas feed and treatment plant and active effluent treatment plant are included in the complex. The two turbo-alternators are set side by side in the adjoining turbine hall on the same centre lines as the reactors and with the steam ends toward them, so as to keep the steam pipe runs short and symmetrical. The central bay of the turbine hall between the two machines houses the water treatment plant, boiler feed pumps, feed heaters, and deaerators. The turbines are single-shaft machines with one single-flow high-pressure cylinder, one double-flow intermediate-pressure cylinder, and three double-flow low-pressure cylinders with integral side condensers. The alternator is hydrogen-cooled with water-cooled stator windings and is designed for 660 MW output at 0.85 power factor. The main steam circuit is shown in Fig. 3.17.

The whole plant is managed from a central control room, with a control desk for each reactor/turbine unit, for the 400 kV system, and for the common services. Computers are used for certain control functions, for data handling, and for presentation and analysis of alarm states.

## 3.2 AGR safety

The AGR system has a number of features which give the plant good safety characteristics in normal operation.

The reactor uses an indirect cycle. That is to say, there are two separate circuits—the enclosed primary gas circuit comprising the reactor core, the gas side of the boilers, and the gas circulators, and the secondary steam circuit taking steam from the boilers through the turbine and recirculating the condensate back to the boilers via the feed pumps. There can be no carry-

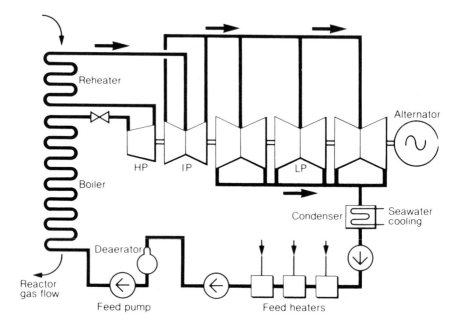

Fig. 3.17 The main steam circuit of the Hinkley AGR station

over of radioactivity from the reactor into the steam circuit in normal operation (or indeed in the event of the fault on the reactor). The turbine plant and its auxiliaries thus do not require shielding and can be operated and maintained in the normal way.

The gaseous coolant carries a relatively low burden of activated corrosion products. The feature of on-load fuelling enables any failed fuel elements to be removed from the core promptly and helps to maintain a clean circuit free from fission product activity. As a result, the radiation doses to the operators are low, the radioactive liquid and gaseous discharges from the plant are small so that doses to the public are also low. Furthermore, the low radiation levels around the circuit ease maintenance, which is itself important in the context of safety.

The plant is designed to protect the public to standards better than the radiation dose limits recommended by the International Commission on Radiological Protection (ICRP).

The direct radiation from the plant is controlled by appropriate shielding. The concrete vessels are more than adequate to shield the reactor cores and coolant, and additional shielding is provided where required for ancillary equipment.

Radioactive and other gaseous effluents are discharged only in controlled quantities from stacks at an appropriate height. Potentially active effluents are discharged through filters and iodine absorption plants with sampling on their downstream side. Major discharges of carbon dioxide occur when the reactors are intentionally depressurized. The coolant includes some gaseous activity in the form of sulphur-35 and argon-41, but may also contain radioactive particulate matter and traces of fission products from uranium contamination of fuel pin surfaces or from the temporary presence of fuel pins with defective cladding. Means are provided to sample the coolant to ensure that releases from vessel leakage or deliberate discharge are within the acceptable limits. Outside the reactor, ventilation extract systems are provided for areas where contamination may occur; these discharge through absolute filters via the stacks.

Radioactive liquid effluents from an AGR are relatively minor. The plant is nevertheless designed to ensure that discharges are within authorized limits. Active and potentially active effluents are carried by a separate drainage system, and tanks for active liquids are provided with leakage monitoring and collection. All active effluents are filtered before passing to the final monitoring and delay tanks before eventual discharge to the cooling water outlet. A relatively major active waste is tritiated water from the coolant dryers which is piped to special closed tanks for storage for a suitable period before discharge. The treatment of liquid effluents, and the control of their discharge and subsequent dilution and disposition, ensures that any resulting dose to the public is well within the ICRP recommended limits.

Solid wastes of high activity are stored on the station in storage vaults of sufficient capacity to take the estimated arisings over the station life. Combustible waste is in some cases incinerated if of low enough activity, otherwise it is drummed and stored temporarily on the station. The precautions taken preclude any significant radiation dose to the public.

The safety of the reactors in potential accidents also springs from features inherent in the system. Prominent among these is the fact that the coolant is a gas. The gas chosen, carbon dioxide, is non-toxic and chemically almost inert. It does not react exothermically with the fuel or cladding even at abnormally high temperatures.

Carbon dioxide is a gas at all temperatures above minus 80 °C (at atmospheric pressure). As therefore it is already a gas at its working temperature it cannot change phase, or 'flash off', as a result of a rising temperature or falling pressure. Thus, there can be no discontinuity in cooling under fault conditions, and flows and temperatures can be predicted with confidence under forced or natural circulation. Indeed, it is virtually impossible to 'lose' the coolant in a gas-cooled reactor: the worst that can happen is that some failure in the pressure circuit could allow the pressure to fall to atmospheric, although it is desirable in such a case to avoid excessive ingress of air which would react with the hot graphite of the core. This is provided for by injecting carbon dioxide as required from multiple storage systems. The operating pressure of an AGR is about 40 bar, so that at atmospheric pressure the density would fall to about 2.5 per cent of its working value. The residual heating in the fuel falls rapidly below 2.5 per cent in the first few minutes following a trip, so that, provided circulation is maintained, there is no problem in removing the heat.

Additionally, being a gas, there is no possibility of explosive evaporation of the coolant, particularly as a result of interaction with molten fuel following some extreme fault condition.

Because the reactor is refuelled continuously, there is at no time any large excess reactivity such as would be needed with batch refuelling. The only excess reactivity required is that to maintain the core critical at steady power and during power manoeuvres. The overall temperature coefficient of re-activity (that is, the way the core power responds to changes of temperature) is positive; it is however dominated by the moderator, which is cooled by a separate re-entrant gas flow and is slow acting, whilst the important fast-acting temperature coefficient of the fuel is negative. The reactivity changes that occur during refuelling and power changing are small so that potential reactivity faults do not have serious consequences. The feature of on-load refuelling requires, of course, equipment of high integrity as the fuelling machine has to be connected to the primary circuit of the operating reactor to change the fuel assembly; several hundred reactor-years of on-load fuelling of the Magnox reactors without mishap gives confidence that this feature will be satisfactory.

The relatively low fuel ratings and core power density, combined with the large heat sink formed by the graphite moderator, tend to make any transients arising from faults slow and easy to control. The high melting points of the steel cladding and ceramic fuel of 1435 °C and 2800 °C respectively, give considerable tolerance in fault conditions before fuel melting can occur. Under normal operating conditions all but less than 1 per cent of the fission products are retained in the uranium dioxide fuel lattice so that can failure in even quite a large number of highly rated pins would not release much activity into the circuit. Natural circulation alone in the compact reactor circuit is sufficient to remove shut-down heating in the pressurized condition provided that water is maintained at a suitable level in the boilers.

A major failure of the vessel is inconceivable. The vessel is prestressed by a multiplicity of steel tendons so that the concrete is in compression up to, and well above, the normal working pressure. Raising the reactor pressure from atmospheric to the working level only insignificantly increases the stress in the steel tendons. The tendons are independent and have considerable redundancy, so that a complete failure of some would not significantly affect the others or the integrity of the vessel as a whole. The tension in the tendons can be checked, and they can be inspected and even replaced if need be.

An important principle in the design is redundancy in the components and systems, so that the failure of one or more items does not leave the reactor in an unsafe condition. This principle applies right through: to the fault sensors, the safety and trip systems, the shutdown systems, the gas circulators (of which there are usually eight), the boilers (four independent boilers in either eight or twelve units), the boiler feed, the auxiliary cooling systems, and all the back-up power supplies.

Any mismatch between reactor power and cooling causes the control rods to motor in. Only if this action is insufficient are the rods required to trip. The required reliability is obtained by using at least two different physical parameters for the trip sensors under any fault, by three separate guard lines operating on a two-from-three basis using solid-state logic, and by two independent interruptors to the electrical clutch supply to each rod. In addition, only a small fraction of the total number of rods is required to initiate a shutdown from the full-power state. This automatic rod system can be backed up by manually initiated injection of neutron-absorbing nitrogen into the gas circuit.

Systematic study of potential faults and their consequences has set the level of protection, backup, and redundancy required to protect the station operators and the public from an accidental release of radioactivity. Possible faults in the pressurized condition include loss of the main boiler feed, failure of a boiler tube, loss of gas flow in one quadrant, loss of main electrical supplies to all circulators, spurious closure of circulator inlet guide vanes, single channel flow faults, and symmetric and asymmetric reactivity faults caused by incorrect rod withdrawal during startup or at power. Depressurization faults are limited to failure at penetrations in the concrete vessel. These penetration closures are made to high integrity, with double closures for the large openings for the circulators so that the risk and size of any failure is limited. A failure of the external steel pipework of the gas treatment plant is possible; this would cause a Hinkley reactor to depressurize with a time constant of about 10 minutes, and protective equipment is provided to deal with the situation. The reactor trips on any significant loss of pressure, and continuity of cooling is ensured by diverse and redundant supplies, independent of the grid, to the main circulators and boiler feed. As the pressure falls, the fission gas in the fuel pin exceeds the coolant pressure and the clad temperature rises transiently, with the result that a few of the more highly rated pins could rupture; the resulting activity release into the gas circuit would be minor, amounting at the most to a hundred curies or so of iodine-131 and other volatile fission products, some of which would escape to the atmosphere.

These characteristics have meant that to meet the very stringent UK safety requirements no secondary containment is required for the AGRs beyond the massive concrete vessel. This has the value that continuous access is possible to the reactor auxiliaries for inspection and maintenance whilst the plant is in normal operation.

## 3.3 Hartlepool and Heysham

The importance of the Hartlepool and Heysham plants can only be judged through the longer perspective of future years. These four reactors represent the third commercial design of AGR, and embody a number of new features that had been under discussion and development for some time.

The reactor[18] has a multi-cavity vessel with its eight boiler units in pods in the vessel wall (Fig. 3.18). This was accompanied by the thought that reactors of greater or smaller outputs could be produced for the home market or export simply by varying the core size and the number of standard boiler units used. The boilers have the feed pipework running down a central core which is surrounded by the helically wound boiler tubing, giving a long cylindrical shape to fit neatly in the circular pods. All feed, superheat steam, and reheater connections are brought out through the top. Thus, the units can be fully works-fabricated ready for insertion into the vessel, site work is reduced, and the concrete vessel can be completed in parallel with the fabrication of the boilers which are not required on site until relatively late. Furthermore, the boilers can in the same way always be removed if major repair or replacement ever became necessary in service. The gas circulators are single-stage machines with direct electric motor drive and inlet guide vane control, and are immersed in the gas. In this case, however, they are unitized with each boiler and mounted vertically in each pod below the boiler. There are thus no major penetrations in the side of the vessel, and the circumferential prestress is applied by tensioned wire winding in horizontal channels in the surface of the vessel wall. The central cavity of the vessel is relatively small, since it contains the core structure only, and the top and bottom slabs are therefore of minimum thickness. A high level 150 t crane is needed to lift the boilers into position at a late stage of construction. This has been turned to advantage by making the crane a permanent feature of a high reactor house erected early on the site, so providing weather protection for much of the reactor construction and enabling the crane to be used for heavy construction lifts.

This style of reactor with pod boilers has met with much favour in studies of gas-cooled thermal and fast reactors in other countries. It overcomes some of the limitations of earlier designs, but, as is the case in many engineering choices, introduces new ones which will be touched upon later. Time will show how the balance lies.

## 3.4 Building the AGRs

It is easy with hindsight to see at least some things that could have been bettered. Certainly it is clear now that these reactors would have benefited from more component development and testing and more objective design analysis before the construction programme started; probably also from more pooling of expertise and ideas and without the intense inter-system and inter-company competition in which the first commercial designs were born.

The later Magnox reactors provided a well-founded engineering base; the

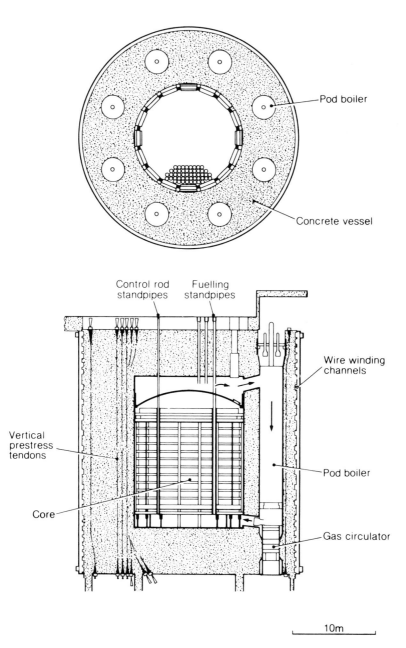

Fig. 3.18 Cross-section and plan of a Hartlepool AGR

30 MW(e) AGR at Windscale was there to demonstrate the new features. The similarity in engineering construction between the Oldbury Magnox reactor and the AGR is clear from Figs 3.5 and 3.6. Yet the translation to the full-scale AGR proved less easy than expected, so much so that construction work at Dungeness came at one point to a halt as the main contractor went into financial difficulties. The reasons are not far to seek. All the AGR stations have been built in a period of rising inflation, of escalating safety requirements, of industrial unrest, and of site labour difficulties to which the AGR with its rather extensive skilled site work was particularly prone; more particularly however the Dungeness main contractor, although bidding what appeared to be a very competitive design, had less experience than the other companies, particularly of the later Magnox concrete-vessel reactors, and underestimated the engineering task.

The next two stations of Hinkley B and Hunterston B had been designed in an atmosphere of calmer collaboration with the clients. They benefited little however from experience at Dungeness, since Hinkley B soon overtook the construction at that site with its engineers meeting and overcoming the technical problems of the new system as they arose.

The demonstration AGR at Windscale was small compared with the commercial plants (Table 3.7), and the evolution of gas-cooled reactors into integral concrete-vessel units had leap-frogged the Windscale plant since its inception. In Hinkley the gas pressure and gamma flux for instance, two key parameters determining the rate of radiolytic corrosion of graphite in carbon dioxide, were respectively twice and four times the Windscale values. Much attention had therefore to be paid to the gas treatment plant at the station, and continuing research and development were needed to establish a coolant mix that would keep corrosion acceptable over the required core life without causing carbon deposition, particularly on the fuel pins where it could impair heat transfer and cause the pins to run hot. Insulation of the inner face of the vessel had, in all previous UK reactors, taken the form of layered stainless steel foil. There was no lack of experience with this and it had been used successfully on the concrete vessels at Oldbury[19] and Wylfa as well as the

Table 3.7 Comparison of the Windscale demonstration AGR and Hinkley B design parameters

| Feature | | Windscale AGR | Hinkley B |
| --- | --- | --- | --- |
| Thermal output | MW | 100 | 1493 |
| Number of channels | | 253 | 308 |
| Core height | m | 4.3 | 8.3 |
| Fuel cluster | | 12 | 36 |
| Fuel pin size | mm | 15.2 dia. × 910 | 15.2 dia. × 1000 |
| Mean core power density | MW/m$^3$ | 1.4 | 2.8 |
| Gas pressure (absolute) bar | | 20 | 42 |
| Core inlet temperature | ° C | 250–325 | 292 |
| Core outlet temperature | ° C | 500–575 | 645 |
| Vessel type | | steel cylinder | prestressed concrete |

Windscale AGR; it was however a slow labour-intensive job applying it *in situ* at the site. In view particularly of the higher pressure and temperatures, a decision was therefore taken to change the insulation at Hinkley to ceramic fibre, a decision which meant that development and testing had to proceed in parallel with the site work, making some delays inevitable.

The telescoping of component development and test into the construction phase has been a feature of the UK nuclear environment, having origins probably in the defence projects of the UKAEA and fostered by the sense of urgency in the early commercial programme; it was difficult to break out of in a competitive bidding situation and without repeat orders to bear speculative research and development spending. This was not however just a feature of the nuclear scene, but one the supply industry had to face in general to meet the rapidly growing load in the 1960s and early 1970s. It applied also to other new features at Hinkley. The immersed-gas circulators for instance, introduced for good reasons to avoid the high-pressure shaft seals of earlier reactors, were of new design. A prototype machine was built early and run under full working conditions in a test rig, but this testing did not start until the third year of the station building programme and the teething troubles naturally enough took time to resolve. On-load fuelling of the AGR is basically very simple: with the fuelling machine latched to the required standpipe the whole fuel assembly has simply to be lifted vertically out of the reactor by a hoist connected to the top of the plug unit, and a new assembly re-inserted in its place. It was recognized at the outset however that the assembly had to stand high gas forces from the enhanced empty-channel flow when passing through the partly inserted position, but extensive tests had to be carried out in a full-scale refuelling rig before the fuel assembly and fuelling tube designs could be regarded as satisfactory. Another aspect is the well-known importance of the right materials for reactor circuits. Proving under realistic conditions takes a long time, however, and it was not until late 1971 that test work showed that the corrosion rate of the 9 per cent chrome steel of the boilers might be excessive in the moist carbon dioxide gas stream. The boiler tube material was left unchanged, but modifications were made to the already installed boilers to duct cool gas to the supports to lessen their corrosion.

Meantime building of the Hartlepool and Heysham stations had started. Bearing in mind the radically new design, much of the work went well. The sophisticated helically tubed boilers required much manufacturing development however before the works was impressively producing the series of 32 units required for the two stations; they represent a large commitment with their behaviour still to be proved in service. Site construction was badly delayed at Hartlepool by a licensing requirement to change the large steel closure at the top of the boiler pods to a prestressed concrete plug considered to have by virtue of its multiple prestressing a larger margin against major failure.

Construction of these reactors soon after Hinkley highlighted some of the

Fig. 3.19 The Hartlepool AGR station during construction, showing the slip-form concrete towers of the reactor building and the temporary extention to the overhead crane tails at right used for construction lifts

differences. The early erection of the reactor house (Fig. 3.19), whilst congesting cranage inside the building, provided good year-round working conditions. The stainless steel foil used for the vessel insulation however took long to apply, requiring much hand work in areas of surface discontinuity such as at the ducts between the core cavity and the pods. Steel foil insulation, used successfully for the vessels of the Oldbury and Wylfa Magnox stations, was retained for the Dungeness, Hartlepool, and Heysham AGRs, where it gave rise to some passing problems owing to the higher gas pressure and temperature in addition to the lengthy site work. These problems are not generic to the AGR; the Hinkley and Hunterston vessels have been successfully insulated and operated with the alternative ceramic fibre insulation, and future stations are planned to follow suit.

Access to the boilers after they had been installed in the vessel at Hinkley had proved invaluable during the construction phase; *in situ* access to the Hartlepool pod boilers is very limited, and whilst they avoid the complication of side penetrations and they can be lifted out completely from their pods, removal and replacement seemed likely to be a longer job than was perhaps at first thought.

## 3.5  Commissioning and running the first AGRs

The comprehensive engineering tests, designed to prove the plant operation or enable defects to be found and put right, were started on the first Hinkley B reactor in mid-1973. For a complex first-of-a-kind plant these tests went well. It was only at a later stage after fuel had been loaded and just before power raising that vibration in the plug units was detected and inspection showed there had been excessive wear in the flow control gag as well as signs of vibration elsewhere in the plug unit. Rectification and re-testing cost a year.   Altogether, the building of the first Hinkley B unit from the start of site work to start of commissioning was extended from a tendered optimistic four-and-a-half years to six, and a further two years were spent on modifications and testing before power was raised. This, for a lead unit with large extrapolations from previous experience, is a time not out of line with many nuclear plants or even conventional ones. It reflects not just the telescoping of the development and test work, but the proper caution of the generating board in clearing the problems in a new plant step by step before going into operation.[20] Hunterston B followed too closely to benefit fully from being a replica. Even so, although started later, its first unit was generating power within days of the Hinkley second reactor, took only three-quarters the man-hours of the first to build, and was up to its interim power rating only six weeks after its alternator was first synchronized to the grid.

The alternators of the first Hinkley B and Hunterston B units were synchronized in February 1976. The success of this power-raising phase went some way to vindicate the time taken in construction and testing. The major plant items, particularly those to do with safety, such as the concrete vessel, insulation, control, and safety equipment worked well. Startup and control of the once-through boilers was satisfactory; they did not leak; the ability of the plant to continue to operate with one boiler quadrant out of action was demonstrated. Over six months the power was raised in steps to 540 MW gross (82 per cent) which was an initial setting-to-work level chosen by the operators to limit the gas temperature in the 9 per cent chrome section of the boilers pending further experience. One reactor was however given a short demonstration which showed it to be capable of its full thermal output.

One feature which has proved more difficult to bring into full operation than expected is the on-load refuelling, despite the rig testing mentioned earlier. There was little pressure initially to put the fuelling plant through its paces since refuelling does not need to start until after the first year or so of operation. When the time came, however, it was clear that further development and proving of the equipment was required, and off-load batch refuelling was temporarily used. Having a single fuelling machine (Fig. 3.20) and a common fuel discharge route between the two reactors, whilst giving an elegant design and simple layout, rather aggravated the problem. A comprehensive programme of work was therefore put in hand, with the generating board taking a cautious approach by working up the equipment initially for part-load fuelling, and with the long and satisfactory experience of non-load fuelling the Magnox reactors giving confidence of a satisfactory outcome.

Not unexpectedly, these plants have taken some time to settle down. Teething troubles that can be found only by running at power (but put right only by shutting down) have so far held down the station load factors. None of the faults has been seen as generic to the AGR system, and many of the outages have been due to the turbo-alternator plant at Hinkley, which is itself in the nature of a prototype (left over from one of the turbine-maker mergers), or to once-off modifications or to statutory inspections.[21] The early load factors are summarized in Table 3.8, in which the turbo-alternator outages are separated out to show the reactor availability. The reasons for non-availability are shown in Table 3.9.[22]

The second Hunterston unit is not in Tables 3.8 or 3.9 as it stopped operating in 1977 when sea-water inadvertently entered the base of the reactor vessel via some temporary pipework when the reactor was shut down and depressurized. The incident caused no safety risk or irreparable plant damage, but inspection and refurbishing took a little over two years. The incident provided an interesting example of the way in which a reactor of this type can be repaired. A major part of the vessel insulation inside the vessel in the annulus below the boilers was removed and replaced, and the radiation

Table 3.8 Early Hinkley B and Hunterston B availabilities (up to March 1978)

| Factor | Hinkley first unit (%) | Hinkley second unit (%) | Hunterston first unit (%) |
|---|---|---|---|
| Unit load factor | 29 | 36 | 39 |
| Unit availability | 33 | 38 | 54 |
| Reactor availability | 49 | 72 | 65 |

Table 3.9 Reasons for early non-availability of Hinkley B and Hunterston B stations

| Reason | Hinkley first unit (%) | Hinkley second unit (%) | Hunterston first unit (%)approx |
|---|---|---|---|
| Block outages: | | | |
| feed system modifications | 4 | — | 3 |
| statutory inspections | 16 | 6 | 18 |
| boiler orifice modifications | 22 | — | — |
| turbo-alternator repairs | — | 21 | 6 |
| Lower power testing | 10 | 10 | 9 |
| Unit trips | 6 | 4 | 6 |
| Other plant faults: | | | |
| reactor/boiler | 1 | 13 | |
| turbo-alternator and balance of unit plant | 8 | 8 | 4 |
| Total loss in unit availability | 67 | 62 | 46 |

Fig. 3.20  Fuelling machine at Hinkley B

level in this area, even with the irradiated fuel still in the core, was in the region of only 0.5 millirem per hour so that it placed no restriction on carrying out the work. After refurbishing, the unit was quickly back at over 500 MW(e) with no apparent ill effects.

More recent operation of these first stations is encouraging. The load factors for Hinkley relative to the full design output are shown in Table 3.10 and are not much different from any other large power stations at this stage. The first Hunterston reactor performed particularly well during 1979, with a load factor, relative to its setting-to-work power level, of 78 per cent. Fuel behaviour has so far been satisfactory. There are 90 000 fuel pins in each reactor and, up to the start of 1980, only two channels in total have had to be discharged from the four operating reactors because of failure.

Table 3.10 Later load factors of Hinkley B

|  | First unit (%) | Second unit (%) |
| --- | --- | --- |
| Financial year 1978/79 | 55 | 35 |
| April 1979 to January 1980 | 54 | 33 |
| Five consecutive months of 1979/80 winter period | 63 | 62 |

The next couple of years are critical. Hinkley and Hunterston should settle down to steady power generation; the different designs of station at Dungeness and at Hartlepool and Heysham should come on line. Whilst these first stations still have to be worked up to give their full performance, there is reason for confidence that they will be a success.

## 3.6 Second-generation AGRs

The government decision in January 1978[23] for work to begin with a view to building two further AGR stations (whilst also developing the option of adopting the pressurized-water reactor in the early 1980s) followed a detailed assessment and recommendation by the industry. The decision was seen by some as a vindication of the AGR, by others as a stopgap to help the industry.

Preliminary site work for these two twin-reactor stations started promptly. This was possible as the chosen sites of Heysham in Lancashire and Torness in Lothian already had planning consent.

With the designers now reorganized into a single company, a common standardized reactor design has been chosen for both stations. The design is based on Hinkley B.[24] Not only is that station operating successfully, but experience has shown the value of having good access to the vessel interior and has caused concern over boiler irreplaceability to recede. The aim is to consolidate the existing design while profiting from the lessons learned. Changes are being made only where their value is judged clearly to outweigh the loss of replication. Particular attention is being given to improving access to the reactor internals and to increasing works fabrication to reduce skilled site work. Construction of these two stations is programmed to start on site in 1980.

## 4 The high-temperature reactor

The HTR differs from other gas-cooled reactors by having an all-refractory core of fuel dispersed in a graphite moderator, higher working temperatures, and inert coolant. Helium is the universal coolant choice. The concept was born at Harwell in the mid-1950s.

The early lure of HTR is easy to see. It promised lower costs through higher core-power, and small plant size and efficiency; in the longer term was the possibility of more savings by cutting out the steam plant and generating by direct-cycle gas turbines; high-temperature process heat was a possibility, and not least, its ceramic core was capable of high conversion ratios and uranium-saving fuel cycles.

Interest was aroused in several countries by the potential of this reactor system. Exploration and development of the core, which could obviously take a variety of forms, was an essential first step and led to the design and building of small experimental reactors in three countries. The first was the Dragon reactor built in the UK, which reached full power operation in 1966, the next the Peach Bottom plant in the USA, which operated at its full output in 1967, and third was the AVR at full power in the Federal Republic of Germany the following year. Table 3.11 gives some details of these plants.

Within the larger family of gas-cooled reactors, the HTR is a family in itself. Because the core is made entirely of ceramic materials, it is less constrained by the exacting heat transfer needs of other reactor systems, more tolerant of local mismatch of power and coolant flow. A large variety of configurations of fuel and moderator is therefore possible, in combination with various alternative fuel cycles.

Table 3.11 Main features of Dragon, Peach Bottom and AVR

| Feature | | Dragon | Peach Bottom | AVR |
|---|---|---|---|---|
| Thermal power | MW(t) | 20 | 115 | 46 |
| Net electrical power | MW(e) | — | 40 | 15 |
| Helium pressure (absolute) bar | | 20 | 24 | 11 |
| Core inlet temperature | °C | 350 | 340 | 260 |
| Core outlet temperature | °C | 750 | 715 | 950 |
| Core effective diameter | m | 1.1 | 2.8 | 3 |
| Core height | m | 1.6 | 2.3 | 3 |
| Core power density | MW m$^{-3}$ | 14 | 8.3 | 2.2 |
| Fuel type | — | prismatic | prismatic | pebbles |
| Fuel cycle | — | various | $^{235}$U–Th | $^{235}$U–Th |

Early studies of the HTR concentrated on the high-enriched uranium–thorium cycle, which the physics of the highly rated homogenous core favoured. The initial fuel loading here consists of uranium enriched to over 90 per cent mixed with fertile thorium to give a $^{235}$U–$^{232}$Th–$^{233}$U cycle with conversion ratios up to around unity. A low-enriched uranium cycle using $^{235}$U–$^{238}$U–Pu similar to the more conventional thermal reactor cycle is however possible. This makes more use of existing process technology. Fuel fabrication costs are fairly high, but in compensation burn-ups proposed are mostly in the high range, 60–120 GW d/t of heavy metal. As with other thermal reactors, the cycle can be once through, but the real value of the uranium–thorium cycle is only realized with reprocessing and recycling of the bred$^{233}$U. HTRs can be used in symbiosis with fast reactors or other HTRs of differing conversion characteristics.

In the early concepts the fuel was dispersed in compacts contained in graphite rods nesting together to form the core and coolant flow passages, with a helium purge system connected to the bottom of the hollow rods to remove the gaseous and volatile fission products from the reactor circuit. A major step forward however came with the development of the coated fuel particle. This is now the basis of all high-temperature-reactor designs. Current coated particles comprise a spherical kernel of enriched uranium oxide or carbide, less than a millimetre in diameter, with a sophisticated coating to retain the fission products. The coating typically starts with a protective buffer layer of pyrolytically deposited carbon, followed by an impervious silicon carbide layer sandwiched between two strong coatings of high-density pyrocarbon (Fig. 3.21). These particles are capable of normal operation at temperatures of 1200 °C to 1300 °C with a release-to-birth ratio (that is the ratio of nuclides released from the particle to nuclides formed by fission) of around $10^{-5}$ for iodine-131 and the metallic fission products. They appear to be able to stand almost endless thermal cycling.

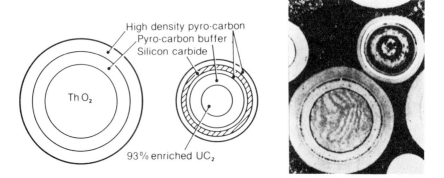

High density pyro-carbon
Pyro-carbon buffer
Silicon carbide

Th O$_2$

93% enriched UC$_2$

Fig. 3.21 Typical HTR coated fertile and fuel particles. On the right is an enlarged photograph of particles after irradiation to over $8 \times 10^{21}$ neutron/cm$^2$ at 1050C

The coated-fuel particle opened the way to greater freedom in core design. Two distinct lines of core development have been followed: the pebble-bed and the prismatic core. Reactor physics and engineering developments of these concepts have gone hand in hand.

In the pebble-bed reactor, pioneered in the Federal Republic of Germany, a fixed graphite reflector forms a container for the core which consists simply of a multitude of graphite spheres. The spheres are typically 60 mm diameter with the required quantities of enriched-uranium and thorium particles embedded within them. In one version of the fuel cycle the spheres are continuously checked for integrity and burn-up before being either re-inserted at the top of the core or rejected.

Many variations of prismatic core have been studied. In these, the core consists of vertical columns of graphite prisms—usually hexagonal, but sometimes square or of more complex section—with various locations of the fuel and coolant passages. The coated-fuel particles are bonded in a carbonaceous matrix, encased in graphite tubes to form fuel rods or annuli supported in the moderator blocks, or inserted directly into blind holes in the blocks. The fuel and moderator blocks are normally replaced together. For the uranium–thorium cycle, the fissile and fertile material can be combined within the particles, mixed in different particles within the rod, or kept separate. The resulting prismatic cores have varying degrees of homogeneity in the reactor physics sense.

Core power densities proposed lie between 5 and 10 MW(t)/m$^3$. Gas outlet temperatures range from about 650 °C to a high 1000 °C proposed for process heat systems.

The engineering design has drawn heavily on the earlier UK and French gas-cooled reactors. All current designs are based on a prestressed concrete vessel; but with the complex multi-cavity styles of some designs and the high temperature and high thermal conductivity of helium, the insulation of the vessel interior is critical. The large surface area of graphite in the core at 500 to 1000 °C reacts readily with oxygen. Steam leakage into the circuit from boilers therefore needs to be kept very low. In addition, to limit core corrosion and carbon deposition elsewhere in the circuit, a gas treatment plant is needed to remove water vapour during start-up and to limit the levels of hydrogen, carbon monoxide, and carbon dioxide formed by reaction with the graphite. Fission product trapping (originally a key feature of the fuel purge system) is now of secondary importance in the gas treatment. The high cost of helium alone requires the circuit to be very leak tight and for rather bulky tanks to be provided for make-up gas and for storing the helium when the circuit is depressurized. Gas flow through the core is either upwards or downwards. The more conventional upward flow is convenient in some respects, but the low density of helium provides little natural circulation and in the event of failure of all circulators reliance has to be placed on the good thermal capacity of the core and eventual resumption of some forced circulation. Downward flow has found much favour: it puts top control rods at the cool end as well as any on-load fuelling gear used with prismatic designs, but, more important, it avoids the problem of levitation of the core graphite.

The safety characteristics of the HTR are similar to those of other gas-cooled reactors in concrete vessels. The combination of good thermal capacity of the core and tolerance to temperature excursions of the fuel is helpful so far as the core itself is concerned; with no forced circulation at all it would take several hours for fuel particles in the shut-down core to reach temperatures around 2000 °C, at which fission product release would increase. In addition, except with the low-enriched uranium–plutonium cycle, both the Doppler and overall (fuel plus moderator) coefficients of reactivity are strongly negative. On the debit side however, the potential reaction rate of the hotter graphite makes the exclusion of air in the event of a pressure circuit fault more important; and the weak natural circulation even in the pressurized state has already been mentioned. These features, together with downward flow and the high starting temperature of the circuit, suggest a greater need for engineered safeguards, particularly in continuity of forced circulation and heat removal.

The bad set-back the HTR had in 1975 on its road to commercialization in the USA was felt round the world. A new reactor system can be introduced commercially only in reasonable numbers and large unit sizes to compete with established systems; marginal benefits of diversity, lower generation costs, and lower cooling water needs are no longer enough to justify either the risks to the utility or the bearing of huge launching costs by a commercial vendor.

There has, as a result, been a shift of interest over the past two or three years to direct process-heat applications. Possibilities being studied include coal gasification, hydrogen production by steam reforming methane, other thermochemical processes, and steel making. Much development is required for these uses. However the market is limited and users tend to be scattered.

There is revived interest also in the direct-cycle system. Apart from the potential cost saving with direct gas-turbine generation in place of the steam cycle, the heat rejection (already smaller than for water reactors) is at a relatively high temperature, over 200 °C being quite reasonable.[25] This makes it feasible to use dry cooling towers, which are becoming necessary in inland industrial areas as well as hot arid countries, without serious thermodynamic loss; it also heightens the prospects for combined heat and power.

Studies and development of the high-temperature reactor are being carried out in several countries besides the UK, the USA, and the Federal Republic of Germany. France for instance continued her early interest in gas-cooled reactors by association with the United States. Although she recently produced a 1200 MW(e) design for steam–electric use,[11] the main HTR interest and research and development effort are directed toward coal gasification using hydrogen provided by a nuclear methane/steam reformer.[26] Switzerland is working closely with Germany in the development of the major components for a demonstration direct-cycle gas turbine plant, including the helium turbine, heat exchangers, and prestressed concrete vessel. The emphasis in Germany is now, however, probably more on coal gasification[27] and the supply of process heat using a chemical heat pipe. In Japan, work has concentrated on nuclear steel-making by a hot reducing gas, with an ambitious programme aimed at having a pilot plant operating in the mid-1980s.[28] Interest and construction of a small demonstration plant in the USSR has been reported.

The major work on the system has however been done in the United Kingdom, the USA, and the Federal Republic of Germany.

### 4.1 St. George and the Dragon: HTR development in Britain

Among the many possible reactor systems studied in the UK in the 1950s, the high-temperature, gas-cooled, graphite-moderated system stood out. Whilst it clearly had great potential it also clearly needed development and demonstration over quite a long timescale. OECD countries were at the time seeking ways to collaborate in the field of experimental reactors. Thus the High-Temperature Reactor Project came into being in 1959, the partners being the United Kingdom, Austria, Denmark, Norway, Switzerland plus the Euratom countries Belgium, France, Germany, Italy, Luxemburg, and the Netherlands. The aims of the project were to build and operate an

Fig. 3.22 The Dragon HTR at Winfrith

experimental reactor and to carry out research and development in the HTR field.

Construction of the 20 MW(t) Dragon reactor was started in 1960[29] at Winfrith in Dorset and was supported by the zero-energy physics reactor Zenith and other facilities (Fig. 3.22, Table 3.11).

The reactor core is 1.07 m diameter and 1.6 m height, and surrounded by a graphite reflector. It is contained in a steel pressure vessel, with six short external loops each comprising a heat exchanger and immersed gas-bearing circulator (Fig. 3.23). The ducts are co-axial and arranged so that the cool gas from the heat exchanger outlet sweeps the pressure circuit before flowing upwards through the reactor core. Control rods are arranged in a ring in the inner reflector around the core. Fuel is changed, with the reactor shut down, by a pantograph machine housed above a shield plug in an extension of the reactor vessel above the core. In view of the early state of the technology at the time of design, the heat is rejected first to a secondary water circuit and thence to atmosphere by forced air coolers in a tertiary circuit; the primary and secondary circuits are both inside a double-walled containment building.

The first core had 37 fuel elements, each of seven hexagonal graphite rods with gas flow passages between. In its original concept the rods contained fission-products-emitting fuel compacts and when placed in the core were connected by a spiked seating to a gas purge system so that the gaseous and volatile products were continuously scavenged from the rods for trapping in a coolant purification circuit. The coated particle was, however, conceived in time for an early version of HTR fuel development.[25,30] When the coated particle was shown to work, purging became unnecessary and a variety of fuel designs were tested, culminating in the simple multi-hole block design, in which the seven-rod clusters were replaced by multi-sided graphite blocks with axial through holes for coolant passages and separate blind holes for the bonded particle fuel rods. This ability to change the complete core design, fuel-to-moderator ratio, and fuel cycle is a characteristic of the HTR. Particles of uranium dioxide, plutonium dioxide and uranium/thorium carbide, were tested with various coatings and over a range of temperatures up to 2000 °C to establish optimum designs and performance limits.

With its original task achieved, the Dragon project ended in 1976 and the reactor was shut down.

Designs for large commercial stations were studied by the Dragon Project and by the UK industry.[31] Among these was the 'feed-breed' reactor. This used two types of fuel-breed elements with a mix of uranium-235 and thorium-232 to give a conversion ratio close to unity so that they needed very infrequent replacement, and a smaller number of feed elements with highly enriched uranium placed so that they could be replaced on-load with direct access. A heterogeneous design close to the AGR, with fixed moderator and low-enriched uranium fuel stringers refuelled on-load, was also examined; it was not pursued owing to doubts over moderator life with the shrinkage and

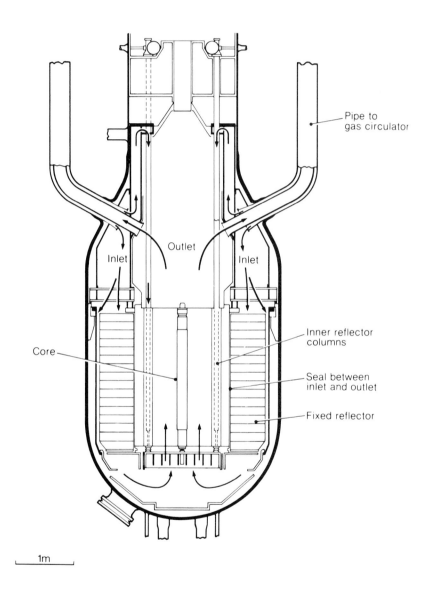

Pipe to
gas circulator

Outlet

Inlet

Inlet

Inner reflector
columns

Core

Seal between
inlet and outlet

Fixed reflector

1m

Fig. 3.23 Cross-section of the Dragon reactor

growth that could occur in the graphite at the higher temperatures and neutron fluxes. A number of homogeneous designs based on low-enriched uranium, the only practicable fuel cycle for commercial use at the time, were also studied,[31,32,33] and a fully detailed design for a 630 MW(e) lead station to be built at Oldbury was tendered in 1970.

With only three small experimental HTRs running however, it seemed clear that much effort and time would be needed before a full-scale plant could be viewed as a reasonable risk. Resources in the UK were aleady committed to other thermal reactor systems. Active work on the HTR therefore, ceased, but its potential was fully recognized and interest in the system, and particularly in the study of its possible future cycles, has continued.

## 4.2 High-temperature reactors in the USA

The HTR idea spread quickly across the Atlantic. Its special characteristics were seen to make it a potential competitor to the light water reactors, and interest was concentrated on the thorium cycle. Research and design work led to the building of the Peach Bottom station[34] by a group of 53 utilities under the Government's power reactor demonstration programme. Its aim was to prove the main elements of the system.

The plant was built beside the Susquehannah River in Pennsylvania (Fig. 3.24, Table 3.11). Work started on site in 1962 and the plant went into full

Fig. 3.24  Peach Bottom HTR station

power operation in 1967. The core, 2.8 m diameter and 2.3 m height, follows the general prismatic design of Dragon with full-length fuel elements surrounded by a fixed graphite reflector. Control and safety rods are inserted into the core from below and are supplemented by thermally released gravity-drop absorbers in the upper reflector. The core sits low in a steel pressure vessel (Fig. 3.25) with space above for fuel handling, and concentric ducts connect the vessel to the two drum-type boilers and gas circulators. Helium flows upwards through the core in the passages formed between the fuel elements and up through the boilers; it returns via the annular ducts and the inner surface of the vessel so as to keep the pressure circuit at the lower temperature. Steam conditions at the turbine alternator are 100 bar and 538 °C, giving the net output of 40 MW(e).

The core is made up of 804 graphite fuel elements of 89 mm diameter. These contain the uranium–thorium fuel particles of about 240 $\mu$m size dispersed in graphite compacts. Coated particles had barely arrived at this time and the fuel in the first core had a simple pyro-carbon coating. A purge system for the fuel elements was therefore an essential part of the design. The second core used improved pyro-carbon coatings with an inner buffer layer and ran for its intended life of almost 900 full-power days with greatly reduced activity release. As with Dragon, the reactor was used also to test various different fuel concepts for the future.

The reactor was shut down in 1974 having given a remarkably good availability over its seven-year test programme.

A year after site work started on Peach Bottom, study of a medium-sized demonstration plant to develop the concept began. In 1968, only a year after Peach Bottom went to power, construction started on Fort St Vrain.[35] This 330 MW(e) station, built in Colorado to provide commercial power, was again part of the federal power reactor demonstration programme (Fig. 3.26).

The reactor is of integral design, with a layout (Fig. 3.27) owing much to the later French Magnox designs. The inside size of the concrete vessel is 9.4 m diameter by 22.9 m height (giving an internal volume little more than one sixth that of the 487 MW(e) St Laurent vessel). The core is located in the upper part of the vessel and consists of 1482 fuel elements stacked in six layers. Each element is a simple hexagonal graphite block with bonded rods of coated fuel particles in drilled holes and separate through holes for coolant flow (Fig. 3.28). The particles are 93 per cent enriched uranium and thorium dicarbides, with multiple pyro-carbon and silicon carbon coatings, in the range 360–860 $\mu$m overall diameter. The fuel loadings are zoned radially and axially. Holes are provided in some of the blocks for the top-operating control rods. The core is designed for refuelling off-load in six annual batches, so that the entire core is replaced over each six-year cycle, using the penetrations in the vessel top normally occupied by the control rod drives.

Helium flow is downward through the core. This gives a hot support

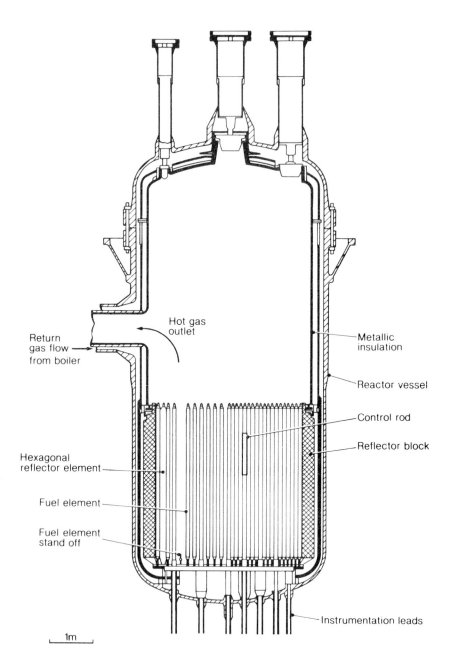

Return gas flow from boiler

Hot gas outlet

Metallic insulation

Reactor vessel

Control rod

Reflector block

Hexagonal reflector element

Fuel element

Fuel element stand off

Instrumentation leads

1m

Fig. 3.25 Cross-section of the Peach Bottom reactor

Fig. 3.26  Fort St. Vrain HTR station

structure for the core; however, it avoids levitation, keeps the hot gas path short, and gives a relatively cool gas flow over the vessel liner and control assemblies above the core. The helium pressure is 48 bar.

The boilers are in twelve cylindrical modules in a ring below the core. They are of once-through design with helical tubing and integral superheaters and reheaters, and the feed and steam penetrations are brought out through the base of the vessel. The four helium circulators are also mounted vertically in the bottom of the vessel. These are single-staged axial compressors driven by single-stage steam turbines.

The prestressed concrete vessel is insulated on the gas side with ceramic fibre and metal cover plates. The penetrations have gas-tight closures at inner and outer ends to provide double closure with the interspace pressurized with gas taken from the helium purification plant.

The steam cycle is the conventional modern one with steam conditions of 159 bar and 538 °C, with a single reheat to 538 °C. In this case however the steam from the turbine high-pressure cylinder passes first through the turbines driving the helium circulators before entering the reheater (Fig. 3.29).

Its construction proceeded well. The success of Peach Bottom and the building of Fort St Vrain coincided with rising concern for the environment. The HTR's good safety, low heat rejection per kilowatt, and conservation of uranium, together with its promising low cost and high availability, attracted US utilities. In the early 1970s standardized designs for 770, 1160, and

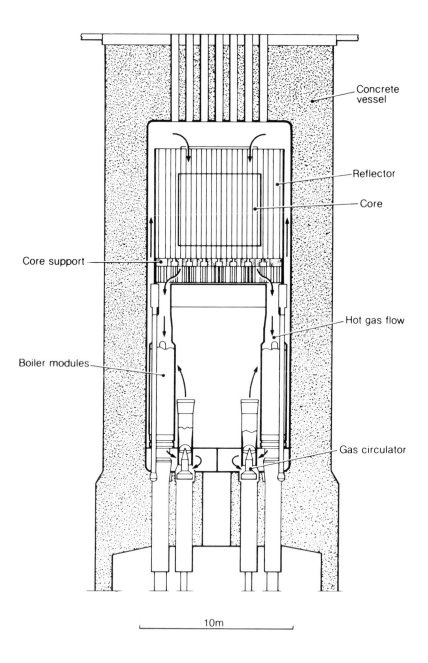

Fig. 3.27 Cross-section of the Fort St. Vrain reactor

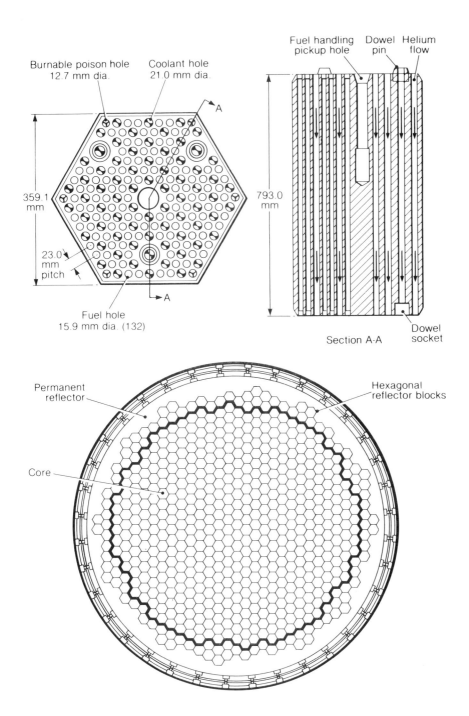

Fig. 3.28  Fuel element and core plan of Fort St. Vrain

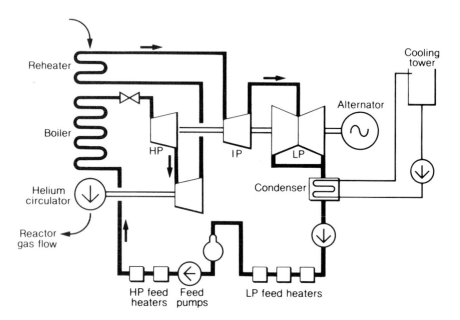

Fig. 3.29 Main steam circuit of Fort St. Vrain

1500 MW(e) outputs were developed.[25] While based largely on Fort St Vrain, these used a multi-cavity wire-wound concrete vessel, similar to the Hartlepool and Heysham AGRs, but inverted to give downward core flow (Figs. 3.30 and 3.31).

Some details of Fort St Vrain and the larger designs are given in Table 3.12. Orders were negotiated for four large commercial twin-unit stations: Summit (2 × 770 MW) in Delaware, Fulton (2 × 1160 MW) in Pennsylvania, St Rosalie (also 2 × 1160) in Louisiana, and Vidal in California.

However, the severe economic problems facing utilities in 1973 and 1974 led to the cancellation of two of these orders, and the general trend of lower growth forecasting brought about the collapse of the market for nuclear plants. As a result of the business turndown the vendor for these plants in the USA cancelled the remaining orders in 1975.

Hints of problems with some Fort St Vrain components had come to light during the latter part of the construction phase. Changing regulatory requirements entailing back-fitting and a succession of problems during requirements commissioning caused some delay in power raising.[36] The reactor was made critical in 1974 six years after construction started; it went to power in 1976 and two years later reached 70 per cent output. None of its problems had thrown doubt on the basic HTR concept, but its chance of immediate commercialization was lost.

HTR technology development has nevertheless continued strongly in the

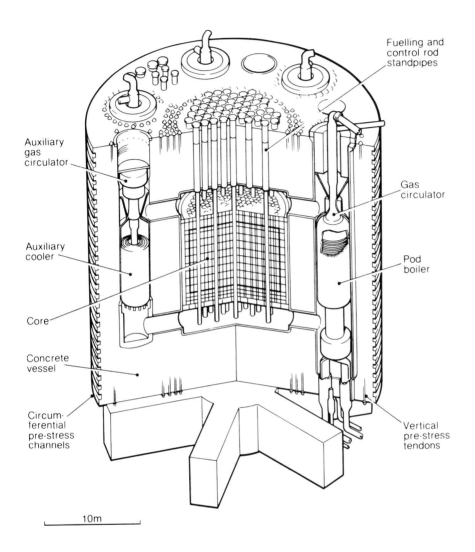

Fuelling and
control rod
standpipes

Auxiliary
gas
circulator

Gas
circulator

Auxiliary
cooler

Pod
boiler

Core

Concrete
vessel

Circum-
ferential
pre-stress
channels

Vertical
pre-stress
tendons

10m

Fig. 3.30 Cutaway drawing of the proposed 1160 MW(e) HTR

Table 3.12  Comparison of Peach Bottom, Fort St Vrain and proposed large HTRs

| | | Peach Bottom | Fort St Vrain | 770 MW(e) | 1160 MW(e) |
|---|---|---|---|---|---|
| Thermal power | MW | 115 | 842 | 2 000 | 3 000 |
| Electrical power | ME(e) | 40 | 330 | 770 | 1 160 |
| Efficiency | % | 35 | 39 | 39 | 39 |
| Inlet gas temperature | °C | 340 | 405 | 320 | 320 |
| Outlet gas temperature | °C | 715 | 780 | 740 | 740 |
| Average power density | MW/m³ | 8.3 | 6.3 | 8.2 | 8.4 |
| Core active height | m | 2.3 | 4.8 | 6.3 | 6.3 |
| Core equivalent diameter | m | 2.8 | 5.9 | 7.1 | 8.5 |
| Fuel lifetime at 80% load factor | years | 3 | 6 | 4 | 4 |
| Refuelling cycle | years | 3 | 1 | 1 | 1 |
| Fraction of core replaced each cycle | | 1 | 1/6 | 1/4 | 1/4 |
| Number of fuel elements | | 804 | 1 462 | 2 744 | 3 944 |
| Number of columns | | — | 247 | 343 | 493 |
| Elements per column | | — | 6 | 8 | 8 |
| Basic fuel component | | Particles in a graphite compact within a cylindrical purged sleeve | Bonded rods of particles within a hexagonal graphite element | | |
| Element height | mm | 3 660 | 793 | 793 | 793 |
| Element width | mm | 89 | 359 (across flats) | 359 | 359 |
| Fuel burn-up average | MW d/t | 73 000 | 100 000 | 95 000 | 95 000 |

Fig. 3.31 Impression of the proposed Fulton HTR station

USA. The aims are to design and test materials and components that are common to the steam cycle, gas turbine, and process heat applications; to develop the fuel; and to study alternative fuel cycles such as the uranium–thorium option with a medium 20 per cent enrichment of non-proliferation interest.

### 4.3 HTR development in Germany

The early interest in developing the HTR in the Federal Republic of Germany has already been mentioned.

Commissioning of the experimental AVR plant at Jülich (Fig. 3.32, Table 3.11) at the end of 1967 marked the start of the practical-test phase. Its design[37] is unique. The reactor and boiler are integrated in a double-walled steel pressure vessel (Fig. 3.33). Helium flows upwards through the core where it is heated to a mixed outlet temperature of 950 °C; this exceptionally high temperature is desired for testing for advanced HTR applications and has been achieved in AVR through successive steps from 750 °C and 850°C. The gas then passes through the boiler situated above the core and returns cool around the inner surface of the vessel to the gas circulators at the vessel base. Steam produced at 75 bar and 500 °C drives the conventional turbo-alternator. The reactor and auxiliaries are housed in a cylindrical steel containment vessel and outer shielded building.

Fig. 3.32  The AVR pebble-bed reactor at Jülich

The exceptional feature of the reactor however is the pebble-bed core. This consists simply of about 100 000 graphite spheres loosely placed in the hopper-like graphite reflector. These balls form the moderator and fuel. They are 60 mm diameter and contain the fissile and fertile material in the form of coated particles. The balls are cycled continuously through the core by gravity at a rate of about 600 per day. Each is checked for damage and for burn-up, and 90 per cent are re-inserted at the top of the core with a make-up of 10 per cent fresh balls replacing those that have reached the required burn-up. The balls are elevated to the reactor top pneumatically using helium from the coolant circulators. About one in 10 000 is rejected for damage. The strong negative temperature coefficient allows the reactor power to be controlled over a wide range simply by varying the gas flow by

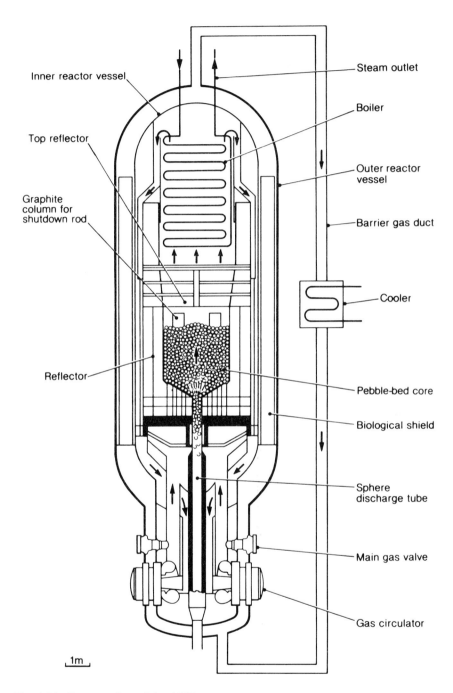

Inner reactor vessel

Top reflector

Graphite column for shutdown rod

Reflector

Steam outlet

Boiler

Outer reactor vessel

Barrier gas duct

Cooler

Pebble-bed core

Biological shield

Sphere discharge tube

Main gas valve

Gas circulator

1m

Fig. 3.33 Cross-section of the AVR

the variable-speed blowers. Vertical-acting rods are used for shutdown, inserted from the bottom in graphite columns near the core edge, together with some secondary shutdown and hold-down arrangements. A helium purification circuit is provided to remove gaseous impurities.

In addition to demonstrating the pebble-bed HTR concept, the reactor has been used as a test bed for developing and proving the fuel.[25,37] The early fuel elements took the form of a hollow graphite ball filled with pyro-carbon coated fuel particles mixed with graphite. Subsequently the balls have been made by isostatic moulding, with a core of fuel particles in a graphite-and-binder matrix enclosed in a fuel-free matrix graphite shell (Fig. 3.34). The fuel content of each element is 1 g of 93 per cent enriched $^{235}$U in the carbide or oxide form, together with 5 g of thorium. Mean burn-ups up to about 150 GWd/t of initial metal atoms (U + Th) have been obtained.

It is planned to continue running the AVR plant into the 1980s, testing a variety of coated particles and fuel cycles, possibly developing increased fuel loadings for near-breeder application.

There was a bold plan to follow the AVR with a demonstration of the direct-cycle gas turbine HTR in a 25 MW(e) plant at Geesthacht near Hamburg.[38] The reactor departed, however, from the initial line of development in the Federal Republic of Germany by using a prismatic core similar to Peach Bottom and, although work began on the project, it stopped, largely for commercial reasons. Subsequently a demonstration helium gas turbine was built with an oil-fired heat input to the closed cycle. This 50 MW(e) plant at Oberhausen started up in 1975 and is providing local power and district heating.

Studies for a large demonstration plant to exploit the pebble-bed concept started in 1963 whilst the AVR was being built. These led to proposals in 1968 for a 300 MW(e) steam–electric plant capable of extrapolation to outputs upto 1200 MW(e).

Construction started in 1972 at Uentrop near Hamm. This thorium high-temperature reactor (THTR) is based on the AVR principles, uses the now standard 60 mm fuel balls already tested at Jülich, but is of advanced engineering design (Fig. 3.35).[39,25]

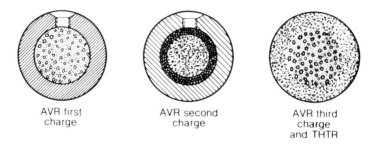

AVR first charge          AVR second charge          AVR third charge and THTR

Fig. 3.34 Successive designs of fuel element of the AVR

Fig. 3.35 The 300 MW(e) THTR at Uentrop nearing completion

About 700 000 fuel balls go to make up the core. Each fresh fuel element contains about one gram of $^{235}U$ and about 10 gram of thorium in solid solution, disposed in 400 $\mu$m particles coated with a double pyro-carbon layer 180 $\mu$m thick. The balls circulate through the core as before, but with a more refined burn-up measuring system, to give continuous on-load refuelling. Absorber rods for control and tripping are top suspended and operate in holes in the graphite reflector. They are supplemented for long-term shutdown by pneumatically driven rods pushed directly into the pebble bed.

The primary circuit is integrated in a relatively simple single-cavity concrete vessel (Fig. 3.36). The inner face of the vessel liner is insulated with steel foil. The reactor is surrounded only by a thermal shield, and the vessel is not designed for entry; the boilers, circulators, and control rods are therefore all made removable. The six helical-tube boiler modules are cylindrical and are inserted into the annulus around the core through openings in the vessel top. Each is unitized with an electrically driven circulator mounted opposite its boiler in the vessel wall. The circulators are constant speed machines with a throttle on the inlet to vary the flow and allow individual boiler operation.

The primary circuit flow is unusual. To avoid the core rating being limited by levitation, the helium flow is downward through the core. It then flows up through the boilers, giving 'downhill' boiling on the secondary side, and

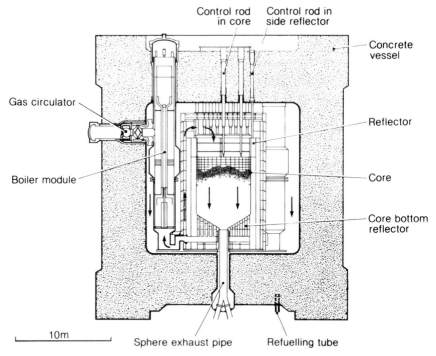

Fig. 3.36 Cross-section of the THTR

after leaving the blowers flows down the annular space outside the boilers, and back up through the thermal shield and side reflector to return to the core inlet plenum. The helium pressure is 40 bar, the core inlet and outlet temperatures about 250 °C and 800 °C. The turbo-alternator operates on the normal single-reheat cycle with inlet steam at 180 bar and 535 °C. Helium purification circuits taking a small flow from the primary circuit remove dust, hydrogen, moisture, carbon monoxide, and carbon dioxide impurities, and in a second low-temperature absorption stage remove fission product gases and methane and nitrogen.

Construction of THTR-300 took longer than planned. Much of the delay has been attributed to stringent licensing requirements imposed on a 'first of a kind'. Some back-fitting required by licensing is now going on, and the plant is in an advanced stage of assembly. It is planned to go into operation in the early 1980s.

Realization of the HTR in Germany has gone more slowly than expected. Plans to introduce a European version of the American 1160 MW(e) design for large steam-cycle plants fell through with events in the USA. Commercialization of a new reactor system to compete with established LWRs was seen as unprofitable. Development aims were therefore restructured[40] and are being pursued vigorously. Work is directed towards a basic reactor concept of a pebble-bed core and concrete vessel; the application targets are electricity production by direct-cycle gas turbine, and, increasingly, for process heat.

Development of the gas turbine plant with Swiss partners has proceeded over a number of years.[41] The objective is a 3000 MW(t) unit delivering 1240 MW(e) from a single turbine. This would be preceded by construction and operation of a smaller demonstration plant, and is supported by some major facilities including a turbine-compressor test loop.

The design output for the demonstration plant is 1640 MW(t); apart from this its technical parameters are intended to be identical to the 3000 MW(t) plant so as to permit extrapolation with minimum further development. It is based on a pebble-bed reactor continuously refuelled on the simpler 'once-through-then-out' cycle, with a single gas-turbine loop with recuperator and single inter-cooling between the low-pressure and high-pressure compressors (Fig. 3.37). The helium enters the turbine from the reactor at 70 bar and 850 °C, the expansion ratio is 2.9 and the low-pressure and high-pressure compressor inlet temperatures are 20 °C. At an efficiency of 41 per cent the unit delivers 675 MW(e) net. The primary circuit is integrated in a concrete vessel, of somewhat complex shape but having a warm liner to avoid insulation on the gas side, with the turbine set horizontally at the base and the alternator outside.

## 5 The past and the future of gas-cooled reactors

'Ideas won't keep: something must be done about them.' So said the creative English philosopher Alfred North Whitehead. The problem with gas-cooled reactors is that they are so versatile. There have been so many ideas leading to so many diverse designs as to make it impossible to exploit even a small fraction of them. As a result no single strong line of gas-cooled reactors has emerged world-wide.

The past 25 years have seen hundreds of concepts studied, experimental reactors of various types built in at least eight different countries, some 50 power reactors built. Of these, two systems have stood the test of time: the AGR in Britain, and the many-faced HTR still on its hard road to commercialization. The bid to introduce gas-cooled reactors in the USA and Germany just failed by a combination of circumstances rather than lack of enthusiasm on the part of the vendor or client. France has no advanced system to take over from Magnox. Few other countries had the skills or resources to pioneer; the easy option of the light water reactor developed by someone else has been irresistible.

All commercial nuclear power in the UK has so far come from gas-cooled reactors. Despite so many variations in design, the Magnox stations have made an important contribution to safe, cheap electricity. During 1976/7 Sizewell-2 for instance generated 3876 million units of electricity in a non-stop run of 653 days; this world record was terminated only by the need to

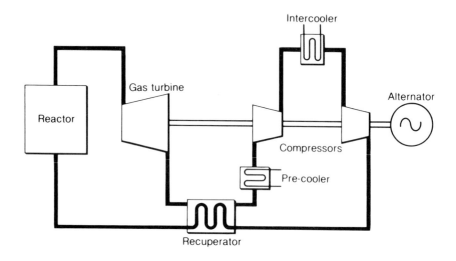

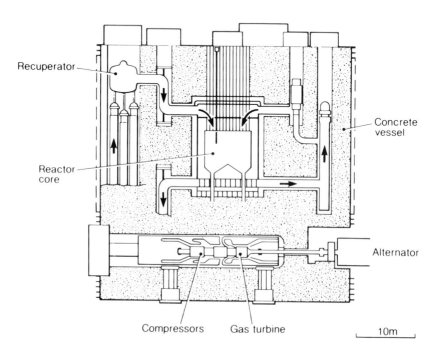

Fig. 3.37 Diagram and layout of proposed demonstration direct-cycle HTR

shut down the plant for its statutory two-year inspection. At the start of 1980 this small Sizewell station still led the world for total power generated at almost 50 billion kWh.[42] And at that time the nuclear station with the best performance in the world was Hunterston A with a cumulative load factor since commissioning of 83.5 per cent. The first of the AGR units coming on line in 1976 is still being worked up; it remains to be seen how well they live up to their promise, but the signs are now encouraging. They are important, for each new unit on stream at full power will save the country £1.5 million a week in fuel costs.

There is no shortage of ideas for developing the AGR further.[43] Direct-cycle gas-turbine operation, exploiting the high-density carbon dioxide and the change in physical properties around its critical point has been shown feasible.[44] The cost and time to develop the system, and the greater lead and potential of the direct-cycle HTR, must however now rule it out. If the AGR line is to be continued the first need is to consolidate and repeat the present design. After that there are a number of possibilities. The best value however might come from making the reactor more reliable, if need be at the expense of technical performance, particularly by simplifying the vessel internals. One possible way would be by eliminating the steam reheaters with their large volume flow penetrations through the vessel wall, and replacing the in-vessel gas-to-steam reheat by steam-to-steam reheat at the turbine. Another possibility is to cut out the re-entrant gas flow cooling the graphite moderator and replace it by simple upward flow in parallel with the fuel channels; this would avoid the need for the internal gas baffle and bring other possible simplifications, but at the expense of a degraded gas outlet temperature and some other minor problems.

This chapter has dealt only with thermal reactors. Gas-cooled fast reactors are a subject in themselves. Studies of gas-cooled fast breeders have been made over a number of years by many countries and international groups, using mostly helium cooling and uranium–plutonium oxide or carbide fuels in the form of metal-clad pins, cermets, or coated particles. All the ingredients of the ultimate reactor system are there: a low fissile inventory, high breeding gain, a coolant which is transparent, not radioactive and (at least nominally) inert, a direct gas-turbine cycle if desired, and low capital costs. Its realization however lies only at the end of a long development programme. Further, although the gas-cooled fast reactor has certain advantages over its sodium-cooled sister, stringent safeguards are needed to protect it against the risk of sudden depressurization; this is enhanced by the absence of a heat-absorbing moderator and by the high gas pressures chosen to give competitive high core ratings and low fissile inventory. A core of metal-clad pins is the currently preferred solution. And with this the engineering advantages of carbon dioxide cooling could in the longer term become of interest again—lower pressures than with helium, more clearly understood coolant-materials chemistry, and of course the large technological base of the AGRs. The way ahead could be eased by going for lower-rated designs as a first step, although these would not match the uranium conservation of other breeder systems. In the USSR work has been done on a fast reactor concept using nitrogen tetroxide in a gas-liquid direct cycle with a top temperature of only 450 °C, making use of the chemical heat of dissociation and recombination of the nitrogen tetroxide to give intense heat transfer in the core and heat exchangers, and of its low latent heat to re-evaporate it in the recuperator before re-entering the reactor.

In the family of gas-cooled thermal reactors however, it is the many-faced HTR that has the unique potential, but most enigmatic future. It has often been said that the HTR is but a step in the evolution of gas-cooled reactors through Magnox and AGRs. This is true in the general engineering. But the change in coolant is a great divide: helium opens up possibilities in fuel and high temperatures denied with carbon dioxide. The HTR characteristics look impressive: replaceable moderator, high safety and low man-rem doses, flexible fuel cycles, limitless power cycling.

Its long-term attractions lie in the gas-turbine for electric power, in process heat and combined heat and power, in power generation without cooling water using a direct cycle with high-temperature heat rejection to the air, possibly even in direct-cycle conversion by magnetohydrodynamics. The high-conversion fuel cycle of uranium–thorium HTRs could also be important. They would not provide a real alternative to the fast reactor as they would need at least an initial charge of enriched uranium. But if for any reason fast reactors failed to arrive, the HTR could have a major role in conserving uranium supplies. The HTR seems increasingly unlikely now, however, to come in for steam–electric use, simply because of the cost and time needed to launch it in the face of established systems. Yet without that prospect, continuity of work on it would be lost. Development of the reactor to the commercial stage together with full-scale thorium and uranium$^{-233}$ reprocessing and fabrication, is likely to take two decades and cost several billion pounds. Two things would help it succeed: international co-operation and concentration on a single design. The trend of gas-cooled reactors to become more diverse and complex needs reversing. As Antoine de Saint-Exupery observed in a different context: 'It seems that perfection is attained not when there is nothing more to add, but when there is nothing more to take away.'

## Acknowledgements

The author is indebted to the following organizations for permission to reproduce illustrations: The IAEA for Figs 8, 23, 25, 27, 32, 34, NPC for Figs 12, 13, 18, Strachan and Henshaw for Fig. 20, General Atomic Company for Fig. 24, Pergamon Press Limited for Fig. 26, Nuclear Engineering International for Fig. 31, Hochtemperatur Reaktorbau GmbH for Fig. 35.

# References

1. HINTON, C. The graphite-moderated gas-cooled pile and its place in power production. In *Proc. Int. Conf. on the Peaceful Uses of Atomic Energy, Geneva, August 1955*, vol. 3, pp. 322–9. United Nations (1955).
2. MOORE, R. V. AND GOODLET, B. L. The 1951–53 Harwell design study. *Br. Nucl. Energy Conf.* **2**(2), 47–60 (1957).
3. MOORE, R. V. The design and construction of the plant. *Br. Nucl. Energy Conf.*, **2**(2), 61–82 (1957).
4. *A programme of nuclear power.* Cmd. 9389 (1955).
5. INTERNATIONAL ATOMIC ENERGY AGENCY. *Directory of nuclear reactors*, vols. IV; VII; IX; X. IAEA, Vienna (1962).
6. SECRETARY OF STATE FOR ENERGY. Statement to Parliament. *Hansard*, **967**(12), col. 187 (1979).
7. DIXON, F. AND SIMONS, H. K. The Central Electricity Generating Board's nuclear power stations: a review of the first 10 years of Magnox reactor plant performance and reliability. *J. Brit. Nucl. Energy Soc.*, **13**(1), 9–38 (1974).
8. PEPPER, R. B. AND SHORT, A. *CEGB nuclear power stations radioactive waste discharges, associated environmental monitoring and personal radiation doses during 1975. NHS* Report 137, Central Electricity Generating Board Nuclear Health and Safety Department (1976).
9. HEALTH AND SAFETY EXECUTIVE. *Health and safety; nuclear establishments 1977–78.* HMSO (1979).
10. PRIOR, W. J. HURTIGER, J., AND INSCH, G. M. Gas-cooled reactor operating experience. *In Nuclear Energy Maturity, European Nuclear Conference, Paris, April 1975*, Plenary Sessions, pp. 217–22. Pergamon, Oxford (1976).
11. RASTOIN, J. AND BRISBOIS, J. Gas-cooled reactor experience and programs in France. *Ann. Nucl. Energy*, **5**, 455–88 (1978).
12. IAEA. *Directory of Nuclear Reactors*, vol. IV, pp. 227–32. IAEA, Vienna (1962).
13. IMAI, R. The Dungeness B appraisal. *Nucl. Engng*, **10**(113), 379–80 (1965).
14. MARSHAM, T. N. AND THORN, J. D. Economic power from gas-cooled reactors. *Nucleonics*, **23**(11), 39–44 (1965).
15. *Report from the Select Committee on Science and Technology, Session 1966–67.* United Kingdom Nuclear Reactor Programme (1967).
16. WARNER P. C. *The nuclear power station at Dungeness B site. In International Nuclear Industries Fair (Nuclex 66), Basle. Technical Meeting no. 3/9* (1966).
17. Special survey of Hinkley Point B. *Nucl. Engng*, **13**(147), 654–68. (1968).
18. Special survey of Hartlepool AGR. *Nucl. Engng. Int.* **14**(162), 973–95 (1969).
19. HUGHES, J. W., FURBER, B. N., LAING, G. W. AND ARMSTRONG. E. Insulation design and development for the Oldbury vessels. In *Proc. of Conf. on Prestressed Concrete Pressure Vessels, London. 13–17 March 1967*, pp. 703–13. Institute of Civil Engineers (1968).
20. MCINERNEY, E. T. AND WESTERMAN, M. Commissioning and operation of Hinkley Point B AGR. In *Proc. of Conf. on the Construction and Operations of Gas-cooled Reactors 26–27 May 1977*, pp. 47–45. Institution of Mechanical Engineers (1977).
21. PASK, D. A. AND HALL, R. A. Review of the commissioning and operational experience of the Hinkley Point 'B' AGR after three years' power operation. *Nucl. Energy*, **18**(4), 237–47 (1979).
22. SOUTHWOOD, J. R. M. The engineering development of thermal reactors in the UK. *Proc. Instn. Mech. Engrs.* **192**, no. 34 (1978).
23. SECRETARY OF STATE FOR ENERGY. Thermal reactor policy. *Hansard*, **942**(45), col. 1391–1408 (1978).

24. SMITH, D. R. AGR design for Heysham II and Torness. *Nucl. Energy,* **18**(4), 251–59 (1979).

25. BOYER, V. S., INSCH, G. M., LANDIS, J. W., HENNIES, H., MULLER, H. W., SHEPHERD, L. R., AND THORN, J. D. High Temperature Reactors. In *Nuclear Energy Maturity, European Conference, Paris, 25 April 1975, Plenary Sessions,* pp. 223–49. Pergamon (1976).

26. MALHERBE, J., *Design of a nuclear steam reforming plant.* In *Specialists' meeting on process heat applications technology, Julich, November 1979.* IAEA (preprint).

27. PRUSCHEK, R. *Basic layout, arrangement and design of heat transfer components of the nuclear coal gasification prototype plant.* In *Specialists' meeting on process heat applications technology, Julich, November 1979.* IAEA (preprint).

28. NAKANISHI, T. *Development of high-temperature heat exchanger on heat utilization from VHTR.* In *5th International Nuclear Industries Fair (Nuclex 78), Basle. Technical Meeting no. A4/17* (1978).

29. ORGANIZATION FOR EUROPEAN ECONOMIC CO-OPERATION. *Dragon High-Temperature Reactor Project. First annual report 1959–1960.* European Nuclear Energy Agency (1961).

30. HUDDLE, R. A. U. Fuel elements for high-temperature reactors: basic materials philosophy of the Dragon programme. In *Proc. Symp. on Advanced and High-Temperature Gas-Cooled Reactors. Julich, 21–25 October 1968,* pp. 631–45. IAEA (1969).

31. LOCKETT, G. E. AND HOSEGOOD, S. B. Engineering principles of high-temperature gas-cooled reactors. In *Proc. Symp. on Advanced and High Temperature Gas-cooled Reactors, Julich, 21-22 October 1968,* pp. 155–79. IAEA (1969).

32. SMITH, D. R., BOHN, E., STORRER, J., AND ACCARIARI, P. Prospects for the HTR with prismatic fuel. In *Proc. Symp. on Advanced and High-temperature Gas-Cooled Reactors, Julich, 21–25 October 1968,* pp. 181–95. IAEA (1969).

33. BROWN, G., BRANSON, P. B., AND MILLS, C. P., Design of a low-enrichment homogeneous, helium-cooled reactor with off-load refuelling. In *Proc. Symp. on Advanced and High-Temperature Gas-Cooled Reactors, Julich, 21–25 October 1968,* pp. 197–221. IAEA (1969).

34. EVERETT III, J. L. AND KOHLER, E. J. Peach Bottom unit No. 1 a high-performance helium-cooled nuclear power plant. *Ann. Nucl. Energy,* **5** 321–25 (1978).

35. HABUSH, A. L. AND HARRIS, A. M. 330–MW(e) Fort Saint Vrain high-temperature gas-cooled reactor. *Amer. Nucl. Soc. Trans.,* **10**(1), 320–21 (1967).

36. WALKER, R. F. Experience with the Fort Saint Vrain reactor. *Ann. Nucl. Energy,* **5,** 337–356 (1978).

37. WIMMERS, M., IVENS, G., LEUSHACKE, D. F., NICKEL, H., AND HUSHA, H. *The AVR–a test bed for high-temperature reactor fuel ball development.* In *5th International Nuclear Industries Fair (Nuclex 78), Basle, Technical Meeting no. A1/5* (1978).

38. BOHM, E., EHRET, A., GEPPERT, H., HAUCK, W., AND KLUSMANN, A. The 25–MW(e) Geesthacht KSH nuclear power plant. In *Proc. Symp. on Advanced and High Temperature Gas-Cooled Reactors. Julich, 21–25 October 1978,* pp. 109–20. IAEA (1969).

39. MULLER, H. W., Design features of the 300 MW THTR power station. In *Proc. Symp. on Advanced and High-Temperature Gas-Cooled Reactors. Julich, 21–25 October 1968,* pp. 135–53. IAEA (1969).

40. FISCHER, P. U. AND OEHME, H. *High-Temperature gas-cooled reactors—experience, development, status and outlook.* In *5th International Nuclear Industries Fair (Nuclex 78), Basle. Technical Meeting, Plenary Session A4* (1968).

41. ARNDT, E., HAFERKAMP, D., HODZIC, A., SCHNEDIER, K. U., BIELE, B., AND SARLOS, G., Development of a plant concept for a demonstration power plant

with high-temperature reactor and helium turbine (*HHT*). In *5th International Nuclear Industries Fair. (Nuclex 78), Basle. 3–7 October 1978. Technical Meeting A2-11* (1978).

42. HOWELLS, L. R. Nuclear station achievement. *Nucl. Engng Int.* **25**(296), 48–50 (1980).

43. DENT, K. H. Where next AGR? *Nucl. Energy* **18**(4), 261–65 (1979).

44. KRONBERGER, H., MAILLET, E., AND COAST, G. Integrated gas turbine plants using carbon dioxide cooled reactors. In *Proc. ENEA Symp. on the Technology of Integrated Primary Circuits for Power Reactors, Paris, 20–22 May 1968* paper 1/45. Organization for Economic Co-operation and Development (1968).

# 4

# Light water reactors

J. G. COLLIER

The light water reactor was a relative latecomer to the nuclear scene, and the first LWR was taken critical only in 1953. Since that time, however, the LWR has developed into the dominant system for commercial nuclear power generation and represents the mainstay of thermal reactor programmes in a majority of countries around the world. This chapter looks at the use of pressurized water as a reactor coolant and traces the two major lines of development that have occurred, leading on the one side to the pressurized-water reactor (PWR) and on the other to the boiling-water reactor (BWR). As well as looking in detail at a typical modern PWR and BWR, to illustrate the main characteristics of the two systems, this chapter also considers a number of design variants that have emerged in the course of commercial development.

The use of a highly rated core and of pressurized or boiling water to cool the reactor has raised important safety and reliability issues, and these are discussed in detail. The final section of the chapter looks forward to possible future developments in light water reactor technology, including the use of plutonium recycle and of a $^{232}$Th-$^{233}$U breeder cycle.

## Contents

# 1   Introduction

This chapter is concerned with light water moderated and cooled nuclear power reactors (LWR). Because they require uranium enriched in the $^{235}U$ isotope they were not amongst the first reactors to be developed. However, they now comprise the majority ot the world's commercial nuclear power stations. After a brief introduction to the general principles of light water reactors, §§3 and 4 trace the development paths of the pressurized-water reactor (PWR) and the boiling-water reactor (BWR) up to the point where the first demonstration plants were operating in the United States, while the latter part of each section deals with the first commercial plant of its type. Typical modern PWR and BWR plants are described in §5 and §6, respectively, and some of the many variations in the design of components and systems are considered. Section 7 deals with safety questions raised by the highly rated water reactors, while §§8 and 9 consider the performance record of the light water reactor and look forward to possible developments over the next twenty or thirty years.

Before considering modern developments, it may be worth glancing back for a moment to prehistoric times. Light water cooled and moderated reactors have been around for rather longer than one might at first imagine. In fact, in the design of such reactors man was not an innovator but an unwitting imitator of nature. In 1972 evidence was found of the dormant remains of a natural fission reactor located at Oklo, in the West African Republic of Gabon. This natural reactor operated about 2 billion years ago for a period extending over hundreds of thousands of years. At that time the relative abundance of $^{235}U$ compared with $^{238}U$ in natural uranium ores was much higher than the present day level of 0.72 per cent. This is because the *half-life* of $^{235}U$ is about 700 million years compared with that of $^{238}U$ which is about 4.5 billion years. When the Oklo reactor was operating the uranium ore contained about 3 per cent $^{235}U$, sufficient to allow a light water reactor to function. Since the reactor was at that stage buried deep underground, the natural ground water, which served both as a moderator and to some extent as a coolant, was under considerable, probably supercritical, pressures. It was also at high temperatures and control of the fission reaction was probably via the significant variation of water density which occurs in the temperature range 350–400 °C. Cooling was provided mainly by conduction with some limited circulation by permeation. The power level was something less than 100 kW(t) and the total energy released was some 15 000 MW(t), representing the fission of about 6 tonnes of $^{235}U$. This amount of energy is about that released in a large modern PWR in just four years.

## 2. The use of light water as a reactor coolant

From the earliest days of the nuclear programme there has been an interest in using ordinary ('light') water as a reactor coolant. The reasons for this are not difficult to understand. Chapter 1 of this book considers the physical characteristics of an effective reactor coolant, and indicates that these include high specific heat (to reduce the rate of mass flow), high density (to reduce the rate of volumetric flow), and low viscosity (to reduce the pumping power required). It is convenient to express the combined effect of these parameters of specific heat at constant pressure, $C_p$, density, $\rho$, and viscosity, $\mu$, by means of a *figure of merit* equal to

$$\frac{C_p^{2.8}\rho^2}{\mu^{0.2}}.$$

Evaluating this parameter for light water, as compared with other coolants, it is found that, in the pressure range of concern in reactor applications, light water offers a figure of merit much higher than that of sodium (because of the better heat *transport* capability) and much higher than $CO_2$ at the same pressure. Light water has moreover the advantage of being easily available, cheap, and, providing any impurities in it are carefully controlled, chemically compatible with core and boiler materials.

The major limitation on the use of water to cool reactors lies in its low boiling point at normal atmospheric pressure. While flowing water is an excellent coolant, steam is evidently not, and water at atmospheric pressure could thus clearly not be used to cool any system operating with a temperature above 100°C. This is, of course, much too low a temperature for a useful heat engine, and would not allow a Carnot thermal efficiency of much above 15 per cent. To allow water to be used in a heat engine operating at a respectable thermodynamic efficiency it is necessary to suppress or limit boiling and maintain the coolant in its single-phase form.

This can be achieved by increasing the pressure of the water, and all applications of light water as a reactor coolant require this overpressurization. Thus, while there is a conventional distinction between the two types of light water reactor, the pressurized-water reactor and the boiling-water reactor, this, to an extent, is a semantic point. Both operate at a pressure well above atmospheric and differ only in the degree of *overpressure* they employ; this aspect will be returned to later in this section.

Essentially, the pressure above which the system needs to be operated to maintain the coolant in a single-phase liquid state is a function of the core temperature. The *saturation* pressure varies with temperature, as shown in Fig. 4.1. The figure shows the familiar *saturation* curve which is of major importance in thermohydraulic and safety studies. The need for high pressures at elevated temperatures carries with it a penalty in terms of more stringent structural integrity requirements, and a balance therefore needs to be struck between considerations of thermodynamic efficiency (which favour increasing the coolant temperatures and pressures) and structural considerations (which favour reducing them). In practice a sensible compromise temperature has been found to be around 300 °C, around half the temperature that is encountered in gas-cooled reactors such as the AGR. These relatively low temperatures in water reactors have some important advantages, allowing, for example, the use of cheaper materials to compensate for their lower efficiency.

At the same time, however, the LWR is, in thermal terms, a highly rated

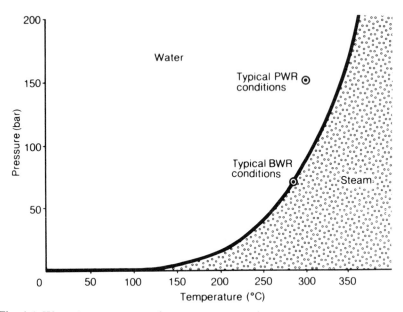

Fig. 4.1 Water temperature and pressure at saturation

system. For reasons of reactor physics the LWR requires a compact core and, in order to help reduce fabrication problems, designers have tended to favour the use of relatively small reactor vessels. In practice LWRs have been designed with a volumetric heat production rate of about 50–100 kW/ litre, the lower figure being appropriate to a BWR, and the higher to a PWR. (By comparison, an AGR has volumetric heat production rate of 2.5–3 kW/litre.) The higher rating of the LWR core and the small heat sink immediately available in the primary circuit under accident conditions, mean that very careful consideration needs to be given to thermohydraulic aspects, and these are discussed in detail in §7.

Any nuclear system operating with uranium at or near natural enrichment requires the use of a moderator (a question more fully discussed in Chapter 1). Light water in terms of its ability to slow down neutrons is an excellent moderator, and the majority of reactor designers considering the use of light water coolants have chosen the simple and attractive solution of using light water as the moderator also. This is the solution which all commercial PWRs and BWRs have adopted. However, light water has a relatively high capture or absorption cross-section and therefore cannot be used with natural uranium fuel. This design concept therefore calls for the use of slightly enriched uranium and the establishment of enrichment facilities. In order to avoid the need for this, some reactor designers have aimed to use light water coolants in association with other moderators, such as graphite (which has a moderating ratio five times that of light water) or heavy water (with a moderating ratio 160 times higher), and these principles underlie, respectively, the Russian graphite-moderated light-boiling-water reactor (briefly mentioned in §5) and the British steam-generating heavy water reactor (SGHWR), (discussed in Chapter 6). In general, however, such alternative approaches have had only a small impact world-wide, and reactor designers and users have considered the use of slightly enriched uranium as more convenient.

As already indicated, two basic lines have emerged in LWR development, this split in design concept centring mainly on the choice of primary circuit pressure. The more widely adopted PWR system involves using a higher primary circuit pressure (say 150 bar at 300 °C), maintaining the entire reactor coolant circuit at a temperature somewhat below the saturation value, and using an indirect cycle to transfer heat from the primary circuit to the turbine, via steam generators (pressurized-water – boiling-water heat exchangers). The primary water circulates through tubes whose outer surfaces are in direct contact with a second stream of water, the feedwater, returning from the turbine condenser. This feedwater is evaporated to form steam to drive the turbo-generator, the feedwater – steam circuit being termed the *secondary* circuit. Figure 4.2 shows a simplified circuit diagram of the PWR. An attractive feature of this system is that only non-radioactive steam passes through the turbine and condenser.

An alternative concept developed slightly later involves operating the primary circuit at the saturation pressure (70 bar at 290 °C), thus allowing limited boiling to occur within the reactor core, so that steam can be passed directly from the reactor to the turbine. This concept, which leads to the idea

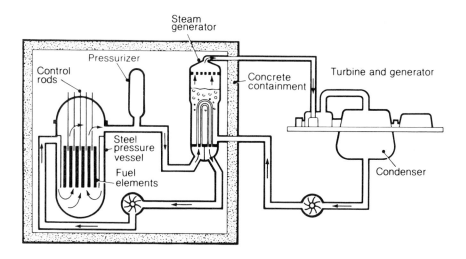

Fig. 4.2 Pressurized Water Reactor (PWR): (schematic)

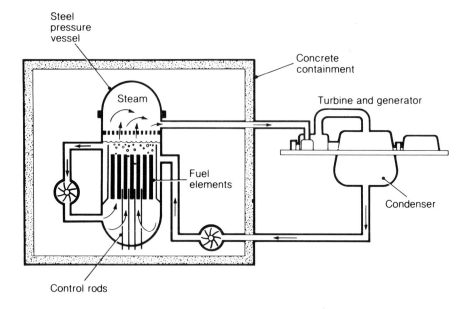

Fig. 4.3  Boiling Water Reactor (BWR): (schematic)

of the direct-cycle BWR (Fig. 4.3), has the advantage of simplicity, though this is an advantage which is purchased at the cost of having to deal with radioactive steam in the turbine, placing a high premium on the integrity of the steam circuit.

In addition to the nuclear boiler and turbo-generator discussed above, important elements within both systems are the coolant pumps and pressure-control systems. In the PWR system, primary circuit pressure is regulated by means of a pressurizer, an independently heated pressure vessel in equilibrium with the primary circuit; in the BWR circuit pressure is controlled by variation of reactor power and by power-operated relief and safety valves in the steam lines. These control systems are further described in §§ 5 and 6.

The next two Sections, §§ 3 and 4, will show how these two basic concepts of LWR have evolved over the past 35 years, leading to the PWR and BWR in their standard present-day forms.

## 3  The development of the pressurized-water reactor

### 3.1  The early years

The story of the successful project to produce atomic weapons during the Second World War, the Manhattan Project, has been recounted many times. The demonstration of a controlled nuclear fission reactor by Enrico

Fermi, at Stagg Field, Chicago in December 1942, using a graphite-moderated pile, opened up a viable route to the production of plutonium from natural uranium. Britain, Canada, and the United Stages pooled their resources on these military projects. A joint British–Canadian team worked on heavy-water-moderated reactors in Canada, whilst in the United States large plutonium-production reactors were constructed at Hanford in Washington State and at Savannah River in South Carolina. The reactors at Hanford were graphite-moderated and cooled by light water taken from the Columbia River, whilst those at Savannah River used heavy water contained in a large tank as both moderator and coolant. These early reactors provided much of the basic technology on water reactors.

At the end of the war there was interest in the United States in using nuclear reactors (or 'atomic piles' as they were then called) for electrical power generation, and a committee was set up to investigate the possibilities. The subsequent project, known as the Daniels Pile was, however, abortive.

At about this time the then US naval captain Hyman G. Rickover recognized the inherent advantages of nuclear reactors for ship and, especially, submarine propulsion. With a determination and a single-mindedness which is now part of the modern folk-lore, Rickover convinced the initially sceptical US Navy that nuclear-powered submarines would have unique capabilities, obtained the necessary authorization and resources, and set about developing and building the propulsion system.

## 3.2 The submarine thermal reactor (STR)

Rickover's project became known as the Naval Reactors Programme. It called for the development of a compact, high-performance reactor. Two separate routes were followed. General Electric received a contract to develop and build a liquid-sodium-cooled system. This ultimately progressed to the stage where such a reactor was installed in the US submarine *Sea Wolf* and saw a number of years of service. However, because of the basic unsuitability of sodium in a marine environment this system was discontinued for sea-going naval vessels.

The second route involved the development of a pressurized-water reactor which became known as the submarine thermal reactor (STR). Westinghouse was given the contract to build the STR, and in 1948 formed its Atomic Power Division (WAPD) at Bettis, Pittsburgh. The STR was a joint effort between the United States Atomic Energy Commission's (USAEC) Naval Reactors Branch, Bettis, and the Argonne National Laboratory.

The range of development work required for the STR was exceptionally broad and was undertaken at a phenomenal rate. Materials such as uranium dioxide fuel and zirconium fuel cladding were developed. Enormous strides were made in the areas of reactor physics, shielding, thermohydraulics, instrumentation, and water chemistry. Perhaps, however, the greatest contributions were in the training of scientists and engineers in these various disciplines, in the development of fabrication and inspection techniques (the foundation of quality assurance), in the constructional experience, and in the commissioning and operation of the plant to the stringent requirements of the US Navy and the USAEC. This achievement owed much to Admiral Rickover personally, who recruited and trained a team of high ability, and who demanded and obtained the very highest qualilty of manufacture to satisfy the standards he specified.

By 1949 the basic features of the STR had been arrived at, and it was decided to construct two plants: STR1, a land-based prototype at the National Reactor Testing Station at Idaho, and STR2, to be installed in the submarine *Nautilus*. STR1 was located in a mock-up of a submarine hull in a large water tank 15.2 m wide and 12.2 m deep. It went critical on 30 March, 1953 and reached full power on 25 June, 1953. *Nautilus* was launched in 1954, and STR2 became the first of a large number of naval propulsion reactors. By 1980 the US Navy had operational naval propulsion reactors in 125 submarines, with a further 21 nuclear submarines under construction, and a number of surface vessels including the 8-reactor *Enterprise* aircraft carrier and the 2-reactor *Nimitz*. A nuclear deep-submergence vessel, the NR-1, has been operational since the early 1970s. A broadly similar number of nuclear submarines have been put in operation by the USSR.

Britain has 15 nuclear submarines with reactors developed by Rolls Royce and Associates based on the same technology with a further submarine of the Swiftsure class under construction. France has developed and put in operation four nuclear-powered submarines, and a fifth, considerably larger submarine is scheduled for completion in 1985. Additionally, there are unconfirmed reports of active development work on nuclear submarines in China.

### 3.3  The Shippingport reactor[1]

In 1953 with the STR project nearing a successful conclusion, attention was turned to other possible applications for nuclear power. Consideration was given to a large military pressurized-water reactor for a naval ship, but this was rejected in favour of the construction of a civil power plant based on the technology acquired during the STR programme. This project was authorized in July 1953, and Admiral Rickover was given responsibility for the project; Westinghouse was the chosen contractor. Proposals were solicited from electric power utilities, and from the nine received by the USAEC, that of the Duquesne Light Company, Pittsburgh was selected. Duquesne provided a site at Shippingport, on the Ohio river about 40 km west of Pittsburgh and contributed to the cost of the plant.

The electrical output of the plant was initially set at 60 MW(e). It was clear that whilst the technology developed during the STR programme could be carried over, a straightforward scale-up of the STR plant was not consistent with the objective of minimum capital and operating cost. Many of the design aspects had to be considered anew. The basic solutions chosen have stood the test of time remarkably well. Only in one major aspect—the core design—and two lesser aspects—the detailed design of the steam generators and the method of reactivity control—did Shippingport fail to anticipate the future technology of the PWR.

The plant basically comprised the reactor vessel, housing the core and four main coolant loops. The reactor vessel (Fig. 4.4) was cylindrical with an inside height of 9.5 m, a diameter of 2.8 m, and a nominal wall thickness of 210 mm. It was fabricated from SA 302 Grade B steel plates and forgings and was clad internally with SA 304 stainless steel cladding. For the vessel shell and bottom hemispherical head, the cladding was roll-bonded. The top head could not be roll-bonded with existing equipment however, owing to its 250 mm thickness, and the cladding was instead deposited by machine welding. The four inlet nozzles penetrated the bottom hemispherical head, whilst the four outlet nozzles were located near the middle of the vessel. The hemispherical closure head was flanged for core replacement although ports were incorporated in the head for the replacement of individual fuel assemblies.

The choice of four coolant loops was a compromise based on considerations of component size and upon the plant output lost with one loop out of service. The system pressure was set at 138 bar, which again was a compromise between allowing as high a secondary steam pressure and temperature as possible and ensuring that the fuel element clad temperature remained below 335 °C, the maximum considered allowable with the then known behaviour of the Zircaloy fuel cladding.

It was originally proposed that the core would be based on uniform slightly enriched uranium. In the event no simple inexpensive fuel element of this type was available at that time, and the so-called *seed*-and-*blanket* concept was adopted. This method of construction, which involved a core consisting of a small region of highly enriched uranium (the *seed*) surrounded by natural uranium fuel elements (the *blanket*), had, at the time, a number of attractive features, although it was discarded for later plants. First, the seed region, which basically dictates the reactivity of the core, could, to a large extent, utilize the technology of highly enriched uranium cores developed for the STR. Secondly, the manufacture of the natural uranium blanket fuel could begin without the need to complete all the physics work required to select an enrichment value. Finally, the core could be refuelled by replacing only the relativity few *seed* fuel assemblies. The highly enriched uranium which formed the seed was contained in 1914 zirconium-clad plates, the active portions of which were 52 mm wide and 1.8 m long. The *blanket* utilized natural uranium in the form of $UO_2$ pellets, stacked in a Zircaloy tube 10 mm diameter and 260 mm long. These pins were mounted in an $11 \times 11$ array in end plates and a stack of seven such bundles made up a *blanket* fuel assembly. A target burn-up of 3000 MW d/tonne ($\approx 8800$ effective full-power hours (EFPH)) was aimed for.

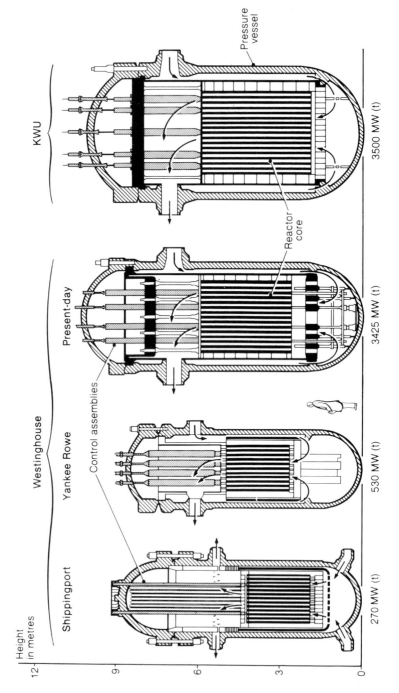

Fig. 4.4 Examples of PWR pressure vessels

Reactivity control was by 32 hafnium control rods, one rod for each seed assembly. The control rod, in the form of a cruciform blade, travelled in the gap formed between four seed subassemblies within the fuel element. The control rod mechanisms were housed on the reactor vessel head and employed a rotating nut and translating screw device which operated in the reactor coolant water.

The Shippingport reactor went critical on 2 December 1957 and reached its full rated power of 60 MW(e) just 21 days later. The power was later increased to 90 MW(e) and the plant remains in operation even now as a test bed for the light water breeder concept (LWBR) which is described in § 9.

### 3.4 The evolution of the pressurized-water reactor

In 1956, even before Shippingport went into operation, the US electric utilities were sufficiently interested in the potential of nuclear power to place orders for the first commercial plants. Consolidated Edison placed a contract with US Babcock & Wilcox for a 265 MW(e) plant at Indian Point and Westinghouse received a contract to build a 4–loop, 175 MW(e) plant, Yankee Rowe. The reactor vessel for Yankee Rowe (Fig. 4.4) was about the same size as the Shippingport vessel, but instead of using the seed and blanket core design this reactor utilized low-enriched uranium oxide fuel clad in stainless steel. The fuelled section of the core was 2.3 m long with the fuel pins assembled in square clusters. The cruciform control rods operated in the gap formed between four subassemblies. The maximum linear rating was quite modest at 32.5 kW/m but the mean core density surprisingly high at 89 kW/litre. This plant also saw the introduction of the vertical inverted U-tube steam generator (see Fig. 4.10) with stainless steel tubes and the use of soluble neutron poisons in the coolant to compensate for both short-term and long-term reactivity changes. Thus Yankee Rowe completed the introduction of the major features of modern-day PWRs not foreseen in Shippingport. Table 4.1 shows the development of the Westinghouse PWR[2] over the 25 years since 1953.

At around this time, the first studies were undertaken in the UK into the feasibility of building pressurized-water reactors for commercial nuclear power generation. A feasibility study, code-named LEO, was initiated by the UKAEA in 1954, with support from English Electric, Babcock & Wilcox, and others, and reported in 1956. The conceptual reactor considered in this study was a 4-loop 440 MW(t) PWR equipped with a 147 MW(t) oil-fired superheating loop to feed superheated steam to two 80 MW(e) turbo-generators. The study assumed a core of length 3.28 m and 2.1–2.5 m diameter, enclosed within a pressure vessel of 2.6–3.0 m inside diameter, and considered a range of possible design variables, with regard to fuel material, enrichment, and cladding. The preliminary conclusions of the study noted the nuclear advantages of metallic fuel and of zirconium cladding. However, considerable economic advantage was seen in the use of stainless

Table 4.1 Westinghouse PWR product lines

| Product line | Year of first order | Characteristic plants and salient features |
|---|---|---|
| PWR/1 | 1953 | Shippingport: demonstration plant, four-loop plant, horizontal steam generators, seed-and-blanket core. |
| PWR/2 | 1956 | Yankee Rowe, Trino, Chooz, BR-3: initial commercial plants, four-loop plants, vertical U-tube steam generators (stainless steel tubes), uniform core low-enriched fuel, zoned core, stainless steel-clad fuel, use of soluble chemical poison for reactivity control. |
| PWR/3 | 1963 | Connecticut Yankee, San Onofre, Zorita: $15 \times 15$ canless fuel assemblies (stainless steel cladding), cluster control rods, Inconel Alloy 600 steam-generator tubing, variable number of loops. |
| PWR/4 | 1966–73 | Zion, Beaver Valley, Beznau: NSSS only, standardization (RESAR), $2-, 3-, 4-$loop plants, $15 \times 15$ Zircaloy fuel assemblies, upper head ECCS, ice containment. |
| PWR/5 | 1973–78 | South Texas, Sayago: increased power output compared with PWR/4 ($\approx 10\%$), $17 \times 17$ extra-long fuel assembly. |

steel and of oxide fuel, whose improved corrosion resistance and consequential higher burn-up were seen as more than compensating for the higher uranium investment. These studies therefore endorsed the general conclusions emerging from US experience at this time.

During the late 1950's Westinghouse initiated construction of the first PWR plants in Europe: Trino in Italy, BR-3 in Belgium, and Chooz in France. By 1963 nuclear plant sizes being ordered had increased to 500–600 MW(e). Connecticut Yankee was a 4-loop plant delivering 575 MW(e) and San Onofre in California, a 3-loop plant delivering 430 MW(e). This latter plant saw the introduction of *finger* or *cluster* type control rods operating within the $15 \times 15$ fuel pin assembly. This reactivity-control arrangement avoided some of the 'peakiness' associated with single large control rods, and helped to produce a flatter power profile across the core. Stainless steel was still being used as cladding material. San Onofre was also operated with plutonium-bearing fuel assemblies. The maximum linear rating in these plants had risen to 46–49 kW/m with mean core power densities of 66–92 kW/litre.

1965 saw a considerable increase in the rate of ordering nuclear plant in the United States and the need for standardization of the product line being offered. The modular concept was introduced with the plant size range 600–1200 MW(e) being covered by 2-, 3- and 4-loop plants. In plants of the same size, the layout of the primary circuit is identical as is the design of the major components such as the reactor pressure vessel, steam generators, pumps, pressurizer, fuel assemblies, and control rod drive mechanisms. For these plants the change from stainless steel to Zircaloy cladding of the fuel was made and the core length standardized at 3.67 m. Peak linear ratings in the 15 × 15 fuel pin array were around 59 kW/m with mean core power densities of 85 kW/litre.

In Europe PWRs were being constructed by Siemens in the Federal Republic of Germany, originally under a Westinghouse licence. However in 1969 Siemens merged its nuclear power interests with Allgemeine Elec-tricitäts–Gesellschaft (AEG) who were already building boiling-water reactors in Germany under licence from General Electric. The combined company Kraftwerk Union (KWU) now sells both PWRs and BWRs and, in particular, has developed its own unique design of PWR further described in §5. About this same time France decided to adopt light water reactors for its future nuclear programme. A company, Framatome, was set up by the French Government in association with French and US industry, to construct PWRs under licence from Westinghouse. The standard 900 MW(e), 3-loop design was adopted for the majority of the French units and a total of 32 almost identical units are either in operation, being commissioned or under construction.

As has been seen, peak linear ratings of PWR fuel pins increased over the first decade of development by almost a factor of two. Additionally, improved control methods (soluble neutron poisons and cluster control rods) allowed for a flattening of the power profile across the core and for a lower peak/mean ratio (*form factor*), so that the core could be operated at powers nearer to the metallurgical limit on individual pins. This again made for high mean core ratings.

At the same time this increase in output was not matched by an increase in vessel size, and revised vessel design codes permitted higher stress levels. By 1972 concern in the United States about the efficiency of the emergency core cooling systems on both the PWR and BWR was growing. The licensing rules which appeared as a result of the 18-month USAEC hearings on this question required a limitation in the power outputs of some of the more highly rated PWR plants.

The response of Westinghouse and the other PWR vendors was to intro-duce fuel assemblies with the fuel more subdivided. Thus Westinghouse replaced the 15 × 15 array by a 17 × 17 fuel subassembly. This allowed the same core power density whilst reducing the peak linear rating down to 39–43 kW/m.

More recently, by lengthening the core to 4.3 m a 10 per cent increase in overall power has been possible without any increase in reactor pressure vessel dimensions and with only limited modifications to reactor vessel internals.

### 3.5 **Marine reactor systems**

From the earliest stage of the nuclear programme, there has been interest in applying nuclear power in surface marine propulsion. Some success has been obtained in this area, and the limitations here have in the last analysis proved to be concerned much more with economics than with questions of technical feasibility.

Virtually every major line of reactor development has at one time or another been associated with a proposal for its use as a marine system. Design studies have been carried out for nuclear ships involving gas-cooled reactors and both main types of light water reactor. Of all these however, only the PWR has in practice been adopted for surface marine propulsion, and all nuclear ships currently operating employ PWRs as their power source.

The first practical success in the civil nuclear propulsion field was obtained by the USSR in the late 1950s with the development of a nuclear-powered icebreaker, the *Lenin*, which was first taken into service in December 1959, as an escort vessel for merchant convoys on the Northern Sea Route. The *Lenin* is powered by two, 90 MW PWRs, with a third reactor in reserve. The core contains 37 fuel assemblies, arranged in a core 0.9 m diameter and 1.4 m high. Other important parameters of this and other marine reactors are shown in Table 4.2. The *Lenin* has demonstrated its capabilities over many years of service, in ice conditions which have hitherto proved impassable to conventional icebreakers, reportedly sustaining speeds of up to 3.7 km/h in 2 m ice. In ice-free waters, the *Lenin* has achieved cruising speeds of 33 km/h, in common with other large nuclear-powered vessels. Two other nuclear icebreakers have since entered service in the USSR.

Nuclear marine developments occurred in parallel in the US, authorization being given in 1956 for the construction of a civil passenger/cargo vessel of deadweight 9990 t, the NS *Savannah*, the contract for the design and manufacture of the reactor and turbine plant being awarded to Babcock & Wilcox. Progress was rapid, and the *Savannah* reactor was first taken critical

Table 4.2 Marine reactor parameters

|  | Lenin | Savannah | Otto Hahn | Mutsu |
|---|---|---|---|---|
| First operation at power | 1959 | 1962 | 1968 | 1974 |
| Deadweight (te) | — | 9990 | 15 250 | 2600 |
| Reactor power (MW) | 2 × 90 | 74 | 35 | 36 |
| Primary circuit av. pressure (MPa) | 18.6 | 12.0 | 6.4 | 10.8 |
| Primary circuit av. temp. (° C) | 286 | 264 | 278 | 278 |
| Secondary coolant mass flow per loop (kg/s) | 45.4 | 32.0 | 17.8 | 17.0 |
| Core diameter (m) | 0.9 | 1.5 | 1.2 | 1.1 |
| Core height (m) | 1.4 | 1.7 | 1.1 | 1.0 |
| PV diameter (m) | 2.0 | 2.4 | 2.4 | — |
| PV height (m) | 5.0 | 8.1 | 8.6 | — |
| Weight of power plant (t) | 2950 | 2230 | 945 | — |
| Average core enrichment (%) | 5 | 4.4 | 3.6 | 3.8 |

in December 1961. In 1965 it was taken into service as a commercial cargo vessel, and continued to give good service in this role until 1970, when Government funding for the project was discontinued.

The *Savannah* was powered by a single 74 MW PWR, which like the *Lenin* follows a 'conventional' layout concept, the steam generators (which are of the recirculating inverted U-tube type) being located external to the reactor pressure vessel. A number of design changes were adopted in the course of early years of operation, most importantly involving modifications of the control rod drive system.

The development of marine PWRs since the early 1960s has concentrated on reducing the overall size and weight of the nuclear system, principally through a reduction in primary circuit pressures (leading to less stringent pressure vessel and containment requirements). Additionally, advantage was seen in the use of a self-pressurized primary circuit, and in the increased use of chemical reactivity control. These design objectives were incorporated in the consolidated nuclear steam generator (CNSG) design, proposed by Babcock & Wilcox in 1962. This differed from the *Lenin* and *Savannah* design of reactor in having an 'integral' layout, all primary circuit components (including pumps and steam generators) being included within the reactor vessel. A modified version of this design was used in the German vessel *Otto Hahn* which first operated at power in 1968. The *Otto Hahn*, of 15 250 tè deadweight, is powered by a 35 MW Babcock & Wilcox PWR with sixteen 226-pin fuel assemblies. Table 4.2 indicates the very much lower primary circuit pressures that are achievable in this design, and the consequential reduction in weight. In assessments carried out in the UK, the CNSG 'integral' layout has been considered as offering considerable advantage over the conventional layout concept, though its compact arrangement may cause difficulties in maintenance.

In Japan, a nuclear-powered vessel, the *Mutsu*, of deadweight 2600 te was brought into active operation in 1974, though it has since been shut down. The *Mutsu* is powered by a 36 MW Mitsubishi PWR, of 'conventional' design with primary circuit pressures lower than on the *Savannah* or *Lenin* though appreciably higher than on the *Otto Hahn*.

Despite early concern about possible problems of roll and shock, the difficulties experienced to date with nuclear ships have by and large been no more severe than those encountered in land-based reactors. Such difficulties as have occurred have been with control rod drives, and with water chemistry, in particular the avoidance of chloride stress-corrosion cracking problems (see § 8), which are more crucial in marine applications, and considerable effort has been deployed on chemistry control. Marine reactors have behaved well under power-cycling conditions, allowing power changes at a rate of 1 per cent per second. Man-rem doses have been low (averaging 200 man-rem/y on the NS Savannah) and well within prescribed limits.

Design studies for marine reactors have been carried out in a number of other countries, including Canada, France, Italy, Sweden, and the UK. In the UK, interest has focused from the early years on the *integral* design, development work during the 50s and early 1960s having centred on the joint Belgonucleaire–UKAEA Vulcain project, an early forebear of the spectral-shift-controlled reactor which is described in § 9. Development funding for several marine applications was reduced in the mid-60s, following the report of the Padmore Committee, and UK work in this field has been maintained since that time at a low level. In the years following the oil crisis however, there has been some reassessment of the long-term economic prospects of nuclear propulsion,[3] and more active consideration is likely to be given to this question over future years.

# 4. The development of boiling-water reactors

## 4.1  The Genesis of the boiling-water reactor[4]

The development of the boiling-water reactor started somewhat later than the pressurized-water reactor. In the early days of power reactor development up to the early 1950s only the non-boiling type of water-cooled reactor received serious attention. The general attitude towards the boiling-water-moderated reactor at that time was that the chances of its feasibility appeared small. As an example of this early attitude in respect of boiling in reactors, the Manhattan District Patent Officer wished to cover all possible types of reactors by patents, so he prepared an application *OSRD*-97 on 'cooling a reactor by latent heat of evaporation of a liquid' filed in the name of one of the principal scientists associated with the project. When apprised of this action however, this distinguished scientist asked that the application be withdrawn, so the Patent Officer wrote a letter withdrawing the application, stating in part: '…in view of the indication that no way had been worked out for controlling a reactor of the type under consideration, the case will be given an inactive status until such time as an operative device can be disclosed.'[4]

Strange as it may seem now, understanding of boiling heat transfer was very rudimentary at the end of the Second World War and there was little or no appreciation of just how much energy could be removed by the boiling process. During the period 1946–51 a considerable amount of research work was initiated first in Canada and subsequently at Massachusetts Institute of Technology (MIT) and the Argonne National Laboratory (ANL), which demonstrated that significant heat flux densities could be accepted with clad surface temperatures only a little above the boiling point. Much of this work was related to the safety aspects of the submarine thermal reactor (STR) and of the high-flux materials-testing reactors (MTRs) then under construction. There was concern that fission product decay heating might cause fuel melting if the water-circulating pumps failed.

There was a further and more important problem which concerned reactor designers at that time: what would happen if all the safety devices failed and the reactor was made supercritical and allowed to 'run away' without any artificial limitation of its power? In particular, how violent would be the energy release before eventual disassembly and shutdown was achieved?

Chapters 1 and 9 of this book discuss in detail the effects of power changes on reactivity. The most important ways a reactor may lose reactivity upon an increase in power are by Doppler broadening of the resonances of $^{238}U$, by expansion of the fuel, and by heating and expansion of the moderator. For water-moderated reactors it was expected that expulsion of the moderator from the reactor core by the formation of steam in the water-filled passages of the fuel element would be an important safety mechanism.

At the same time it was realized that the efficiency of this mechanism was strongly dependent on the rate of steam production. Light-water-moderated reactors have a neutron lifetime of around $10^{-4}$ seconds. A sudden increase in reactivity to 1 per cent beyond prompt criticality would therefore cause the reactor power to increase by a factor $1.01^{100}$, which is 2.7, in 0.01 s. If such reactivity excursions are to be terminated before melting of the fuel occurs, steam bubbles have to appear within a few hundredths of a second.

Transient experiments at ANL using water-filled metal tubes heated by the momentary passage of low-voltage alternating current indicated that this timescale could be readily achieved. The next stage was to put this finding to the test by observing experimental runaways of actual reactors. In early 1950, ANL submitted a proposal to the USAEC for a reactor experiment to be conducted at an isolated location. This location was originally Frenchman Flats in Nevada but was later changed to the National Reactor Testing Station in Idaho, and the reactor experiments were code-named BORAX (boiling reactor experiments). The purpose of the experiments was to answer the following questions:

1. whether steam formation would actually limit the extent of a reactor power excursion and so act as an inherent safety mechanism for light water reactors, and
2. whether a reactor in which boiling was occurring was controllable and could produce useful amounts of power.

## 4.2 **The BORAX experiments**

The first of the BORAX experiments known as BORAX-I was conceived as a method of establishing whether or not steam formation in the reactor would reduce reactivity and so allow the system to be self-regulating. BORAX-I was of very simple construction, consisting of an unpressurized tank containing the core which was made up of an adjustable number of MTR plate-type fuel elements. The reactor tank was contained in a larger shield tank which, in turn, was sunk part way into the ground. The shield tank was flooded with water when the reactor was shut down. The reactor was operated remotely from a control station.

During the summer of 1953 BORAX-I demonstrated not only that a boiling-water reactor has a high degree of inherent safety in its ability to shut itself down in the event of a reactivity excursion, but also that it could operate stably over a range of conditions. The final objective of the programme was however to test the system to destruction, and, after additional tests in the summer of 1954, BORAX-I was deliberately destroyed during a violent subcooled power excursion. The reactor was made supercritical and the power excursion melted most of the fuel plates. The resulting metal–water reaction burst the reactor tank and ejected the contents of the shield tank into the air (Fig. 4.5).

Shortly afterwards a new reactor, BORAX-II, was constructed, utilizing the same control equipment and many of the actual parts of BORAX-I salvaged after the explosion. The new reactor was capable of operation at 26 bar and, with twice the core volume of BORAX-I, delivered about 6 MW(t).

In 1955 BORAX-III was installed. This utilized a new core in the same pressure vessel as that used for BORAX-II, and at the same time a turbine and generator of 3500 kW(e) capacity were added to provide a complete power-generating system. Thus it became the first direct-cycle boiling-water-reactor power plant. The amount of reactivity which could be safely added to this reactor varied with system pressure. Above a certain excess reactivity, an instability set in, which manifested itself as an oscillation in

Fig. 4.5 Reactor destroyed by melting of fuel plates after a significant reactivity insertion

power level. Subsequently, during the later development of the boiling-water reactor, considerable attention was directed to the prediction of the threshold of this type of instability—a coupled hydrodynamic-neutronic instability—which was found to be strongly pressure-dependent. Experiments with BORAX-III also provided the first information on the amount of radioactive carry over in the steam to the turbine—it was lower than anticipated—and the extent of the radiolytic decomposition of the water.

Following the success of the BORAX experiments it was decided to separate the reactor safety work from the further development of the boiling-water reactor. The latter was pursued vigorously, and in 1954 construction was started on the 20 MW(t) experimental boiling-water reactor (EBWR) at Argonne. The safety research was in turn continued and extended in a new experimental programme code named SPERT (special power excursion reactor tests) undertaken at NRTS Idaho.

## 4.3  The experimental boiling-water reactor (EBWR)

The BORAX experiments had demonstrated the basic feasibility of the boiling-water reactor; EBWR was built with the aim of providing a flexible demonstration reactor in which information for use in designing large central station units could be obtained. Construction started in 1954 and it reached its rated power of 20 MW(t) in late 1956. The power output was, however, subsequently raised to 60 MW(t) and then to 100 MW(t). Heat transfer and fluid flow characteristics of the EBWR are given in Table 4.3.

The reactor was designed to provide steam to the 5000 kW(e) turbo-generator at 41 bar and 254°C. The reactor vessel was 2.1 m in diameter and 7.0 m high. The fuel assemblies were of the plate type using 1.4 per cent $^{235}$U formed in a uranium–zirconium–niobium alloy clad in Zircaloy 2. The plant included such safety features as an all-welded steel outer containment, a 57 m$^3$ tank of water for emergency use suspended at the top of the containment shell, two emergency boron-solution injection systems, and provision for completely remote operation from outside the containment shell. In this sense it set the pattern for the later commercial boiling-water reactor plants.

One of the problems that concerned designers at that time was the mode of control of a boiling-water reactor. An increase in steam demand from the turbine tended to reduce the reactor pressure, and hence the reactivity and power level (owing to the higher steam/water ratio and consequently reduced moderation). The EBWR designers got around this problem by installing a special bypass throttle in the steam line which connected directly with the main condenser bypassing the turbo-generator. The plant was normally operated with 5 per cent of the steam being bypassed to the condenser. This provided a margin for increasing load without reducing reactor pressure, by closing the bypass valve and opening the turbine valve. It was soon realized that the reactor could be operated satisfactorily with the bypass valve closed.

Early EBWR operating experience indicated that the design power rating

Table 4.3 EBWR heat transfer and flow characteristics

| | |
|---|---|
| Reactor power | 20 000 kW |
| Pressure | 41 bar |
| Steam flow | 7.56 kg/s |
| Total coolant heat transfer volume (consists of fuel channel, surrounding water within guide structure but does not include the control rod guide channels) | 955 litre |
| Coolant volume contained within fuel plate channels | 790 litre |
| Average power density based on total coolant volume | 21.0 kW/litre |
| Average power density based on coolant contained within heated channels | 25.3 kW/litre |
| Water volume in average fuel element to water volume in thick plate assembly | 1.07 |
| Maximum flux radially to average radial flux | 1.58 |
| Local maximum power in thick enriched assembly to average fuel assembly power | 1.12 |
| Flux in water channel near control rod to average flux in assembly | 1.30 |
| Power density in hottest channel radially | 62.5 kW/litre |
| Maximum axial flux to average axial flux | 1.49 |
| Total heat transfer surface, based on fuel area | 49 m$^2$ |
| Core average heat flux | 143.5 kW/m$^2$ |
| Average surface temperature | 258 °C |
| Mean fuel centreline temperature enriched thick plate | 282° C |
| Mean fuel centreline temperature enriched thin plate | 276° C |
| Mean fuel centreline temperature natural thick plate | 274° C |
| Mean fuel centreline temperature natural thin plate | 267° C |
| Max. heat flux in thick enriched fuel plate near control rod | 489 kW/m$^2$ |
| Max. fuel centreline temp. in enriched thick plate | 349° C |
| Average core height non-boiling ratio | 0.40 |
| Average core height boiling | 0.73 m |
| Mean core exit quality based on total coolant circulated | 1.56 per cent |
| Average voids at core exit | 28.5 per cent |
| Average voids in core height (1.22 m) | 10.5 per cent |
| Average voids in chimney | 16 per cent |
| Subcooling at core inlet | − 15° C |
| Estimated burnout heat flux | 2370 kW/m$^2$ |
| Total recirculation flow ratio | 63.0/1.0 |
| Steam to water velocity ratio | 1.5/1.0 |
| Downcomer flow area | 2.1 m$^2$ |
| Velocity in downcomer | 0.27 m/s |
| Coolant flow area in fuel zone | 0.78 m$^2$ |
| Coolant inlet velocity in fuel cluster | 0.76 m/s |

of 20 MW(t) was conservative and that the plant had a considerable amount in reserve. Experimental determination of the reactor power-transfer function showed that the reactor would be stable at power up to 60 MW(t) and this was demonstrated in late 1957.

## 4.4 The evolution of the boiling-water reactor[5]

Following the successful BORAX tests and the initiation of the EBWR project, the US General Electric Company became involved with the development of the BWR. Early in 1955 General Electric signed a contract with Commonwealth Edison of Chicago and a number of other utilities to construct a 180 MW(e) nuclear station at Dresden near Chicago. General Electric had complete turnkey responsibility for the project at a lump sum contract price of $45 million. It was decided that a small development plant was needed to assure the success of the full-scale Dresden plant and in the spring of 1955 a proposal was made to construct a small reactor (VBWR) at Vallecitos near Pleasenton, California. The speed with which this reactor was constructed was extraordinary even for the 1950s. The purchase of the land was negotiated late 1955, the USAEC issued a construction permit in June 1956, construction was complete by mid-June 1957, the reactor went critical on 3 August 1957, and delivered electric power to the grid on 24 October 1957.

The Vallecitos reactor was basically an experimental reactor used to develop the necessary fuel technology and to demonstrate the alternative modes of operation of a BWR, although a small amount of electric power was supplied to the Pacific Gas and Electric Company. The reactor initially used highly enriched uranium plate-type fuel elements but was also fuelled with the low-enrichment (2.3 per cent $^{235}$U) $UO_2$ fuel rod cluster-type elements now standard in light water reactors. Assemblies containing both 9 rods and 16 rods were loaded. The Zircaloy clad fuel rods were 13 mm in diameter and up to 710 mm in length. The assembly was enclosed in a shroud or channel made from Zircaloy and just under 76 mm square. The fuel rods were located in spacer blocks top and bottom. These fuel assemblies were thus smaller versions of the present day BWR fuel assembly (see §6).

The operating experience gained with the VBWR, whose operating pressure of 69 bar was much higher than that of earlier BWRs, showed that the turbine plant had low radiation levels and that the BWR was stable in the various operating modes, although the natural-circulation mode was less stable than the forced-circulation mode.

The first BWR specifically constructed as a commercial reactor was Dresden-1 (see Table 4.4). It used a dual cycle and was the prototype for several other dual-cycle plants including Tarapur in India, Garigliano (Senn) in Italy and Gundremmingen (KRB) in the Federal Republic of Germany. In this system (Fig. 4/6) two separate cycles are used to supply steam to the turbine. In the first cycle, steam generated in the reactor vessel passes directly to the turbine. The second cycle is indirect: steam is produced in a steam generator or heat exchanger by cooling a portion of the water taken from the reactor. This type of reactor is thus a half-way house between the direct-cycle BWR and the indirect-cycle PWR. The dual-cycle approach had as its objective the removal of as much heat as possible by evaporation in the upper part of the reactor core without exceeding the limits of reactivity for good stability. With the dual cycle the reactor response to load following is straightforward because the reactor pressure is not changed greatly by movements of the turbine regulating valve.

These dual-cycle plants were designated the BWR/1 design, indicating they were the first product line marketed by the Nuclear Energy Division of General Electric. Each BWR/1 plant was, however, custom built to the specifications of the individual utility for which it was designated.

Table  4.4  General Electric BWR product lines

| Product line number | Year of first order | Characteristic plants and salient features |
|---|---|---|
| BWR/1 | 1955 | Dresden-1, Big Rock Point, Humboldt Bay, Tarapur, Senn, KRB: initial commercial BWRs, dual cycle, direct-cycle prototypes, natural circulation (Humboldt Bay), pressure-suppression containment, first internal steam separation (KRB). |
| BWR/2 | 1963 | Oyster Creek, Nine Mile Point: first turnkey plants, direct cycle, motor–generator flow control, 7 × 7 fuel bundle. |
| BWR/3 | 1965 | Dresden-2: internal jet pumps, improved emergency core-cooling system (ECCS). |
| BWR/4 | 1966 | Browns Ferry: increased power density (20%). |
| BWR/5 | 1969 | Zimmer: improved ECCS, valve flow control. |
| BWR/6 | 1972 | Grand Gulf: 8 × 8 fuel bundle, additional assemblies, increased power (20%), improved ECCS performance, reduced peak linear rating (lower kW/m). |

The layout for the Dresden plant followed the pattern set by EBWR by enclosing the reactor within a 58 m spherical containment and having separate buildings for the turbine and fuel handling. This general pattern has been followed to the present day. The reactor vessel itself was made of low-alloy steel clad inside with stainless steel. The vessel thickness was 143 mm. The design pressure was 86 bar, sufficiently above the operating pressure of 69 bar to allow for operational transients. The core comprised 488 fuel assemblies, each in the form of a channel or shroud 109 mm square containing 36 fuel pins. The total weight of uranium in the core was 57 tonnes. Eighty control rods moved in the interspaces between the square fuel assemblies. The control rod drives, as is common with BWRs, were located on the bottom of the reactor vessel. The steam–water mixture produced in the reactor flowed through a number of riser pipes to a primary steam drum above the reactor vessel; the separated steam was passed to the turbine and the water pumped to the secondary steam generators, which were of the natural-circulation inverted-U type used on PWRs. There were four separate recirculation loops, each consisting of a secondary steam generator, recirculating pump and valving. Each loop could be isolated from the primary circuit.

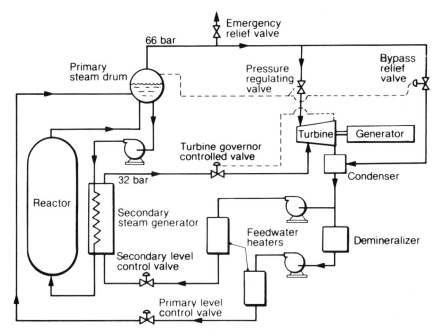

Fig. 4.6 Dual cycle system (Dresden–1)

The operation of Dresden-1 proved that the basic concept of the BWR was sound, and it is still in operation at the present time. However, the dual-cycle plants with their external steam generators and other components were more expensive to build and to operate than the direct-cycle plants. At the same time, direct-cycle plants like Big Rock Point and Humboldt Bay came into operation soon after Dresden-1 and demonstrated significant advances over the earlier demonstration direct-cycle units, setting the pattern for subsequent development of the BWR system.

The initial plan followed by General Electric for developing the BWR was known as Operation Sunrise[6]. This plan called for the simultaneous development of both natural- and forced-circulation concepts as well as 'nuclear superheat' technology. For some years around the late 1950s and early 1960s a number of major development programmes were carried out to evaluate the attractive concept of directly generating superheated steam within the reactor, which offered increased efficiency and reduced plant sizes, and two experimental superheat reactors were used for fuel development work, one at Vallecitos in California and one in Germany. The programme was finally discontinued owing to problems with the integrity of the stainless steel cladding, the relatively low power density achievable, and the marginal economic advantages to be gained through nuclear superheat.

The first real attempt by General Electric to establish a standard BWR product line was Oyster Creek—650 MW(e)—which was introduced in 1963. These BWR/2 plants had internal steam separation within the reactor vessel and were fitted with a $7 \times 7$ fuel rod bundle. They were of the forced-circulation type with all the core flow passing through five external recirculation flow loops. The recirculation pumps had variable speed capability via motor–generator sets to allow load-following via flow control. This concept of flow control is common to all direct-cycle BWRs.

In 1965 a second reactor was ordered by Commonwealth Edison for the Dresden site—Dresden-2—followed a year later by a third—Dresden-3. These 809 MW(e) reactors were typical of the BWR/3 product line. Basically, the design was similar to the BWR/2 concept except that it featured internal jet pumps instead of the external recirculation loops of the earlier design. The principle of the jet pump is discussed in §6; their use reduced the size and number of the external recirculation loops from five to two since only about one third of the core flow is needed to drive the pumps.

General Electric accepted the contracts for these early plants on a turnkey fixed price basis—12 such plants were sold between 1962 and 1966. However, changing construction and licensing requirements meant that this was not a profit-making exercise. After 1966 General Electric supplied only the nuclear steam supply system (NSSS), leaving the architect-engineers or the utilities themselves to undertake the procurement of the balance of plant and site construction.

In 1966 also the power density of the core was increased by 20 per cent to 51 kW/litre. In the then standard $7 \times 7$ fuel assembly this meant a peak linear rating of 60.7 kW/m. The first of this new line—BWR/4—which took the total power output over 1000 MW(e) was Browns Ferry. In 1969 the BWR/5 or Zimmer class of plants were introduced. Basically similar to the BWR/4, they had an improved emergency core-cooling system, and flow control was accomplished by valves rather than by altering the speed of the recirculation pumps. This enabled the plants to follow more rapid load changes and at the same time reduced capital costs.

In 1972, as earlier indicated, concern was growing about the efficiency of the emergency core-cooling systems installed on water reactors in the United States. The hearings on this question came up with a series of licensing rules (10*CFR*50 Appendix K, *Federal Register*, 39, 3 (1974)) which required a reduction in the peak linear rating of some plants. General Electric responded to these events by introducing the BWR/6. The most noticeable change from BWR/5 is the use of an $8 \times 8$ fuel rod bundle to replace the previous $7 \times 7$ array. This change allowed the power density to be increased to 56 kW/litre whilst the increased heat transfer area allowed the peak linear rating to be dropped to 44 kW/m. The increased power density and the loading of more sub-assemblies into a given size of pressure vessel means that the BWR/6 can deliver up to 20 per cent more power than the BWR/5. The opportunity was also taken to make other improvements including improved steam separators, high-efficiency jet pumps, better power flattening across the core through better coolant distribution, and the use of a burnable gadolinia poison. The BWR/6 is described in some detail in §6.

# 5 Present-day pressurized-water reactors

## 5.1 **Introduction**

This section describes the nuclear steam supply system (NSSS) of modern pressurized-water reactors. Such systems are designed by a number of industrial companies around the world and there exist differences in the details of the design and the engineering approach used in the various systems. Moreover, engineering details in successive generations of plants evolve rapidly in response to increases in size, improvements in performance, and increased stringency in environmental and safety standards.

The PWR plant described in the first part of this section is basically a typical Westinghouse four-loop plant of the type installed at the Trojan site, in Oregon. Some of the many variants from this basic design are discussed in the latter part of the section. The BWR system is considered separately in § 6.

## 5.2 **The pressurized-water reactor**

### 5.2.1 *Reactor vessel*
The reactor vessel (Fig. 4.7) is cylindrical with an overall height of 13.4 m and an internal diameter of 4.4 m. It has a hemispherical bottom head, and a flanged and gasketed removable upper head. The vessel contains the core, the core support structures, the control rod clusters, the thermal shield, and other parts directly associated with the core. Inlet and outlet nozzles are located at an elevation between the flange and the core. In the US the vessel is designed and manufactured to the requirements of Section III of the ASME *Boiler and pressure vessel code*. The basic design parameters are listed in Table 4.7. The vessel is constructed from low-alloy carbon steel SA 533 Grade B plate and/or SA 508 Cl.2 or Cl.3 forgings. The inside surfaces of the vessel in contact with the primary coolant are clad with a layer approximately 3.2 mm thick of austenitic stainless steel to reduce the amounts of corrosion products which could contaminate the cooling water. The vessel is supported by steel pads integral with the coolant nozzles. The pads rest on steel base-plates mounted in the concrete support structure.

The reactor core is located below the nozzles and above the bottom hemispherical head. The reactor internals are designed to support and locate the reactor core fuel assemblies and control rod assemblies and to transmit static and dynamic loads to the reactor vessel flange. The internals comprise three components: the lower-core support structure (including the core barrel and the thermal shield), the upper-core support structure and the in-core instrumentation support structure. An annular passage called the downcomer is formed between the core barrel and the vessel wall. Water from the inlet nozzles passes down this passage to the lower plenum formed by the bottom head, then up through the core to the upper plenum, and

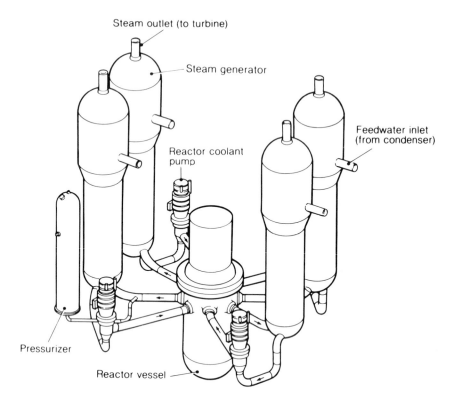

Fig. 4.7 Schematic of reactor coolant system for PWR

thence out through the outlet nozzles. The thermal shield, which is integral with the core barrel, protects the vessel by attenuating much of the gamma radiation and some of the fast neutrons that escape from the core thus reducing the irradiation damage and thermal stress in the pressure vessel steel.

The upper-core support structure consists of the upper support plate, the support columns, the control rod guide tubes and the upper core plate. The in-core instrumentation support structure consists of an upper system to convey and support core thermocouples which penetrate the vessel through the upper head, and a lower system which conveys and supports neutron-flux-measuring gauges (*flux-thimbles*) penetrating the bottom hemispherical head. The control rod drive mechanisms are mounted on the removable upper head of the vessel.

### 5.2.2 *The Reactor core*

The reactor core consists of about 100 tonnes of $UO_2$, enriched in $^{235}U$ to about 3 per cent. The $UO_2$ is in the form of pressed and sintered pellets and these are loaded into Zircaloy-4 tubes and sealed by welding on an end-cap. The pins are usually prepressurized with helium before being sealed and a compression spring is fitted between the pellet stack and the end-cap to prevent movement of the 3.7 m long stack during handling. Pressurization reduces mechanical interaction between the fuel and the cladding and has important safety implications. The fuel rods are assembled in square arrays (Fig. 4.8) to form a fuel assembly. Fuel assemblies may consist of (1) 15 × 15 arrays, or (as indicated in § 3) more recent Westinghouse designs use (2) 17 × 17 arrays of slightly smaller-diameter pins so as to improve performance in the event of a loss-of-coolant accident. Details of these alternative fuel element designs are given in Table 4.5. The assembly has a top and bottom nozzle which serve as structural elements. Control rod guide thimbles replace fuel rods at selected spaces in the array and join the top and bottom nozzles. Spring clip grids mounted off the guide thimbles support the fuel rods at different elevations. All fuel assemblies are of the same basic design though not all have control rod clusters. Selected fuel assemblies have neutron sources or burnable poison rods installed in the control rod guide thimbles. Fuel assemblies not containing control rod clusters or other devices are filled with plugs in the upper nozzle to restrict the water flow through the vacant areas.

PWR fuel assemblies are not surrounded by any form of wrapper. The open lattice tends to provide uniformity of flow and radial mixing reduces temperature gradients between fuel assemblies. The control rod cluster assemblies are used for reactor control and consist of clusters of cylindrical absorber rods which move within the guide thimbles. Above the core, each cluster of absorber rods is attached to a spider connector and drive shaft, which is raised and lowered by the drive mechanism mounted on the vessel head. Under trip conditions the control rods are delatched and fall into the core under gravity.

Control of the reactivity of the core is accomplished in a number of ways. A soluble chemical neutron absorber (boric acid) is used in the water to control long-term reactivity changes such as fuel depletion and fission product build-up, cold to hot conditions and other effects. The control rods are used to provide rapid reactivity control for shutdown, temperature changes at load, and reactivity changes due to changes in power or steam formation in

Table 4.5 Alternative fuel element designs for a PWR reactor of 3411 MW(t)

|  | (1) | (2) |
|---|---|---|
| Fuel rod array in assembly | 15 × 15 | 17 × 17 |
| No. of rods in assembly | 204 | 264 |
| Total no. of rods in core | 39 372 | 50 952 |
| Outside diameter, mm | 10.7 | 9.5 |
| Cladding thickness, mm | 0.62 | 0.57 |
| Pitch, mm | 14.3 | 12.6 |
| Surface area per assembly, $m^2$ | 25.1 | 28.8 |
| Maximum heat flux, $MW/m^2$ | 1.83 | 1.50 |
| Average linear heat rate, kW/m | 23.0 | 17.7 |
| Average channel power, MW(t) | 17.7 | 17.7 |
| Maximum linear heat rate, kW/m | 61.7 | 44.6 |

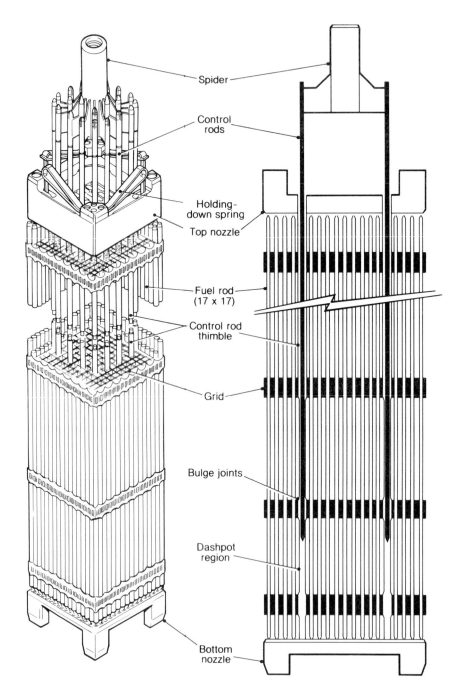

Fig. 4.8 Pressurized water reactor fuel assembly

the core. On large LWRs, local flux perturbations can occur for complex reasons, and the use of local reactivity control by means of sector control rods is important in counteracting the instabilities this can cause.

The core is usually divided into three regions and, in the initial core loading, three different fuel enrichments are used. Fuel asemblies with the highest enrichments are placed around the outside of the core and two groups of assemblies of lower enrichment are arranged in a selected pattern in the central region. Because the first fuel cycle contains more excess reactivity than subsequent fuel cycles as a result of the loading of all-fresh fuel, a so-called burnable poison is introduced in the form of boron-10 incorporated in the form of borosilicate glass tubes. The boron-10 is transmuted during the course of the irradiation. Refuelling takes place annually when one third of the fuel is discharged and fresh fuel loaded into the outer region of the core, burnable poisons in some cases also being introduced in these subsequent fuel loadings. The remaining fuel is moved to the central region in such a way as to maintain an optimum power distribution.

### 5.2.3 *Reactor coolant pumps*

The primary coolant pumps circulate the water through the primary circuit. Each loop contains a verticle single-stage shaft seal pump. For a 1000 MW(e) plant the pumps are about 7.9 m high, are driven by a 4.9 MW electric motor, and have a design flow capacity of 5.87 m³ per second at a design head of 93 m. A cut-away view of a primary pump is shown in Fig. 4.9.

### 5.2.4 *Steam generators*

Each loop of the primary system also contains a vertically mounted inverted U-tube steam generator (Fig. 4.10). The steam generator consists of two separate sections: an evaporator section containing the tube bundle and the steam-drum section where the steam is separated and dried. Table 4.6 gives typical parameters for a Westinghouse steam generator.

The high-temperature reactor coolant flows into the channel head at the base of the unit, through the inside of the Inconel U-tube bundle containing some 72 km of tubing, and back to the channel head. A partition plate divides the channel head into inlet and outlet sections. The channel head is fabricated from ferritic steel and clad internally with stainless steel. The Inconel tubes are mounted on a thick ferritic steel tube plate also clad on the primary side with Inconel. The Inconel tubes are rolled into the tube plate and welded to the primary side cladding. The tubes are supported at intervals by tube support plates (see Fig. 4.10). Feedwater from the turbine condenser enters the steam generator in the upper shell and mixes with water separated from the steam by the swirl vane separators. This water flows down the annulus between the steam generator shell and a baffle surrounding the tube bundle. When the water reaches the tube plate, it flows radially across the upper surface of the plate into the tube nest. Boiling occurs on the outside surfaces of the tubes within the bundle, and the steam–water mixture is

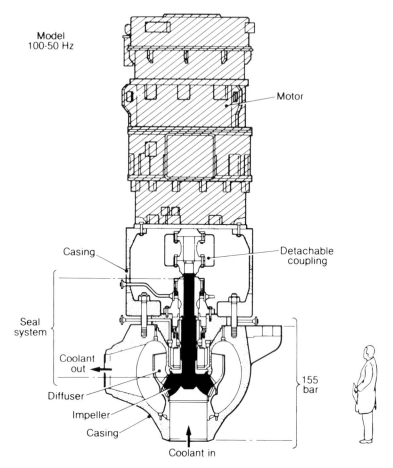

Fig. 4.9  PWR (Westinghouse) primary circuit loop pump

passed into the swirl vane separators. Natural circulation is induced as a result of the density difference within the bundle and the annular downcomer. The steam from the separators passes over corrugated *impingement-type* driers, designed to trap entrained water droplets, and exits from the top of the shell. The steam generator is a most important component of the PWR and alternative designs have been established, as described later in this section. Difficulties have been experienced with maintaining the integrity of the primary to secondary circuit boundary due to corrosion on the secondary side (see §8).

### 5.2.5 *The pressurizer*
PWR steam supply systems are equipped with pressurizers to maintain the required primary coolant pressure during steady-state operation, and to prevent the primary system pressure from exceeding the design pressure of

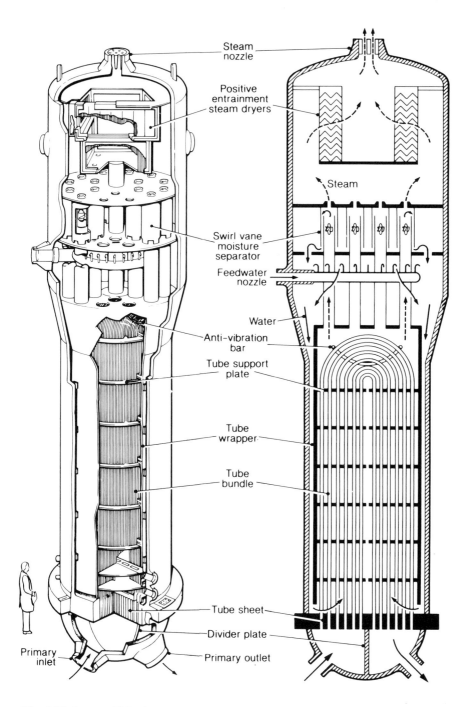

Steam
nozzle

Positive
entrainment
steam dryers

Steam

Swirl vane
moisture
separator

Feedwater
nozzle

Water

Anti-vibration
bar

Tube support
plate

Tube
wrapper

Tube
bundle

Tube sheet

Divider plate

Primary
inlet

Primary outlet

Fig. 4.10 Inverted U-tube type steam generator (Westinghouse)

the system. The pressurizer is connected to the rest of the primary system via piping referred to as the surge line.

Typically a pressurizer for a 4-loop plant will be a pressure vessel some 16 m high and 2.8 m in diameter. It operates about 60 per cent full of water and is heated by immersion-type electric heaters of about 1800 kW capacity located in the lower section of the vessel. These keep the water at saturation conditions making it, therefore, the hottest part of the primary circuit. The vessel is also equipped with multiple safety and relief valves and a spray system. When the plant load is decreased, the temperature of the primary coolant rises and there is a pressure surge in the circuit which results in the automatic operation of the spray system in the top of the pressurizer; this condenses some of the steam to keep the pressure below that which will actuate the relief valves. If the spray system should prove inadequate to limit the pressure rise, then two power-operated relief valves mounted on the top of the vessel automatically open at a pressure below the system design pressure. If the pressure continues to rise, self-actuating ASME-code safety valves will open. Steam from the relief valves is piped to the pressurizer relief tank which contains sufficient water to condense the steam. A rupture disc on this tank vents to the containment building sump if the design pressure of the tank is exceeded.

### 5.2.6 *Chemical and volume control system*
During power operation a continuous feed and bleed stream is maintained

Table 4.6 Typical design data for Westinghouse steam generator

| | |
|---|---|
| Number and Type | 4 vertical U-tube steam generators with integral steam drum |
| Height overall | 20.6 m |
| Upper shell o.d. | 4.48 m |
| Lower shell o.d. | 3.44 m |
| Operating pressure (tube side) | 155 bar |
| Design pressure (tube side) | 172 bar |
| Design temperature (tube side) | 343° C |
| Full load pressure (shell side) | 69 bar |
| Max. moisture at outlet (full load) | 0.25% |
| Design pressure (shell side) | 83 bar |
| Reactor coolant flowrate | 4360 kg/s |
| Reactor coolant inlet temperature | 325.8° C |
| Reactor coolant outlet temperature | 291.8° C |
| Sheet material | SA533B or SA508 |
| Channel head | clad internally with Inconel |
| Tube sheet material | SA508 clad with Inconel on primary face |
| Tube material | Inconel |
| Tube o.d. | 22.2 mm |
| Average tube wall thickness | 1.3 mm |
| Steam generator weight: | |
|     dry in place | 312210 kg |
|     normal operation | 376030 kg |
|     flooded cold | 509380 kg |

to and from the reactor coolant system. The bleed stream or let-down flow passes to the chemical and volume control system (CVCS), which performs a number of functions: it initially fills the reactor coolant systems, it provides a source of high-pressure water for pressurizing the reactor coolant system when cold, it maintains the water level in the pressurizer when the system is hot, it reduces the concentration of corrosion and fission products in the system, it adjusts the boric acid concentration in the primary coolant, and it provides high-pressure water for the reactor pump seals.

### 5.2.7 *Residual heat-removal system*
A low-pressure (28 bar) low-temperature (177 °C) residual heat removal system is provided for the purposes of removing the decay heat from the core during plant shutdown and refuelling operations. It consists of dual heat exchangers and circulating pumps together with associated piping, valves, and controls.

### 5.2.8 *Emergency core-cooling system*
In addition to the equipment needed for normal operation of the plant, a number of systems and devices are provided with the express purpose of protecting the plant against damage during accident or fault conditions. Since even when the reactor is shut down the radioactive decay of the fission products still provides substantial quantities of heat, a system must be provided to cool the core and prevent the fuel pins from melting if normal cooling is interrupted. The emergency core cooling (ECCS) is thus one of the more important protective systems.

The ECCS differs in detail from plant to plant but, in general, consists of several independent sub-systems characterized by equipment and flow path redundancy which assures a very high reliability of operation and continued core cooling even in the event of a failure of any one component to carry out its design functions. The first sub-system is an accumulator injection system. It consists of a number of large tanks (one per loop in the Westinghouse design) that are connected through non-return valves and piping to the main coolant loops between the pump and the reactor vessel. These tanks contain cool borated water stored under nitrogen gas at a pressure of 45 bar. If a large break were to occur followed by a rapid depressurization when the circuit pressure dropped below the accumulator gas pressure, the non-return valves would open to discharge rapidly a large volume of water into the reactor vessel and core. The accumulator injection system is termed a *passive* system because it functions automatically without any activation of pumps, motor-driven valves, or other equipment.

Two *active* sub-systems are also incorporated in the ECCS. One is a low-pressure system for use in the event of large breaks which would result in rapid depressurization and discharge of the primary coolant. This low-pressure system injects water taken from the refuelling water storage tank into the system over long time periods. Recirculation from the containment sump via residual heat exchangers is also possible for longer-term cooling of the core after an accident.

The other active sub-system is a high head system which will supply borated cooling water if the break is small and the primary system pressure remains high. Use is initially made of the chemical and volume control system charging pumps to force highly concentrated boric acid solution into the cold legs of each loop. Subsequently, the high-pressure safety injection pumps are activated, drawing water from the refuelling water storage tank, and these supplement the charging pumps. The high-pressure ECCS is activated when the primary circuit pressure falls to 110 bar.

### 5.2.9 *Containment*

Pressurized-water reactors are designed so that the primary system is entirely within the containment building. The containment buildings are built to contain all of the water that would discharge from the primary system if a loss of coolant accident (LOCA) occurred. There are several different containment designs.

Figure 4.11 shows a cross-sectional view of a Westinghouse PWR dry containment. The building is usually designed to withstand a pressure of 3.5

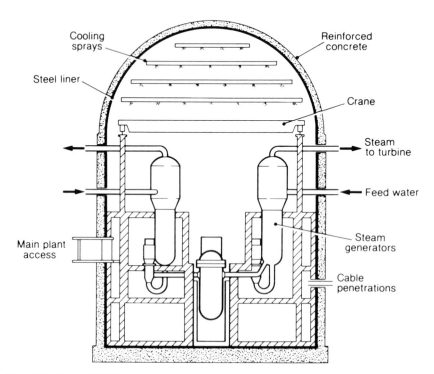

Fig. 4.11 Typical PWR containment

bar. It is constructed from reinforced concrete 1.1 m thick surrounding an inner steel liner. Typically the height would be 60 m and the diameter 38 m. A modification of this concept is the so-called *dual* or *annular* design in which there is a second concrete wall outside the first forming an annular space. This annular space is held at a pressure below the internal cavity and external atmospheric pressures. Any leakage from the inner volume is then into the annular space rather than to the environment. In many plants cooling sprays, containing dilute sodium hydroxide to remove any fission product iodine, are provided to condense the steam resulting from a major escape of primary system coolant into the containment. A fan-driven air-conditioning system is also provided to remove heat from the building during normal and accident conditions.

In some plants stored ice is used to supplement the spray system and keep the peak pressure low. The ice forms a 'passive' heat sink requiring no external power systems during the course of the accident. Systems for the control of hydrogen from both metal–water reactions and radiolytic decomposition of the water are also provided to ensure that flammable concentrations are not reached in the containment.

### 5.2.10 *Fuel-handling system*
The reactor is refuelled with equipment which handles spent fuel under water from the time it leaves the reactor vessel until it is placed in a transport flask for shipment from the site. The reactor cavity is flooded for refuelling and is connected to the spent fuel pond by a transfer canal. Refuelling normally requires the disassembly of all the components mounted on the vessel head. Typically the refuelling outage occupies about 27 days and is the dominant item during the annual outage. New methods have, however, been introduced for rapid refuelling. These include the replacement of the conventional studs and nuts on the pressure vessel by quick-acting studs of the breach block type and the development of a two-lift refuelling concept. In this latter concept the refuelling operations have been simplified to just two activities: removal of the upper package, which consists of the missile shield connected to the vessel head, together with the cooling system for the control rod drive mechanisms, and the removal of the reactor upper internal assembly. This improvement reduces the refuelling outage to about 11 days.

### 5.2.11 *Control*
The PWR is normally operated with constant mass flowrate and constant reactor pressure. Therefore as the reactor power increases so the temperature rise across the core increases. The average coolant temperature is arranged to change linearly from a no-load value of about 286 °C to a full-load value of about 330 °C. At the beginning of core life the effect of coolant/moderator temperature upon reactivity is small. Any increase in temperature reduces reactivity. At the end of core life the effect is substantially greater, largely because of the removal of the boric acid chemical

shim. At the same time, because the upper part of the core sees hotter water, it becomes less reactive compared with the lower half. However, this occurs sufficiently slowly that burn-up of fissile material acts to keep the power distribution flat by preferentially consuming fissile material in the lower half of the core. The cumulative effect is a flat power distribution at the end of core life with full-power operation.

### 5.2.12 *Effluents*

There are about 50 000 fuel pins in a large PWR. A small number of these pins, typically 25–50, might be leaking fission products into the primary circuit. The noble gases (krypton and xenon) soon separate from the coolant and enter the plant off-gas system. Other fission products become dissolved in the coolant or become finely suspended solids. These are largely removed by the coolant purification system. Isotopes of nitrogen and oxygen are also formed by neutron activation of the coolant and dissolved air. However, because there is an excess of dissolved hydrogen in the reactor coolant any oxygen formed recombines immediately. In addition, dissolved and particulate activation products also appear in the coolant as a result of corrosion and erosion of structural and cladding materials in the circuit, and in part precipitate around the circuit. The gaseous effluents are removed via the chemical and volume control system and stored for a period of 30–60 days prior to release to atmosphere. Typically a maximum of 300 curies/y of krypton and xenon might be released in this way. Liquid wastes from vents, drains, valve leaks, etc. are treated by demineralization, filtration, and evaporation as appropriate to reduce the amount of radioactive material in the liquid effluent. Apart from tritium, very small amounts of activity are released in the liquid effluent usually less than 1 curie per year. Tritium is however produced in PWRs by ternary fission (a process which occurs when a uranium nucleus fissions to form three rather than the more normal two fission products); tritium is also produced from the boric acid through a fast neutron reaction with $^{10}$B and by other reactions. About 15 000–25 000 Ci GW(e)$^{-1}$ y$^{-1}$ are produced in the fuel, where most of it remains to be discharged and either decay during storage or be released during reprocessing. Some however finds its way into the coolant and about 600 Ci GW(e)$^{-1}$ y$^{-1}$ are discharged as part of the liquid effluent.

Solid wastes produced from the liquid treatment systems and other solids such as the demineralized resins are packaged in steel drums for shipment to low-level radioactive waste disposal areas. Typically these wastes will contain about 2500–5000 Ci GW(e)$^{-1}$ y$^{-1}$ of mixed radioactive species.

### 5.3 **Variations on a theme**

Apart from Westinghouse there are a number of other companies in the United States, Europe, and Japan offering pressurized-water reactor plants. Whilst the basic concepts of these PWRs are all broadly similar, they differ from each other in many points of detail. The remainder of this section reviews some of the more important variations between these various plants.

### 5.3.1 USA

In the United States apart from Westinghouse, two other companies market PWRs—Combustion Engineering and Babcock & Wilcox (see Table 4.7). The Babcock & Wilcox design is unique in making use of a once-through steam generator. This component is described in more detail later. The layout of a 900 MW(e) reactor consists of two steam-generator units each joined to the reactor vessel by a single pipe which carries the hot primary coolant to the steam generator (the hot leg), and two pipes, with a pump mounted in each, carrying the cooler primary water back to the reactor (the cold legs).

Table 4.7 Thermal and hydraulic parameters for typical pressurized-water reactors

| Vendor: | Westinghouse Electric Corporation | | | Combustion Engineering Inc. | Babcock & Wilcox | |
|---|---|---|---|---|---|---|
| Reactor: | Trojan | Turkey Point 3 & 4 | Kewaunee | Forked River | Oconee | Davis-Besse |
| Thermal power rating MW(t) | 3411 | 2200 | 1650 | 3390 | 2568 | 2633 |
| Number of loops | 4 | 3 | 2 | 2(2×4) | 2(2×4) | 2(2×4) |
| Number of pumps | 4 | 3 | 2 | 4 | 4 | 4 |
| System pressure, bar | 155 | 155 | 155 | 155 | 152 | 152 |
| Coolant flow: | | | | | | |
| total flowrate ($10^3$ kg/s) | 16.7 | 12.8 | 8.6 | 18.6 | 16.5 | 16.5 |
| effective flowrate for heat transfer ($10^3$ kg/s) | 15.9 | 12.2 | 8.2 | 18.0 | 15.6 | 15.6 |
| average mass velocity (kg/ $m^2$s ) | 3354 | 3150 | 3286 | 3639 | 3422 | 3422 |
| Coolant temperatures: | | | | | | |
| inlet (°C) | 289.1 | 285.6 | 279.7 | 289.4 | 290.0 | 291.7 |
| vessel outlet (°C) | 324.8 | 316.7 | 315.0 | 321.7 | 310.1 | 320.0 |
| average core outlet (°C) | 326.4 | 318.0 | 316.6 | 322.7 | 310.6 | 321.0 |
| hot channel outlet (°C) | 341.7 | 338.9 | 335.0 | 340.5 | 339.3 | 342.7 |
| Heat transfer at rated power: | | | | | | |
| active heat transfer surface area ($m^2$) | 4852 | 3943 | 2667 | 5112 | 4622 | 4622 |
| average heat flux (MW/$m^2$) | 0.685 | 0.541 | 0.602 | 0.647 | 0.541 | 0.555 |
| maximum heat flux (MW/$m^2$) | 1.83 | 1.75 | 1.69 | 1.74 | 1.68 | 1.70 |
| average linear heat rate (kW/m) | 23.0 | 18.0 | 20.3 | 22.6 | 18.5 | 19.0 |
| maximum linear heat rate (kW/m) | 61.7 | 58.7 | 56.7 | 60.7 | 57.8 | 58.4 |
| Fuel rods: | | | | | | |
| number | 39 372 | 32 028 | 21 659 | 38 192 | 36 816 | 36 816 |
| outside diameter (cm) | 1.07 | 1.07 | 1.07 | 1.12 | 1.09 | 1.09 |
| cladding thickness (mm) | 0.62 | 0.62 | 0.62 | 0.66 | 0.67 | 0.67 |
| diametral gap (mm) | 0.16 | 0.16 | 0.19 | 0.21 | 0.18 | 0.18 |
| fuel pellet diameter (cm) | 0.932 | 0.932 | 0.929 | 0.964 | 0.940 | 0.940 |
| active length (m) | 3.66 | 3.66 | 3.66 | 3.66 | 3.66 | 3.66 |

The latest Combustion Engineering design—the so-called System 80—produces 1200 MW(e) and utilizes just two very large recirculating-type inverted-U-tube steam generators each rated at nearly 2000 MW(t).

### 5.3.2 *Germany*

In the Federal Republic of Germany the Siemens-owned Kraftwerk Union (KWU) company has developed a standardized 1300 MW(e) four-loop PWR design. The first such plant was installed at Biblis. The KWU design differs from that offered by Westinghouse in a large number of ways of which a few will be mentioned here. The KWU pressure vessel (Fig. 4.4) is made from large ring forgings with no longitudinal welds. In contrast to the Westinghouse design, the KWU vessel design has no penetrations below the primary circuit nozzles and the in-core instrumentation is inserted through the top of the vessel.

The KWU steam generator utilizes Incoloy Alloy 800 (33.5% Ni, 21.5% Cr, 42.7% Fe, 0.7% Mn, 0.5% Si, 1.1% other) for the tubing rather than Inconel Alloy 600 (76.5% Ni, 15.8% Cr, 7.2% Fe, 0.2% Mn, 0.2% Si, 0.1% other). The reasons for this are examined in §8. The ECCS system consists of four identical sub-systems, each having one HPIS (high-pressure injection system), two accumulators, and one LPIS (low-pressure injection system). The system feeds cooling water to the vessel via both the cold leg and the hot leg of each primary loop. An emergency feedwater system is also provided in the form of four completely independent sub-systems to allow heat removal from the steam generators under accident situations where the primary circuit remains intact and at pressure. The KWU fuel has a somewhat thicker cladding than the Westinghouse fuel, the clad thickness being 0.72 mm, compared with 0.54 mm in the Westinghouse design. Finally, the spent fuel storage ponds in the case of the KWU design are located inside the spherical containment building.

### 5.3.3 *USSR*

In the Soviet Union, PWR development has followed a relatively independent path. Starting in 1954 the USSR has developed two types of nuclear reactor for electricity production—a graphite-moderated, boiling-water-cooled pressure tube reactor unique to the Soviet Union, and a pressurized-water reactor basically similar to that used in the United States, Europe, and Japan. The first PWR put into operation in the USSR was at Novovoronezh in 1964. This reactor had six loops and produced 210 MW(e). It is equipped with horizontal steam generators, which are described later. The design was uprated and a second unit producing 364 MW(e) went into operation at Novovoronezh in 1969. The design was uprated again to 440 MW(e), and two further reactors of this power came into operation at Novovoronezh in 1971 and 1972 respectively. The 440 MW(e) PWR has since become a standard design (VVER), and a number of other stations are either operating or under construction in the USSR. The components for these reactors have been manufactured at the Izhorsk works near Leningrad. In future they will be manufactured at a new plant at Skoda in Czechoslovakia. This plant will also make the 440 MW(e) reactor for export to other Comecon countries. More recently the Soviet Union has developed a larger 1000 MW(e), 4-loop PWR. The first such reactor at Novovoronezh is being commissioned and a number of other stations are in various stages of construction. The components for these larger reactors are being manufactured at the Atommash plant on the River Don in southern Russia. It has the capacity to produce pressure vessels and steam generators for between six and eight 1000 MW(e) reactors per year. Soviet reactors are equipped with similar reactor control and protection systems and with engineered safety features such as emergency core cooling. Although the earlier PWR plants are not equipped with concrete containment buildings, there are indications that the large 1000 MW(e) plants will have a conventional reactor containment.

## 5.4 Pressure vessel variants

Early light water reactors had pressure vessels which were constructed mainly from forgings. As reactor sizes increased, however, the larger diameter of the reactor vessel necessitated the use of forged rings for the flanges and either forged rings or rolled plates for the shell sections. These parts had to be joined together, and the total length of weld in a modern reactor pressure vessel is considerable. In addition, the nozzles connecting the vessel to the reactor pipework required welding into the shell section.

Over the past few years a number of companies have increased their capacity to manufacture very large forgings up to 160 tonnes or more in a finished state. As a result designs of reactor pressure vessel have been produced using these large forgings. In the Westinghouse design the lower part of the vessel shell is made from just three major pieces (see Fig. 4.12). These pieces are an integral flange–nozzle course, the parallel section course, and the hemispherical bottom head. The material used is ASTM A 508 Class 3 steel and the initial ingots weigh between 465 and 510 tonnes. The attachment of the nozzle is by means of a butt weld. This form of construction has the advantage of reducing considerably the length of weld with consequent savings in both time and cost of manufacture and in-service inspection.

Apart from ferritic steel, alternative materials of construction proposed for PWR pressure vessels include cast iron and prestressed concrete.

## 5.5 Steam generator variants

Whilst the majority of PWR power plants are equipped with steam generators of the vertical shell, inverted U-tube recirculating design described in §5.2.4, other steam generator designs are employed in some plants.

In the original Shippingport plant,[1] the steam-generating equipment consisted of four units each comprising a heat exchanger, a steam drum, and

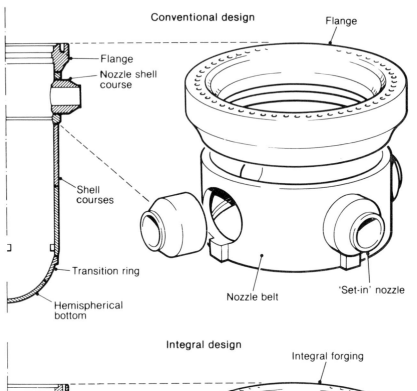

Conventional design

Flange

Flange
Nozzle shell course

Shell courses

Transition ring

Hemispherical bottom

Nozzle belt

'Set-in' nozzle

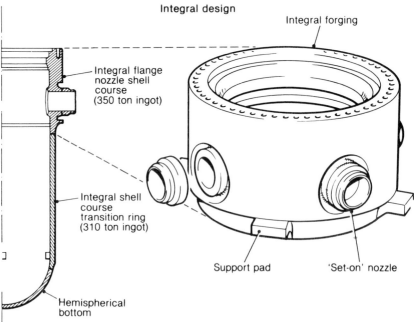

Integral design

Integral forging

Integral flange nozzle shell course (350 ton ingot)

Integral shell course transition ring (310 ton ingot)

Hemispherical bottom

Support pad

'Set-on' nozzle

Fig. 4.12 Comparison between conventional and integral pressure vessel designs

connecting piping. Two different types of heat exchanger, a horizontal U-tube design supplied by Babcock & Wilcox and a horizontal straight-tube design supplied by Foster Wheeler, were installed to evaluate the relative performance of the two designs (Fig. 4.13). Each heat exchanger was of the shell-and-tube type, in which primary coolant flowed through the tubes and steam was generated on the shell side. The steam–water mixture passed up the risers to the steam drum where standard separators and driers were used to separate the water from the steam. The water returned to the lower heat exchanger via the downcomers. The Babcock & Wilcox design was rated at 75 MW(t) and contained 921 19 mm stainless steel tubes 15.2 m long. The Foster Wheeler design was also rated at 75 MW(t) and contained 2096 12.7 mm stainless steel tubes 9.5 m long.

Horizontal natural-circulation steam generators are also widely used in PWRs constructed in the USSR[10]. The units for the 440 MW(e) plant consist of a horizontal shell 11.5 m long and 3 m in diameter. Vertical tubular headers located half way along the shell act as the inlet and outlet for the primary coolant. Horizontal bundles of U-tubes mounted on these headers provide the heat transfer surface. The particular units shown in Fig. 4.14 are rated at 250 MW(t), but units of 800 MW(t) have been manufactured for a 1000 MW(e) plant.

In the United States, one vendor, Babcock & Wilcox, equips its reactors with a vertical shell, straight-tube, once-through steam generator (Fig. 4.15). The primary coolant enters the header at the top of the unit and flows down through the tubes to exit at the base. On the secondary side, the feed-water is boiled in the interspace between the tubes, totally evaporated, and slightly superheated (by ≈ 30 °C). The positioning of the feed nozzles and steam outlet on the shell and the use of some of the steam to preheat the feedwater in the annulus around the tube bundle overcomes the problem of differential thermal expansion of the tubes and shell. A feature of the once-through steam generator which was important in the accident at Three Mile Island in 1979 is the reduced water inventory in the unit compared with the recirculating design which, in turn, leads to a shorter time before the unit dries out in the event of a loss of feedwater.

Even with the vertical shell, inverted U-tube recirculating units, there are significant differences between vendors in respect of design details such as thermohydraulics parameters, methods of construction, tube supports and materials which profoundly influence their performance.

Some units are equipped with a feedwater preheating section or economizer located just above the tube plate on the cold leg side of the U-tubes. The feedwater enters the preheater and is heated almost to saturation temperature by countercurrent heat transfer from the reactor coolant within the tubes, flowing countercurrent to the direction of feedwater flow.

In the design of unit offered by Foster Wheeler Limited,[8] the massive thick-tube plate and channel head is dispensed with and is replaced by two cylindrical horizontal headers upon which the tube bundle is mounted

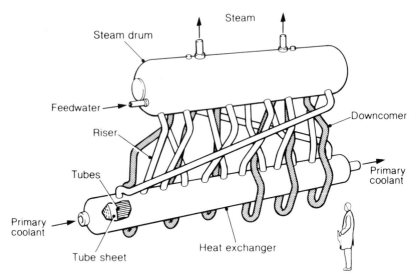

Foster Wheeler steam generator

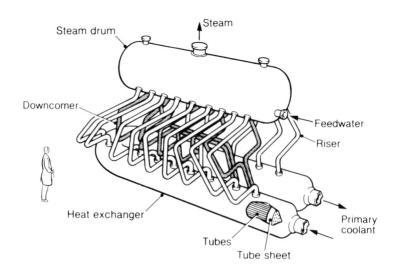

Babcock & Wilcox steam generator

Fig. 4.13 The steam generators used at Shippingport

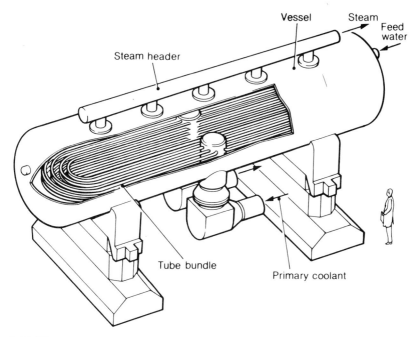

Fig. 4.14 The steam generators used on the 440 MW(e) units constructed in the USSR

directly. The primary reactor coolant passes through a vertical penetration in the steam generator shell to this horizontal header feeding the tube tank and exits by way of a similar header and shell penetration (Fig. 4.16). The advantages claimed for this design include avoidance of sludge deposition on the tube plate and the elimination of tube-to-tube plate crevices.

Tube support designs are particularly important because of the consequences of corrosion of the tube support material.[9] Since the corrosion products of carbon steel occupy about twice the volume of the original metal, it is possible, with some designs of tube support, for the corrosion product to dent the tubes and to distort the support plate itself. Corrosion in PWR steam generators will be discussed in §8. Figure 4.17 shows a variety of tube support arrangements used in the units offered by the various vendors.

Whilst all the shells of steam-generator units are invariably one or other grade of ferritic steel with the surface exposed to the primary coolant clad in stainless steel, a wide range of tubing materials have been used including 1Cr–½Mo, 300-series stainless steels, Incoloy Alloy 800, and Inconel Alloy 600. The material most widely used is Inconel Alloy 600 but the German plants constructed by Kraftwerk Union (KWU) use the high-iron, low-nickel Incoloy Alloy 800 for tubing.

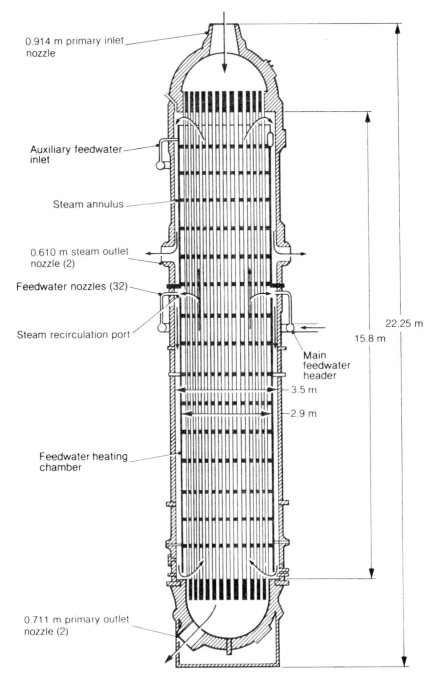

Fig. 4.15 OCONEE 1 Babcock and Wilcox nuclear once through steam generator

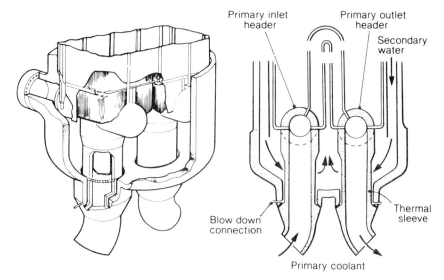

Bottom end arrangement

Fig. 4.16 Header arrangement in the Foster–Wheeler steam generator

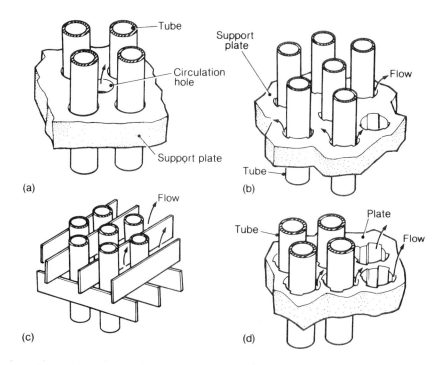

Fig. 4.17 Tube support plate designs
    a) Westinghouse drilled hole    b) B & W Tri-foil plate for OTSG units
    c) KWU Egg-crate    d) Westinghouse Quartrafoil for Model F units

# 6 Present day boiling-water reactors

## 6.1 **Introduction**

The nuclear steam supply system for a boiling-water reactor (BWR) consists primarily of the reactor vessel and the components inside the vessel. Because the energy from the nuclear chain reaction is used to evaporate water to form steam which is then transported directly to the turbine, the BWR is called a direct-cycle system. The pressure in a BWR system is maintained at about 69 bar; at this pressure water boils at about 290 °C. The following paragraphs describe the current General Electric BWR/6 nuclear boiler. Typical parameters for large BWRs are given in Table 4.8.

## 6.2 **The boiling-water reactor**

### 6.2.1 *Reactor vessel and internals*

The reactor itself consists of the reactor vessel, the internal components of the core, jet pumps for reactor water recirculation, steam separators, and steam driers. The reactor vessel is a pressure vessel about 6.4 m diameter, 22 m in height with a wall thickness of 150–180 mm. It has a single full-diameter removable head and is constructed in low-alloy steel. The interior, except for the top head and nozzles, is clad with stainless steel weld overlay to reduce corrosion. The vessel is mounted on a support skirt which, in turn, is bolted to a concrete and steel cylindrical pedestal integral with the reactor building foundation (Fig. 4.18).

Inside the vessel, and surrounding the reactor core, is the core shroud, which is a cylindrical, stainless steel structure that separates the up flow through the core from the down flow of water in the annulus between the shroud and the vessel itself. The shroud head extends like a dome over the top of the core. Mounted on the shroud head is an array of standpipes with a three-stage steam separator at the top of each standpipe.

Water boils in the core and a mixture of steam and water flows out of the top of the core into the plenum below the shroud head and through the steam separators. Steam from the separators passes through a drier assembly mounted in the vessel above the shroud head and leaves the vessel through four nozzles located below the vessel flange. Water from the driers and from the centrifugal separators is returned to the recirculation annulus and is joined by the feedwater which enters through nozzles in the vessel wall. The head of the vessel, the steam separators, and the driers are removable for refuelling the core.

In the pressurized-water reactor, the flow of coolant through the core is provided by reactor coolant pumps which circulate the heated coolant to the steam generators. In the BWR, coolant flow derives from the internal circulation of the coolant within the reactor vessel. High-performance jet pumps located in the annulus between the vessel and the core shroud

Table 4.8 Thermal and hydraulic parameters for typical boiling-water reactors (vendor: General Electric)

| | Duane Arnold | Zimmer | La Salle County |
|---|---|---|---|
| Thermal power rating (MW(t) ) | 1593 | 2436 | 3293 |
| Electrical power rating (MW(e) ) | 545 | 840 | 1052 |
| Reactor vessel: | | | |
| inside diameter (m) | 4.65 | 5.54 | 8.15 |
| inside length (m) | 20.2 | 20.9 | 23.2 |
| design pressure (bar) | 86.2 | 86.2 | 86.2 |
| design temperature (°C) | 301.7 | 301.7 | 301.7 |
| Number of control rods | 89 | 137 | 185 |
| Recirculation system: | | | |
| number of loops, 1 pump/loop | 2 | 2 | 2 |
| pipe size, suction and discharge (cm) | 55.9 | 50.8 | 61.0 |
| pipe size, manifold and discharge (cm) | 40.6 | 35.6 | 40.6 |
| pipe size, inlet, number and size (cm) | 8–30.5 | 10–25.4 | 10–30.5 |
| pump flow, each ($10^3$ kg/s) | 1.3 | 1.57 | 2.25 |
| total jet pump flow ($10^3$ kg/s) | 6.1 | 9.9 | 13.4 |
| Number of steam lines and pipe size (cm) | 4–50.8 | 4–61.0 | 4–66.0 |
| Steam flow rate, maximum ($10^3$ kg/s) | 0.86 | 1.32 | 1.78 |
| Number of feedwater lines and pipe size (cm) | 4–25.4 | 4–30.5 | 6–30.5 |
| Core coolant flowrate ($10^3$ kg/s) | 6.1 | 9.9 | 15.5 |
| Coolant conditions, pressure and temperature: | | | |
| out of core (steam dome) (bar/°C) | 70.3/286 | 70.3/286 | 70.3/286 |
| feedwater inlet (bar/°C) | 75.0/215 | 75.0/215 | 75.0/215 |
| inlet core subcooling (kJ/kg) | 57 | 53 | 57 |
| Heat transfer at rated power: | | | |
| active core heat transfer area ($m^2$) | 2959 | 4503 | 6144 |
| average heat flux, ($MW/m^2$) | 0.517 | 0.520 | 0.519 |
| maximum heat flux ($MW/m^2$) | 1.35 | 1.35 | 1.34 |
| average linear heat rate (kW/m) | 23.2 | 23.3 | 23.1 |
| maximum linear heat rate (kW/m) | 60.7 | 60.7 | 60.7 |
| Fuel data: | | | |
| number of fuel assemblies | 368 | 560 | 764 |
| number of rods | 18 023 | 27 400 | 36 672 |
| fuel cell spacing (control rod pitch) (cm) | 30.5 | 30.5 | 30.5 |
| active length (m) | 3.66 | 3.66 | 3.66 |
| cladding outside diameter (cm) | 1.43 | 1.43 | 1.43 |
| cladding thickness (mm) | 0.81 | 0.81 | 0.81 |
| pellet diameter (cm) | 1.24 | 1.24 | 1.24 |
| rod pitch (cm) | 1.87 | 1.87 | 1.87 |
| channel thickness (mm) | 2.03 | 2.03 | 2.03 |
| channel outside across flats, dimensions (cm) | 13.8 | 13.8 | 13.8 |
| total weight of $UO_2$ (tonnes) | 81.3 | 123.8 | 168.9 |
| total weight of U (tonnes) | 71.7 | 109.1 | 148.9 |
| core power density ($MW/m^3$) | 50.9 | 51.2 | 51 |
| CSCS model (year) | 1967 | 1969 | 1969 |
| Reference | PSAR | PSAR | PSAR |

(Fig. 4.18) circulate the water through the reactor core. The jet pumps, which have no moving parts, provide a continuous circulation path for the majority of the coolant flow.

The recirculation system employs 20 to 24 such jet pumps each about 5.8 m in length. The internal jet pumps are driven from two external recirculation loops, each containing a pump with a directly coupled water-cooled motor, a flow control valve, two shut-off valves, and a by-pass valve. Approximately one-third of the core flow is taken from the vessel through the two recirculation nozzles, pumped to a higher pressure, distributed through a manifold to which a number of external riser pipes are connected, and returned to the vessel through inlet nozzles. Each inlet nozzle is connected

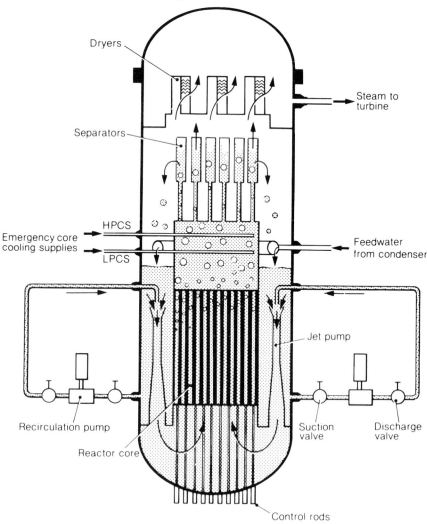

Fig. 4.18 Schematic of BWR reactor coolant system

by an internal riser pipe to a pair of jet pumps. The high-pressure water flow is discharged into the throat of the jet pump, thus inducing the surrounding water in the downcomer region to be drawn into the pump. The two flows mix and pass via a diffuser into the lower plenum of the reactor vessel. Because the reactor heat output is sensitive to the rate of flow of coolant through the core, partial control of reactor power is effected by varying the driving flow to the jet pumps and thereby the rate that the jet pumps recirculate water through the core.

Carbon-steel steam lines are welded to the vessel outlet nozzles and run parallel to the vertical axis of the vessel, downward to an elevation where they emerge from the containment. Two air-operated isolation valves are installed on each steam line, one inside and one outside the primary containment. In addition a combination safety-relief valve and a flow-restricting nozzle (to protect against rapid uncovering of the core in the case of a main steam line break) are installed in the steam line. The safety-relief valves discharge directly into the pressure-suppression pool of the containment. The safety function includes protection against overpressure of the reactor primary system. The relief function is provided by a power-operated valve to relieve steam during operational transients resulting in high system pressure. In certain accident sequences it can also be used to depressurize the reactor primary system.

### 6.2.2 *The reactor core*
The fuel pins used in a BWR core are similar to those for PWRs. The Zircaloy-2 clad fuel pins are about 13 mm diameter and 3.7 m long, assembled on an $8 \times 8$ square array into fuel bundles (Table 4.9). Each bundle or assembly has, in fact, 62 active fuel pins and two Zircaloy-2 tubes which contain water. These 'water rods' introduce additional moderator into the bundle interior and also act as a support for the seven grid spacers. Eight of the fuel pins in each bundle act as tie rods to reinforce the assembly. They have threaded end-plugs which screw into the lower tie plate and extend through the upper tie plate. A stainless steel nut and locking tab hold the assembly together. A square Zircaloy-4 casing called a fuel channel encloses the fuel bundle; in this respect the assembly differs from the PWR fuel

Table 4.9 Alternative fuel element designs for BWR reactors of 3293 MW(t)

| | $7 \times 7$ | $8 \times 8$ |
|---|---|---|
| Fuel rod array in assembly | $7 \times 7$ | $8 \times 8$ |
| No. of rods in assembly | 48 | 63 |
| Total no. of rods in core | 36 672 | 48 132 |
| Outside diameter (mm) | 14.3 | 12.5 |
| Cladding thickness (mm) | 0.81 | 0.86 |
| Pitch (mm) | 18.7 | — |
| Surface area per assembly (m²) | 8.0 | 9.3 |
| Maximum heat flux (MW/m²) | 1.34 | 1.12 |
| Average linear heat rate (kW/m) | 23.0 | 19.7 |
| Average channel power (MW(t) ) | 4.3 | 4.3 |
| Maximum linear heat rate (kW/m) | 60.7 | 44.0 |

assembly in that the closed sides prevent lateral flow of the coolant between adjacent bundles, thus avoiding oscillations that could prove troublesome in a two-phase system.

Different $^{235}$U enrichments are used in fuel bundles to reduce local power peaking and in addition selected rods contain gadolinium, a burnable poison, to achieve optimum power flattening. Low-enrichment uranium pins are used for corner rods and for rods near the water gaps. The end fittings for each type of fuel pin are such that it is not mechanically possible to mis-assemble a fuel bundle.

Changes in reactivity are made by movable control rods, cruciform in shape, which move in the interspace between four fuel bundles, and are interspersed throughout the core. The rods contain sealed stainless steel tubes filled with boron carbide power. These control rods act on the overall reactor power level and provide the principal method of shutting the reactor down. They are moved vertically by hydraulically actuated drive mechanisms mounted on the bottom hemispherical head of the vessel. Control rod entry from below provides the best power shaping and thus fuel economy for a BWR and moreover causes no interference during refuelling.

### 6.2.3 *Emergency core-cooling system (ECCS)*

BWRs, like PWRs, have multiple provisions for cooling the core in the event of an unplanned depressurization or loss of coolant from the reactor. The BWR/6 ECCS is composed of four separate sub-systems: the high-pressure core spray (HPCS) system, the automatic-depressurization system (ADS), the low-pressure core spray (LPCS) system, and the low-pressure coolant injection (LPCI) system. These systems are shown in Fig. 4.19 for a BWR/6 plant installed in a Mark III containment.

The HPCS pump obtains water from the condensate storage tank and/or the pressure-suppression pool. Injection piping enters the vessel near the top of the shroud and feeds two semicircular spargers that are designed to spray water radially over the core into the fuel assemblies. The system works over the full range of reactor pressure and is activated by a low reactor water-level signal or a high containment pressure.

If the HPCS cannot maintain the water level or fails to operate, the reactor pressure is reduced automatically by operation of the automatic-depressurization system (ADS), which discharges to the pressure-suppression pool. This then allows the LPCI and LPCS to provide sufficient cooling.

The LPCS pump takes its water from the suppression pool and discharges from a circular spray sparger in the top of the reactor vessel above the core. Low water level or high containment pressure activates this system which begins injection when the reactor pressure is low enough.

The LPCI is an operating mode of the residual-heat removal system. It also is activated by low water level or high containment pressure and when used with the other ECC sub-systems can reflood the core and thereafter maintain a sufficient water level in the core.

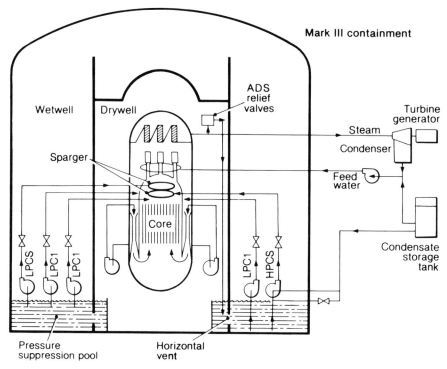

Fig. 4.19 A BWR ECCS system arranged in a Mark III containment

### 6.2.4 *Containment*

Containment systems for current design of BWR generally provide both primary and secondary containment. The former is a steel pressure vessel surrounded by reinforced concrete and is designed to withstand the peak transient pressures which might occur in the most severe LOCA. This primary containment employs a *dry well* enclosing the entire reactor vessel and its recirculation pumps and piping. This dry well is connected through large ducts to a lower level pressure-suppression chamber which stores a large pool of water (Fig. 4.20). Under accident conditions the main steam isolation valves close and any steam escaping from the reactor primary system would be released entirely within the dry well. The resulting increase in dry well pressure would force the air-steam mixture in the dry well down into and through the water in the pressure-suppression chambers where the steam would be completely condensed. Similarly, steam released through the pressure relief valve would also be condensed in the pool. This pool serves as one source of water for the ECCS. Systems for the control of hydrogen from metal–water reactions and radiolytic decompositon of the water are also provided to ensure that flammable concentrations are not reached in the containment.

The secondary containment building is the reactor building which houses

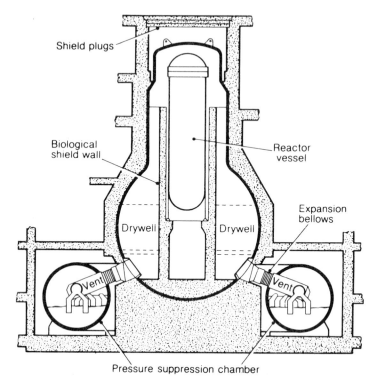

Fig. 4.20 BWR light bulb torus containment

the reactor and its primary containment. The Mark III containment system (Fig. 4.19) uses a separate free-standing leak-tight containment shell inside a sealed building, which provides a further barrier to the escape of gaseous effluents, as well as shielding to reduce the escape of radiation emanating from the reactor itself.

### 6.2.5 *Control*

The overall control concept of the BWR is that the reactor pressure is maintained approximately constant by means of a pressure-regulating valve upstream of the turbine. To follow load changes, reactor power level is varied by reactor coolant flow control. Thus, if more power is demanded at the turbo-generator, an increase in reactor power is necessary. This is achieved by increasing the flow through the core, which, in turn, initially reduces the steam-to-water ratio (the *average void fraction*) in the core and thereby increases the amount of neutron moderation. An equilibrium is, however, reached at the higher power level and flowrate in which the void fraction returns to approximately its former value.

For a drop in load, the turbine tends to speed up, sending a signal to close the flow control valves in the recirculation loops to a new position. The flow through the reactor core is reduced increasing the core average void fraction. Reduced moderation lowers the reactor power consistent with the reduced load. Most operational transients on a BWR are accommodated in this manner, whilst maintaining constant system pressure and reducing control rod movements to a minimum.

### 6.2.6 *Effluents*

In the BWR, gamma radiation decomposes a small amount of the coolant water to produce hydrogen and oxygen gases. These together with gaseous fission products from leaking fuel pins pass to the condenser, where they are removed via the air ejector. A device for recombining the gaseous hydrogen and oxygen to form water is located in the line from the air ejector to the stack. A typical recombiner consists of a replaceable cartridge containing a catalyst in a steel vessel. The catalyst is in the form of alumina pellets precoated with platinum or palladium. The catalyst bed is heated to a temperature around 200–400 °C. Because the catalyst becomes poisoned by materials such as iodine, multiple units are used and periodic replacement of the cartridge is necessary. The amount of radioactive gases released from a BWR can be reduced by storing the gases for a length of time sufficient to allow the short-lived isotopes to decay to low levels. This may be done in hold-up tanks or by retaining the gases on large charcoal beds for periods up to 24 hours for the krypton and 15 days for the xenon. The gases are finally released from a tall stack to obtain maximum dilution. Early BWRs equipped only with a 30 min hold-up tank released up to $2 \times 10^6$ Ci $GW(e)^{-1} y^{-1}$ of xenon and krypton. By using charcoal delay beds with the delay times quoted above this is reduced by a factor of about 100–1000.

Although gaseous releases from BWRs are significantly higher than for PWRs, the liquid effluents are comparable, except that the tritium content is reduced by a factor of about ten. This is because boric acid is not used in the BWR coolant.

## 6.3 Design variants

A double-ended break of the recirculation line of the General Electric BWR/6 is the worst pipe break that can be postulated for such a plant. In the recent designs offered by Asea-Atom in Sweden and KWU in Germany (the Type 72), the recirculation line has been eliminated and replaced by a number of glandless pumps mounted directly on the lower head of the reactor pressure vessel in a ring at the base of the downcomer annulus. It is claimed that this arrangement improves reactor coolant circulation, reduces containment vessel size, and dispenses with a number of penetrations, thereby considerably reducing the initial capital cost and increasing safety.

The arrangement of the pressure-suppression primary containment for the KWU BWR is shown in Fig. 4.21. This in turn is surrounded by a reinforced concrete containment building with a reduced pressure in the

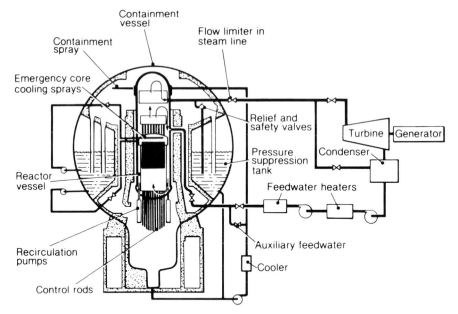

Fig. 4.21 German BWR with engineered safeguards (KWU)

interspace between the steel inner containment and the concrete outer containment.

# 7 Safety aspects

### 7.1 Introduction

Earlier sections, dealing with the basic design features of the PWR and BWR, have emphasized the extent to which safety is an integral part of the design concept of both systems. It is now necessary to look in more detail at the types of possible maloperation which require particular consideration in assessing LWR systems, and the principal types of accident sequence needing to be guarded against.

This section will look in some detail at the measures taken to forestall damaging losses of cooling or containment, concentrating on two requirements of particular significance in LWR design: the effectiveness of the emergency core-cooling systems and the integrity of the reactor vessel. It then considers some accident sequences and their implications for the safety of the system.

### 7.2 Thermohydraulics

As §2 has indicated, a light water reactor core when considered as a heat source is relatively highly rated. Volumetric heat production rates are in the range 50–100 kW/litre, a factor of 20–40 higher than for an AGR. The surface heat flux—that is, the amount of heat passing through unit surface area of the fuel element in unit time—in an operating water reactor is of the order of 1 MW/m². Such heat fluxes can be removed quite satisfactorily using flowing water at velocities of a few metres per second. In these circumstances the difference in temperature between the fuel element surface and the cooling water will be quite low—a few tens of degrees. Should the fuel element surface temperature exceed the boiling point, then boiling will occur. Again heat fluxes of the order of 1 MW/m² can be safely dissipated by flowing boiling water, provided that the system is pressurized.

If, however, the surface heat flux becomes too high or alternatively the water flow too low (or the steam content too high in a boiling system), then overheating of the fuel pin can occur. This overheating occurs quite suddenly at a particular set of thermal and hydraulic conditions. The threshold at which it occurs is usually described in terms of the maximum heat flux which can be sustained without overheating at a particular set of conditions—the *critical heat flux* (CHF), *departure from nucleate boiling* (DNB), or (colloquially) the *burnout* heat flux, because damage to the heat-transfer surface can occur if the heat is not removed. If no action is taken, for example, to shut down the reactor, then the fuel element will heat up until the heat flux can be dissipated through the blanket of steam which forms adjacent to the surface and tends to act as an insulator. At the surface temperature corresponding to this new equilibrium state the Zircaloy cladding will become hydrided and brittle, and may react exothermically with the steam to form hydrogen. This chemical reaction starts at about 950 °C. As the reaction progresses, the fuel element can become badly damaged, and NRC licensing requirements stipulate *inter alia* that the ECCS should guarantee that peak clad temperatures are held below 1200 °C.[11]

It is therefore important to know the *critical heat flux* (or burnout heat flux) for every likely condition within the reactor. This is achieved by carrying out experiments with full-scale electrically heated models of the PWR and BWR fuel elements over a range of pressures, flowrates, and water temperatures. The measure of safety within an operating light water reactor core is the CHF or DNB ratio, defined as the ratio of the predicted critical heat flux to the actual local heat flux in the reactor. Typically, in a PWR the DNB ratio in the worst overpower situation with all the uncertainties on flow rates, temperatures, and fuel assembly dimensions working in worst possible combination will be about 1.5. Perhaps two or three fuel pins only within a core of 50 000 will approach the critical heat flux closely. In the case of a BWR, General Electric use a *critical power ratio* (CPR), which is the maximum power in the fuel assembly before overheating occurs divided by the operating power of the subassembly. Typically, the use of a CPR of 1.2 would mean that there is a 50 per cent probability of a critical heat flux condition occurring on some (0.1 per cent) of the fuel pins.

The reactor must be able to withstand transients without any core damage. In the design of water reactors it is usual to distinguish between different types of transient in terms of their frequency of occurrence. Thus tables of various foreseeable transients are drawn up under the following headings.

1. *Normal Operation.* This includes conditions that occur frequently or regularly in the course of normal operation, refuelling, or maintenance, e.g. isolated failure of a fuel pin, changes in load, startup, shutdown, etc.

2. *Upset Conditions.* These include all faults not expected during normal operation, but which can be reasonably expected during the lifetime of the plant. Examples are loss of power to reactor components or control systems, loss of condenser cooling water, loss of feedwater, inadvertent control rod withdrawal, etc. The plant design must ensure that no such fault which might occur with a frequency of one or two events per year can cause either a more serious fault or damage to the fuel.

3. *Emergency Conditions.* These are departures from normal conditions that are not expected to occur in the lifetime of any particular plant, but which may be expected to occur a few times within a large number of plants over a 30–40 year period. Examples would be minor loss of primary coolant, minor breaks in the secondary steam side, or inadvertent boric acid dilution. The plant design should be such that safe shutdown is achieved without any off-site consequences.

4. *Fault Conditions.* These are faults considered very improbable, i.e. less than once in 1000 reactor operating years, but whose consequences include the potential for serious plant damage or public injury. Thus despite their low probability protective measures must be taken. The possibility of serious plant damage is accepted, but it is required that in the event of such a fault the off-site radioactivity release must be contained within limits specified by national authorities.

The major transients which a water reactor designer has to consider are (1) loss of load due to a turbine trip, (2) loss of flow due to loss of pump electrics, (3) reactivity insertion, (4) breaks in the secondary steam line to the turbine, and (5) a loss of coolant accident due to a rupture of the primary circuit. Because (1), (2) and (3) are considered likely to occur they are classed as 'upset' conditions, whilst (4) and (5) are either 'emergency' or 'fault' conditions, depending upon the magnitude of the break. Some of these transients will be described in more detail later in this section.

A further characteristic of water reactors which is important, and which was demonstrated by the BORAX tests described in §4, is that loss of coolant/moderator shuts down the reactor. This is a very important fail-safe characteristic.

Even when the reactor is shut down, decay of fission products still generates significant amounts of heat. For a reactor operating at 2800 MW(t) the decay heat is still 100 MW(t) 1 minute after shutdown, 37 MW(t) 1 hour after shutdown, 10 MW(t) 1 day after shutdown, 7 MW(t) 1 week after shutdown, and 3 MW(t) 1 month after shutdown. These power levels are such that the core must be kept covered by water to prevent it overheating.

The considerations discussed so far point to two fundamental safety issues: first, the need to provide core-cooling systems which will be adequate for all foreseeable situations including the improbable 'fault' situations, and secondly, the need to ensure that the container for the core, that is the pressure vessel, maintains its integrity and that adequate make-up is available (from the ECCS) to compensate for any foreseeable leakage. Hence the preoccupation in discussions on light water reactor safety with two issues: (1) the effectiveness of the ECCS and the (2) integrity of the reactor pressure vessel. These issues are dealt with in turn in §§ 7.3 and 7.4.[7]

### 7.3 Loss-of-coolant accidents (LOCA)

The worst accident sequence with a light water reactor which is considered credible and as needing to be designed against is a complete circumferential break—'guillotine' break—of one of the main primary coolant pipes resulting in coolant discharging from both ends of the pipe. The relevant pipe for the PWR is one of the main inlet (*cold leg*) 710 mm diameter pipes to the reactor vessel, and, for the BWR, the suction side of the jet pump recirculation line. In either case, in the absence of any protective measures, the contents of the primary circuit of the reactor will discharge through the breach in a few tens of seconds and although the nuclear reaction is shut down, the core will become uncovered and the decay heat will cause an uncontrolled rise in temperature resulting in core damage and possible melting of the fuel.

Prior to 1966 LWRs were not provided with the high-capacity emergency core-cooling systems described in §5. These early plants had make-up systems which could cope with relatively small leaks and had decay heat removal systems for long-term cooling of the core. They did not, however, have the capacity to handle a major rupture of the primary circuit, and for these relatively small early LWRs reliance was placed primarily on the containment.

In 1966 a USAEC task force under William K. Ergen, set up to examine the problem, concluded that not only could the sequence of events following a LOCA not be predicted with any accuracy, but there were also serious concerns that plant containment integrity would not be maintained in these circumstances. In particular the Ergen report mentioned the possibility that substantial portions of the core could melt through the bottom of the pressure vessel and thence through the containment and into the earth below. This event has since become colloquially known as the 'China syndrome' since, for reactors sited in the United States, the molten core is supposed to be headed in the general direction of China. In actual fact numerous studies have since shown that should this unlikely event occur the molten material would become diluted and would become stabilized as a pool with a maximum depth of perhaps 15–18 m.

During 1966 reactor vendors in the United States started to introduce high-capacity emergency core-cooling systems (ECCS) to cope with the large-break LOCA. Because the overall ECCS had to be very reliable, it became the practice to use a number of different systems to provide diversity and redundancy. A typical arrangement would comprise the high-pressure injection system (HPIS), the low-pressure injection system (LPIS), and core-flooding accumulators, each individual system being arranged in up to three and sometimes four parallel and identical lines or *trains*. Some earlier plants have been modified (*back-fitted*) by the addition of ECC systems.

About this time the USAEC initiated an extensive programme of analytical and experimental work to predict the course of a loss-of-coolant accident and the performance of the emergency core-cooling systems. The main focus of this programme was the so-called Loss-of-Fluid Test (LOFT) Facility, a 55 MW(t) model of a PWR under construction at NRTS Idaho. This reactor was originally intended to study a complete core meltdown and the associated release of fission products. However, in 1967, the objective was altered to the investigation of the ability of the emergency core-cooling systems to prevent a core meltdown. At the same time the reactor vendors also undertook extensive development of computer codes to predict the course of events in their particular reactors.

As part of the USAEC studies a semi-scale simulation of the complete blowdown transient followed by operation of the ECCS was also set up at Idaho Falls. It represented a single PWR primary loop equipped with an electrically heated core of 120 fuel pins, having a 0.228 m heated length and a total power output of 1.1 MW. The results of these tests became available during 1970–71. They showed that the injected water failed to flow into the simulated core and cool the pins as expected. Although it is now known that the experiments were not representative of the behaviour of large systems, the USAEC felt obliged in 1971 to issue its *Interim ECCS acceptance criteria* which specified in considerable detail the techniques it would accept for calculating ECCS performance. When the results of the Idaho Falls tests became widely known, environmentalist groups in the US intervened in reactor licence hearings with criticisms that the performance of the ECCS was unsatisfactory and the reactors were unsafe. Consequently the USAEC arranged public hearings to allow a full discussion of the subject before issuing *Final ECCS acceptance criteria*. The hearings started in January 1972 and lasted over 18 months. They resulted in increased conservatism both in the allowable peak clad temperature (1200°C) and in the methods to be used to calculate peak clad temperatures.

It may be helpful at this stage to describe briefly the sequence of events in a LOCA for a PWR and a BWR respectively.

### 7.3.1  *LOCA in a PWR*
The sequence of events for a PWR can be divided into four distinct phases:
    (1)  the blowdown phase,
    (2)  the refill phase,
    (3)  the reflood phase, and
    (4)  the long-term cooling phase.

Immediately following any rupture, the high-pressure water undergoes a rapid decompression from a subcooled to a saturated state. This process is controlled by the velocity of transmission of pressure waves in the fluid and by the areas and lengths of the piping. The break in the inlet piping would mean that the decompression wave would reach the lower plenum first. The core flow could show violent oscillations followed by a reversal in direction. In addition, interaction between the pressure waves and the structure could result in movements of the vessel and its internals. The rapid production of steam within the core would shut down the reactor, leaving the fuel however inadequately cooled. A *critical heat flux* situation would be reached after about two seconds and the fuel clad temperature would start to rise. The fuel temperature itself would drop sharply since the reactor would be shut down but the large amount of stored energy in the fuel and the radial redistribution of temperature would cause the clad temperature to rise. At about four seconds there would be a relatively abrupt reduction in the mass flowrate out of the break as a result of a transition from subcooled to saturated fluid discharging at the break. This corresponds to the point where the hot fluid from the reactor core reaches the break. As a consequence there would be a corresponding net increase in flow down the downcomer and into the core which would be sufficient to stop further increase in peak clad temperature when this had reached about 700 °C. Both the HPIS and LPIS would be activated on receipt of the safety injection signal (about 2 seconds) and all pumps would start. At about 14 seconds the HPIS would be initiated automatically followed immediately by the accumulator system as the system pressure fell below 41 bar. The LPIS would start to inject water at about 30 seconds as the circuit pressure continued to fall and the circuit and containment pressures finally equalized, this representing the end of the blowdown phase and the start of the refill phase. During blowdown the lower plenum would have been partially voided but water would now return to the lower plenum filling it. However during this refill phase, the fuel remains relatively poorly cooled and despite the low heat generation, the clad temperature and the fuel temperature would rise appreciably with possible bursting of the clad.

The third phase of the accident, the *reflood phase*, would start once the ECCS water reached the bottom of the core and started to quench the over-heated fuel pins. Quenching of the complete core might take up to 5 minutes, for example in the core where the reflooding rate occurred at 25 mm elevation per second. At the end of the quenching period the fuel and clad temperatures would have been reduced to very low values, 100–200 °C. The last phase of the accident would now begin corresponding to the long-term steady-state condition of removal of decay heat in the reactor.

### 7.3.2 *LOCA in a BWR*
The sequence of events for a BWR differs somewhat from the PWR. An initial rupture in one of the recirculation loops would produce a more

gradual depressurization compared with a PWR. The flow in the damaged loop would reverse draining water from mixing plenum and downcomer. However, the core flow would be maintained during the early part of the accident by the coastdown of the feed pump and by the undamaged recirculation loop. Finally the suction of the jet pumps would become uncovered and the core flow would drop to zero. A *critical heat flux* situation would occur at about 10 seconds as a result of the core flow stagnation and the fuel clad temperature would start to rise. The flow at the break would switch mainly to steam, the water in the downcomer would be completely discharged, and steam formation would occur in the lower plenum as the system pressure decreased more rapidly. This in turn would force a two-phase mixture up through the jet pump diffusers and the core, resulting in enhanced core heat transfer.

After about 30 seconds the ECCS would be triggered either by low water level or high dry-well pressure. The Automatic Depressurization System (ADS) would operate reducing the vessel pressure and allowing the LPCI and LPCS to come into operation. Initiation of the core spray system would result in wetting of the shrouds surrounding each fuel element. This in turn would limit the temperature rise of the cladding. Accumulation of water in the lower plenum leading to reflooding would provide a second mechanism of quenching. Quenching would be complete after about 80–100 seconds and the long-term decay heat cooling phase would then begin.

### 7.3.3 *LOCA Studies*

Since 1972 the USAEC followed by the NRC (Nuclear Regulatory Commission), in collaboration with many other organizations both inside the US and in other countries operating light water reactors, has mounted a massive programme of work to understand all aspects of the course of a large LOCA. During 1979 the first tests on the LOFT reactor were performed with a nuclear heated core. The results of these tests confirmed firstly, that a good quantitative understanding of the complex thermohydraulic and fuel behaviour conditions of the accident had been established and, secondly, that the criteria adopted for licensing purposes in the United States were significantly conservative.

Consideration is given not only to the very improbable large-break type of LOCA described in the previous paragraphs, but also to the much more likely small-break accidents (a category including, for example, the Three Mile Island accident discussed later in this section). Small breaks, which have received increasing attention recently, are usually characterized as those with discharge areas up to about $0.01\ m^2$ corresponding to about 2 per cent of the cross-section of one of the main coolant pipes on a PWR. The rate of depressurization is dictated by both the break size and the break location. Such events differ from the large-break LOCA in requiring heat removal from the system via the steam generators or some other path over an extended time period, whilst the primary system remains at pressure. The important variable during this extended depressurization is the level of the steam–water mixture in the reactor pressure vessel. Every effort must be made to ensure that the core does not become uncovered during this period.

## 7.4 Pressure vessel integrity

The integrity of the reactor pressure vessel (RPV) for a light water reactor is vital from a safety viewpoint for two separate reasons. Firstly, the reactor pressure vessel is the container in which the core sits. If a leak should develop below the level of the core and the flow through the leak path be greater than the maximum flow capable of being supplied by the ECCS, then the core may become uncovered and consequently overheat. Secondly, a massive failure of the reactor pressure vessel can both seriously damage the core and break the containment. Thus a single event could outflank the various sequential barriers which prevent the escape of fission products in other accident sequences: namely the fuel cladding, the primary circuit and, should these fail, the containment building. Clearly, it is necessary to demonstrate conclusively that a disruptive failure of the reactor pressure vessel has a very low probability indeed.

Failure of the RPV could conceivably occur because of an inherent weakness in the construction of the vessel itself, or alternatively as a result of an internal or external event which is outside the design basis of the plant. Such events could possibly be a molten fuel–coolant explosion inside the vessel or alternatively a gross failure of the vessel support system. Provided such events can be shown to have a very low probability then the main consideration must be of the vessel structure itself.

As far as can be ascertained, no large pressure vessel built to the standards appropriate for fossil or nuclear power stations has failed catastrophically in service over the past few decades, although a number of vessels have failed whilst undergoing hydrostatic testing. Thus, prior to 1965, it was maintained that failure of the RPV of a light water reactor was incredible. However, during the late 1960s programmes of work, notably the Heavy Section Steel Technology (HSST) Programme in the US, were initiated in a number of areas to improve confidence in pressure-vessel integrity. The primary areas covered were materials properties, fracture, fatigue crack growth, irradiation effects on steels, thermal shock, and model pressure vessel tests. One of the major advances over this period was the development of *fracture mechanics*, a method of analysis which can establish the response of a loaded structure to flaws or cracks postulated to be present at various locations in the material of the structure. The local stresses in the vicinity of a crack are greatly magnified over those in unflawed material and indeed can exceed the fundamental cohesion of the material. In these circumstances the crack extends indefinitely and the section fractures. For a given section, material and loading there is a limiting or *critical* crack size, below which the flaw will be stable and above which the flaw will either extend or become unstable. Fracture mechanics allows this critical size to be estimated.

The physical property which is a measure of the resistance of a material to crack extension under stress is known as the *toughness* or *fracture toughness*. It is important that the materials used in pressure vessels possess high fracture toughness. This is true not only of the forging or plate materials themselves, but also of the weldments joining them together, since it is in these regions that flaws might be expected to occur most frequently.

Despite the presence of a thermal shield which is designed to reduce the neutron flux reaching the pressure vessel wall, that region of the RPV immediately opposite the core is exposed to neutron irradiation. As a result the fracture toughness of the material is affected. The extent to which the toughness is reduced depends critically upon the levels of certain chemical elements in the steels like copper, vanadium, and phosphorus.

During the lifetime of the vessel, a defect in the material will be subjected to cyclic pressure and thermal stresses resulting from the normal operational transients experienced by the vessel. As a consequence the defect will grow by a small amount, and it is important to establish that this accumulated growth throughout the lifetime of the vessel is not such that the defect will become critical. Extensive experimental testing programmes have been undertaken to establish crack growth-rates as a function of the change in stress per cycle, the cycle frequency, and whether the crack is 'dry', i.e. in air or 'wet', i.e. exposed to reactor coolant.

It is also important that the integrity of the reactor pressure vessel should be maintained during a serious accident sequence such as a loss of coolant (LOCA) or rupture of the secondary steam line between the steam generator and the turbine. This latter accident would result in an uncontrolled removal of heat from the intact primary circuit, causing both the system temperature and pressure to drop sharply over a short time period. Subsequently the system pressure returns to its normal level as the HPIS is activated. During a LOCA, cold ECCS water would come into contact with the vessel wall. Thus, in both accident sequences thermal shocking of the inside surfaces of the vessel occurs. This is compounded in the case of the steam line break by the additional stresses resulting from the repressurization. As a result critical crack sizes for accident conditions are significantly smaller than for normal operation.

Model pressure vessels tested under a variety of conditions with large flaws present in either the membrane or nozzle sections have confirmed the general applicability of the methods used to establish reactor pressure-vessel integrity and have demonstrated the tolerance of such vessels to the presence of defects. In one test a vessel with a crack whose depth was over 88 per cent of the wall thickness and with a length over three times the wall thickness sustained twice the design pressure before leaking—rather than failing catastrophically.

The integrity of any particular vessel depends upon achieving uniformly high standards of materials selection and fabrication. To achieve this uniformity the fabricator of the vessel applies a set of procedures and methods with the aim of ensuring that every vessel meets the standards required. These procedures and methods are written down in a quality-assurance manual which must be strictly followed. A series of checks and audits are carried out by independent inspectors, although the ultimate responsibility for the quality of the vessel remains firmly with the vessel fabricator. An important feature of the vessel fabrication is the use of non-destructive examination techniques including ultrasonics and radiography to ensure the absence of significant defects in the vessel when it enters service. Flaws initially found to be present in, say, a weldment must be removed by grinding out and a repair made by welding prior to a final post-weld heat treatment.

Not only must the vessel be free of significant defects when it enters service, but it must remain so throughout its service life. To ensure this, in-service inspections involving visual and ultrasonic examinations are undertaken at intervals throughout the vessel's life. The ultrasonic inspection method consists basically of injecting a beam of ultrasound into the metal and displaying the reflection of this beam from any discontinuities in the metal. Because the RPV becomes radioactive during the course of its operation, use is made of automated equipment which can be inserted into the flooded vessel once the core internals have been removed (Fig. 4.22). Access to the entire inside surface of the vessel including the nozzles is possible in this way.

A number of detailed examinations of the integrity of light water reactor pressure vessels have been undertaken, including one under the chairmanship of Dr. Walter Marshall whose first report was published in 1976. These studies conclude that, provided the design intentions are met, and that the vessels are fabricated from well-characterized materials using well-established procedures backed up by good quality-assurance, inspection, and testing, then the vessel will have a high integrity and reliability at the start of its life and its failure probability during service will be less than or equal to one in a million reactor operating years.

## 7.5 Accidents involving light water reactors

### 7.5.1 *The SL-1 accident*
On 3 January 1961 the small experimental reactor SL-1 installed at the National Reactor Testing Station Idaho was destroyed as a result of a large insertion of reactivity caused by manual withdrawal of a control rod whilst the reactor was shut down. The input of 100–200 megawatt-seconds of energy melted the plate-type fuel elements. This was followed by a molten fuel–coolant explosion, which caused bulging of the pressure vessel and the vessel to jump several feet off its mountings. The three operators were killed.

Large reactivity insertions could occur in modern LWRs as a result of such low-probability events as a failure to trip the reactor, operator error, sudden

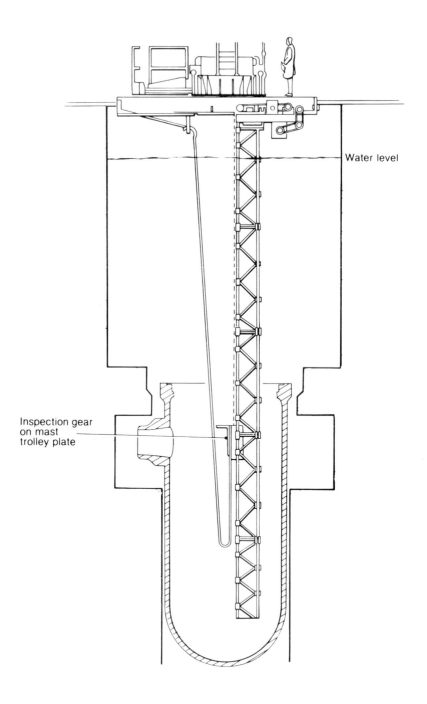

Water level

Inspection gear
on mast
trolley plate

Fig. 4.22 Diagram showing an arrangement for the automated ultrasonic and visual
inspection of PWR vessels

changes in core configuration, control rod ejection etc. The immediate consequence of a reactivity insertion is an increase in pressure, an effect which may reinforce the increase in reactivity if the reactor has a strong positive pressure coefficient of reactivity, as in a BWR. The behaviour of the fuel under such conditions depends upon the amount of energy released from the core and upon the level of its irradiation. Following the BORAX experiments, a continuing series of tests on oxide fuel rods has been carried out in the USA, first in SPERT, then in TREAT, and finally under representative full-scale pressure and temperature conditions in the Power Burst Facility (PBF). Failure of fuel cladding occurs with unirradiated fuel at about 942 kJ/kg $UO_2$ energy release, whilst irradiated fuel fails somewhat earlier at about 586 kJ/kg $UO_2$. At higher energy inputs there is fuel pin swelling followed by extensive cracking and crumbling of the fuel. At very high energy inputs, 2000–3000 kJ/kg $UO_2$, fuel melting occurs. It is not yet clear whether the conditions exist to allow molten fuel–coolant explosions with oxide fuels as well as with molten metallic fuels as occurred in the SL-1 accident.

### 7.5.2 *Browns Ferry fire*

The Browns Ferry plant, near Decatur, Alabama is owned by TVA and consists of three 1100 MW(e) BWR/4 units. On 22 March 1975 a fire originated in the electrical cable trays and burned for seven hours before it was extinguished. Units 1 and 2 were both at full power when the fire started. Unit 3 was under construction and was not affected by the accident. The fire was initiated by a small lighted candle that was being used to check for air leakage from the containment. The initial location was in an electrical cable penetration between the cable spreading room and the reactor building. However the fire spread both horizontally and vertically, affecting about 2000 cables and causing about $10 million of damage. There was a reluctance to use water on the fire until both reactors were in a stable shutdown condition, because of the possibility of short circuiting. Once water was used the fire was rapidly put out.

Both reactors were safely shut down. However, because of the fire, both the shutdown cooling system and the emergency core-cooling system for Unit 1 were inoperable for several hours. Alternative means of injecting water into the reactor were used, including a control rod drive system pump and condensate booster pumps. The use of these alternative water supplies required depressurization of the reactor, and during this manoeuvre the water level over the core dropped to a point 1.2 m above the top of the fuel. Sufficient cooling was however provided throughout the incident to prevent the core from overheatig. No significant problems were encountered with the cooling of Unit 2 and the high-pressure cooling system was successfully initiated. There was no release of radioactivity off-site and no-one on the site was seriously injured. Both units were however out of operation for over a year whilst the damage was repaired.

### 7.5.3 *Three Mile Island*

On 28 March 1979 a serious accident occurred at the Three Mile Island Unit 2 (TMI-2) plant near Harrisburg, Pennsylvania. The plant consists of two Babcock & Wilcox PWR reactors having an electrical capacity of 961 MW(e) each. At the start of the incident the TMI-2 reactor was operating at close to its design output. Difficulties with the feedwater system caused the main feedwater pumps to trip, which, in turn, resulted in a main turbine trip. The plant responded as expected: the primary circuit pressure increased, the reactor tripped on a high-primary-circuit-pressure signal, and the power-operated relief valve on the pressurizer opened to release the excess pressure. The auxiliary feed pumps started, but failed to deliver water to the steam generators because isolating valves had been inadvertently left shut following maintenance and testing. When the circuit pressure fell back to the normal level, unknown to the operators the relief valve failed to close, allowing a continuous discharge of water and steam, first to the reactor coolant drain tank and subsequently to the containment building. The high-pressure emergency injection system (HPIS) was activated when the primary circuit pressure fell to 110 bar, but because of a high-water-level signal in the pressurizer, the operator thought the circuit was full of water and throttled back the HPIS flow. Auxiliary feedwater was restored to the once-through steam generators but not before they had boiled dry.

Over the next 100 minutes about two-thirds of the total water inventory of the primary circuit was discharged to the reactor containment. The containment was not isolated and some of this water was pumped to the auxiliary building. During this time significant boiling was occurring in the primary circuit, but the decay heat in the core was being removed without overheating of the fuel by this boiling process. However cavitation in and vibration of the primary pumps meant that they had to be stopped, and when this happened the water levels in the circuit equalized at an elevation about 300 mm above the top of the fuel. The operators finally realized that the relief valve was stuck open and took action to stop the discharge.

However, due to operator error, inadequate water flow was being maintained to the system, and the level of water in the core fell exposing fuel. Three hours after the start of the incident, most of the core was uncovered, had heated up to high temperatures (up to 1800 °C), and a chemical reaction between the zirconium alloy cladding and the steam had been initiated, forming large quantities of hydrogen. High radiation levels were indicated and a general emergency declared. The heat-up transient was terminated by the restoration of the full HPIS flow.

During the next ten hours various efforts were made to return the plant to a stable cooling mode. Finally, repressurization allowed the reactor coolant pumps to be restarted and circulation through the primary circuit to be re-established. A large bubble of incondensible gas had accumulated in the head of the reactor pressure vessel and there were fears, later proved to be based on erroneous technical information, that there was a danger of a hydrogen–oxygen explosion. Such a reaction had occurred earlier in the containment building itself. Over the next five days the bubble of hydrogen was removed by degassing the circuit through the pressurizer relief valve and the chemical and volume-control system.

Significant damage was done to the reactor core; about 40 per cent of the zirconium alloy was consumed in the steam reaction. Fuel in the upper regions of the core disintegrated and fell down on to the intact lower parts blocking the coolant flow paths. The majority of the noble gases (xenon and krypton) were released from the fuel, most of these short-lived gases being discharged from the site. However very little iodine was released. An evacuation of the area immediately surrounding the site was considered, and there was a recommendation for expectant mothers and children to be evacuated. In the event, however, exposure of the general public to radio-activity was very small indeed and the future consequences (statistically less than one additional cancer death) undetectable in the surrounding population.

A Presidential Commision into the causes of the incident found that operator error was the direct cause of the accident. Contributory factors were operator training, control room design, and the attitude to safety by the US industry. The Commission was also very critical of the NRC. The plant is badly damaged and recovery operations are likely to take a number of years and cost $500–1000 million. It is not clear whether the unit can be repaired and put back into operation or whether it will be written off.

## 7.6 **The reactor safety (Rasmussen) study**[12]

In 1974 the United States Nuclear Regulatory Commission published a detailed study of the safety of light water reactors. It made use of probabilistic risk-assessment techniques (see Chapter 22), and examined the consequences of a wide spectrum of initiating events and subsequent plant responses, including the failure of one or more of the safety features installed on the plant. Two plants were used as the basis of the study, a PWR—the Surry No. 1 unit, 788 MW(e)—and a BWR—Peach Bottom No. 2 unit, 1065 MW(e). They were selected on the basis that they were the largest plants of each type about to start operation in the United States.

It was concluded that the risks from accidents to light water reactor nuclear power plants were smaller than many other man-made and natural risks to which the human race is exposed to as a society and as individuals. These other risks include fires, explosions, dam failures, air travel, toxic chemicals, tornadoes, hurricanes, and earthquakes. The differences between the PWR and the BWR were not significant. The chance of a meltdown of the reactor core was estimated at 1 in 17 000 reactor years, but the probabil-ities were that such a meltdown would have relatively minor consequences in terms of acute illnesses, latent cancers, or property damage. The small-breach loss-of-coolant accident was identified as contributing. the largest probability to PWR core melting, and it is interesting to note that the accident at Three Mile Island was such an accident sequence. In the case of the BWR it is failure of the decay heat removal systems coupled with the large number of transient events requiring activation of the trip system (ten times/year) which dominates the core melt sequence. The study also concluded that failure of the pressure vessel would have to be about ten times more likely than the value estimated by the Marshall Study Group if it were to make a significant addition to the risk of large accidents.

A report produced under the chairmanship of Professor Lewis criticised some aspects of Rasmussen's conclusions, in particular the confidence limits attached to the assigned probabilities. The Lewis Report confirmed however, that the methodology employed by Rasmussen was fundamentally sound.

Essentially similar results to Rasmussen's were obtained in Phase A of the German Risk Study, carried out by the Gesellschaft für Reaktorisicherheit.[13] While departing from Rasmussen in the dose-risk relationship assumed, as well as in a number of aspects specific to local site features, the German study again confirmed the basic soundness of the event-tree/fault-tree approach.

# 8 Performance

## 8.1 Introduction

Of the world's operating commercial nuclear power plants, over 70 per cent by number and more than 80 per cent in terms of power output are light water reactors. At the end of 1979, with 177 commercial reactors in service, 77 were PWRs (corresponding to a total power output of 59.6 GW(e) ), 51 were BWRs (with a total power output of 35.3 GW(e) ), and the rest either heavy water reactors or gas-cooled reactors (with a combined power output of 14.8 GW(e) ).

The only performance index which is generally available and which can readily be used for comparison purposes is the load factor. This is defined as the ratio between the gross electricity production achieved over a time period and the theoretical maximum electricity production assuming continuous operation at the design output. Care must be exercised in comparing load factors between plants, because sometimes the ratio is based not on design output but on some other figure such as *licensed power level* or *guaranteed power levels*, which may be below the intended design output. In addition, average load factor does not take into account either the number of stations that have just entered service and are in their shakedown period, or whether stations are shut down because of licensing problems.

The performance of light water reactors varies from country to country, with the type of reactor and with the manufacturer. In particular, the PWR stations in Switzerland and the Federal Republic of Germany have achieved remarkably high load factors. In the year up to the end of March 1979, the

PWRs around the world achieved an average load factor of 66.8 per cent, whilst the BWRs reached 61.2 per cent. Cumulative load factors give a picture of the complete history of the reliability of a particular reactor system, and up to the end of 1979 the PWR had achieved an overall lifetime figure of 55.8 per cent and the BWR 52.8 per cent.

Thus there is some evidence that over the early years at least, the BWR had a somewhat lower performance than the PWR. This difference may be due to basic problems with the BWR in the early years such as cracked feedwater spargers and vibration in the core, repairs to the turbine and steam plant possibly being impeded by radiation, and the fact that the smaller fuel assemblies of the BWR mean that a greater number of movements are necessary during refuelling. The amount of accumulated operating experience with the PWR is also very much greater than with the BWR if the experience with the naval submarine reactors is taken into account. Examination of the variation of load factor with age of a reactor after commissioning suggests that performance improves somewhat with age, certainly over the first five years of a plant's life.

## 8.2 Fuel performance[14]

The early water reactors attempted to use Zircaloy cladding for the fuel, but difficulties were experienced with achieving the necessary quality. Many of the plants constructed during the early 1960s therefore utilized stainless steel cladding and accepted the consequent neutronic penalty. Stainless steel cladding performed adequately in the PWR but failures were experienced in the BWR cores as a result of intergranular cracking in the oxygenated boiling-water environment. By the late 1960s the quality of Zircaloy tubing had improved significantly, and the change back to this material as cladding was made in both the BWR and the PWR.

Problems were experienced with particular batches of Zircaloy-clad fuel, and early failures occurred owing to internal contamination of the cladding. The offending impurities appeared to be moisture, hydrocarbons, or halogens, particularly fluorine, in the fuel pellets. The first two impurities can liberate hydrogen which may locally cause hydriding, whilst the halogens appeared to produce a type of stress-corrosion cracking. These problems affected the Dresden-2 core in particular. The failure rate around this time (1972–3) was about 0.5 per cent (Fig. 4.23) corresponding to 250 failed fuel pins in a 1000 MW(e) reactor.

The design basis for clean-up of fission products in the primary coolant is, however, equivalent to a 1 per cent fuel failure rate, although limits on the concentration of particular fission products in the primary coolant, particularly $^{131}I$, do prove to be somewhat more restrictive.

A further problem which occurred around this time was that of densification of the relatively low-density $UO_2$ pellets then in use. As the pellet stack shrank so axial gaps tended to open up, and in the PWR flattening of the

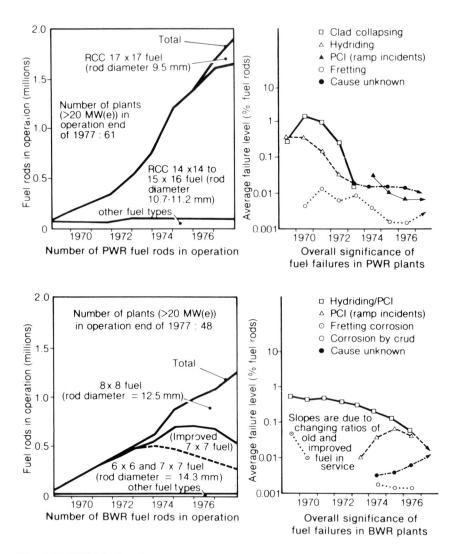

Fig. 4.23  LWR fuel performance

unsupported cladding occurred owing to the external pressure. This problem was alleviated initially by restricting maximum power levels and later by pre-pressurizing the fuel pins with helium before sealing, thus increasing the internal gas pressure and reducing the compressive stress in the cladding. At the same time higher-density $UO_2$ pellets were substituted.

Some power increases are also a cause of cladding failure. A power increase generates a radial temperature gradient within the fuel pellet and the cladding becomes stressed as a result of the fuel expansion. This stress, coupled with embrittling fission products on the inside surface of the clad, is responsible for the formation of stress-corrosion cracks, which occur at pellet-to-pellet interfaces where there are ridges in the cladding. Current operational practice is to limit the rate of any imposed power increases; for example, PWRs are limited to increases of 5 per cent per hour up to 93 per cent power, then progressively 3 per cent per hour, and finally 1 per cent per hour up to full power.

Over the period since 1972/3 water reactor fuel performance has improved significantly, such that the failure rate is now around 0.05 per cent or lower (corresponding to 25 failed pins for a 1000 MW(e) reactor). This has come about as a result of improved fuel-manufacturing techniques to avoid hydriding early in life, operating restrictions particularly on the rate of power increases, lower fuel ratings, and special techniques to minimize overstressing of the cladding, e.g. pre-pressurization, chamfered pellets, and high-density $UO_2$.

## 8.3 Steam generators

The component with which there has been most problems is the steam generator. About 25 per cent of all major outages (100 hours or over) on PWRs are caused by problems with this unit. The mean duration of such an outage for repairs is about 500 hours and, for a 1000 MW(e) plant, the replacement power during this time, assuming it is generated by fossil fuels, requires 700 000 barrels of oil or 150 000 tonnes of coal. In addition repair of steam generators accounts for a significant fraction of the radiation doses to workers at PWR power plants.

Each year a survey is published of experience with this particular component in respect of tube failures. The survey covers all types of water reactors, but of the 86 plants surveyed in 1978[15] 67 were PWRs (the remainder being mainly CANDUs). Table 4.10 reproduces information taken from this survey. This shows that 31 of the reactors surveyed or 36 per cent have

Table 4.10 Steam generator performance (taken from reference (14) )

(a) Overall statistics

| Year | Reactors | | | Tubes | | |
|---|---|---|---|---|---|---|
| | in survey | with defects in SGs | % with defects | in survey | with defects | % with defects |
| 1971 | 34 | 19 | 56 | 337 808 | 1305 | 0.39 |
| 1972 | 36 | 13 | 36 | 364 691 | 1066 | 0.29 |
| 1973 | 48 | 11 | 23 | 553 883 | 3942 | 0.71 |
| 1974 | 59 | 25 | 42 | 742 623 | 1990 | 0.27 |
| 1975 | 62 | 22 | 35 | 783 433 | 1671 | 0.21 |
| 1976 | 68 | 25 | 37 | 879 333 | 3763 | 0.43 |
| 1977 | 79 | 34 | 43 | 1 008 893 | 4355 | 0.43 |
| 1978 | 86 | 31 | 36 | 1 119 728 | 1242 | 0.11 |

Table 4.10 (*cont.*):

(b) Timing (cumulative tube defects to end 1978)

| Effective full power days | Reactors | | | Tubes | | |
|---|---|---|---|---|---|---|
| | in survey | with defects in SGs | % with defects | in survey | with defects | % with defects |
| 500 | 16 | 4 | 25 | 243 084 | 19 | 0.008 |
| 500–1000 | 21 | 11 | 52 | 317 941 | 3511 | 1.1 |
| 1000 | 49 | 36 | 73 | 558 703 | 14489 | 2.6 |

(c) Location of 1978 tube defects

| Location | Number of reactors affected | Number of tubes plugged | % of tubes plugged |
|---|---|---|---|
| Within tubesheet | 4 | 63 | 5.1 |
| Near tubesheet | 13 | 131 | 10.5 |
| Support plates | 13 | 964 | 77.6 |
| U-bend | 5 | 20 | 1.6 |
| Undetermined | 5 | 64 | 5.2 |
| Total | — | 1242 | 100.0 |

(d) Causes

| Cause | Number of reactors affected | Number of tube defects | % of tube defects |
|---|---|---|---|
| Denting (D) | 9 | 926 | 74.6 |
| Wastage (W) | 8 | 90 | 7.2 |
| Stress-corrosion cracking (SCC) | 8 | 81 | 6.5 |
| Fretting (Fr) | 3 | 14 | 1.1 |
| Fatigue (F) | 1 | 5 | 0.4 |
| Other | 5 | 39 | 3.1 |
| Unknown | 11 | 87 | 7.0 |
| Total | — | 1242 | 100.0 |

(e) Materials (defects to end 1978)

| Tube material | Number of reactors | Number of tubes | Number of tube defects | % with defects | Failure mechanism |
|---|---|---|---|---|---|
| SS | 11 | 69 853 | 1 019 | 1.5 | SCC,W |
| Inconel 600 | 63 | 809 458 | 16 793 | 2.1 | SCC,W,D,Fr,F |
| Monel 400 | 8 | 167 700 | 335 | 0.2 | SCC,Fr |
| Incoloy 800 | 6 | 72 717 | 0 | 0 | — |

reported tube failures. These tube defects however take some time to develop after commissioning of the reactor and the table shows that for reactors which have been operating longer than 1000 effective full-power days the figure rises to 73 per cent. The total number of defects was some 70 to 80 times higher than would on the record have been expected in an equal overall length of conventional boiler tubing.

The remaining parts of Table 4.10 show that 'denting' of the tubes at the tube support plates is a major cause of failure. Experience, however, varies greatly between one design of steam generator and another, between one reactor and another, and between one operating utility and another. Water chemistry on the secondary side, design details, e.g. tube support-plate design (see Fig. 4.18), and materials used all contribute to this variation in performance.[16] The interested reader is referred to the excellent review by Garnsey[9] for fuller details.

The majority of PWR steam generators have Inconel Alloy 600 tubes. Early designs of recirculating steam generators allowed a considerable accumulation of corrosion-product sludge to occur on the horizontal tube sheet. At tube sheets and tube support plates there also exist crevices where mixing with the bulk liquid is restricted and the heat added from the hot tubes can evaporate water to generate highly concentrated solutions of aggressive salts. The same effect can occur within the porous deposit lying on the tube sheet. These early plants used phosphate additions to control the secondary-side water chemistry. By 1971/2 caustic stress-corrosion cracking of the Inconel Alloy 600 tubes had been identified at the tube sheet or just above the tube sheet within the sludge layer. Westinghouse, in particular, decided to continue with phosphate control, but to restrict the ratio of sodium to phosphate within the range 2.0–2.6. This appeared to suppress the caustic stress-corrosion cracking, but inspections made towards the end of 1973 revealed that extensive tube thinning was occuring in nearly all plants. The exact mechanism of this thinning has not yet been established, but Garnsey[9] has suggested that it is caused by the separation of a phosphate-rich liquid at temperatures above 275 °C.

By August 1974 Westinghouse had abandoned phosphate dosing and was recommending to its customers the use of an all-volatile alkali (AVT) water chemistry on the secondary side. However, within 3–6 months of the changeover a new phenomenon known as 'denting' was observed on an increasing number of stations. It was apparent that tube deformation in the form of a reduction in tube diameter was occurring at tube support-plate locations. This was occurring as a result of corrosion of the support-plate material. Since the corrosion product occupies approximately twice the volume of the metal consumed, this expansion crushes the tubes and distorts the support plate when the design of the latter is such that the corrosion product connot escape and there is insufficient flexibility in the tube support system. The corrosion itself is rapid acid chloride attack resulting from ingress of sea water. Phosphate residues can enhance the attack. Other agents such as copper from the feed train or air ingress can also promote denting. One other consequence of the physical distortion caused by denting has been the increased strain at the apex of the U-tube. The tubes have been distorted into an oval shape and some stress-corrosion cracking has occurred in this region.

Damage at two US plants, Surry Units 1 and 2 and Turkey Point 3 and 4, has been such that replacement of the steam generator units has been considered necessary. The existing steam generator will be cut apart at the transition piece to the upper section of the shell. The upper section of the steam generators will be stored inside the containment and joined to the new lower steam generator assembly which includes the new tube bundle. The damaged lower assembly, including the old tube bundle, will leave the containment via the equipment hatch. An alternative which has also been developed by Westinghouse is a method of removing only the steam generator tubes as opposed to replacing the entire lower section. With the exception of the station at Obrigheim the PWRs constructed by KWU in the Federal Republic of Germany have not experienced caustic cracking. KWU steam generators use Incoloy Alloy 800 tubing. Table 4.10 indicates the apparent superiority of this tubing material. The German units continue to operate using a phosphate treatment and some tube thinning has been experienced in the plants. None of the KWU stations has experienced denting.

The once-through steam generators built by Babcock & Wilcox have not suffered from the problems identified above, but have experienced fretting, fatigue failures, tube pitting, and erosion-corrosion.

Vulnerability to denting attack can be reduced by a number of precautions. All-volatile water treatments give no protection against ingress of sea-water and therefore it is vital to ensure that the condenser has the highest integrity. Full-flow condensate polishing, and full-flow steam deaeration (using direct contact thermal deaeration, and oxygen-scavenging agents such as hydrazine) offer additional protection. The use of copper in the feed train should be avoided. In the more recent designs of recirculating steam generator changes have been made in the material of the tube support plate from carbon steel to 12 Cr steel—or austenitic steel as in the KWU units—and a change to a more open design.

## 8.4  Nozzle cracking in boiling-water reactors

Many boiling-water reactors have experienced extensive cracking of austenitic stainless steel pipework and nozzles. Such cracking was first observed on BWRs as early as 1965. Dresden-1 in particular experienced 13 separate cracks in the period up to 1974. In September 1974 cracks were found in the pipework for the recirculation-loop valve bypass and in the reactor core spray lines of the Dresden-2 reactor. Similar cracking has occurred on most other BWR/3 and BWR/4 plants in the USA and elsewhere. Three factors combine to cause this cracking: the choice of material, the stress level, and the water chemistry. The material used for this pipework is SA 304 stainless steel. It becomes sensitized to intergranular stress-corrosion cracking either during its initial heat treatment or during welding. High initial stresses usually occur in piping when it is installed, and cyclic thermal and vibration stresses often occur in service. The stress levels are usually of little concern except when an environment also exists which can cause stress-corrosion cracking. Such an environment is however provided by the water chemistry of the BWR coolant. In particular, high dissolved oxygen levels occur in BWR water. One of the major potential concerns is that they may trigger off an extensive failure, leading to a loss-of-coolant accident, although this has not occurred thus far. When cracks occur they are located and the affected pipework repaired or replaced. Failure rates have been of the order of $10^{-2}$ per reactor year. This is much higher than the failure rates normally expected for small pipes; the normal rate is some $10^{-3}$ per reactor year, and for larger pipes it is $10^{-4}$–$10^{-6}$ per reactor year. PWRs have also experienced various types of stress-corrosion and fatigue cracking of small-bore pipework associated with the ECCS, boric acid, and feedwater systems.

Cracking has also occurred on the inner clad surface of the feedwater nozzles of most BWRs. This has been caused by thermal stresses set up by the leakage of cold feedwater between the internal sparger and the nozzle bore. A modified design of sparger and removal of the stainless steel cladding in this area of the nozzle is the proposed solution to this problem.

## 8.5 Other causes of lost output

### 8.5.1 *Vibration*

Another problem which has been experienced by a number of water-cooled reactors to a greater or lesser extent is that of flow-induced vibration of components. This phenomenon can manifest itself in different ways in the various parts of the system. The primary sources of local hydraulic pressure oscillations occur in every reactor system—pump noise, eddy shedding, cavitation, boundary layer separation, stalls in diffusers, and so on.

Fuel pins in PWR systems can be excited by random pressure fluctuations in the turbulent boundary layer on the rod and in BWRs by the unsteady momentum fluxes inherently associated with two-phase vapour–liquid flows. A small but consistent number of fuel failures occur as a result of wear and fretting corrosion induced by vibration (see Fig. 4.23). Fretting has also been a minor cause of tube failures in PWR steam generators (see Table 4.10).

Rather more variable and much more difficult to analyse at the design stage is the vibration behaviour of the reactor's internal equipment. A particular example was the damage to the flow channel around the fuel assemblies in some BWR/4 designs. The amount of damage varied from reactor to reactor, but was extreme in the Swiss Müleberg reactor and in the

US Pilgrim reactor, where the shroud was perforated in eight and six channels, respectively, resulting in a narrow vertical slit. The wear was caused by vibration of temporary control curtains, in-core flux monitors, and secondary neutron sources, excited by the bypass flow in the gap between subassemblies. The problem was solved by plugging the bypass flow holes in the core support plate. Vibration of thermal shields and shrouds has occurred, leading, in the case of incidents at the Trino and Chooz PWRs, to collapse of the neutron shield and lengthy outage times. In the case of Big Rock Point, a self-excitation mechanism was set up between the shield and the reactor vessel wall.

### 8.5.2 *Turbine overspeeding*
Certain difficulties have been experienced with the overspeeding of turbines on some BWRs following a reactor trip. One reactor affected in this way was Garigliano. The problem arises on account of residual water in the turbine steam–water separators flashing-off, and the main remedial measure has involved introducing intercept valves which close on a reactor trip.

### 8.5.3 *Water hammer in feed water piping*
The feedwater system piping in some PWRs has experienced a phenomenon described as *water hammer*. The significance of the events varies from plant to plant, but components have been damaged in some cases, i.e. excessive pipe movements and damage to the valve. Most occurrences have taken place after trying to restart feed flow following an operational transient such as loss of steam-generator water level. The events are of significance because of concern that both the main feed flow and the auxiliary feedwater systems might be simultaneously severed to several steam generators and plant cooling capability affected. Detailed modifications to the pipework and feedwater distribution system within the steam generator has reduced the incidence of such events.

### 8.5.4 *Valves and piping*
Minor problems with valves including packing, gasket or seat leakage, galling, and erosion of internals contribute significantly to lost output for both PWRs and BWRs. In addition erosion problems due to two-phase flow in both process and drain piping are a second major cause of lost output.

## 8.6 Occupational radiation exposure

Radiation exposure to operators of light water reactors is significantly above that for gas-cooled reactors but rather less than that for heavy water reactors. An analysis of the data available in the US over an eight year period gives the average occupational exposure for BWRs at 1.5 man-rem per megawatt-year and for PWRs 1.1 man-rem per megawatt-year. As Fig. 4.24 shows, reactor operation accounted for only 10 per cent of this exposure, main-

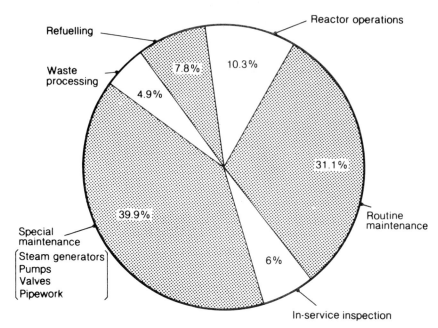

Fig. 4.24 Distribution of dose with task (LWR, 1976 data)

tenance work for 70 per cent, and in-service inspection, refuelling, and waste processing for the remainder.

The radioactivity producing this dose arises in the main from the activation of corrosion products in the primary circuit, particularly structural cobalt and nickel, which are entrained by the coolant and activated within the core. A significant proportion of the dose is a result of unscheduled maintenance operations (e.g. for the replacement or plugging of defective steam generators). It is clear, then, that this dose will to some extent be a function of overall system performance, and the design and operational improvements discussed earlier in this section as likely to improve performance (e.g. improved water chemistry) may serve also to reduce this man-rem commitment.

Other design improvements that may prove important include questions of plant layout, and it may for example be helpful if active plant can be located away from inactive plant so as to minimize operator dose during inactive maintenance. Further reductions may be achievable by increased use of remote inspection and remote maintenance techniques. Materials selection may also have an important role to play here, and there may be advantage in employing primary circuit materials with a reduced cobalt and nickel content (e.g. Incoloy instead of Inconel in steam-generator tubing).

Finally, consideration has been given to the periodic decontamination of the primary circuit to reduce operator dose during subsequent maintenance. The advantages of this need to be assessed on an individual case by case basis, and detailed scrutiny has to be given to the system in assessing whether this would be advantageous.

## 8.7  Construction

Mention has already been made of the relatively rapid construction times achieved for LWR plants during the late 1950s and throughout the 1960s. For example, the Beznau 1 and 2 units constructed in Switzerland were completed in a little over four years each. The LWR has the benefit that much of the nuclear steam supply system (NSSS) is shop fabricated. Because of the very tight specifications and manufacturing procedures required and the high levels of quality assurance necessary as a result, most reactor vendors have set up dedicated manufacturing facilities.

Although the NSSS is largely shop fabricated there is still a considerable amount of site work associated with the balance of plant on an LWR. Site work on all major civil projects around the world tends increasingly to be disrupted by delays due to strikes, industrial problems, difficult working conditions, and other reasons. Throughout the 1970s construction times for LWR stations have lengthened progressively. Some of the reasons, but by no means all, are associated with increased standards and specifications necessary as part of increased safety requirements of national licensing bodies. In the United States and Germany the licensing process itself can take two or more years. Typically the amount of engineering effort required on a project measured in such items as man-hours, drawings produced, calculations performed, etc. has increased by a factor of five despite strenuous efforts to achieve standardization.

Typically a modern LWR might require eight to ten years from ordering to commercial operation. However, this can be improved by standardizing on a particular off-the-shelf design. For example, in France, four standard three-loop Westinghouse plants were ordered by EdF from Framatome in August 1974 for installation at Gravelines, and all four units were in operation by 1981 just 69 months later.

## 8.8  Power plant costs

The costs of light water reactor power plants vary from site to site, from vendor to vendor and from country to country. A review of US power plant costs have been carried out by Sherman of Ebasco. Table 4.11 summarizes his findings. In particular nuclear capital costs have increased by a factor of about seven over the period 1969–78 or by 24 per cent per annum. The capital costs of coal-fired plants have also increased by about the same factor. This represents a considerable real price increase, over and above the effects of inflation, which accounts for only about 21 per cent of the overall increase from \$160/kW(e) in 1969 to \$913/kW(e) in 1978.

This real increase is due less to increases in hardware costs, which have been fairly modest, than to changes in plant design and increased construction

Table 4.11 US power plant costs

| Power plant | | Nuclear 2 × 1200 MW(e) | | Coal 3 × 800 MW(e) | |
|---|---|---|---|---|---|
| Year of estimate | | 1969 | 1978 | 1969 | 1978 |
| Capital cost | $/kW(e) | 160 | 913 | 122 | 639 |
| Expected inflation during construction | $/kW(e) | 62 | 694 | 61 | 627 |
| Total capital cost | $/kW(e) | 226 | 1607 | 183 | 1266 |
| Increase 1969–78 | | — | 7.1 | — | 6.9 |

| Power plant | | Nuclear 2 × 1200 MW(e) | | Coal 3 × 800 MW(e) | |
|---|---|---|---|---|---|
| Capital cost in 1969 | $/kW(e) | 160 | | 122 | |
| Increase due to inflation during construction | $/kW(e) | 162 | (21%) | 120 | (23%) |
| Increase due to licensing changes | $/kW(e) | 591 | (79%) | 397 | (77%) |
| Capital cost in 1978 | $/kW(e) | 913 | | 639 | |

Each plant requires about 7–9 years to commercial operation

times, which are a result of licensing changes following additional safety requirements.

# 9. Looking to the future

## 9.1 Improvements in the LWR

The electrical power generated from a single reactor core has increased from a few tens of megawatts in the mid-fifties to over 1200 MW currently. There is little doubt that, given the incentive, this could be increased further. It would seem that with modern pressure vessel practice, 1800 MW is an attainable output for the PWR and probably also for the BWR. Indeed some vendors have outlined designs for such sizes. However, there is some evidence that the larger units have proved less reliable than the smaller units, or at least have taken more time to achieve high reliability (see § 8). The cost of and the need to provide replacement power for outages with such large plants indicates that it is unlikely there will be significant increases in size over the present units, at least for the foreseeable future.

Likewise, although it has been shown that the present peak linear rating of 39–43 kW/m is conservative and that peak clad temperatures during a LOCA are well below the 1200 °C limit, nevertheless it is unlikely that there will be any move to increase fuel ratings significantly.

Over the next decade there will be a series of improvements to individual components to improve their performance, their reliability, their inspection, and their maintainability. Particular attention will be given to the steam generators. Attention will also be paid to detailed design layouts to reduce the radiation exposure of operators during plant maintenance. The refuelling operation will be streamlined to reduce the annual outage time. Systems will be simplified wherever possible.

In addition, attempts to improve the efficiency of utilization of the uranium fuel may be expected. At present a light water reactor operating on a once-through cycle with a 70 per cent load factor has a lifetime (30 year) requirement of between 4347 and 4610 tonnes per GW(e) of natural uranium (assuming the tailings from the enrichment plant contain 0.2 per cent $^{235}U$). Various improvements have been proposed. If the length of time the fuel remains in the reactor could be increased from three years—equivalent to a burn-up of approximately 30 000 MW d/t—to four or even five years (i.e. burn-ups of 40 000 or 50 000 MW d/t) then an 8 to 12 per cent saving in uranium would be obtained. However, various factors could limit the maximum fuel burn-up possible including pellet–clad interaction, fuel assembly dimensional stability, external clad corrosion, and fuel rod internal pressure (see § 8).

It is estimated that increased burn-up coupled with other measures which could add up to a total of 10–15 per cent savings in uranium, might be demonstrated over the next decade provided current development programmes are successful. These initial savings can probably be achieved without significant economic or safety consequences. Further improvements might be feasible on a somewhat longer timescale. However, it is not yet clear to what extent and at what rate these improvements will become accepted by manufacturers and users. The rate of implementation is very dependent upon licensing conditions, availability of resources, and the particular state of development of the system in the country concerned.

### 9.2 Plutonium recycle

The fuel discharged from a light water reactor contains significant quantities of both unused uranium and plutonium. If these materials are separated from the fission products in a reprocessing plant and recycled via a fuel fabrication plant to the reactor, significant savings in natural uranium requirements of as much as 40 per cent can result. During the early 1970s many countries including the United States undertook the development of the technology to fabricate and use plutonium-bearing fuels within light-water reactors. In the US a series of studies and hearings were held with the object of deciding whether to allow the operation in the United States of light water reactors with plutonium-bearing fuels. However, these hearings—Generic Environmental Statement on the Use of Recycled Plutonium in Mixed Oxide Fuel in Light Water Cooled Reactors (GESMO)—were terminated when President Carter in May 1977 expressed his concern that the widespread use of plutonium could lead to the spread of nuclear weapons, and, as a result, deferred indefinitely the reprocessing of spent fuel in the United States. Other countries, however, assessing the position differently, have experimental irradiations of plutonium-bearing fuels in LWRs in progress as a preliminary to the wider use of plutonium recycle.

If plutonium is successively recycled in light water reactors (i.e. the process of irradiation, reprocessing, and refabrication is repeated), a state of equilibrium will be reached. This occurs when the amount of plutonium recovered from the spent fuel is equal to the amount of plutonium in the new fuel rods loaded into the reactor three or four years earlier. Typically only about one quarter of these fuel rods will contain plutonium as $PuO_2$; the remainder will still contain 3 per cent $^{235}U$ as $UO_2$. A PWR of current design operating for 30 years with the recycling of uranium and plutonium would have a gross natural uranium requirement of between 3850 and 4300 tonnes per GW(e)y (assuming enrichment tails of 0.2 per cent $^{235}U$). Thus the saving is 35–40 per cent compared with the uranium requirement when the spent fuel is not reprocessed.

The physical disposition of the plutonium-bearing fuel pins in the sub-assemblies is of importance. The higher thermal neutron absorption cross-section of plutonium compared with $^{235}U$ means that if the plutonium-bearing fuel pins are loaded near to control rods, those control rods will be less effective. The location of the control rods in a PWR is such that one possibility is to have the fuel assemblies without control rod clusters consisting entirely of plutonium-bearing fuel pins, and the fuel assemblies with control rod clusters to be fuelled entirely with uranium-bearing fuel pins. In a BWR, on the other hand, all sub-assemblies would have to contain a central core of plutonium-bearing fuel pins surrounded on the outside by uranium-bearing fuel pins.

## 9.3 Spectral-shift-controlled reactor

As has been shown in § 5, at the beginning of each reactor cycle neutron poisons have to be added to the core to compensate for the excess reactivity necessary to cover the consumption of fissile material during the cycle. If the average neutron energy could be increased at the start of the cycle (giving a *hard* neutron spectrum), the fuel would be less reactive, owing to increased capture in the $^{238}U$ resonances and a decrease in the rate of $^{235}U$ fission. Likewise if the neutron spectrum could be *softened* towards the end of the cycle, then the capture in the $^{238}U$ resonances can be reduced. In a spectral-shift-controlled reactor, this control of neutron spectrum and hence reactivity is achieved by the adjustment of the light/heavy water ratio in the reactor coolant. At the beginning of the core life the heavy water content could be up to 85 per cent of the primary coolant giving a *hard* spectrum in this design of reactor. During the operating cycle the concentration of heavy water is progressively reduced until at the end of the cycle it may be as low as 2 per cent or less, *softening* the spectrum and increasing reactivity. The preferential absorption of neutrons in fertile material—$^{238}$U—rather than in poisons means that additional fissile material is produced. The net effect is to reduce the amount of $^{235}$U and thus natural uranium needed as compared with the standard cycle. Savings of up to 12 per cent are possible with this concept.

A PWR design of this type was offered by the US Babcock & Wilcox Company in the early 1960s. A joint UK/Belgian experimental core was successfully operated in the Belgium BR3 reactor around 1965.

The special shift reactor requires a plant to upgrade the heavy water at the end of each cycle. Modifications to the reactor itself are necessary to control the heavy water concentration and to ensure that losses of coolant at pumps, valves, etc. are minimized and the heavy water recovered. Thus the capital cost of the reactor would be higher. This is offset however by the lower fuel costs.

Other simpler ways of changing the neutron spectrum are possible. In some BWRs it may be achieved by changing the coolant steam content in the core by altering the coolant flow and/or inlet temperature. The uranium savings depend on the excess capacity of the main circulating pumps and the capability to utilize reduced feedwater temperatures. A similar but smaller effect can be achieved in a PWR by varying the coolant temperature during the operating cycle.

## 9.4  Light water breeder reactor

The light water breeder reactor (LWBR) is basically a PWR with a core designed to run on an optimized $^{233}$U–$^{232}$Th fuel cycle. In current LWRs the *conversion ratio* of the fissile material produced (plutonium) to the fissile material consumed ($^{235}$U) is about 0.6. One of the reasons for this relatively low value is that the water itself absorbs a significant fraction of the neutrons from fission. Reducing the amount of water present makes possible a higher conversion ratio at the expense of increasing the required inventory of fissile material. This under-moderation tends in fact to increase the neutron capture in either $^{238}$U or $^{232}$Th, but in other respects the balance of advantage lies with the thorium system. Economic conditions are conceivable in which the use of $^{233}$U–$^{232}$Th in a much undermoderated lattice may be desirable. The reactor design aims to maximize conversion ratio, and a $^{233}$U–$^{232}$Th system, with its much improved characteristics at thermal neutron energies, opens up the prospect of achieving a conversion ratio close to unity. In this case the reactor makes or breeds as much fissile material as it consumes.

However, with the LWBR the moderator and coolant are still light water and that represents a significant neutronic penalty. There is no prospect of breeding ratios much in excess of unity and, in fact, a breeding ratio of 1.02 is the best that can be calculated. On the basis of 2 per cent unrecoverable residues in the fuel processing and refabrication, it is apparent that self-sufficiency is all that can be expected. Moreover, this also presumes the availability of an adequate supply of $^{233}$U and, in an overall analysis, its production must be taken into account. For a fuller discussion of the neutron physics aspects of thorium systems the reader is referred to Chapter 17.

Because of the need to conserve neutrons in the core, neutron poisons such as boron cannot be used for control under normal operating conditions, although they are used as a back-up shutdown system. The reactor core uses the seed-and-blanket concept of the original Shippingport core (see §3), but in this case the reactor is controlled by actually moving the seed fuel assemblies out of the core. This is equivalent to changing the core configuration and hence the neutron multiplication factor. However, because of flux peaking problems, the power densities in the LWBR are only about 60–70 per cent of those in a modern PWR.

Preliminary work on the LWBR started in the 1960s, and a demonstration of the concept in the Shippingport nuclear reactor is in progress. Full power operation at 62 MW(e) gross was reached in December 1977. Figure 4.25 and Table 4.12 give the major core and plant parameters. Because of the small size of the plant only the central three modules are typical of a larger breeder core. The demonstration core has incorporated a number of compromises and conservatisms which mean that the performance of the present core is not necessarily indicative of the ultimate potential of the concept. For example the fuel grids are fabricated from AM-350 steel, which imposes a heavy penalty in neutron absorption. If zirconium grids could be employed, the breeding gain would be increased.

$^{233}$U does not occur naturally; it has to be made by irradiating $^{232}$Th. For the Shippingport demonstration the necessary $^{233}$U was supplied by the Savannah River production reactors. If the LWBR were to be introduced, it would be necessary to adopt two distinct reactor types: a prebreeder in which the $^{233}$U is created and a breeder in which the $^{233}$U can be used in a self-sustaining mode. Most attention has focused on the latter, but the design of the prebreeder is crucial to the commercial success of the concept.

Present day PWRs could be used as prebreeders simply by replacing the core. The fissile material for these prebreeder cores could be either highly or slightly enriched $^{235}$U with pure $ThO_2$ rods for $^{233}$U production, located in a separate region, or plutonium separated from PWR spent fuel mixed as $PuO_2$ with thorium oxide (see Figs 4.26 and 4.27). Since a 1000 MW(e) PWR can produce about 290–580 kg of $^{233}$U annually, and since about 4000 kg of $^{233}$U would be needed to start up a LWBR, the $^{233}$U would have to be accumulated for perhaps 7–14 reactor years.

The breeder reactor core could also be fitted into an existing PWR if the vessel head and internals were replaced. However, this is not likely to happen because of the lower power density of the LWBR. It is more likely that a new plant would be constructed with a larger core. This would however increase the cost of the NSSS. The LWBR prebreeder/breeder combination would require more uranium input for the first 12–14 years and

Table 4.12 Major core and plant parameters for the LWBR demonstration core at Shippingport

| | |
|---|---|
| Power plant | |
| Gross electrical output | 62 MW(e)† |
| Net station output | 50 MW(e)† |
| Net station heat rate | 14.2 MJ/(kW h) |
| Steam pressure: | |
| full load at generator | 51.3 bar |
| no load at generator | 61.7 bar |
| Number of loops | 4 |
| Reactor pressure drop | 4.8 bar |
| Coolant piping, o.d. | 457 mm |
| Coolant piping, i.d. | 381 mm |
| Coolant velocity, main piping | 10.7 m/s |
| Reactor core type | Pressurized light water cooled and moderated seed and blanket |
| Total reactor heat output | 204 MW(t)† |
| Total coolant flow rate | 3856 kg/s |
| Reactor coolant inlet temperature at 236.6 MW(t) | 271° C |
| Reactor coolant outlet temperature at 236.6 MW(t) | 283° C |
| Average coolant temperature, nominal | 277° C |
| Primary system pressure, nominal | 138 bar |
| Nominal core height, including $ThO_2$ reflector | 3.1 m |
| Mean core diameter | 2.3 m |
| Fuel loading (thorium and uranium) | 42 t |
| Lifetime, effective full-power hours (EFPH)† | 15 000 |
| Fuel material: | |
| movable seed | $^{233}UO_2$ –$ThO_2$; with $ThO_2$ end-reflectors |
| stationary blanket | $^{233}UO_2$–$ThO_2$; with $ThO_2$ end-reflectors |
| reflector blanket | $ThO_2$ |
| Fuel cladding material: | |
| seed, blanket, and reflector | Zircaloy-4, low hafnium |

Rod diameter and cladding thickness†:

| | Outside diameter (mm) | Cladding thickness (mm) |
|---|---|---|
| seed rod | 7.8 | 0.56 |
| blanket rod | 14.5 | 0.70 |
| power-flattening blanket rod | 13.4 | 0.65 |
| reflector blanket rod | 21.1 | 1.07 |

Number of rods per module†:

| Type of module | Seed | Standard blanket | Power-flattening blanket | Reflector blanket |
|---|---|---|---|---|
| Type I | 619 | 443 | | |
| Type II | 619 | 261 | 303 | |
| Type III | 619 | 187 | 446 | |
| Type IV | | | | 228 |
| Type V | | | | 166 |

† These are the minimum expected performance values for the LWBR operation at Shippingport. To ensure that environmental impacts have been conservatively evaluated, the following parameters have been analysed in this environmental statement: gross electrical output, 72 MW(e); net station output, 60 MW(e); total reactor heat output, 236.6 MW(t); lifetime, 18 000 EFPH.

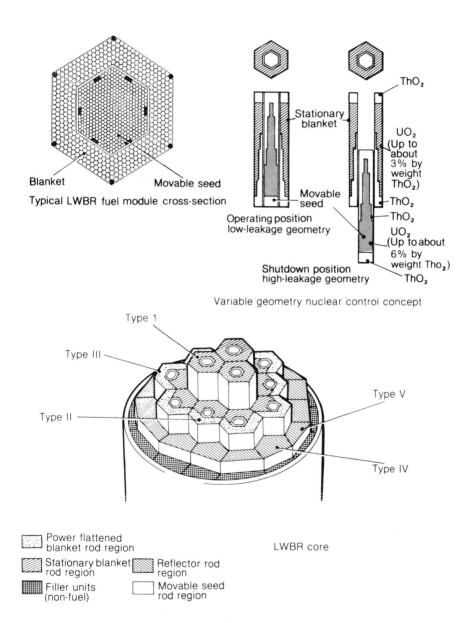

Blanket                    Movable seed

Typical LWBR fuel module cross-section

ThO$_2$

Stationary blanket

UO$_2$ (Up to about 3% by weight ThO$_2$)

Movable seed

Operating position low-leakage geometry

ThO$_2$

ThO$_2$

UO$_2$ (Up to about 6% by weight Tho$_2$)

Shutdown position high-leakage geometry

ThO$_2$

Variable geometry nuclear control concept

Type 1

Type III

Type II

Type V

Type IV

LWBR core

Power flattened blanket rod region

Stationary blanket rod region

Reflector rod region

Filler units (non-fuel)

Movable seed rod region

Fig. 4.25 LWBR concept in Shippingport

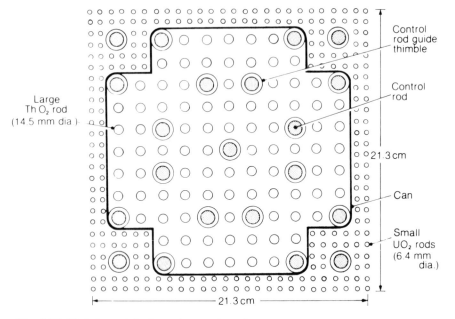

Fig. 4.26 Slightly enriched uranium prebreeder geometry

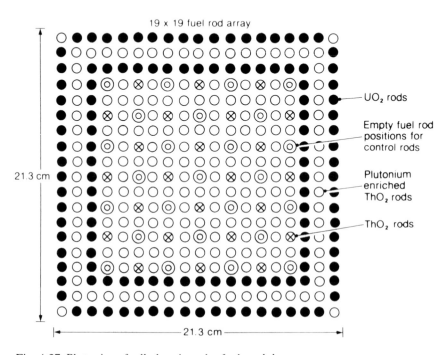

Fig. 4.27 Plutonium-fuelled prebreeder fuel module

more total heavy metal (U and Th) for about 18 years than a modern PWR. The same conclusion also applies to separative work requirements.

Reference 7 gives further details of the LWBR and the reader is also referred to Chapter 17 for a general discussion on the state of the technology.

To summarize, it seems unlikely that future LWRs will either get larger in size or that increased fuel ratings will be attempted. However, improvements will be made in the performance and reliability of individual components. Operating procedures will be modified to reduce operator radiation exposures and annual refuelling and maintenance outages. Major efforts will be made to improve the efficiency of utilization of uranium by a series of fuel management improvements, including increased dwell times for the fuel in the reactor. The recycling of uranium and plutonium contained in the spent fuel discharged from the reactor would make significant savings in uranium, but the extent to which it will be adopted is uncertain given the current concerns about the widespread use of plutonium and the proliferation of nuclear weapons. Other techniques such as altering the neutron spectrum during the course of the fuel cycle have some capacity to save uranium, but involve complications and additional costs and may not be attractive. Finally, efforts are being made to produce a reactor design—the light water breeder reactor—which, if it were developed, might afford substantial uranium savings. Although development work is underway, its introduction into commercial service is not foreseen before the year 2000 at the earliest.

## Acknowledgements

I would like to acknowledge the many helpful suggestions for the improvement of this article made by my colleagues in the UKAEA and elsewhere. In particular, I would like to express my appreciation to Mr. Jonathan Heller, who helped me considerably with the various reviews of the draft.

# References

1. *The Shippingport pressurized water reactor.* Addison Wesley Publishing Company Inc., Reading, Mass (1958).
2. WEAVING A. H. Development of the pressurized water reactor *Nucl. Energy,* **18**(2), 101–115 (1979).
3. DEPARTMENT OF INDUSTRY. *Second report on the nuclear ship study.* HMSO, London, (1975).
4. KRAMER A. W. *Boiling water reactors.* Addison Wesley Publishing Company Inc., Reading, Mass (1958).
5. LAHEY, R. T. AND MOODY, F. J. *The thermal-hydraulics of a boiling water nuclear reactor.* The American Nuclear Society (1977).
6. COHEN, K. AND ZEBROSKI, E. Operation Sunrise, *Nucleonics* (17 March 1959).
7. INTERNATIONAL ATOMIC ENERGY AGENCY. *Status and prospects of thermal breeders and their effect on fuel utilization.* IAEA Technical Report Series No. 195 (1979).
8. DAVIS, R. J. AND HIRST, B. Twin header bore welded steam generator for pressurized water reactors. *Nucl. Energy,* **18**(2), 133–40 (1979).
9. GARNSEY, R. Corrosion of PWR steam generators. *Nucl. Energy,* **18**(2), 117–132 (1979).
10. STYRIKOVICH, M. Mass transfer in two-phase flows and its importance in nuclear power plants. In *Proc. Int. Seminar Momentum, Heat and Mass Transfer in Two-Phase Energy and Chemical Systems. Dubrovnik, Yugoslavia 4–9 September, 1978.* International Centre for Heat and Mass Transfer (1978).
11. *US Code of Federal Regulations,* Part 50, §50.46 (b).
12. US NUCLEAR REGULATORY COMMISSION. *Reactor safety study: an assessment of accident risks in US commercial nuclear power plants. WASH*–1400 (*NUREG* 75/014) (1975).
13. FEDERAL MINISTER OF RESEARCH AND TECHNOLOGY. *The German Risk Study— summary.* Bundesministerium Für Forschung und Technologie Bonn. Cologne (1979).
14. GARZAROLLI, F., VON JAN, R., STEHLE, H. The main causes of fuel element failure in water cooled power reactors. *Atom. Energy Rev.* **17**(1), 31–128 (1979).
15. PATHANIA, R. S. AND TATONE, O. S. *Steam generator tube performance: experience with water-cooled nuclear power reactors during 1978.* Atomic Energy of Canada Ltd. Report No. AECL 6852 (1980).
16. EISENHUT, D. G., LAW, B. D., STROSNIDER, J. *Summary of operating experience with recirculating steam generators NUREG*–0523. US Nuclear Regulatory Commission, Washington (1979).

# 5

## Fast reactors
### A. M. JUDD

Fast reactors—that is reactors which make use of high energy or *fast* neutrons—are able to breed fissile material and thus extract energy from the otherwise almost useless fertile isotope of uranium, to make available an enormous source of energy.

This chapter starts by explaining this breeding process, and goes on to describe typical fast reactors. The starting point is the physics of the nuclear reactions which take place in the reactor, because these go a long way towards determining the nature of a fast reactor and the way it behaves. Equally important, however, is the chemistry of the fuel elements and, in particular, the behaviour of the fission products, which determines the *burn-up*, the extent to which the fuel can be used before it is reprocessed.

The engineering of the reactor core is determined mainly by the need to transport a large amount of heat from a small volume. The other important consideration is the behaviour of the structural materials under neutron irradiation; important choices have to be made about the layout of the primary coolant circuit and the design of the steam generators.

A sodium-cooled fast reactor has many inherent features, such as the low coolant pressure, which make it very safe. These are backed up with protective safety systems such as automatic shutdown equipment and decay heat rejection circuits, so that the risk of injury to the public or to the operating staff from accidents is small enough to satisfy the licensing authorities.

## Contents

# 1  Introduction

The significance of fast reactors is underlined by the following considerations.
1.  The world's energy reserves from fossil fuels are about $80 \times 10^{21}$ joules of heat.
2.  The world's economically recoverable resources of uranium including speculative deposits are about $10^7$ tonnes, and in thermal reactors would yield about $4 \times 10^{21}$ joules of heat; in fast reactors this same $10^7$ tonnes would yield about $200 \times 10^{21}$ joules of heat (i.e. more than twice as much as could be gained by burning the total fossil fuel reserves).
3.  As fast reactors can extract more energy from uranium they make it possible to pay more for it. If the price rises it may be worth extracting uranium from lower grade sources such as, for example, sea-water, in which the equivalent of some $10^{26}$ joules is present.
    Thus fast reactors can make a very important contribution to the world's energy needs. The reason for this, and the way it can be done, form the subject matter of this chapter.

# 2   Fast and thermal reactors

When the fission of uranium nuclei was discovered, and it was observed that a fission caused by one neutron liberated energy and two or three neutrons, the possibility of a chain reaction generating power was apparent. There was, however, a serious difficulty in the way.

Natural uranium consists of two isotopes: (1) the very rare *fissile* $^{235}$U, in which fission can be caused by neutrons of any energy, and (2) $^{238}$U, which can be fissioned by *fast neutrons* (with kinetic energy above about 1 MeV), but captures neutrons of all energies readily. In natural uranium the probability of capture in $^{238}$U outweighs that of fission for neutrons of all but the lowest energies (below about 0.1 eV), whereas the neutrons are born in fission with an energy of about 2 MeV.

For a chain reaction to work, therefore, either the energy of the neutrons causing fission has to be around the *thermal* level of 0.025 eV (in which case natural uranium can be used), or the proportion of $^{235}$U or some other fissile isotope has to be increased substantially. Both of these routes were followed from the early days: the first led to the development of thermal reactors, and the second to fast reactors.

The term fast reactor does not define a specific style of design any more than does the term thermal reactor. It simply means that the reactor is constructed without a moderator so that the average neutron energy is much higher than in a thermal reactor. In a typical fast reactor most of the neutrons are in the energy range 0.1–0.8 MeV, whereas in a thermal reactor they are roughly equally divided above and below 1 eV.

# 3 Breeding

To return to the neutrons captured in $^{238}$U, each neutron captured turns a $^{238}$U nucleus into $^{239}$U, which decays as follows:

$$^{239}\text{U} \xrightarrow[\text{23.5 min}]{\beta^-} {}^{239}\text{Np} \xrightarrow[\text{2.35 days}]{\beta^-} {}^{239}\text{Pu} \xrightarrow[\text{24 360 years}]{\alpha} {}^{235}\text{U, etc.}$$

The times are the half-lives of the decay processes. As far as reactor operation is concerned $^{239}$Pu is the end product of the chain. Its nuclear properties are similar to those of $^{235}$U, and in particular it is *fissile* (meaning that it can be fissioned by neutrons of any energy). $^{238}$U is called a *fertile* material because, although it is not fissile, it can be turned into a fissile material by neutron capture.

The quantity of fissile material which can be generated depends on the material in which the fissions take place and the energy of the neutrons causing them. The important parameter is $\eta$, the number of fission neutrons generated per neutron absorbed. Figure 5.1 shows the variation of $\eta$ with neutron energy for two fissile isotopes, $^{235}$U and $^{239}$Pu.

Of the $\eta$ neutrons, one must be absorbed in fissile material to maintain the chain reaction. Some of the rest are lost because they diffuse out of the reactor or are absorbed in other materials such as structure or coolant. The

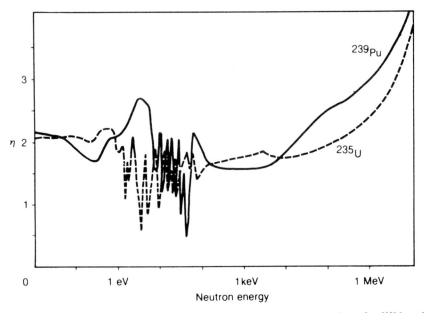

Fig. 5.1 The number of neutrons generated per neutron absorbed, $\eta$, for $^{235}$U and $^{239}$Pu

remainder, $C$, are available for capture in fertile material to turn it into fissile material. If $L$ is the number of neutrons lost and $\epsilon$ is a factor near unity to allow for fissions caused by fast neutrons, then

$$C = \eta \epsilon - 1 - L.$$

If $C>1$, more fissile nuclei are created than consumed; $C$ is then called the *breeding ratio* and the reactor is a *breeder*. If $C<1$, as it is in most thermal reactors, it is called the *conversion ratio*.

In practice $L$ cannot be reduced below about 0.2, so breeding is possible only if $\eta>2.2$. Figure 5.1 shows that this is possible in a fast reactor using either $^{235}$U or $^{239}$Pu, but $^{239}$Pu gives the wider margin; $^{235}$U will breed only if the energy of the neutrons causing fission is above 1 MeV or so. In either case, the higher the neutron energy, the higher the breeding ratio. Thermal reactors using $^{235}$U have conversion ratios in the range of 0.6 (light water reactors) to 0.8 (heavy water and gas-cooled reactors).

If $N$ atoms of $^{235}$U are fissioned, $CN$ atoms of $^{239}$Pu are generated. If these in turn are fissioned in a similar reactor, $C^2N$ atoms of $^{239}$Pu are formed (assuming for illustration that $C$ is unchanged), and so on. The total number of atoms fissioned if this process goes on indefinitely is $N(1 + C + C^2 + \ldots)$. If $C < 1$ the series converges to $N/(1 - C)$. If the fuel is natural uranium $N$ is 0.7 per cent of the total number of uranium atoms supplied, and the maximum number of atoms which can be fissioned in a thermal reactor with $C = 0.7$ is $0.7/(1 - 0.7) \approx 2$ per cent of the uranium supplied.

For a breeder reactor with $C > 1$ the series diverges, and in principle all the uranium supplied can be fissioned, the $^{235}$U directly, and the $^{238}$U by first converting it to $^{239}$Pu. Some of the $^{239}$Pu is however lost to residues when the fuel is reprocessed, and some is converted to higher isotopes (see §5.1), so that in practice about 50–70 per cent of the total uranium feed can be fissioned in fast reactors as against 1–2 per cent in thermal reactors. Thus fast reactors can extract 25–50 times as much energy from uranium as thermal reactors. This is why fast reactors have such an important effect on the world's energy reserves. However, it must be remembered that the need to accumulate enough fissile material to start up each new fast reactor imposes a limit on the rate at which the $^{238}$U can be exploited.

## 4 Description of a fast reactor

Although many variations are possible, the basic physics imposes constraints, so that all practical fast reactors are very similar in many ways. The essential parts of fast reactor are shown in Fig. 5.2.

The core of the reactor contains the fissile material which maintains the chain reaction, and also fertile $^{238}$U. It is surrounded by a *breeder* or *blanket* of $^{238}$U to capture neutrons escaping from the core. Because there is no need

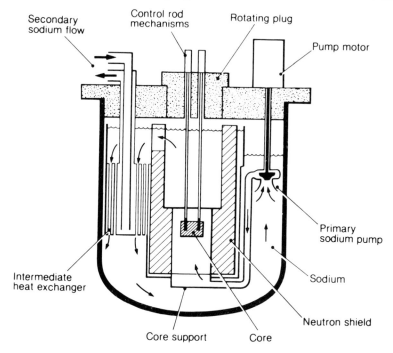

Fig. 5.2 Layout of a typical fast reactor

for any moderator, the core is much smaller than that of a thermal reactor. The core of a 2500 MW(t) fast reactor would be a cylinder about 1 m high and 2 m in diameter.

In order to remove the heat from such a small volume, a coolant with very good heat-transfer properties is needed. Water is ruled out because it acts as a neutron moderator. In almost all fast reactors so far a liquid metal has been used, usually sodium. In principle it is possible to use a gas such as helium at high pressure as a coolant, but the balance of advantage lies with sodium and no high-power gas-cooled fast reactor has yet been built.

After a period of operation, the fuel has to be removed and reprocessed (1) to remove the accumulated fission products (because they absorb neutrons and reduce the breeding ratio), (2) to adjust the amount of fissile material in the core, and (3) to reconstitute it (because of the changes in the structure of the fuel material brought about by exposure to the neutron flux).

The extent to which the fuel can be used before it has to be reprocessed is called the *burn-up*. It is measured in terms of the proportion of the total number of *heavy atoms* (i.e. of all isotopes of uranium or plutonium) which are fissioned. The higher the burn-up the better. Fuel which is being reprocessed is temporarily unproductive, whereas fuel in a reactor is earning revenue. What is more, reprocessing is expensive, and the less frequently it is done the less the total fuel cost.

High burn-up cannot be achieved with metal fuel. The limit is about 1 per cent before its structure is so damaged that it has to be reprocessed. If the fuel is in the form of oxides (a mixture of $UO_2$ and $PuO_2$), experience shows that up to 10 per cent burn-up can be reached. Oxide fuel also has the advantage that it can be operated at higher temperatures than metal so as to allow the power plant to operate at higher thermodynamic efficiency.

In practice then the main constituents of the core of a fast reactor are fixed: mixed oxide fuel and sodium coolant. The third constituent is the structural material, usually stainless steel, used to clad the fuel and hold it in place. The volumes occupied by fuel, coolant and structure are usually roughly in the ratio 30:50:20.

The following sections provide a rather more detailed description of the most important aspects of a fast reactor. These are reactor physics, the behaviour of the fuel, the engineering of the core and the coolant circuits, and safety. The chapter concludes with a brief survey of the development of fast reactors.

# 5  Reactor physics

## 5.1  **Breeding gain**

Breeding is basically the conversion of $^{238}U$ into $^{239}Pu$ by neutron capture, but there are other factors to be considered. When a $^{239}$Pu nucleus absorbs a neutron, it does not necessarily undergo fission. It may capture the neutron to form a $^{240}$Pu nucleus instead.

$^{240}$Pu is not fissile like $^{239}$Pu, but fertile. (In fact all the even-numbered isotopes of uranium and plutonium are fertile, while the odd-numbered isotopes are fissile.) $^{240}$Pu behaves very much like $^{238}$U, and it captures neutrons to form $^{241}$Pu which is fissile. If the $^{241}$Pu is not fissioned, it can capture neutrons to form $^{242}$Pu, and so on to form heavier isotopes and elements such as americium and curium which lie beyond plutonium in the periodic table.

Because of these complications the simple concept of breeding ratio ($C$) given in §4 is not really appropriate. For example $^{241}$Pu fissions to produce more neutrons more readily than $^{239}$Pu and is therefore more useful as a reactor fuel. A better measure of the ability of a given reactor to breed is given by the *breeding gain*, $G$, the definition of which takes into account the production and destruction rates of all the individual heavy isotopes and the worth of each as a reactor fuel. This ratio measures directly the rate at which fuel for a new reactor is made. If $G > O$ the reactor breeds. If, for example, $G = 0.2$, for every 100 fissile atoms destroyed (by fission or neutron capture) the equivalent of 120 new fissile atoms are generated. After the equivalent of five complete reactor cores have been irradiated to complete burn-up, enough excess fuel has been accumulated to start a new reactor.

Fissile material which is bred in the core of a reactor tends to compensate for the fissile material which is destroyed and helps to keep the reactor critical. Fissile material bred in the breeder, however, is mainly used after the breeder has been reprocessed, when it can be incorporated in new core fuel. A distinction has to be made therefore between breeding gain $G$, which measures the total gain of new fuel, and the *internal breeding gain*, $G_I$, which measures the gain of fuel in the reactor core. $G_I$ is negative for typical oxide-fuelled fast reactors, which means that reactivity is steadily lost as the reactor operates, and can be regained only when the fuel is reprocessed and some of the fissile material from the breeder is put into the core.

The actual value of $G$ depends on the isotopic composition of the plutonium and on the design of the reactor, and it is possible by altering the design to reduce $G$ to zero, or even to make it negative. This means then that a fast reactor is not necessarily a breeder; it can be made to consume fissile material. The value of $G$ can be reduced by making the breeder smaller or by reducing the mean neutron energy - *softening the spectrum* (see §5.2). This reduces the effective value of $\eta$ (see Fig. 5.2).

## 5.2 Neutron flux and importance

Neutrons are born from fissioning nuclei, and at birth have a kinetic energy of about 2 MeV. They diffuse through the reactor until they cause another fission, or are captured, or leak out into the surrounding shielding. As they diffuse they lose their energy, mainly by inelastic scattering by $^{238}$U, $^{56}$Fe, and $^{23}$Na nuclei (above 0.5 MeV), or to a lesser extent by elastic scattering by $^{23}$Na or $^{16}$O (at lower energies). In an inelastic-scattering interaction the neutron loses a large fraction of its energy and leaves the nucleus in an excited state, which subsequently decays by $\gamma$-emission, whereas in an elastic-scattering event it loses only a little of its energy, which appears as an increase of the kinetic energy of the nucleus.

The resulting variation of the neutron flux $\phi$, with energy, the *neutron spectrum* as it is called, is shown in Fig. 5.3. The peak is at about 0.2 MeV because neutrons with higher energy are likely to be scattered into this energy range by inelastic scattering, and those with lower energy are likely to be lost by causing fission or being captured. Very few neutrons get below 1 keV, but the few that do are significant because they contribute to the *Doppler effect*, which is explained in §5.3.

In assessing the effects of changes in the reactor composition or dimensions (see §5.3), it is very useful to know the contribution of a neutron of a particular energy to the reactivity of the whole reactor. This is measured by the *neutron importance*, $\phi^*$, which is also shown in Fig. 5.3 and is actually proportional to the change in the overall flux level that would result from introducing a new neutron of that energy into the reactor.

Above 1 MeV $\phi^*$ is high because of the possibility of fission in $^{238}$U. It is low at about 0.2 MeV because neutrons of this energy have the greatest

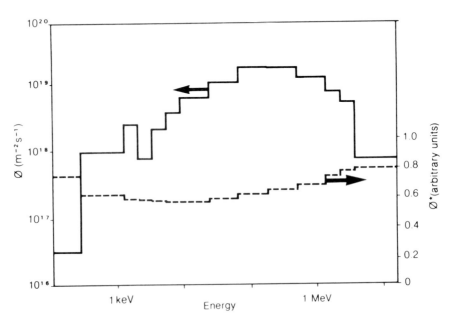

Fig. 5.3 Neutron flux $\phi$ and importance $\phi^*$ in a typical fast reactor

chance of being captured or leaking rather than causing fission, but as the energy decreases ø* increases, because the chance of causing fission increases more than the chance of being captured.

The neutron flux within a fast reactor is not uniform; Fig. 5.4 shows the spatial variation of the total flux (i.e. neutrons of all energies) across a reactor core and breeder. At the centre of a typical power reactor core the flux is about $10^{20}$ m$^{-2}$ s$^{-1}$. If the fuel composition is uniform, the distribution of power follows that of neutron flux fairly closely, but it is more convenient, and efficient, if the power is distributed more uniformly. This can be done by increasing the concentration of plutonium in the outer parts of the core to compensate for the lower flux (Fig. 5.4).

## 5.3 Temperature and reactivity

Figure 5.5 shows the fission cross-section of $^{235}$U. Below 100 eV it is dominated by many very sharp peaks or *resonances*, and in fact there are resonances at higher energies, but these are so fine and close that they cannot be resolved experimentally. The shape of the fission cross-section curve for $^{239}$Pu is very similar.

A resonance depresses the neutron flux because a neutron of that energy interacts very quickly and is removed. The effect can be seen in Fig. 5.3, where the dip in the spectrum at 3 keV is caused by a large scattering resonance in $^{23}$Na.

The width (in energy) of the dip in the spectrum is determined by the *effective width* of the resonance. This has two components: one is the actual width of the resonance in the cross-section, as shown, for example, in Fig. 5.5, while the other arises from the thermal motion of the nucleus, which has the effect of broadening the resonance. The higher the temperature the greater the effective width of the resonance and the wider the dip in the neutron spectrum.

The net result of this is that, as the temperature increases, the relative frequency of neutron interactions also increases. This is true of any of the neutron interactions which are affected by resonances (fission, capture, or scattering). The most important effects are upon the rates of fission in $^{239}$Pu and capture in $^{238}$U. An increase in the fission rate increases reactivity, whereas it is decreased by an increase in the capture rate. As there is more $^{238}$U in the core than $^{239}$Pu, the latter effect dominates, and there is an overall loss in reactivity with increasing temperature. This is known as the Doppler effect because it depends on changes in the effective energy of a neutron due to the relative motion of neutron and nucleus.

Temperature effects reactivity in other ways than by the Doppler effect. An increase in the temperature of the structure makes the whole core expand, increasing the probability of neutron leakage and so reducing the reactivity. This is however a relatively small effect. More important are the change in the composition of the core caused by the fact that the sodium

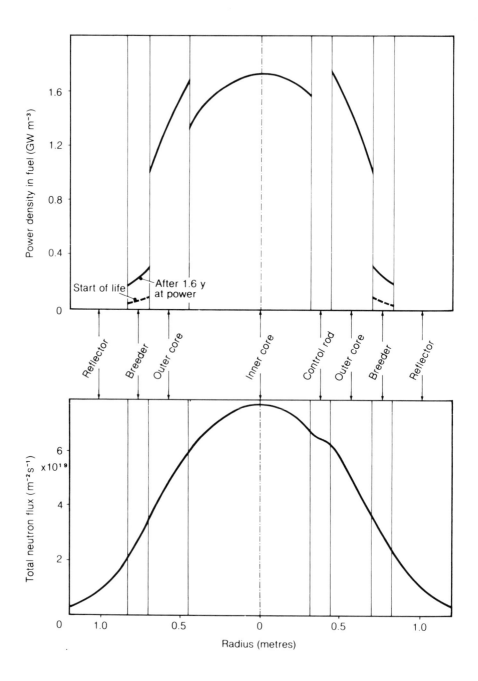

Fig. 5.4 Variation of flux and power density across a reactor core shown along a diameter which happens to pass through a control rod

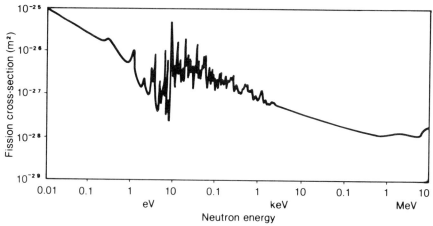

Fig. 5.5 The fission cross-section of $^{235}$U

coolant expands more than the fuel or the structure, and the *bowing* or distortion of the fuel elements in temperature gradients.

The sodium interacts with neutrons by capturing them (only a few, because it is a weak absorber), scattering them (which tends to prevent their leaking out of the core), and reducing their energy or *moderating* them. As the temperature increases the density of the sodium falls. As a result there are fewer sodium nuclei in the core, and each of these three effects consequently diminishes. The decreased scattering allows more leakage and therefore decreases reactivity. The effect of the decreased moderation, as shown in Fig. 5.3, is a slight increase in the mean neutron energy, so the peak of the spectrum moves to the right. This is a region of higher neutron importance, so there is an increase in reactivity.

The effect of increasing sodium temperature is thus twofold: there is a positive effect from absorption and moderation, and a negative effect from scattering. The net result in a small reactor, generating 600 MW(t) for example, is negative (that is, increasing sodium temperature reduces reactivity), although it may be positive in a larger reactor. Since however the net effect results from large positive and negative components, it is sensitive to detailed features of design.

The bowing effect on reactivity arises as follows. The fuel is mounted in steel tubes or *subassemblies* (see §7.4), which stand vertically in the core. Because of the variation of flux across the core (Fig. 5.4), the power and hence the temperature of a peripheral subassembly is higher on the side nearer the centre of the core. As a result it takes up a bowed shape convex towards the centre. The effect on reactivity depends on how it is supported, as shown in Fig. 5.6. If it is pin-jointed above and below the core, it moves inwards as the power increases causing it to bow more, thus increasing the reactivity. If, however, it is cantilevered below the core, it moves outwards and the reactivity decreases. For this reason the cantilevered arrangement is preferred.

Pin joint                                    Cantilever

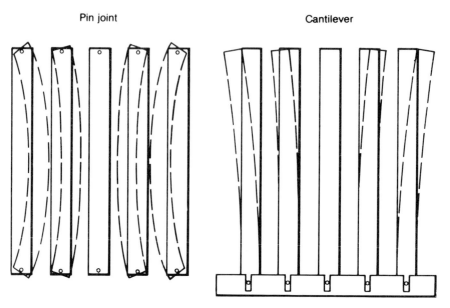

Fig. 5.6 The effect of bowing fuel elements

## 5.4 Reactivity coefficients

It is best to consider these various effects of temperature on reactivity not separately but together as they actually affect operation of the reactor. Two quantities are important: the *isothermal temperature coefficient* and the *power coefficient*. The isothermal temperature coefficient is the rate of change of reactivity with temperature, assuming temperatures throughout the core change by the same amount. This coefficient depends then on all the effects described previously: Doppler, sodium expansion, expansion of the structure, and bowing.

The power coefficient is the rate of change of reactivity with reactor power, assuming the coolant flowrate moves in step with the power. This means that coolant and structure temperatures are nearly constant and only the fuel temperature changes significantly. The power coefficient therefore depends mainly on the Doppler effect.

If the reactor is to be stable in operation, it is important for both of these coefficients to be negative. In other words, an increase in power or in temperature tends to reduce the reactivity so that power and temperatures fall, producing a negative feedback. As will be shown (§8.1), it is particularly important that the Doppler effect is negative. This effect is very reliable in that it does not depend on the transfer of heat from fuel to coolant or structure, and as it acts instantaneously, it is a valuable safety feature.

The values of the power and temperature coefficients can be manipulated by altering the design. If the ratio of $^{239}$Pu to $^{238}$U in the core is increased so that the core becomes smaller, neutrons have a better chance of diffusing out of it. At low energy ∅* increases (because if a low-energy neutron does remain in the core it has more chance of causing a fission), and as a result, the moderating effect of loss of sodium becomes less positive while, because there is less $^{238}U$ the Doppler effect becomes less negative. As a result, the temperature coefficient becomes more negative and the power coefficient less negative. Both coefficients can be made more negative by introducing a little moderator, such as beryllium oxide, into the core, but this has the undesirable effects of reducing the breeding gain and increasing the amount of fuel needed. Both coefficients are made more negative if the fraction of $^{241}Pu$ in the fuel increases. If carbide fuel were to be used instead of oxide (a possible development in the future), the mean neutron energy would increase and the power coefficient would become more negative, while the effect of sodium expansion would be more positive.

# 6 Fuel elements

## 6.1 Design

The fuel for current fast reactors is a mixture of uranium and plutonium dioxides, containing usually 15–30 per cent $PuO_2$. It is made up into fuel elements, each of which consists of a stainless steel tube, usually about 6 mm in diameter, containing the fuel mixture.

In most reactors the fuel is in the form of sintered pellets. These are made of $UO_2$ and $PuO_2$ powders mixed in the correct proportions with a binder, formed to shape, baked, and then ground to size. When finished the pellets are porous, and the porosity can be controlled by careful choice of the binder, and the baking time and temperature. The pellets are then placed in the steel tube as shown in Fig. 5.7.

An alternative is to fill the tube with sintered particles of oxide and then vibrate it until they are compacted. Particles of at least two sizes, typically 800 and 80$\mu$m in diameter, have to be used if a high packing fraction is to be attained, but even then it is difficult to get the porosity below about 18 per cent of the overall fuel volume. Nevertheless, this method of manufacture, which is called *vibro-compaction*, is attractive because it is cheap and simple.

The purpose of the steel tube is to hold the fuel and also—most important— to keep the radioactive fission products from getting into the coolant. It is usually called the *cladding*, and it is important that it retains its integrity— i.e. it does not leak before the fuel reaches up to 10 per cent burn-up. There are three main difficulties in ensuring this: the fuel swells, it releases gaseous fission products, and some of the fission products tend to corrode the cladding.

Each fission replaces one atom by two fission product atoms (or in rare cases three). Now in broad terms all atoms occupy roughly the same volume in a solid (about $10^{-29}$ m$^3$). Thus when the fuel is fissioned it swells. Some of the fission products, however, are gases and these escape into the fuel pin. The increase in volume amounts to about $0.8B$, where $B$ is the burn-up. This means that if the fuel experiences 10 per cent burn-up, it undergoes nearly 3 per cent linear expansion. If it is touching the cladding, this is stretched by the same amount. Steel tends to lose its ductility when it is irradiated, and so the material for the cladding has to be chosen carefully to withstand the swelling of the fuel without cracking.

Some of the fuel swelling is accomodated by filling the pores in the sintered fuel. As an alternative the pellets can be made with a hole in the centre. The fuel is usually made so that pores or a central hole occupy some 10–20 per cent of the total volume. It is not certain, however, that either pores or a central hole work in the intended manner to accommodate swelling. What actually occurs in practice is a combination of creep and swelling of the cladding, as well as changes in the porosity of the fuel. The various mechanisms interact in a very complex way, which is not easy to model satisfactorily, and ultimately the main guarantee that the cladding will not leak is provided by extensive testing.

Various isotopes of krypton and xenon are produced by fission—in all about 0.26 atoms per fission. As these are inert gases they do not combine chemically with the fuel or other fission products, but form tiny bubbles between the individual crystals in the fuel material. Eventually these link up forming channels to the pellet surface, and the gas is released.

The quantity of gas produced is very large. At 400 °C and 1 bar the gas generated by 10 per cent burn-up would occupy about 50 times the fuel volume. As it is radioactive it is usually arranged to retain it within the cladding. An empty volume or *plenum* is provided at one end of the fuel element for this purpose. If the volume of the plenum is about equal to that of the fuel the pressure inside the fuel element after 10 per cent burn-up is about 5 MPa (50 bar).

Figure 5.7 shows a typical fuel element. Each element passes right through the core, so it contains a region of core fuel sandwiched between regions of breeder fuel (containing natural or depleted $UO_2$) above and below.

## 6.2 Effects of irradiation

When the reactor is operating, each fuel element may be generating heat at a rate of up to 50 kW per metre length. The thermal conductivity of the fuel is very low (about 2 W m$^{-1}$ K$^{-1}$), and so the temperature in the centre of the fuel element is very high, up to 2500 °C, while the outer surface of the fuel may be at about 1000 °C. (This is not to be confused with the temperature at the outside of the cladding, which does not exceed about 600 °C. There is a large temperature difference across the gas gap between fuel and cladding.)

These high temperatures and high temperature gradients have the effect of completely altering the microscopic structure of the fuel. Figure 5.8 shows polished and etched cross-sections of fuel elements before and after irradiation.

The most striking changes are the appearance of a hole in the centre of the fuel, even if there was no hole to start with, and of a region surrounding the hole made up of long *columnar* grains lying along radii of the cylinder. These columnar grains are much larger than the grains of which the fuel was composed before it was irradiated.

This alteration of the structure is due to two effects. Firstly, in the 1300–1600 °C temperature range, the sintering process started during manufacture continues, and individual grains coalesce and grow. Secondly, above 1600 °C (i.e. in the central part of the fuel) the pores migrate towards the higher temprature. This happens as individual atoms of the fuel material diffuse from the hot side to the cold side of each pore, so that the pores move to the centre, coalescing to form a central hole. The fuel then recrystallizes in the wake of each pore to form a single crystal or columnar grain. This can be seen in progress in Fig. 5.9. The columnar grain region is continually being recrystallized as first the original pores and later bubbles of fission product gas migrate to the central hole.

The temperature difference between centre and surface results in the fuel cracking. When the fuel is first put in the reactor and starts generating heat, the centre expands more than the circumference, and radial cracks tapering towards the centre are formed. In the central region, however, they soon 'heal' as the fuel is recrystallized. Once the reactor is shut down the centre contracts and new cracks tapering towards the circumference are formed, and these are the cracks which can be seen in Fig. 5.8.

The mechanical interaction between fuel and cladding is very complex. Initially the fuel swells and presses against the cladding. The stress is limited however because the fuel deforms plastically or *creeps*. In the centre of the fuel the high temperature results in a high rate of thermal creep which relaxes any stress within a time of the order of minutes or hours. At the outside where it is cooler, thermal creep is very slow, and the main effect contributing to the relaxation of stresses is so-called *irradiation creep* (caused by the destruction of the atomic structure of individual crystals by fissions). In spite of this the cladding is strained, and the strain is concentrated at irregularities such as cracks in the surface of the fuel. Later in life the rate of swelling of the cladding increases and may exceed that of the fuel, so that a gap opens between fuel and cladding.

As the fuel is irradiated there is a tendency for the plutonium to migrate relative to the uranium. This migration depends on the amount of oxygen in the fuel, which may not be of the exact stoichiometric composition. The composition can be represented by $(U, Pu)O_2 + x$, where $x$ may be either positive or negative. It is found that for $x > -0.04$ the plutonium in the hottest part of the fuel tends to move towards the centre, while for $x < -0.04$ it moves outwards. The plutonium in the cooler outer part of the fuel is hardly affected. Typical plutonium distributions after irradiation are shown in Fig. 5.10.

The phenomenon seems to depend on migration of volatile fuel species. If

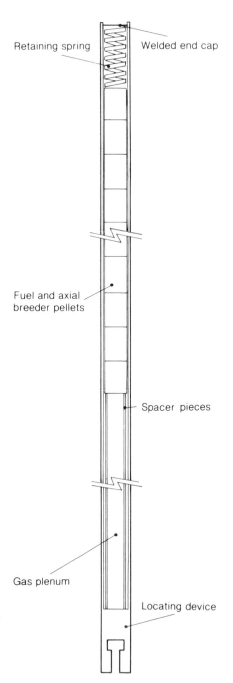

Retaining spring

Welded end cap

Fuel and axial
breeder pellets

Spacer pieces

Gas plenum

Locating device

Fig. 5.7 A typical fuel element

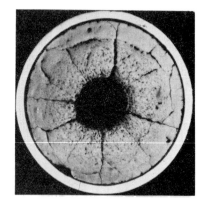

Fig. 5.8 Cross-section of an irradiated fuel element; the outer diameter of the cladding is 6 mm.

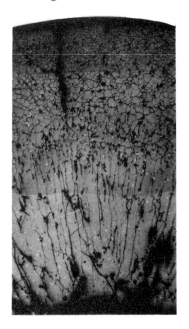

Fig. 5.9 Recrystallization of fuel; the section shown is 2 mm across

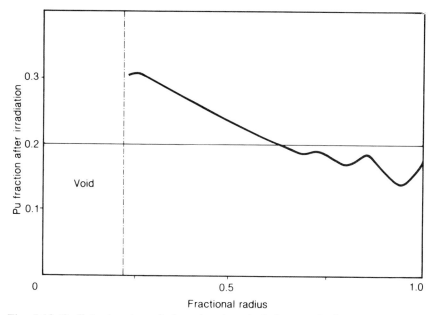

Fig. 5.10 Radial migration of plutonium in a fuel element. Before irradiation the plutonium fraction was 0.2 throughout the fuel

there is excess oxygen, the vapour in equilibrium with the fuel contains more $UO_3$, which migrates to the cooler regions, displacing plutonium to the hotter, while if oxygen is deficient the vapour contains more PuO, which migrates outwards.

## 6.3 Behaviour of fission products

A wide range of isotopes of many different elements is produced as fission products, and they form a very complex chemical system. It is possible to give only the briefest outline of the behaviour of this system here.

1. The inert gases (Kr, Xe) are progressively released from the fuel and go to the plenum at the end of the fuel element. A small quantity remains in solution or as small bubbles in the fuel.

2. The alkali metals (Rb, Cs) are volatile and tend to migrate to the cool periphery of the fuel. This is illustrated in Fig. 5.11, which shows the distribution of $^{137}$Cs after irradiation. Isotopes which come from $\beta$–decay of inert gases, such as $^{133}$Cs produced from $^{133}$Xe, appear in the fission product gas plenum.

3. Halogens (I, Br) are produced in small quantities. It is thought that iodine collects near the cladding where it forms caesium iodide, CsI, which may under some circumstances be involved in corrosion of the cladding. This is conjecture because it is impossible to observe what is happening inside a fuel element while it is in the reactor, and observations made on fuel

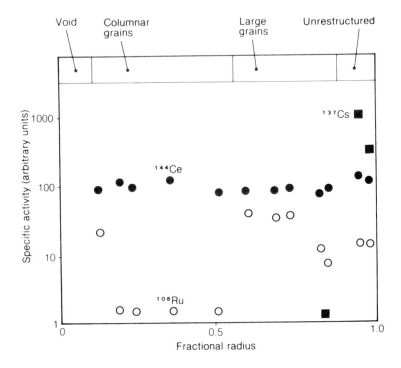

Fig. 5.11 Distribution of fission products after irradiation

after it has been removed from the reactor may not reflect correctly what happens under irradiation.

4. Metals forming refractory oxides (Sr, Y, Zr, Ba, La, Ce, Pr, Nd, Pm, Sm) stay more or less where they are formed (see Fig. 5.11). Some like $^{89}$Sr are daughters of volatile precursors which may migrate, so they do not necessarily appear where the fissions take place.

5. Metal not forming oxides (Tc, Ru, Rh, Pd, Ag, Te) are found as metallic inclusions, sometimes dispersed in the fuel, sometimes (if the central temperature is high) as metallic droplets in the central hole (Fig. 5.11).

6. One metal (Mo), which sometimes forms an oxide, appears in the metallic inclusions if the oxygen potential is low and as an oxide if it is high.

The oxygen potential (which determines the tendency of materials to be oxidized) is determined initially in manufacture, but it changes as burn-up proceeds. Each fission frees two oxygen atoms which go to oxidize the fission products, but the number needed depends on the yields of the various elements, which are different for fission of different isotopes. Fission of $^{238}$U produces more zirconium and strontium (which form oxides) and less ruthenium and palladium (which do not) than fission of $^{239}$Pu. Thus in the core, where most of the fissions occur in $^{239}$Pu, the oxygen potential rises as irradiation proceeds, while in the breeder, where fission in $^{238}$U predominates, it falls. In the core an upper limit to the oxygen potential is set by the molybdenum, which acts as a buffer by combining with excess oxygen and prevents $x$, where the effective fuel composition is $(U, PU)O_2 + x$, rising above about 0.015.

# 7 Reactor engineering

## 7.1 **Heat transfer in the core**

It is often thought that the size of the core is determined by reactor physics—by the need to achieve a critical mass—but this is not in fact the case. An assembly of more or less any size can be made critical by adjusting the ratio of $^{239}$Pu to $^{238}$U in the fuel.

The most important dimensions of the core and of the fuel elements are in fact set by considerations of heat transfer—by the need to get the heat from the fuel into the coolant and to provide space for enough coolant to flow through the core. This has to be done without allowing the fuel elements to get too hot.

The crucial points are the maximum temperatures of the fuel and of the cladding. The cladding temperature cannot be allowed to exceed about 700 °C, because if it does the cladding may be broken open by the stresses imposed on it by the fuel and the fission-product gas. If this were to happen, radioactive fuel and fission products might be released into the coolant. The fuel temperature cannot be allowed to exceed the melting point, which is about 2800 °C, because liquid fuel would be free to run down the central void.

The temperature difference between the centre and the surface of the fuel depends on the power generated per unit length (called the *linear rating*) and on the thermal conductivity of the fuel. It is independent of the radius of the fuel element. Fixing the maximum fuel and cladding temperatures therefore effectively fixes the linear rating. Allowing for uncertainties in the thermal conductivity and in the temperature drop across the gap between fuel and cladding, 50 kW m$^{-1}$ is about the maximum possible.

Because the neutron flux varies across the core (see Fig. 5.4), and with it the linear rating of the fuel elements, the average linear rating is about 0.6 of the maximum, or 30 kW m$^{-1}$. This means that a reactor generating 2500 MW of heat has to have a total of about 80 km of fuel elements.

As the fuel is expensive there is an incentive to keep the total amount of fuel in the core small. In principle it can be made as small as desired by making the fuel elements very thin, provided that their overall length equals the required 80 km. In practice the cost of manufacturing and handling the elements rises if they are too small, and as a compromise a radius of about 3 mm is chosen. This implies that there is a total fuel volume of about 2.3 m$^3$, weighing about 20 tonnes, in the 2500 MW(t) reactor. The maximum heat flux from cladding is about 2.6 MW m⁻².

Because sodium has very good heat-transfer properties, this enormous heat flux can be transferred with only a small temperature difference between cladding and coolant. A sodium speed of about 10 m s⁻¹ gives a heat transfer coefficient of about 200 kW m⁻² K⁻¹, so 2.5 MW m⁻² can be transferred with a temperature difference of 13 K.

Unfortunately this is an optimistic estimate because substantial additional allowances have to be made for irregularities in the heat transfer or 'hot spots'; for example, there are temperature variations round the fuel elements, some elements receive more coolant than others, the fuel element supports interfere with the coolant flow, and there are random variations in the dimensions. As a result a hot-spot allowance of about 100 K has to be made, so that if the maximum cladding temperature is 700 °C the maximum coolant temperature needs to be somewhat less than 600 °C.

The rate at which coolant flows through the core is determined by the rate at which heat is to be removed and the temperature rise the coolant experiences. If the outlet temperature is fixed at 560 °C, say, it is thermodynamically advantageous to keep the inlet temperature as high as possible, because this maximises the temperatures at which heat is transferred to the steam and with it the efficiency of the power plant. The higher the inlet temperature, however, the smaller the coolant temperature rise and so the higher the total flowrate, which requires either

(a) higher coolant speeds and higher pressures, increasing the possibility of the fuel elements and other components vibrating, or

(b) a larger volume of coolant in the core which would absorb more neutrons, reducing the breeding gain and increasing the quantity of fuel needed to make the reactor critical.

Again a compromise has to be reached. A typical coolant inlet temperature is 400 °C, so with a temperature rise of 160 K and a specific heat capacity of 1.2 kJ kg⁻¹ K⁻¹, a total sodium flowrate of 13 000 kg s⁻¹ is required for the 2500 MW(t) reactor. To avoid problems with vibration the average sodium speed must not exceed about 8 m s⁻¹. This speed, together with the sodium flowrate of 13 000 kg s⁻¹ and density of 900 kg m⁻³, leads to a total flow area of 1.8 m².

The final consideration in determining the dimensions of the reactor core is the pressure drop required to drive all this coolant through the core. This depends on the height of the core. If the core is low, the pressure drop is also low, and the circulating pumps and structural members can be simpler, but because of the necessity of accomodating 2.3 m³ of fuel the diameter of the core has to be large. A core 1 m high is a reasonable compromise, giving a manageable pressure drop of 400 kPa or so. The 1.8 m² required for coolant flow thus sets the coolant volume at 1.8 m³, so that with 2.3 m³ of fuel the total core volume is 4.1 m³, and its radius is about 1.2 m.

The calculations in this section are a much-simplified version of the calculations on which the design of a real reactor core is based. In reality things are far more complicated and there are many other considerations to be taken into account. In spite of this the conclusions are broadly correct, and they show how heat transfer is fundamental to the design. They also show why the cores of all oxide-fuelled sodium-cooled fast reactors look very much the same.

## 7.2 Effects of irradiation

The structural materials used in the reactor core, which are either steels or nickel alloys, have to undergo irradiation by the neutron flux, and this has several different effects on their properties. The most important is that it makes them swell. The way this happens is as follows.

The neutrons in the core are scattered several times between being born in fission and being absorbed, and they may be scattered by iron nuclei in the structural steel as well as by uranium, oxygen, or sodium. Each time it is scattered the neutron gives up some of its kinetic energy to the scattering nucleus. A 100 keV neutron imparts up to 7 keV to an iron atom in an elastic-scattering interaction.

Each iron atom is held in position in the array of atoms that makes up one of the crystals of which the metal is composed, but if it is given enough energy (i.e. if it is knocked hard enough) it leaves its position. It takes about 25 eV to remove an atom from its lattice site, so the atom which scatters the neutron is easily removed. It goes careering through the crystal and, as it goes, it interacts with many other atoms, knocking them out of position as well. In this way each neutron-scattering event results in the displacement of a hundred or more atoms.

It is possible to estimate the total number of atom displacements which take place. While the fuel in a typical reactor undergoes 10 per cent burn-up, each atom in the steel cladding is displaced from its site in the crystal lattice about 70 times on average. There are about $2 \times 10^{14}$ displacements per second in each cubic millimetre of steel at the centre of the core of a reactor at full power. If it seems amazing that the material is not completely destroyed by this, it should be remembered that each cubic millimetre contains nearly $10^{20}$ atoms.

Each time an atom is displaced it leaves a hole, or *vacancy*, in the lattice together with an *interstitial*, where the atom itself comes to rest jammed into the lattice at a point where there is no proper lattice site for it. These vacancies and interstitials diffuse through the crystal under the influence of the thermal agitation of the atoms. Eventually each interstitial meets a vacancy and they annihilate each other (i.e. the atom falls back into a hole).

Unfortunately this process is complicated by other reactions which go on in the steel as it is irradiated. Some neutrons are captured by $(n, \alpha)$ reactions, mainly in $^{10}B$ which is present as an impurity. The neutron is captured by the boron nucleus and liberates an $\alpha$–particle, which is a helium nucleus, and forms an interstitial in the crystal. It so happens that an interstitial helium atom attracts vacancies as they diffuse, and forms a microscopic polyhedral void, typically $0.1\,\mu$ m across. An electron micrograph of such voids is shown in Fig. 5.12. Meanwhile the interstitials tend to accumulate at the grain boundaries, which are the surfaces of individual crystals. The result is that each crystal grows as its average density decreases. In extreme cases volume increases of 10 per cent are possible.

Irradiation has other effects as well as causing swelling. Because it makes atoms mobile, it causes creep at low temperatures at which ordinary thermal creep is not significant. This can be important because it tends to relieve the stresses caused by swelling. In addition the defects—interstitials or vacancies—make dislocations less mobile so that plastic strain is inhibited. The yield stress is increased and the material becomes harder. At temperatures above 700 °C or so helium tends to diffuse to grain boundaries, where it causes a loss of cohesion so that the material can be torn apart more easily. This reduces the ductility and makes it brittle.

## 7.3  Structural materials

In addition to the effects of the neutron flux, structural materials used in the core have to stand up to the coolant. Sodium is slightly corrosive to stainless

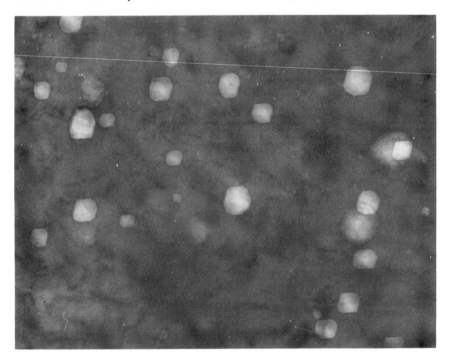

Fig. 5.12 Voids in stainless steel irradiated at 510 °C until each atom has been displaced about 42 times; the larger voids are about 60 nm across

steel and dissolves it at a rate which depends on the oxygen concentration and on the temperature. At 500 °C sodium containing 10 p.p.m. of oxygen corrodes type 316 steel at a rate of about 2 $\mu$m per year, but if it contains 30 p.p.m. of oxygen the rate rises to 10 $\mu$m per year.

Sodium also serves to transport carbon between different materials exposed to it. Depending on the temperatures, carbon tends to be lost from ferritic steels, causing them to lose strength, and to be gained by austenitic steels, reducing their ductility. Carbon deposition can be controlled by limiting the carbon concentration in the coolant.

Finally, sodium tends to subject materials to unusually high thermal stresses because of its high thermal conductivity. If the sodium temperature fluctuates, these fluctuations are transmitted to the surface of the steel, causing high thermal stresses that may cause fatigue damage and make any cracks which are present grow.

The materials available to meet these demanding requirements fall into three classes: austenitic stainless steels, nickel alloys, and ferritic steels.

Austenitic steels of the AISI 304, 316 or 321 types are widely used. They have relatively low yield strength, but they are ductile and they resist corrosion. They suffer, however, by becoming brittle under irradiation, being susceptible to thermal shock, and in particular by swelling under irradiation. *Cold-working*, producing permanent plastic strain during manufacture, reduces but does not eliminate the swelling. In some cases irradiation creep tends to offset the effects of swelling, but in general the need to accommodate 5–10 per cent increases in volume has to be taken into account in design.

High-nickel alloys such as PE16 (about 40 per cent nickel) are stronger and more resistant to swelling. Nickel absorbs neutrons to a greater extent than iron, but this can be offset by its greater strength, which allows less of the material to be used. However, nickel is quite soluble in sodium, so alloys with more nickel, such as Nimonic 80 A (75 per cent nickel), corrode too fast to be acceptable.

Ferritic steels are rarely used in reactor primary circuits because of the effects of decarburization, but it may be possible to allow for this in design. A low-alloy steel, for example, containing 2.25 per cent chromium and 1 per cent molybdenum, then has the significant advantage over a stainless steel that, because its coefficient of thermal expansion is lower, it is less susceptible to damage by thermal shock.

## 7.4 The structure of the core

As explained in §7.1, there are about 80 000 fuel elements in a 2500 MW(t) reactor core, and so, obviously, they cannot be handled one by one. They are made up into *subassemblies*, each consisting of 200–300 elements in a hexagonal steel tube or *wrapper*, as shown in Fig. 5.13. This arrangement allows the coolant flow to each subassembly to be matched to the power generated in it, which varies with neutron flux across the core. The fuel elements are located in the wrapper either by transverse grids or by wires wrapped round each element.

Probably the most awkward consequence of irradiation swelling is that it distorts the subassemblies. The neutron flux is highest at the centre of the core, so each wrapper swells more on that side and becomes curved or *bowed* with the convex side inwards. The extent of bowing depends on the dimensions of the subassembly and its temperature, but if not restrained could result in the top being displaced some 10–50 mm outwards during its life.

Bowing can be avoided by rotating the subassemblies from time to time to equalize the swelling on opposite sides, but it is more usual to prevent it. This can be done by a *passive* restraint structure consisting of a rigid ring round the core that prevents any outward movement, or by an *active* restraint mechanism consisting of clamps, which are tightened after the core has been assembled and slackened to allow the fuel to be changed. The restraint mechanism itself does not swell, because it is outside the core in a region which is cool and where the neutron flux is low.

Figure 5.14 shows how the subassemblies are arranged in a typical reactor. The fuel in the outer part of the core contains more plutonium than that in the inner to compensate for the lower neutron flux and make the power density more uniform. The control rods occupy subassembly positions. They consist of neutron absorbers, usually boron in the form of boron carbide, inserted and withdrawn by mechanisms above the core. The core is surrounded by a radial breeder in which the fuel elements contain natural or depleted uranium dioxide. This in turn is surrounded by a reflector consisting mainly of steel, which serves to scatter some of the neutrons that would otherwise be lost back into the breeder. Outside this is a neutron shield. The whole of the core is supported by a *diagrid*, which is a box-like structure serving to locate the subassemblies and to distribute coolant to them.

The coolant exerts a substantial upward drag on a subassembly as it flows through it, and provision has to be made to hold it in place, usually by balancing the hydraulic forces on the base of the subassembly where it fits into the diagrid.

## 7.5 Primary coolant

The biggest problem in using sodium as a coolant is that it becomes radioactive in the core. $^{23}$Na captures neutrons, and the resulting $^{24}$Na is $\beta$ –and $\gamma$ –active with a half-life of 14.7 hours. With adequate shielding and containment it poses no undue risks, although under some circumstances it makes operation and maintenance a little more complicated.

The difficulty arises when transferring heat from the sodium to water and steam. Heat exchangers of any type are inevitably complex structures, subject to thermal stresses, and carry a risk of failure. A leak in a sodium–

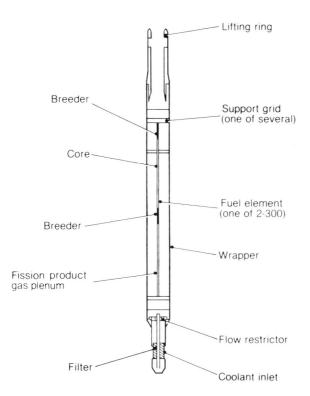

Fig. 5.13 A typical fuel subassembly

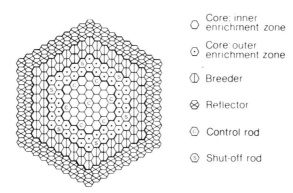

Fig. 5.14 Layout of a typical fast reactor core

water heat exchanger results in the well-known exothermic chemical reaction. It is possible to cope with such a leak (see §7.6) by venting the reaction products to the atmosphere, but this cannot be done if they are strongly radioactive. This is why all sodium-cooled fast power reactors built so far have incorporated a secondary sodium coolant circuit. The primary sodium, which cools the core and becomes radioactive, transfers its heat in intermediate heat exchangers to the secondary sodium, which does not becomes radioactive. This in turn gives up its heat to raise steam in a further set of heat exchangers which are usually called *steam generators.*

Previously it has been stressed that the designer of a reactor is constrained so much by the properties of the materials that all fast reactor cores are very similar. This does not apply, however, outside the reactor core, and the differences between different designs are obvious in the layout of the primary coolant circuit.

This consists of the core, the intermediate heat exchangers, and the circulating pumps. The main choice is whether the pumps and heat exchangers should be in the same vessel as the reactor core (to give what is called a *pool* reactor) or in separate vessels (a *loop* reactor). Figure 5.2 shows a typical pool layout. It has the advantage that the whole primary circuit is contained in a vessel of simple form. It has no pipes connected to it to cause stress concentrations at the nozzles, and it is exposed only to sodium at the relatively cool core inlet temperature.

The vessel has to be large enough to contain the core, the pumps, and the intermediate heat exchangers (at least three of each), and also a neutron shield, usually mainly of steel about 1 m thick, to protect the secondary sodium in the heat exchangers from neutrons from the core and thus prevent it from becoming radioactive.

The simplicity of the vessel is paid for by complexity of the roof. This is large because it has to span the vessel which may be 15 m in diameter, and it has to support the weight of the core, the pumps and their motors, the heat exchangers, and the neutron shield.

The individual vessels of a loop reactor are much smaller and the roof of the reactor vessel is much simpler (Fig. 5.15). It is much easier to manufacture the smaller vessels, and very much easier to inspect them than the load-bearing members of a pool reactor, which are immersed in the primary sodium. The disadvantages of the loop layout are that the vessels are of complex form and are partly exposed to hot coolant and partly to cool, and so there are inevitable thermal stresses and stress concentrations.

In either pool or loop reactors the coolant is circulated usually by single-stage centrifugal pumps at the bottom of long vertical shafts. This arrangement requires a single sodium-lubricated bearing beneath each pump and a single seal where the shaft goes through the roof. The seal has to prevent egress of sodium vapour, but it does not have to cope with liquid sodium.

The sodium is covered with a blanket of argon at a pressure slightly above atmospheric to prevent ingress of air. The reactor vessels and pipework

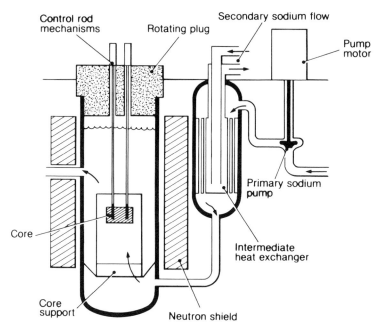

Fig. 5.15 Layout of a loop fast reactor

therefore have to withstand only a moderate pressure of a few hundred kPa. They are usually made of stainless steel and are double-walled with leak detectors in the space between the walls.

The oxygen concentration in the sodium has to be kept below about 10 p.p.m. to control corrosion. This is done by a *cold trap*, in which use it made of the fact that the solubility of $Na_2O$ in sodium decreases with decreasing temperature. A stream of sodium from the primary circuit is passed through a vessel, where it is cooled to about 160 °C or so, and the oxide is precipitated.

Access to the reactor core is through the roof of the reactor vessel. While the reactor is operating, the control-rod mechanisms on top of the roof move the control rods by means of rods passing down into the core. While it is shut down, access is needed to remove and replace the fuel. There are many ways in which this can be done. A typical method is to use a *charge machine*, consisting of a shielded flask which can be placed over a hole in a rotating section of the roof. This can be moved so that an extraction tool can be lowered from the charge machine to remove any of the subassemblies in the core. The control rods have to be disconnected from their mechanisms and left in the core before the roof is rotated.

## 7.6 Steam generators

An important advantage of sodium compared with water as a reactor coolant is that sodium can be heated to 560 °C at atmospheric pressure and used to raise steam at 520 °C. This is the same steam temperature that is achieved in the most modern fossil-fuelled power plants, so the steam cycle of a fast reactor power station differs only in a few respects from that of a conventional power station.

By far the most important differences are the size and form of the steam generators. The boiler of a typical 300 MW(e) coal-fired power station has tubes with a total heat-transfer area of about 30 000 m². By comparison the 250 MW(e) PFR has some 4000 tubes, each about 10 m long, arranged in nine separate units with a total heat-transfer area of 3000 m².

This much smaller size is made possible by the superior heat-transfer properties of sodium compared with those of the hot gases in a fossil-burning boiler, but it also has some disadvantages. The heat fluxes through the walls of the tubes are much higher so that thermal stresses are higher, and the conditions of service are much more rigorous. In addition the smaller size has an effect on the operation of the whole steam plant, because the *thermal inertia* of the smaller mass of steel and water in the steam generator enables it to respond much more quickly to changes in the operating conditions such as variation of the electrical load, and this imposes more stringent requirements on the control system.

The steam generators are usually, but not always, shell-and-tube heat exchangers with the water or steam in the tubes and the sodium in the shell. Various configurations are possible—U-tubes, straight tubes, helical tubes, for example—but in all cases allowance has to be made for differential thermal expansion between shell and tubes. Figure 5.16 shows typical steam

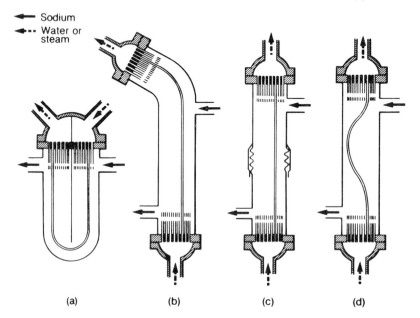

Fig. 5.16 Steam generator designs

generator designs. In design and manufacture particular attention has to be paid to welds between the tubes and the tube plate, because they suffer severe thermal stresses. As an alternative, it is possible to continue the tubes right through the tube plate and joint them together in a manifold, but with many hundreds of tubes to join such a manifold is very complex.

Often the steam generators consist of three separate sections—evaporator, superheater, and reheater. The advantage of this system is that the water does not have to be evaporated to dryness in the evaporator. A two-phase mixture of steam and water is produced, and the two phases are separated in the steam drum. The water is recirculated to the evaporator, while the dry saturated steam passes to the superheater. As there is always water present throughout the evaporator, the tubes never dry out, so that concentration of impurities, which may be corrosive, at this dry-out point is prevented. Also, thermal stresses due to the change in tube wall temperature at the dry-out point are avoided. If this arrangement is used, the superheater and reheater, which do not come into contact with water in the liquid phase that might contain impurities, can be made of austentic stainless steel, but the evaporator has to be made of a ferritic steel which is more resistant to stress corrosion.

The alternative is a *once-through* steam generator, in which water is evaporated to dryness and then superheated in a single heat exchanger. This system is much simpler, but there is much greater difficulty in designing the tubes to tolerate dryout, and in controlling the purity of the boiler water.

If a leak occurs in a steam generator, steam or water is injected at high pressure into the sodium. The resulting chemical reaction produces sodium oxide and hydroxide and hydrogen, and is exothermic. The hot reaction products are very corrosive, and a reacting jet from a leak in one tube could act like a flame playing on an adjacent tube and eventually cause it to burst in turn.

This has to be avoided. It is quite easy to detect a leak in a steam generator, either by means of the hydrogen generated by the reaction or by the increase in pressure on the sodium side. If a small leak is detected, the steam generator can be shut down and emptied and the leak repaired (or the damaged tube plugged and taken out of service). In addition, an automatic protective system has to be incorporated to protect the steam generator from damage if there is a large leak. It acts to close valves isolating the damaged unit on both the steam and sodium sides, and open vents to allow the sodium, water, and reaction products to flow into effluent vessels, from which the hydrogen can be released in a controlled and safe way.

# 8 Safety

## 8.1 Inherent features

The designer of a fast reactor, or of any other piece of plant or machinery, has to make sure that it is safe both when it is operating normally and when something goes wrong. He has to make sure that the risk of injury, either to the operating staff or to the general public, is acceptably small, i.e. the risks have to be shown to conform to the standards imposed by the authorities, and these may be different in different countries.

There are two ways to make a reactor safe: firstly a design concept can be chosen which is inherently safe, and secondly protective systems can be incorporated to prevent accidents having damaging consequences. The four inherent features of an oxide-fuelled sodium-cooled fast reactor that have bearing on safety are discussed in the following paragraphs.

1. The first is that the radioactive materials are separated from the environment by three barriers: the cladding of the fuel, the primary coolant containment (i.e. the reactor vessel if the pool layout is used or the vessels and the connecting pipework if it is the loop layout), and the reactor building. All three contain the fuel and the fission products, while the last two contain the $^{24}$Na in the primary sodium.

There are so many fuel elements that the possibility of one or two failing from time to time has to be catered for. Failures are usually in the form of narrow cracks, however, which allow only some of the gaseous and volatile fission products to escape. These are either retained in the primary coolant or go to the argon gas blanket, from which they can be removed by suitable traps and filters. There may be slight leakage through pump shaft seals for example into the atmosphere of the reactor building, but filters in the ventilation system prevent escape to the outside.

2. Another inherent safety feature is the low pressure of the coolant. This means both that the primary containment is lightly stressed so that it is not likely to fail, and also that even if it does fail the coolant does not immediately escape (as it does from a gas- or water-cooled reactor). If the reactor vessel is situated in a hole in the ground it can be arranged so that the core and the intermediate heat exchangers stay covered with sodium even if the vessel and its leak jacket should both leak. This means that even in this unlikely event the decay heat can still be removed safely.

3. The third inherent feature is the large mass of the primary coolant. This is particularly important in the case of a pool reactor, which may have a few thousand tonnes of primary sodium. It means that even if for some reason the emergency cooling system does not work and no decay heat at all can be removed from the primary circuit, it takes several hours for temperatures at which failure of the cladding becomes likely (800–1000 °C) to be reached. Figure 5.17 shows the rise in mean temperature of a typical pool reactor assuming no heat is lost after the reactor is shut down. The significance of this is that even in this case several hours are available to provide emergency cooling.

4. The final inherent safety feature is the Doppler effect, described earlier, which provides a reliable prompt negative feedback, resulting in a reduction of reactivity if the fuel temperature rises. This means that if

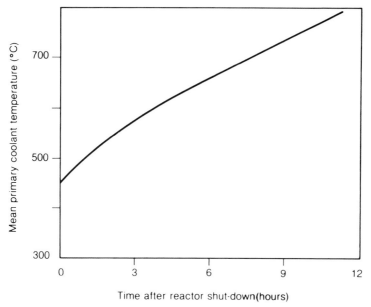

Fig. 5.17 Total cooling failure in a pool fast reactor

reactivity is added accidentally in some way, the power does not rise exponentially with a short doubling time but slowly in a limited manner.

As well as these features which tend to make the reactor inherently safe,
there are other characteristics which might produce difficult situations if
suitable precautions were not taken. One is the high temperature of the fuel
in the centre of a fuel element. This is well above the boiling point of the
coolant, and raises the possibility that, if for some reason a large number of
the fuel elements should suddenly break, molten fuel and liquid coolant
might come into contact. Under some special circumstances a *sodium vapour
explosion* might take place, leading to damage to the rest of the core. This
has never been observed to happen with any significant violence, but it is
nevertheless known that an analogous *steam explosion* can take place if
water is mixed suddenly with certain molten metals, and until it has been
proved to be impossible in sodium it has to be allowed for and guarded
against.

There is also the fact that the fuel in a fast reactor is not in its most reactive
configuration, and in principle it is true that if the fuel were free to move
(because of widespread cladding failure, for example) it might add reactivity
and cause the power to rise to very high levels. In practice, however, it is
very difficult to envisage realistic ways this could happen. Collapse of part of
the core under gravity, for example, reduces reactivity. If the core were
crushed radially the reactivity would increase, but there is no serious risk of
this happening. Loss of coolant from part of the core can also, in principle,
cause an increase in reactivity.

## 8.2  **Protective systems**

In spite of the inherent safety features described in the previous section it is not possible to allow the reactor to look after itself if something goes wrong. Protective systems have to be provided which give warning if a minor mishap occurs, and which shut the reactor down automatically if there is anything more serious. Automatic shutdown is usually called *tripping* the reactor. Instruments are provided which measure certain critical parameters such as the coolant temperature. If the maximum safe value, or *trip level*, is exceeded, the control rods and shut-off rods are automatically inserted into the core.

It is important that the trip system is very reliable. This is achieved in two main ways: by redundancy, which means that several quite separate systems are provided to perform essential functions, and by making the system *fail-safe*. For example, there are more than enough control rods, so that only a fraction of the total is needed to shut the reactor down, and they may be attached to the actuators by electromagnets which are automatically switched off when the reactor is tripped, allowing the rods to drop into the core (this is a fail-safe system because the reactor shuts down automatically if the electricity supply fails).

It is relatively easy to detect something going wrong if it affects the whole of the reactor core. The reactor is tripped if, for example, the neutron flux is too high, the coolant is too hot, the flowrate is too low, or the primary coolant pumps stop. To avoid tripping the reactor unnecessarily, the essential instruments are usually at least triplicated. If there are three thermocouples to measure the coolant temperature and only one indicates that it is too high, an alarm is sounded, but the reactor is not tripped because it is probably the thermocouple which is faulty. If two or more indicate that the coolant is too hot however the reactor is tripped.

While it is fairly easy to detect something going wrong if it affects the whole core, more care has to be taken to make sure that local malfunctions, affecting only a part of the core, do not go unnoticed. Because there are so many fuel elements and the coolant passages between them are so narrow, it is particularly necessary to guard against blockages in the coolant flow. The coolant inlet to each subassembly is protected by a filter, but precautions have to be taken against the filter breaking and allowing some foreign material to be swept in and lodged between the fuel elements, where it might cause overheating and failure of the cladding.

A local blockage can be detected in a variety of ways. Thermocouples at the subassembly outlets are simple and reliable, but not particularly sensitive; a small blockage would cause local overheating without affecting the outlet temperature very much. If the overheating is severe enough to boil the coolant, acoustic detection—by listening to the noise—may be possible, but the method has not been proved. The most reliable method is to detect failure of the cladding by looking for radioactive fission products from the fuel in the coolant.

A system to do this is known as a *burst cladding detection* system or BCD. It often consists of several components, searching for radioactive fission-product gases in the gas blanket above the sodium and in the sodium itself, and also for delayed neutron precursors in the sodium. This last is a particularly sensitive means of detecting failure: if a sample of coolant is taken well away from the core and neutrons are found coming from it, there must be neutron-emitting fission products, which must have come from a failed fuel element.

When a malfunction is detected and the reactor is tripped, adequate cooling has to be guaranteed so that the decay heat can be removed and the fuel elements kept cool, the cladding intact, and the fuel and fission products contained. Decay heat can be rejected by means of the normal cooling system, the primary and secondary sodium rejecting heat to the steam generators and thence to the condenser. Reliable cooling is so important however, that an alternative is provided in case something should go wrong with the main system. A *decay-heat rejection system* may consist of one or more separate heat exchangers, by means of which the primary coolant can reject heat to a coolant circuit filled with an alloy of sodium and potassium which is liquid at atmospheric temperature. This circulates by natural convection and rejects its heat to the air. In this way, whatever happens the fuel can be kept cool indefinitely, even in the unlikely event of complete failure of the normal cooling system.

## 8.3 The risks

In the previous two sections discussion has centred on the various features, both inherent and engineered, which would have an effect on the behaviour of a fast reactor if an accident should happen. What would actually take place depends on the details of the design, in particular of the protective systems. In order to determine whether a proposed reactor design if safe enough, it has to be analysed by tracing the consequences of a range of different accidents, taking account of the frequency with which each is expected to be initiated and the probability of the various ways it might develop. The resulting range of possible accident consequences and the frequencies with which each is expected can be compared with a criterion of acceptability. If the reactor does not meet the criterion the protective systems have to be improved until it does.

The methods of conducting this type of analysis and of setting criteria for acceptability are described elsewhere. None is universal: licensing authorities in different countries set different standards and require different methods of analysis. As a result, it is not possible to make a general statement about the protective systems needed or the standards of safety achieved by fast reactors; systems suitable to meet whatever standards are specified can, however, be incorporated.

It is nevertheless possible to indicate the types of accidents which have to be guarded against. There are the usual range of operator errors or equipment malfunctions which, as with any other sort of reactor, have to be shown to present no unacceptable risk. There are also a few accidents that are specific to fast reactors, to which attention has to be given.

Local blockages to the coolant flow were mentioned in §8.2. It is necessary to guard against these because of the fine clearances between the fuel elements, the difficulty of detection, and the high heat flux from the fuel to the coolant. There is also an historical reason in that in 1966 the Enrico Fermi reactor in the USA was severely damaged by an accidental blockage of a subassembly. Other reactors have been designed with two or more coolant inlets to each subassembly, so that a repeat of the same accident is impossible, but the lesson that local blockages have always to be considered has been well and truly learnt.

An earlier accident to the EBR–1 reactor, also in the USA, in 1955, took place during tests involving adding reactivity, and serves as a reminder that an accidental addition of reactivity must also be considered. To guard against such an accident it is usual to analyse the consequences of a reactivity addition from an unspecified cause. The effect is to raise the power, and so this accident is often called a *transient over-power*. It is necessary to make sure that the reliable negative feedback from the Doppler effect and other sources will shut down the reactor safely.

The third important class of accident relates to a loss of primary circuit pumping power. It has always been clear that there is a possibility of overheating if something happens to interrupt the flow of coolant, and so an accidental stoppage of the primary coolant pumps needs to be analysed to make sure that emergency cooling systems can control the overheating and prevent widespread damage to the fuel.

Another accident which has to be considered is a sodium fire. This by itself is unlikely to affect the core, but if the primary coolant should leak to the atmosphere it will burn, releasing radioactive sodium oxide. The rector containment building has to be designed to prevent release of this to the environment.

Any accident that can be foreseen can be designed against. But clearly not every accident can be foreseen. It is to prevent the unforeseen accident that initiating events, such as uncontrolled addition of reactivity, are assumed to happen even though no way is known in which they can actually come about. Because of rigorous analysis of this type and the application of stringent safety criteria, fast reactors have achieved an excellent safety record.

# 9  Development of fast reactors

Before about 1960 a high breeding ratio was thought to be the most important quality of a fast reactor. This meant that the mean neutron energy had to be kept high: thus extraneous materials, especially moderators, had to be excluded from the core. As a result the early reactors had metal fuel, either enriched uranium or plutonium.

The cores of these reactors were small, and for high-power operation they had to be cooled with high-density coolants. Water was precluded because it is a moderator, and so liquid metals–either sodium or sodium potassium alloy–were used. Many neutrons leaked from the small cores and were captured in surrounding natural uranium breeders.

Low-power experimental reactors of this type were built in the late 1940s and early 1950s. EBR–1 (USA) was the first step towards a power reactor. It produced 1.2 MW and was the first nuclear reactor of any type to generate electricity.

For full-scale power production the core had to be enlarged to allow for the extra coolant flow. The result was EBR–II and the Enrico Fermi FBR (EFFBR) in the USA and DFR in the UK. When they were constructed these were seen as prototype power station reactors, but in the event they were the end point of metal-fuelled reactor development, and EBR–II and DFR were used mainly to test oxide fuel for the next generation of reactors.

About 1960 it became clear that there is more to a profitable fast reactor than good breeding. The necessity for high burn-up became clear as well. It was found that metal fuel could only withstand about 1 per cent burn-up before it became so affected by irradiation that it had to be reprocessed. The costs associated with such frequent reprocessing were excessive, and so an alternative was found in the form of oxides, either $UO_2$ or $PuO_2$ or a mixture of the two. Oxide fuel can stand 10 per cent burn-up or more, thus giving lower reprocessing costs.

Oxide fuel has other advantages over metal in that it can be operated at high temperatures, and is compatible with stainless steel cladding (whereas refractory metal cladding has to be used with metal fuel). The main disadvantage is the introduction of oxygen into the core, which acts as a partial moderator, reducing the mean neutron energy, and therefore reducing the breeding ratio.

Carbide fuel (UC and PuC) should be better than oxide because it has a higher thermal conductivity and also allows a higher breeding gain, but suffers the disadvantage that it is not so well understood as oxide, which is widely used in thermal reactors as well as fast. The need for reliability often dictates a conservative approach, so oxide fuel is currently used almost universally. In the future, however, it is possible that carbide will be preferred.

Mixed-oxide fuel, stainless steel structure, and sodium coolant have become accepted widely as the route for the development of fast breeder reactors. The use of these materials restricts the engineering design so that all current reactors, from whatever country, show marked similarities. The core of PFR (UK), Phenix (France), BN350 (USSR), SNR 300 (Germany), Monju (Japan) and the proposed CRBR (USA) are very similar. These are all prototypes to be followed by commercial reactors for the generation of electricity. It is expected that the start-up of the first commercial-size demonstration fast breeder reactor will be in France in 1984.

## Suggestions for further reading

### Physics

BAKER, A. R. AND ROSS, R. W. Comparison of the value of plutonium and uranium isotopes in fast reactors. In USAEC *Breeding, economics and safety in large fast power reactors. ANL* 6792, pp. 265–329. Washington, United States Atomic Energy Commission (1963).

DUDERSTADT, J. J. AND HAMILTON, L.J. *Nuclear reactor analysis.* Wiley, New York (1976).

HUMMEL, H.H. AND OKRENT, D., *Reactivity coefficients in large fast power reactors.* American Nuclear Society, Hinsdale, Ill. (1970).

OKRENT, D., Performance of large fast power reactors including effects of higher isotopes, recycling and fission products. In IAEA. *Physics of fast and intermediate reactors,* vol. 2, pp. 271–297. International Atomic Energy Agency, Vienna (1961).

OKRENT, D., COHEN, K. P. AND LOEWENSTEIN, W. B. Some nuclear and safety considerations in the design of large fast power reactors. In UN. *Peaceful uses of atomic energy,* vol. 6, pp. 137–148. United Nations, New York (1964).

YIFTAH, S. Effect of the plutonium isotopic composition on the performance of fast reactors. In IAEA. *Physics of fast and intermediate reactors,* vol. 2, pp. 257–270. International Atomic Energy Agency, Vienna (1961).

### Fuel

FINDLAY, J. R. The composition and chemical state of irradiated oxide reactor fuel material. In IAEA. *Behaviour and chemical state of irradiated ceramic fuels,* pp. 31–39. International Atomic Energy Agency, Vienna (1974).

MEYER, R. O., O'BOYLE, D. R., AND BUTLER, E. M. Effect of oxygen-to-metal ratio on plutonium redistribution in irradiated mixed-oxide fuels. *J. Nucl. Mat.,* **47,** 265–267 (1973).

OLANDER, D. R. *Fundamental aspects of nuclear reactor fuel elements.* Energy Research and Development Administration, Washington (1976).

PERRIN, J. S. Effect of irradiation on creep of $UO_2$—$PuO_2$. *J. Nucl. Mat.* **42,** 101–104 (1972).

POWELL, H. J. Fission product distribution in fast reactor oxide fuels. In IAEA, *Behaviour and chemical state of irradiated ceramic fuels,* pp. 379–392. International Atomic Energy Agency, Vienna (1974).

### Core materials

BAGLEY, K. Q., BRAMMAN, J. L., AND CAWTHORNE, C. Fast neutron induced voidage in non-fissile metals and alloys: a review. In BNES. *Voids formed by irradiation of reactor materials,* pp. 1–26. AERE Harwell for British Nuclear Energy Society, London (1971).

BAGLEY, K. Q., BARNABY, J. W., AND FRASER, A. S. Irradiation embrittlement of austenitic stainless steels. In BNES. *Irradiation embrittlement and creep in fuel cladding and core components,* pp. 143–153. British Nuclear Energy Society, London (1973).

BRAMMAN, J. I., BROWN, C., WATKIN, J. S., CAWTHORNE, C., FULTON, E. J., BARTON, P. J. AND LITTLE, E. A. Void swelling and microstructural changes in fuel pin cladding and unstressed specimens irradiated in DFR. In AIME. *Radiation effects in breeder reactor structural materials*, pp. 479–508. American Institute of Mining Engineers, New York (1978).

BROWN, C., BUTLER, J. K. AND FULTON, E. J. Void swelling studies of austenitic fuel pin cladding material irradiated in the Dounreay Fast Reactor. In CEA. *Irradiation behaviour of metallic materials for reactor core components*, pp. 129–136. Commissariat a L'Energie Atomique, Gif-sur-Yvette (1979).

MOSEDALE, D. AND LEWTHWAITE, G. W. Irradiation creep in some austenitic stainless steels, Nimonic PE16 alloy, and nickel. In Iron and Steel Institute. *Creep strength in steel and high-temperature alloys*. pp. 169–188. The Metals Society, London (1974).

THORLEY, A. W. AND TYZACK, C. Corrosion and mass transport in steel and nickel alloys in sodium systems. In BNES. *Liquid alkali metals*, pp 257-273. British Nuclear Energy Society, London (1973).

## Engineering

BROOMFIELD, A. M. AND SMEDLEY, J. A. Operating experience with tube to tubeplate welds in PFR steam generators. In BNES. *Welding and fabrication in the nuclear industry*, pp. 3–18. British Nuclear Energy Society, London (1979).

CLAXTON, K. T. Solubility of oxygen in liquid sodium—effects on interpretation of corrosion data. In ANS. *Liquid metal technology in energy production*, Vol. 1, pp. 407–414. American Nuclear Society, Hinsdale, Ill. (1976).

FRAME, A. G., HUTCHINSON, W. G., LAITHWAITE, J. M. AND PARKER, H. F. Design of the prototype fast reactor. In British Nuclear Energy Society. *Fast breeder reactors*, pp. 291–315. Pergamon, Oxford (1967).

HANS, R. AND DUMM, K. Leak detection of steam or water into sodium in steam generators for LMFBRs. *Atom. En. Rev.*, **15**, 611–699 (1977).

HAYDEN, O. Design and construction of past and present steam generators for the UK fast reactors. *J. Brit. Nucl. En. Soc.*, **15**, 129–145 (1976).

LEWINS, J. *Nuclear reactor kinetics and control*. Pergamon, Oxford (1978).

YANG, Y. S., COFFIELD, R. D. AND MARKLEY, R. A. *Thermal analysis of liquid-metal fast breeder reactors*. American Nuclear Society, Hinsdale, Ill. (1978).

# 6

# A review of the UKAEA interest in heavy water reactors

R. J. SYMES

Heavy water has been an important material throughout the whole history of nuclear reactor technology because of its ability to slow down neutrons resulting from fission without absorbing them. The possibility of using heavy water as a moderator in thermal reactors has been understood from the early days of research into nuclear energy, and this chapter describes the evolution of the heavy water power reactor in the context of the development of the UK's thermal reactor programme. The viewpoint expressed by the author is that of a power engineer who became involved in this evolutionary process from its outset. A review is also made of the quite considerable range of research and experimental reactors throughout the world which have incorporated heavy water in their cores.

The chapter commences with a brief account of the history of heavy water production and then begins the story of the British use of this moderator in power reactors. This is equated with the introduction and development of the tube reactor as a distinct and important form of reactor construction in contrast with the perhaps better known vessel design that has tended to dominate reactor engineering to date. The account thus includes a succession of reactor designs including the gas- and steam-cooled heavy water systems in addition to the steam-generating heavy water reactor. The SGHWR was demonstrated by the construction of a substantial prototype, which continues in operation as a flexible and reliable electricity-generating plant. It was also, for a time, identified as the system to be used for Britain's third reactor programme. Today the successful Canadian CANDU power reactors represent the only penetration of heavy water reactor technology into large scale electricity generation.

## Contents

# 1 Introduction to the British heavy water reactor

The physicist Frederic Joliot, whilst he was a professor at the College de France, concluded that two substances might work as practical moderators for a nuclear reactor: pure carbon in the form of graphite and heavy water ($D_2O$), a compound of an isotope of hydrogen (deuterium) with oxygen. Deuterium absorbs neutrons to a much smaller extent than ordinary hydrogen, making heavy water a more attractive moderator than 'light' water ($H_2O$). The abundance of 'light' hydrogen is much greater than that of deuterium. Naturally occurring water consists principally of light water with only about one part in 5000–6500 of heavy water.

The American scientist, Harold Urey, had identified in 1932 a denser fraction in water after a very large number of successive distillations. Until the outbreak of World War II in 1939, however, heavy water was used only in scientific research. It was produced in relatively large quantities by a

Franco-Norwegian company involved in the manufacture of synthetic ammonia, who took advantage of the low cost of electricity in Norway and linked a system of fractional electrolysis with the ammonia production process. In this way, heavy water was made at a cost of some $14 an ounce, equivalent to $500 000 per tonne.

Subsequently, various isotope-separation processes have been used or studied for concentrating deuterium above its natural abundance level. No single process has been identified which is superior over the whole range of enhancement generally required, i.e. from a starting concentration of 0.015 per cent to a production concentration of 99.7 per cent or higher.

After the war, there was an expansion in the production of plutonium for the US Atomic Energy Commission, and reactors moderated by heavy water were chosen for the Savannah River production plant. It therefore became necessary to construct large-scale heavy water manufacturing factories, and it was decided to use a dual-temperature chemical exchange system as the initial process with water distillation and electrolysis for the intermediate and final stages of concentration. The American government built two large plants of this pattern—one at Newport, Indiana, known as the Dana Plant and another at Savannah River itself. The Dana Plant was operated for the AEC by duPont until 1957, when it was first reduced to a stand-by status and then transferred to the US Army Chemical Corps to be rebuilt for a different purpose. The Savannah River plant, which has a capacity of some 400 tonnes of $D_2O$ per year (99.75 per cent concentration), continues to operate and has supplied the bulk of world requirements for heavy water. More recently, Canada has built its own large-output $D_2O$ production plant in support of its large and successful national nuclear power programme based on the CANDU reactor system, which employs heavy water both as moderator and as coolant. In 1963 heavy water assayed as 99.75 per cent $D_2O$ could be purchased at a price of around $60 000 per tonne in minimum quantities of 57 kg—a very substantial reduction compared with the early Norwegian production.

This country's achievement in heavy water power reactor technology consists essentially of the design of the prototype steam-generating heavy water reactor (SGHWR) and its application to the construction of a 100 MW(e) power station (1963-7), incorporating a prototype reactor which embodied 'full-sized' components. In the course of arriving at the SGHWR design, many alternative heavy water reactor layouts were examined and much was learned that contributed to thermal reactor design in the UK. The SGHWR was constructed to timescale and within its estimated cost. It has operated successfully for sustained periods whilst connected to the national electricity transmission grid (1968–today) under all the various operating regimes that a practical power plant has to accommodate during its working life.

It was claimed that the SGHWR was a system whose main components— certainly its coolant circuit components and fuel—could remain of fixed dimensions while individual plant sizes ranged from 100–1000 MW electrical output. This was demonstrated by producing a series of designs spaced within this band of outputs that contained standard modules and applied the same design philosophy as the 100 MW prototype. It was similarly claimed that the SGHWR design permitted the principal reactor components to be regularly examined and monitored during each refuelling programme and that even major circuit components could be replaced with a minimum of downtime if any unacceptable damage or deterioration was discovered. These claims have been demonstrated on the prototype plant by the re-placement of seven experimental *pressure-tube* assemblies after the reactor had been in commission for two years. The extensive sub-division of the primary coolant circuit, achieved in the pressure-tube type of reactor, facilitates the provision of an independent emergency cooling supply to each fuel assembly. The design of the fuel element was developed to exploit this feature to the full so that the SGHWR possesses some safety advantages over other types of water-cooled reactor.

Finally, it was foreseen that the SGHWR could be built economically in sizes appropriate to central power station use in the UK. This was acknowl-edged by the long-considered decision to adopt the system for the third phase of the British nuclear power programme. However, in the event, too long had elapsed between the construction of the prototype plant and the authorization to build the commercial version, so the re-establishment of the specific industrial capability was found to be too expensive. The decision was accordingly reversed after not inconsiderable commitment and, effec-tively, it must be reckoned that Britain's heavy water power reactor develop-ment stopped at that point. So much for an initial appraisal of the end-game which took place in 1976/7. The prototype SGHWR continues to operate at Winfrith as a power producer for the national grid.

## 2 Evolution of the SGHWR and its principal design features

A good starting point in looking at the SGHWR is to consider what is perhaps its principal design future, the use of pressure-tube construction.

In basic terms, every fuel assembly of a pressure-tube reactor is located for its useful lifetime within an individual pressure envelope which forms an integral and therefore nominally permanent part of the reactor structure. By arranging a collection of these cylindrical conduits of essentially small dia-meter, side by side, with suitable external pipe connections, the total coolant flow to the core may be divided so that the requisite amount passes through each fuel assembly. Heat is thereby removed from the fuel and transported to the turbines.

The pressure-tube form of construction can be directly contrasted with so-called pressure vessel or tank systems. The contrast is between a large, single, undivided pressure vessel accommodating all the fuel assemblies and any associated core structure and a collection of individual small-diameter pressure-tubes, each capable of holding just one fuel assembly, which pass in parallel array between an inlet and an outlet manifold system.

## 2.1  Evolution of the tube system in the UK

The first major study of tube reactor systems by the UKAEA was in 1956, two years after the Authority had come into being.

The early 1950s were a time of rapid progress on a number of fronts. The fast reactor development programme had been launched with the design of the Dounreay Fast Reactor (DFR); construction was already nearing completion of the first of a series of new production reactors at Calder Hall—adjacent to the Windscale Establishment. These were the first reactors in the UK to make use of the heat of nuclear fission and were effectively prototypes of thermal reactor power stations that could produce both plutonium for fuelling fast reactors and electricity for the national network. Both thermal and fast power reactors were, then, in their infancy and, not surprisingly, it was felt at the time that it would be expedient to make common as many features as possible between the parallel programmes. Since the only coolant then regarded as feasible for fast reactors was liquid metal, this implied consideration of a sodium-cooled thermal reactor which would employ similar coolant circulators, cold traps, heat exchangers, steam generators, etc., as the contemporary fast systems. The only practical thermal reactor meeting this specification appeared to be the sodium-graphite reactor (SGR), and this concept was studied at about the same time in both the UK and the USA.

Hot sodium damages graphite by spalling and so cannot be permitted to contact the graphite directly. The American solution was to arrange for hexagonal columns of the moderator to be *canned* in zirconium alloy sheet, to nest these together with a regular vacancy, and to use these vacancies to constitute fuel channels. In the British design, a stack of graphite bricks and tiles, generally similar to the moderating structure of a Calder Hall reactor, with regular, vertical through-holes would have been provided with a circular zirconium tube in each of these holes, each tube accommodating a simple design of multi-pin fuel assembly and a channel for the sodium coolant. This was in retrospect an important step, and even bolder at the time when one appreciates that it represented the first proposal to use zirconium alloys on a commercial scale and also one of the first suggestions to use multi-pin ceramic fuel in a thermal reactor.

The design withstood the test of most of the searching economic and engineering questioning to which it was subjected; it brought to bear a wide forum of technical effort on such items as heat exchangers and steam generators, and the proposals in these areas for the SGR resemble quite closely those now being associated with the commercial fast reactor (CFR). In the final analysis, however, the reactor would have required more enriched material than it was then considered could be afforded for the thermal reactor programme, and the study was terminated in 1957. This placed thermal reactor design emphasis on the gas-cooled, graphite-moderated systems that are now generically described as Magnox reactors.

Development of thermal reactors in the UK accordingly became more strongly oriented towards the improvement of gas-cooled systems, and the ceramic fuel considered for the SGR was quickly explored in this context. The advantages soon became clear. Higher coolant temperatures leading to superior steam conditions for the generating plant were possible, and so, in 1957, the design of a prototype advanced gas-cooled reactor (AGR) was begun. This used the only form of construction then being commercially exploited, i.e. with the whole core structure located inside a steel pressure vessel. The AGR encountered a number of problems. The 'Windscale Accident' occurred (October 1957), casting doubts on the suitability of graphite as a moderator for power reactors, since it could so readily become a chemical fuel. Beryllium, chosen as the fuel cladding material, to minimize absorption while permitting higher temperatures than Magnox, proved to be intractable and incapable of being extruded to form thin-walled tubes. Finally, the prospect of constructing, on-site, the steel pressure vessel needed to house an AGR core of the size envisaged for ultimate application was daunting. Although, ultimately, all these difficulties were resolved or designed out of the AGR design, it took time to find the solutions and to demonstrate their efficacy. In the meantime, the future of thermal reactor design seemed again open and uncertain.

This period of uncertainty in late 1957 convinced the management of the UKAEA that an 'insurance policy' thermal reactor development option was essential. Diversity of approach was therefore sought in any of the aspects then suggested for the AGR which might prove 'difficult'. Thus, while the fuel form, uranium dioxide ($UO_2$), and the coolant, carbon dioxide ($CO_2$), were accounted to be wholly satisfactory, alternatives were considered prudent for the fuel cladding, the moderator, and the pressure vessel. The substitutes suggested were: stainless steel for the cladding; heavy water in place of graphite for the moderator; and, in place of the normal AGR pressure vessel, the tube construction basically demonstrated in the SGR feasibility study. The name given to this alternative reactor did not appear unduly cumbersome—the gas-cooled heavy water reactor—and not everyone in those days converted this to a string of initial capitals—GCHWR.

The GCHWR was studied in considerable detail and many of the features and characteristics now associated with the SGHWR were firmly established. First, a design of pressure tube was proposed which, dimensionally, resembled very closely that currently incorporated in the SGHWR and which included similar materials and forms of construction. (Essentially, this dimensional similarity followed because the coolant pressures associated with British tube reactor studies have always been in the same area of design— between 41 and 69 bar.) From the first, pressure-tube designers have been aware of the difficulties associated with using zirconium alloys for this component. It has always been accepted that zirconium should be used for the portion of the pressure-tube assembly which actually passes through the core region since acceptable fuel economics depend on the use of a workable structural material with a low neutron absorption cross-section. However, as early as 1957, zirconium was already commanding a high price, and so it was vital from capital cost considerations to minimize the amount employed. Cheaper materials are therefore needed in the coolant circuit immediately the limits of the core itself have been passed. It is the much smaller coefficient of linear expansion exhibited by zirconium compared with more common pressure-circuit materials that creates difficulty. Direct-joining techniques such as welding are not practicable at the sizes required and no simple form of flanged connection would have been reliable.

The GCHWR study suggested two solutions: a complex form of flange joint described as a *compensated Boltzmann joint* involving relatively elaborate flanges and special metal gaskets, and a *rolled joint*, in which the zirconium alloy tube was roller-expanded into a thick-walled tubular section described as a *hub*, machined with a number of separate grooves and made of the structural material used for the rest of the coolant circuit. Ultimately, the latter was demonstrated to be completely practicable and very reliable even under the most rigorous simulation of possible working conditions. This was particularly fortunate since it was by far the less complex approach and permitted the use of an extremely simple form for the zirconium alloy components.

Secondly, the GCHWR study showed that the tube reactor could be a truly modular concept provided a liquid moderator were employed in a core with a vertical axis. The heat output of such a reactor could be varied by altering the effective height of the core—by raising or lowering the level of the moderator in the tank—and this control effect would occur uniformly across the whole diameter of the core. By a modular concept, it is understood that the design and dimensions of the fuel assembly and the associated pressure-tube are standardized and held constant. The overall power of the reactor is then determined by the number of these fuel–tube modules incorporated in a particular core lattice. This possibility of varying only the diameter of a reactor core to achieve different core heat outputs was then completely novel, since, previously, all cores had been typified by a characteristic height/diameter ratio appropriate to the type of reactor involved. Further, in the tube reactor, the complete channel assembly is sufficiently small for it to be wholly manufactured in a factory and sent to the reactor site as a completed component. Not only does this reduce the capital cost, but also—and this is important—a significant portion of the reactor assembly is given a cost that is almost directly proportional to reactor heat output. The specific capital cost (£/kW) of a tube reactor design with constant fuel–tube dimensions is therefore appreciably flatter when plotted against reactor output than the cost/size curves for other reactor systems which have no such comparable feature to produce linearity. Accordingly, the possibility was seen that the type of reactor being discussed could uniquely display lower capital costs at the smaller sizes of real interest to emerging countries with small electricity-generating networks. The British tube reactor was therefore regarded as offering prospects of being a highly competitive export system.

These aspects were identified in the GCHWR study and have characterized all subsequent design of British tube reactors. No other tube reactor exploits this idea of constant channel–fuel dimensions, and no pressure–vessel reactor can reproduce the feature over a wide range of reactor outputs.

The possibility of using a different material for moderating a reactor from that used to cool it was seen as conferring substantial freedom in the quest for good neutron economy. British gas-cooled systems in fact have this choice even when they are built with pressure-vessel construction since they use a solid moderator, graphite. In these, however, the operating temperature of the moderator is necessarily closely related to the temperature range of the coolant gas. In a water-cooled reactor constructed inside a pressure vessel, it would seem that no such separation can be attained without quite excessive complications. Pressure-tube construction by contrast gives the technologist a substantially free choice of material and also of the operating conditions for what are quite obviously very different functions. In the case of the GCHWR, the carbon dioxide used for the coolant was to be operated at high pressure and through a similar temperature range to that in the AGR. The heavy water moderator was, however, to be employed at low pressure, and it was held at under 80 °C by circulating it through suitable coolers arranged close to the core to minimize hold-up of this expensive commodity. Use of low pressure, low temperature for the heavy water system has enabled experience derived from the British research reactors of the DIDO class to be applied directly to the UKAEA's development of pressure-tube systems. It is clearly an important feature enabling the design of a flexible, low-loss heavy water circuit to be carried out with simple, low-cost engineering techniques.

The heavy water moderator is typically contained in a separate low-pressure tank known as a *calandria*. This results in there being two coaxial tubes surrounding each fuel assembly: the pressure tube itself and then, with some radial clearance from it, the associated tube of the calandria tank, this second tube providing the local boundary of the moderator in the lattice cell. This leads to a number of important design features. First, while the pressure tube is intended to remain in the core during the whole lifetime of the reactor, damage may occur during refuelling or a defect may develop. The compartmenting introduced into the core structure by this double-tube system makes it comparatively simple to design such a reactor so that a complete, individual pressure-tube assembly may be replaced without removing fuel from adjacent channels or draining down the moderator. The ability to rebuild the pressure circuit within the core region is unique in reactor technology. Secondly, the interspace between the tubes can be charged with a gas that is compatible with adjacent structural materials under reactor working conditions and of low thermal conductivity. This effectively prevents any excessive loss of heat from the 'hot' coolant circuit to the 'cold' moderator without requiring the complication of a 'solid' thermal-insulating barrier. If the gas ($CO_2$) is introduced in a dry state and is allowed to drift slowly down between the tubes—whilst notionally stagnant in the annular interspace—it also provides a localized and direct means for identifying the first signs of any leakage from either the coolant circuit or the moderator because of a tube or joint defect. This is an essential capability if the application of the leak-before-break fracture mechanics criterion to the coolant circuit design is to be meaningful in actual operating practice.

As a final comment on the convenience of the lattice arrangement of a tube reactor, it should be mentioned that it is an especially simple system to survey, so that information on the condition of the pressure parts in the core region can be obtained directly throughout the lifetime of the plant. This monitoring is most conveniently undertaken as part of the refuelling sequence. Once the spent fuel assembly has been lifted from its channel, the whole internal surface of this major pressure component, which is normally adjacent to the fuel and being swept by the coolant, is available for examination by television scanning heads or for ultrasonic testing or direct measurement, which can include taking impressions of surface defects. Eddy current devices can also be lowered into the channel to determine whether the pressure tube is maintaining its position relative to the surrounding calandria tube. This capability for monitoring the actual pressure parts is superior to that achievable with present vessel designs of reactor, where the inside surfaces of the vessel are screened by stainless steel lining, thermal shields and core support structures.

## 2.2  From GCHWR to SGHWR

The GCHWR feasibility study established most of the characteristic features of vertical pressure-tube reactors using techniques available in 1957/8. The route taken by the development team from this time forward in evolving the design was novel in that the main study was on overall power station designs of commercial size. Small, prototype designs were only considered if it was judged that the evaluation of a demonstration unit would be worthwhile. This was in direct contrast to the main Magnox/AGR development route or even the fast reactor at this time, where the prime consideration was given to small-scale prototype plant which could be used to demonstrate the main features characterizing a new stage of development and help provide firm data for it.

It was a result of studying large-output tube reactors that the prospect of standardizing fuel–tube modules emerged, and it meant that some of the early UKAEA designs were for sizes close to those that were ultimately worked upon for commercial application.

It also led to the abrupt termination of the GCHWR study in 1958 when it was appreciated that the cost of steam-raising towers for a large unit would be too high to permit competitive generating costs to be achieved. This arose because of the following. Since the carbon dioxide coolant would have been raised only to AGR values, then around 660 °C, a steam cycle was implicit. However, whilst the design of the small Windscale prototype (WAGR) was then being formulated with a steel reactor pressure vessel and separate steam-generating towers to be operated at a gas pressure of around 20 bar, optimization of the pressure-tube configurations of the GCHWR pointed to an operating pressure of about 40 bar. Thus, the problems of designing a larger steel pressure vessel—one of the reasons for considering the GCHWR—was transferred from the vessel for the core to the vessels for the steam generators! The study still had some funds and time to run, so a modification was made by changing the coolant from a permanent gas to a vapour—steam. This was to enable an ingenious recirculation reactor coolant system to be applied, the so-called *Murgatroyd system*. The scheme sub-stituted small units exchanging heat from steam superheated in the reactor to boiling feedwater for the very large steam generators of the GCHWR, and also exchanged feed pumps for gas circulators thereby markedly reducing the auxiliary power load.

The reactor was effectively a multipass steam superheater and was there-fore called the Steam-Cooled Heavy Water Reactor or SCHWR. Essentially, the same fuel element was used as in the GCHWR—enriched $UO_2$ pellets clad in stainless steel in a multi-rod assembly not unlike that now enclosed in a graphite sleeve for the AGR. The outer surface of the cladding was 'roughened' to improve the heat-transfer performance by producing tur-bulence in the boundary layer of the coolant as it flowed over the fuel element. This roughening took the form of regular, low-height circum-ferential fins machined from the thickness of the fuel can and represented a very early proposal to employ such a heat-transfer surface. Similar channel tube assemblies were employed as those designed for the GCHWR save that the zirconium alloy pressure-tube section was protected internally by a thin stainless steel liner to prevent excessive hydride formation. Outside the channel tubes, the system followed the design of the GCHWR closely.

During this time, the British work on pressure-tube reactors was discussed with AECL during one of the regular UK–Canada exchanges on nuclear matters. Canadian engineers judged that the pressure tube offered advan-tages in the design of their CANDU system. Thus CANDU evolved from a pressure-vessel design (NPD) to a horizontal pressure-tube system (NPD-2).

The SCHWR was effectively a direct-cycle system providing superheated steam at the turbine stop-valve without the complication of having two distinct zones within the reactor core. This was a great advance over con-temporary pressure-vessel designs that were proposing to incorporate both boiling and superheating sections in a single core with consequent problems of control and matching. The SCHWR had its coolant circuit arranged as a number of passes in series with progressively diminishing pressure in each pass, but it employed a single cooling regime throughout—saturated steam which received superheat as it passed down the channels—and there was a standard temperature at the outlet from each pass. Ultimately, the steam arrived at the turbine at 900 lb/in², 900 °F (62.1 bar, 482 °C), so the design pressure for channel tube assemblies became similar to those now employed in the SGHWR.

Historically, it is important to realize that the evolution of the SCHWR coincided in time with a period of improved confidence in the AGR. In fact, a notable contribution was made to the AGR by concepts first introduced in the tube reactor studies. The prospects for the SCHWR to have overtaken and displaced the established gas-cooled systems was therefore probably never as good as its protagonists thought, though its ability to show favourable economics at small sizes resulted in it being considered for marine application. In 1959, it was set aside because of the difficulties in achieving the required performance from the stainless steel cladding when this was cooled by steam.

The remaining stages of the story of pressure-tube reactor evolution in the UK are better known. They are part of more recent history and many more people tend to have been involved. In retrospect, the tenacity of those engaged upon the early UKAEA HWR studies has to be admired. The concept of the 'insurance policy reactor', to safeguard the thermal programme, was largely a thing of the past. The AGR work in the Authority was moving continuously towards a satisfactory conclusion, and such interest as there was in alternative systems was directed more at achieving enhanced performance in the future, rather than contemporary, parallel paths. The high temperature gas-cooled reactor (HTGCR—later HTR) therefore began to be considered as the prime contender to be successor to the AGR, leaving the study of tube systems very much in the background. The retention of some effort on tube reactors probably owed as much to foreign activities as to those inside the UK. The Canadians were maintaining their enthusiasm for CANDU, and the American pressure-vessel LWRs were beginning to make solid progress in the land power generation business. Interest in water reactor systems therefore seemed to be increasing and, as a consequence, the UKAEA was reluctant to drop the tube concept completely.

The idea of using boiling light water in a direct-cycle, pressure-tube configuration was seen as a convenient way of maintaining British interest in a very flexible form of construction—which, in any case, had been a UK innovation—and also keeping abreast of matters pertaining to fuel, cladding, and coolant performance and behaviour that were implicit in the water-cooled systems of overseas competitors.

The case for the SGHWR—as the system was called—are not perhaps as clear-cut as had been imagined, and its position was, at times, tenuous. At one stage in 1960, it was thought that a pressure-tube reactor experiment (PTRE) would be a more practicable way of maintaining close contact with other water reactors, and this was studied in detail. (This would have been an assembly of vertical pressure tubes with a few channels of virtually every type of water reactor that could possibly be represented in tube form.) However, this rather complicated concept was eventually superseded. A massive comparative economic study, which the UKAEA conducted with the assistance of the British nuclear plant consortia of the day (1961), rationalized and analysed 29 possible water reactor systems, with common performance criteria and laid out to a consistent set of design standards, and estimated capital and fuel costs. The saturated-steam-producing SGHWR came first in this mammoth survey.

This vindication of the Authority's pressure-tube design was only five years after work started on the sodium graphite reactor (SGR) and coincided with a time of economic restraint. A limited development programme of two years' duration was therefore mounted to probe specific aspects of the SGHWR design, following which the 1961 comparative study was re-appraised by a joint assessment panel of the Authority and the Central Electricity Generating Board (CEGB). This also reviewed gas-cooled systems and, while it endorsed the AGR for immediate application, it also confirmed the future prospects of the SGHWR. The decision was almost immediately taken by the UKAEA to build a prototype at the Winfrith site of 100 MW(e). It was commenced in May 1963, and the plant was commissioned at the end of 1967.

# 3  Essential circuitry of the SGHWR

It may be helpful to present the basic arrangement of the plant by reference to one reactor circuit at a time, itemizing the functions, main components, materials of construction, and operating and design conditions in the process. These circuits are as follows:

(1)  Reactor coolant circuit,
(2)  Steam, condensate, and feed circuits,
(3)  Moderator and helium-blanket circuits,
(4)  Liquid shutdown circuit, and
(5)  Emergency core-cooling circuit.

## 3.1  Reactor coolant circuit

This closely resembles the evaporator section of a recirculation, water-tube boiler both as regards the function and the disposition of its principal components. The primary circuit actually consists of two similar, left- and right-handed half-circuits, coupled in a nuclear sense by arranging all the pressure or channel tubes so they form a regular, square lattice within a single *calandria*, the tank holding the low-pressure, low-temperature heavy water bulk moderator. This configuration was common to most commercial

SGHWR designs up to the largest unit output, though modular designs incorporating a relatively small number of pressure tubes associated with a steam drum-circulator unit and arranged in a common design of sub-calandria tank have also been proposed. In the Winfrith prototype, the two half-circuits each contain 52 channels.

Slightly sub-cooled light water ($H_2O$ some 10 °C below its boiling temperature), at a pressure of 65 bar, is taken from the water space of each steam drum through seven drum downcomers to the downcomer header—also called the pump inlet or top header—from which two larger downcomers each connect to a main circulating pump. The discharges from the two pumps of each half-circuit are joined by the pump outlet header—also called the feeder manifold or bottom header—from which rise six vertical sub-headers or feeder headers. Separate feeder pipes connect these to the bottom of each channel tube. Each circulating pump is fitted with an inlet and discharge valve, but the reactor is only operated with all the pumps fully opened to the primary circuit and running. All this pipework from drum to channel tube assembly is in stainless steel to specification EN58B and varies in inside diameter from 79.5 mm of the individual feeders up to 375 mm in the case of the main downcomers, these having a wall thickness of 16 mm.

A channel tube assembly is a composite item with a bore of 132 mm for the majority of its length which varies between 9.1 m and 10.7 m according to its location within the reactor lattice. The feeder pipework connects directly with the tailpiece of the assembly. This is also made of EN58B stainless steel and is fitted externally with components forming a sliding seal between the pipe and the lower neutron shield tank. The tailpiece itself is welded to a transition section of stainless iron (to specification A1331.410), which has the hub of the lower rolled joint formed at its upper end. The bottom end of the Zircaloy-2 pressure tube is expanded into this hub and the pressure-tube, with a wall thickness of 5 mm, extends up through the calandria and the upper auxiliary neutron shield to another rolled joint into a hub formed at the lower end of a further transition section of stainless iron. This upper transition tube incorporates a flange which sits on the upper neutron shield tank and, thereby, supports the weight of the whole channel assembly together with that of a portion of the feeder pipework. It is welded to the uppermost section of the assembly, the standpipe. This is of stainless steel and accommodates the seal plug at the channel top. Within the section housing the seal plug is an individual connection to the channel from the emergency core-cooling circuit, while the main coolant outlet branch is below this but above the support flange. It will be appreciated that the length of the tailpipe and the level of the coolant outlet branch within the common height of the standpipe depend on the position of the particular channel tube assembly in the core. It is therefore convenient to visualize the production of a 'universal' portion of a channel tube assembly which is then made into a 'specific' assembly for a particular channel by the addition of the appropriate tailpipe and standpipe, even though these also incorporate a number of

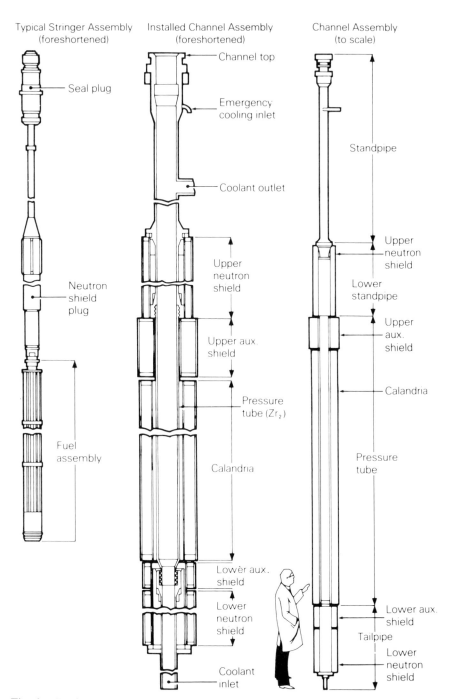

Fig. 6.1 Typical stringer and channel assembly (SGHWR)

standard components. Figure 6.1 illustrates a typical stringer and channel assembly. Figures 6.2 and 6.3 illustrate the primary circuit components of the SGHWR.

From the coolant outlet branch of each standpipe, an individual riser pipe takes the mixture of boiling water and steam to a connection on the adjacent steam drum thereby completing the primary coolant circuit. Each riser is fitted with a quality meter—a flow meter of the venturi-tube type calibrated to measure flow of the two-phase mixture. This reading taken in conjunction with that derived from a similar flow meter located in the associated inlet feeder provides an indication of channel heat output. On the Winfrith plant, the riser pipework is in stainless steel, while the steam drums are of carbon steel lined with stainless steel. Each circuit was subject to a cold hydraulic overpressure test to 110 bar and a hot hydraulic overpressure test to 90 bar, 303 °C, before power generation commenced.

### 3.2  Steam, condensate, and feed circuits

Just as the primary coolant circuit of the SGHWR has a close resemblance to a common form of water-tube boiler, so its steam, condensate, and feed circuits (Fig. 6.4) are very similar to the corresponding sections of a conventional power plant. Dry saturated steam from each steam drum passes in

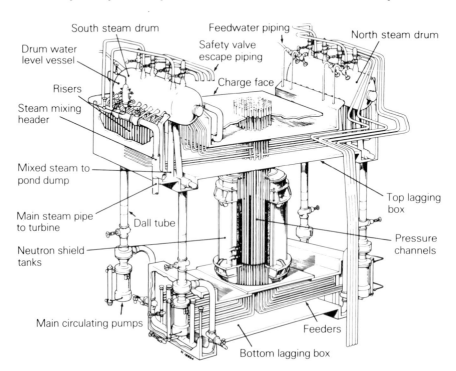

Fig. 6.2  Plant in primary containment (SGHWR)

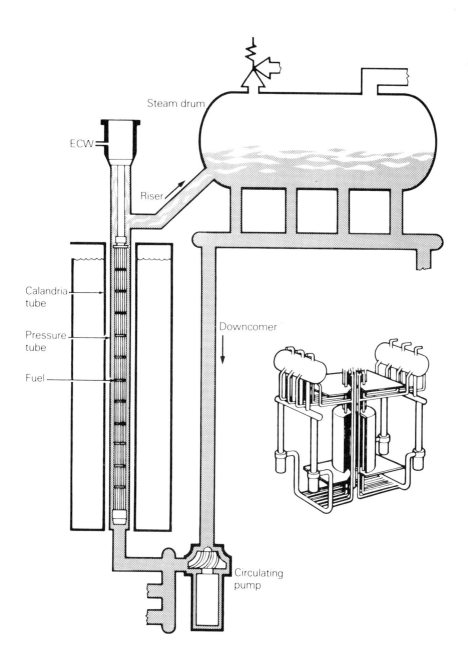

Fig. 6.3 Primary circuit—simplified diagram (SGHWR)

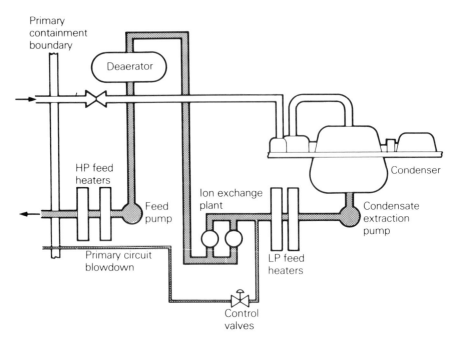

Fig. 6.4 Feed train (SGHWR)

a normal manner to the turbine emergency stop valves, the only notable features being that quick-acting shut-off valves are provided in each line immediately before, and immediately after it passes through the primary containment wall to preserve the integrity of this containment, and that the steam from each drum passes separately to the turbine admission—there is no interconnection between the two half-circuits until the turbine itself is reached.

Similarly, the condensate and feed circuit of the Winfrith plant also follow closely standard turbine house practice of comparable period and cycle conditions. Thus, the bled steam feedheating train consists of five stages: two low-pressure (l.p.) feedheaters of tube-and-shell construction followed by a direct-contact unit, arranged as a deaerating head on the surge tank which rides on the suction of the two feed pumps, followed by two further tube-and-shell units, designated high-pressure (h.p.) heaters. This train produces a final feed temperature of 199 °C, which is appropriate to the steam conditions at turbine inlet (62 kg/cm², 278 °C). Such a performance would be typical of any size of SGHWR and the modifications needed to bring this configuration in line with current power plant practice are probably restricted to the use of direct-contact heaters throughout the condensate section.

One substantial departure from standard practice for sub-critical steam conditions is, however, evident: the inclusion of a full-flow water conditioning plant of the ion-exchange type between the second low-pressure feedheater and the deaerator. Another is the way the primary circuit blow-down is taken to the inlet of this conditioning plant so that the 'boiler' blow-down is also treated. Magnetic traps are incorporated in this blow-down line which is typical of direct-cycle water reactor practice and their operation is now a routine function. However, the implication of radioactivity at this point in the circuit and the effectiveness of polishing plant in removing various impurities at the somewhat limited choice of water temperatures available with practicable arrangements of feedheating plant—coupled with the economics of resin useful lifetime—are matters deserving due attention when turbine house layout is being considered.

Carbon steel is the construction material used for the steam, condensate, and feed circuits of the Prototype SGHWR. Experience with this has been generally satisfactory save perhaps for the bled steam lines. A consequence of the radiolysis of the light water coolant is that oxygen levels at the turbine tappings points, from which the bled steam is extracted, are higher than in plant where the steam is not generated inside a reactor core. Corrosion of this pipework is, as a result, at an enhanced rate. This has been common experience on direct-cycle water reactors, and stainless steel is now the favoured material for bled steam piping.

### 3.3 Moderator and helium blanket circuits

As has been shown, heavy water has an appreciably smaller neutron absorption cross-section than light water, so that a practicable lattice providing sufficient space between pressure tubes to insert the coolant feeder pipework becomes possible. The SGHWR has its channels arranged on a square pitch of 260 mm, resulting in a void coefficient which, although still negative, is close to zero. Consequently, inherent safety of control is retained but, at the same time, the axial flux form remains effectively unperturbed, which enables efficient fuel utilization to be achieved.

The arrangement of the heavy water moderator system of the SGHWR (Fig. 6.5) takes account of many years of practical experience of similar circuits in the British-designed materials-testing reactors of the DIDO class. These reactors had demonstrated—and continue to demonstrate—that a $D_2O$ system can be trouble-free provided it operates at essentially atmospheric pressure and at low temperature. The successful operation of the Winfrith SGHWR has fully endorsed this approach, and the normal rate of make-up to replace all losses, including routine daily chemical sampling, is only some 200 litres per year in a total inventory of about 40 000 litres. Most of this make-up requirement results from this chemical sampling and the actual leakage of heavy water from the circuit averages only one litre per month.

During the operation of the Winfrith reactor, methods have been developed for improving the effectiveness of the clean-up plant which regulates

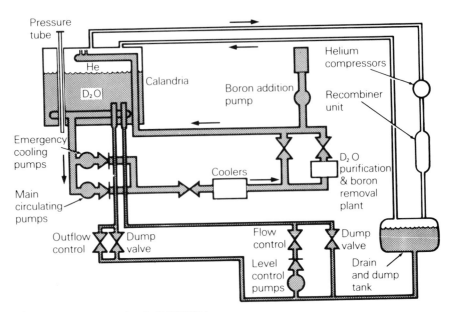

Fig. 6.5 Moderator circuit (SGHWR)

both general purity of the heavy water and the boron concentration used to compensate for long-term core reactivity changes. Over the first seven years operation, the isotopic purity of the heavy water fell from 99.75 per cent to 99.54 per cent, equivalent to the degradation of the total inventory by only some 80 litres of light water. Irradiation results in the formation of tritium in the heavy water, the level increasing at a rate of about one curie per litre per year of operation. Spillage or exposure of $D_2O$ during maintenance results therefore in local airborne radioactivity, but application of well-established techniques reduces such problems to acceptable levels. In fact, the tritium provides an extremely sensitive method for detecting any leakage of heavy water.

Some 4 per cent of the fission energy appears in the heavy water moderator primarily as a result of gamma heating, with a smaller contribution from neutron moderation and simple conduction from the pressure tubes. Nuclear activity also causes heat to be generated in the calandria material itself. Since the heavy water is held at low pressure and any boiling of the moderator with the formation of bubbles would not be conducive to stable operation, the bulk moderator is circulated from the calandria through coolers to maintain its temperature below 80 °C. This function determines an important section of the moderator circuit, which is carried out in stainless steel external to the calandria itself. At the Winfrith plant, the calandria is made of aluminium, though more recent designs have been based on zirconium alloy tubes and stainless steel tube-plates and shell. The main heavy water circulating pumps, backed up by emergency pump units, draw from a circular ring main near the bottom of the calandria, and the $D_2O$, after being cooled, passes to a similar main below the top of the calandria. From this top main, connection is made to the interspace between the double plates of a composite tube sheet forming the top of the calandria. Slots are arranged in the lower plate providing an exit from this interspace around each calandria tube so that the cooled heavy water returns to the core volume by flowing as a curtain down the outer diameter of every calandria tube.

A feature of British tube reactor design, which the SGHWR continues, is to make reactivity control a function of the heavy water moderator circuit. Thus, fine reactivity control on a moment-to-moment basis is achieved by variation of the $D_2O$ moderator level in the calandria. Large changes of level at full power would detract from the achievable thermal performance, since they would reduce the dryout margin but, in practice, the load changes required can be made without moving the level below the top 30 cm of the calandria. The moderator filling and emptying valves were so dimensioned that excessive rates of increase of reactivity are not possible and, in the prototype, the maximum rate-of-change of reactivity is limited to about 2 milli-niles/second with a full calandria.

Long-term reactivity changes are compensated by varying the concentration of the boric acid enriched in boron-10 which is present in the bulk moderator. Boric acid can be added using the boron-addition pumps and can be removed by diverting some of the moderator being circulated to the cooler through a strong-base ion exchange bed. The maximum rate-of-removal of boric acid corresponds to a rate-of-removal of reactivity of 4 milli-niles/second at a maximum boron-10 concentration of 15 p.p.m. by weight. The process has the advantage that the large amounts of excess reactivity which occur with fresh, unpoisoned fuel are absorbed in a uniform manner. This avoids the localized flux distortions associated with solid absorber rods and the attendant safety problems arising from the possible accidental withdrawal of such rods.

The moderator circuit also has to be arranged for admission of the $D_2O$ moderator at start-up and to permit the bulk moderator to be *dumped* to a drain tank as a back-up to the rapid shutdown system which is the next circuit to be discussed. Since $D_2O$ in a reactor environment is subject to radiolysis, recombination units are incorporated in the moderator circuit, and helium is used throughout to provide an inert 'blanket' to any free surfaces of heavy water.

## 3.4 Liquid shutdown circuit

The long-term shutdown of the SGHWR, by draining heavy water moderator to a point where criticality cannot be inadvertently achieved, has already been briefly mentioned. The primary shutdown system (Fig. 6.6) inserts negative reactivity rapidly into the core by introducing neutron-absorbing

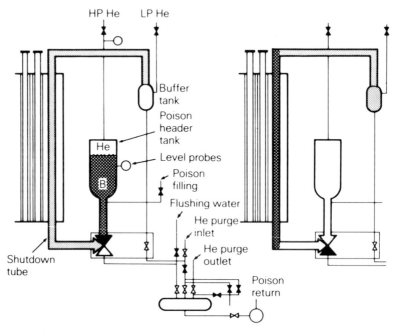

Fig. 6.6 Shutdown loop (SGHWR)

liquid into a separate series of tubes located interstitially within the main lattice of the pressure tubes and distributed throughout the core. Both shutdown systems are brought into action simultaneously for every reactor trip, the liquid shutdown system becoming effective within one second of the trip initiation, the dump hold-down within 1.5 minutes.

The prototype reactor has twelve 63.5 mm bore tubes positioned symmetrically within its array of 104 pressure tubes, each shutdown tube being provided with its own independent reservoir of lithium borate solution. The reservoirs are pressurized by helium and are connected to their associated core tube through a three-way trip valve. The upper end of each core tube is connected by a vent pipe to a buffer tank. Since each loop is thus completely independent, it is considered extremely unlikely that a fault on any one loop could affect another. The trip valves are held closed against a powerful spring by action of a solenoid energized through the protective system. They are located outside the core area where they are protected from damage and can operate in a relatively cool environment. During normal operation, all the shutdown tubes are purged continuously by helium. A reactor trip signal de-energizes the electromagnets, and the valves are opened by the springs, allowing the boron solution to be forced into the core by the pressure of the helium in the header tanks. The absorber fluid is brought to rest within each shutdown tube by the rising pressure of the gas trapped in the associated buffer tank. Even if two valves supplying adjacent tubes fail to operate, there is still sufficient negative reactivity injected into the Winfrith core to hold the reactor shut down until the moderator is fully effective.

### 3.5 Emergency core-cooling circuit

In the unlikely event of a rupture of the primary circuit and, with it, a loss of the normal coolant, the SGHWR ECC circuit (Fig. 6.7) permits an independent supply of water to be injected into each fuel element along its whole length. The water is initially supplied from a high-pressure tank at 14 bar above circuit pressure—the tank is pressurized by nitrogen gas—and connects by permanent pipework to each pressure tube. It then passes to the centre of the seal plug, down the centre of the hanger 'bar' and neutron scatter plug at the top of the fuel stringer, to connect with the sparge tube forming the central feature of each fuel assembly.

The capacity of the high-pressure tank is sufficient to provide adequate cooling to the tripped reactor until the primary coolant circuit has been depressurized. Thereafter, the cooling is supplied under gravity from a low-pressure tank. The emergency core-cooling system of the prototype is automatically initiated by detection of a pressure rise or rate-of-change of pressure in the primary containment. Apart from trip valves, it is entirely independent of moving parts or prime movers.

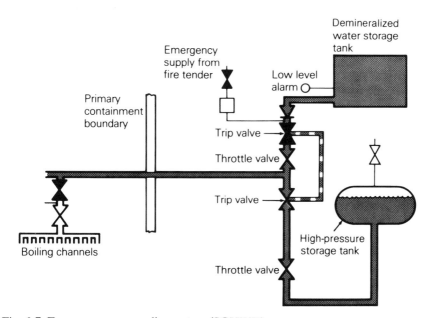

Fig. 6.7 Emergency core cooling system (SGHWR)

# 4  The construction of the Winfrith Prototype SGHWR

Government sanction for the construction of a prototype SGHWR was given in February 1963, and site work began at Winfrith in May of that year. Building construction and mechanical testing were planned for completion May 1967, and the station was due to produce power by the end of the same year. The total cost was to be £16M (1963 prices). Both the timescale and the cost targets were met. The reactor became critical on 14 September 1967, the first electricity was exported on Christmas Eve, and full power output— 100 MW electrical—was achieved on 24 January 1968. The plant was officially opened by HRH Prince Philip on 23 February 1968.

The Winfrith reactor was designed and built by the UKAEA. Application of the pressure-tube system, with its absence of large pressure vessels, allowed the construction to become little more than a routine assembly job on site, the more precise engineering operations being carried out at manufacturers' works. Thus, it was possible for civil work and plant installation to be largely segregated, and extensive need to use 'clean' conditions was thereby avoided. None of the components was heavier than the corresponding item of a conventional boiler, thus obviating the need for the erection and use of a massive crane, so much a feature of the Magnox power stations which the generating boards were having built at the time.

Critical path control methods, then in their early days were used throughout the four-year construction period with very satisfactory results. Flexibility of the planning methods was demonstrated when significant damage caused by the introduction of a mercury compound into the calandria had to be rectified. The damage was discovered in February 1967, and remedial work involving a variety of novel techniques began immediately. This was completed by the end of June 1967 and, in spite of the extra work involved, full power was achieved in Week 243 instead of Week 242 as had been planned four years earlier.

## 4.1  Siting the reactor

Considerable care has been taken by the Authority to ensure that its Winfrith Establishment blends as unobtrusively as possible with the surrounding countryside. An advantage enjoyed by Winfrith over all other major UKAEA sites is that it is the only one to be built on previously undeveloped land. It occupies about one thousand acres of Dorset heath near the small River Frome and is centred roughly two miles west of the village of Wool, which has a station on the Weymouth to Waterloo line of British Rail. A few miles to the south are the Chaldon Hills, which overlook Lulworth Cove and the English Channel; the setting is immediately adjacent to an area of outstanding natural beauty and is, in fact, the centre of Thomas Hardy's Wessex.

When seen from some viewpoints, the SGHWR buildings are silhouetted against the skyline whilst, from others, the undulating heathland forms a backcloth. To blend with both backgrounds, a pale, green-blue colour was chosen for the cladding of the main hall but, to provide some relief from the large, regularly shaped masses of the building, an abstract mural of contrasting colours and measuring 28 m by 10 m was painted on the concrete wall adjacent to the deaerator tank. This has subsequently become a well-known feature of the plant.

Natural-draught cooling towers could not be used because of a height limitation of 18 m then imposed by the planning authority. The most acceptable alternative was judged to be the provision of two batteries of cross-flow, fan-induced-draught cooling towers, the cooling air being drawn through the full height of the louvred sides of the towers.

## 4.2 Construction

Formation of the major concrete foundation raft was started in September 1963. The layout chosen for the main structure with the pond complex, reactor, and turbine on a common axis, made construction fairly straightforward. The reactor plant, its primary containment, and the fuel handling and pressure suppression ponds are carried on the foundation raft together with the superstructure of the reactor hall in that area. The reactor itself is supported on the concrete primary shield, which is carried by concrete columns from the raft, allowing maximum accessibility to the pipework. After completion of the raft, civil engineering work associated with the reactor was undertaken and erection of the main structural steelwork for the power hall began. Early in 1965, the structural steelwork of the main buildings and also of the associated annexes on the northern and southern flanks had been completed. The roof decking was put into position and the structure made weatherproof by attachment of the cladding. A suitable stage had then been reached for reactor components to arrive for installation in semi-clean conditions. Since the start of site work, 87 weeks had elapsed.

A critical activity in the overall construction schedule was the manufacture of the calandria, the tank that holds the heavy water moderator. A high standard of dimensional accuracy was imperative in both manufacture and erection to ensure that the annular gas gap between each calandria tube and pressure tube would be correctly maintained at all lattice positions when the channel tube assemblies had been inserted into the calandria. Installation of the factory-manufactured channel tube assemblies into the core was basically a matter of lowering the completed units into the holes in the upper neutron shield tank and through the calandria and lower shield.

During 1966, civil engineering construction of the main buildings and most of the auxiliary structures was largely completed. Included were the CEGB sub-station at the western end of the site, the two cooling tower batteries to the south, the overhead conductors from the power house to the transformer and switch yard, and the off-gas exhaust stack. Components of reactor systems such as the heavy water cooling circuits and the ion exchange beds for the control of boron in the moderator were assembled on site. As previously mentioned, the presence of mercury was detected in the aluminium calandria in February 1967. Repair work which involved cutting out the corroded metal and its replacement began at once and was completed by the end of June. Some changes to the programme were necessitated by this unscheduled work but, by the end of July 1967, virtually all of the construction had been finished except for the experimental loops and the on-load refuelling machine.

The overall commissioning programme of the Winfrith SGHWR occupied a period of 18 weeks, full power output being reached in January 1968. Table 6.1 summarizes the construction and commissioning schedule which was achieved.

Table 6.1  Key construction and commissioning dates of the Winfrith steam-generating heavy water reactor

| | Programme week number | | Approximate date |
|---|---|---|---|
| | Schedule (1962) | Actual achievement | |
| Start on site | 0 | 0 | May 1963 |
| Complete primary containment concrete | 78 | 80 | October 1964 |
| Start installation—shields and calandria | 86 | 86 | January 1965 |
| Start installation—channel tubes | 111 | 114 | July 1965 |
| Complete installation—channel tubes | 126 | 129 | October 1965 |
| Start installation—turbo-generator | 129 | 129 | October 1965 |
| Start installation—rotating shields | 153 | 153 | April 1966 |
| Start hot tests | 183 | 188 | December 1966 |
| Commence loading heavy water | 216 | 225 | September 1967 |
| Reactor made critical | — | 227 | September 1967 |
| Electricity first fed to National Grid | — | 241 | December 1967 |
| Reactor at full power | 242 | 243 | January 1968 |

# 5  Operating experience at Winfrith since January 1968

When introducing a new technology, construction of a prototype allows representative components to be subjected to the full range of actual operating conditions likely to be met in commercial installations. Prototype operation, at minimal cost, probes behaviour both of individual items and of the integrated plant. If realistically specified, such a plant affords early opportunities of dealing with any operational problems; solution of such problems makes a vital contribution to the economic viability and safe operation of the subsequent commercial design. The operation of the Winfrith SGHWR since January 1968 has fully endorsed the basic features and performance parameters of the steam-generating heavy water reactor concept.

A research and development programme has been vigorously prosecuted since the prototype was commissioned, in spite of which the load factor from 25 January 1968 to mid-1980 has been 54.8 per cent. During the winter period (which runs from November/December to March/April of the following year) station load factors have been high, varying between 83 per cent in 1970/1 to 98 per cent in 1973/4. These figures do not allow for potential generation lost through that part of the development and experimental work that was continuing.

## 5.1 **Plant operational control**

Since the system has desirable operating economics and is revenue earning, it has primarily been operated as a base-load plant, although periods have been devoted to demonstrating load-following and optimizing the control system for this important regime. By employing the *coupled control system*, the Winfrith plant can respond automatically to the demands of the national electrical network, varying its output to contribute to the stability of system frequency.

Two-shift working was demonstrated between 1970 and 1972 by undertaking a programme of 204 daily power cycles, using a manoeuvring rate of 3 per cent full power per minute. Fuel behaviour remained highly satisfactory throughout these power cycles: no fuel defect was caused and there was no significant deterioration in defective fuel already in the core.

Fast load pick-up tests have demonstrated that the prototype can accommodate a step increase in demand of at least 10 per cent full power. In response to a typical demand change of 10 per cent, the plant output has increased by 8 per cent in 1.5 s. Tests have also shown that, with the present simple plant control system, load variation of 30 per cent full power can be achieved in 1 minute, and of 80 per cent full power in 20 minutes. In addition to this considerable flexibility under automatic control, it has been shown that satisfactory manual control of the plant is always possible. Of particular interest has been the successful demonstration of the Winfrith prototype's ability to supply its own station load of about 6.5 MW when disconnected from the stabilizing influence of the national grid.

## 5.2 **Fuel experience**

It was planned from the outset that fuel element behaviour would be monitored by detailed post-irradiation examination. This required systematic discharge of a substantial proportion of the initial core loading before the clusters could achieve their designed burn-up. Examination in shielded cells at Windscale and Winfrith have indicated that the incidence of such phenomena as hydriding of the Zircaloy-2 cladding, ridging of this cladding, and swelling of the oxide fuel pellets has been minimal and consistent with design predictions, though changes in fuel pin length with irradiation have been somewhat greater than expected. A smaller modification of initial end clearance led, in 1969, to the formulation of the present standard design which can readily accept such increases in length as may occur. Shortly after the prototype was commissioned, a fault in the feedwater purification plant caused build-up of impervious deposits on the fuel pins resulting in defects in some thirty fuel clusters. This was subsequently rectified. Since 1969, only a very few isolated pins have shown defects in standard fuel clusters. Such faults appear to be associated with quality control during manufacture, and recent experience with small discontinuous batch production has indicated an expected defect rate of about one pin per year, though such failures would not normally lead to premature shutdown.

During 1970—for the first time in an operating power reactor—a fully instrumented experimental fuel assembly was subjected to a series of transient and steady-state dryout tests using the cluster loop in the central region of the core. This investigation of fuel behaviour under abnormal coolant conditions clearly has most significant operational implications. The results provided direct in-pile confirmation of data previously obtained out-of-pile in the 9 MW heat-transfer rig at the Winfrith Establishment, and also demonstrated that the fuel had not been damaged and could continue in normal operation after completion of the tests.

Safety considerations on light water reactors led to the introduction of fuel assemblies with a larger number of smaller-diameter pins so that, for a similar channel heat output, individual pins now have much lower heat ratings than those of earlier designs. Similar considerations for the SGHWR have produced a new standard design with 60 pins, and this fuel has been routinely loaded since 1975. The centre temperature of the newer fuel designs is also lower so that the release of fission gas from the fuel pellets— and hence the gas pressure inside the cladding—is much reduced.

The opportunity has also been taken to improve the emergency core-cooling distribution system within each individual fuel assembly. This no longer relies upon a single sparge tube along the centreline of the fuel assembly as in the 36-pin design, but includes further secondary tubes within the fuel bundle to improve the distribution of the emergency coolant.

### 5.2.1  *Operating at power with faulty fuel*

Whilst the possibility of pin failure is very small, it is not negligible, and in the case of the prototype, experimental programmes have also involved deliberate loading of defective fuel. The health physics implications of fuel faults therefore demand close attention, the main considerations being those of gaseous discharges to the atmosphere and water or steam leaks into the secondary containment.

If a defect develops in the fuel cladding, some fission products will escape into the coolant, the most likely to do so being isotopes of inert gases, xenon and krypton. Volatile fission products, predominantly radioiodine, are almost completely retained within the uranium dioxide fuel pellets so that the level of radioactivity has not risen above a few microcuries per litre of coolant even during power-cycling operations when there have been large variations in reactor load. These variations in power level have tended to cause surges of inert gases to enter the coolant, but it has not been until the reactor was shut down and the primary coolant circuit depressurized that significant quantities of radioiodine were able to escape.

Under steady power operation with failed fuel present, inert gases are released continuously into the coolant and, preferentially favouring the steam phase, pass through to the turbine and condenser from which they are removed with other incondensibles via the off-gas system. Such gases contribute little to radiation levels around the turbine—which continue as in normal operation to result principally from nitrogen-16—and, before discharge, pass through large charcoal beds, introducing a delay of several hours for krypton isotopes and several days for those of xenon. Since most of the isotopes of krypton and xenon emanating from defective fuel are relatively short-lived, delays of this order result in an appreciable decay of the radioactivity, so that discharges to atmosphere as noble gases via the ventilation stack of the prototype SGHWR have always been within the limiting discharge rate agreed with the appropriate national controlling bodies. Because of their solubility, any iodine isotopes—particularly iodine-131 and iodine-133—entering the coolant remain essentially dissolved in the water phase and within the primary circuit. The very small quantities which do pass to the turbine with the steam are held in solution in the condensate and, so far, no iodine has been detected in the condenser off-gas line of the prototype, so there has never been any discharge of radioiodine to atmosphere by this route.

While there is defective fuel in the core, any leakage of water from the primary circuit could lead to high concentrations of iodine-131 in the atmosphere of the primary containment, but this has no significance since there is no possibility of personnel access to this location when the reactor is in operation. To prevent release of the iodine-131 to atmosphere with the primary containment ventilation air, it is possible to direct this ventilating flow through one of the large charcoal delay beds which are highly efficient absorbers of iodine. A leakage from the primary circuit blowdown pipework, which is in part external to the primary containment, could cause a release of iodine-131 into the turbine hall, but the level of iodine activity in the air of this building—which constitutes the secondary containment—has never exceeded the maximum permissible concentration.

Steam leaks are almost inevitable in any large turbo-alternator system. Should these be significant when there is faulty fuel in the reactor, radioactive inert gases escape into the secondary containment. These present an operational inconvenience rather than a radiological hazard since the isotopes decay fairly rapidly to short-lived solid radioactive daughter products—krypton-88 to rubidium-88 and xenon-138 to caesium-138—which become deposited on surfaces. Rubidium-88 and caesium-138 have half-lives of only 18 minutes and 32 minutes, respectively, and are radiologically innocuous. They can however mask the presence of longer-lived contamination which could be radiologically significant. Because so little iodine is present in the steam phase, concentrations of iodine-131 in the secondary air attributable to steam leaks from the turbine are negligible.

Figure 6.8(a) shows the *clean-up plant* flowsheet and (b) illustrates the method of monitoring gaseous discharges.

### 5.3  **Fuel handling**

To form a complete fuel stringer, a new fuel assembly is coupled to a suspension section comprising a seal plug, a support tube and a neutron shield plug, which, in most cases, has already seen service in the reactor and is, therefore, radioactive. The standard route for loading new fuel into the core is thus under water through an intermediate facility. This is located at the end of a fuel-transfer tunnel, which extends from the main fuel pond, in the secondary containment, under a portion of the primary containment structure and, in conjunction with a charge–discharge facility, provides a pressure seal or lock between the two containments. Irradiated fuel is moved back into the secondary containment under water via these facilities and the transfer tunnel.

On return to the main pond, the irradiated fuel stringer is divided into its two main sections. The suspension section is held for inspection and reuse, while the spent fuel assembly is stored until its radioactivity has decayed sufficiently to permit it to be transported for post-irradiation examination or reprocessing. Preliminary inspection and monitoring of irradiated fuel—as well as some modification of assemblies—has been regularly undertaken in the underwater storage area.

Apart from shift periods following depressurization of the primary coolant circuit, water clarity has been sufficient for observation throughout the fuel pond, in which the water depth is approximately 10 m and the total volume some 1250 m$^3$. Water clarity reduces towards the end of a circuit depressurization because, when the circuit pressure has fallen to around three bar, low-pressure steam is diverted for condensation in the suppression ponds. These are at either end of the fuel pond and connected to it by ports or openings below the surface of the water. The depth of this water is also 10 m and the volume of these ponds a further 1250 m$^3$.

Until the advent of the SGHWR, British nuclear power-generating experience had largely come from operation of gas-cooled reactors employing metallic uranium fuel. The storage in ponds of fuel discharged from such systems after irradiation poses a number of problems, since corrosion of the fuel cladding commences as soon as it is immersed in water. This is particularly so if the fuel cladding is defective. Fission products in company with corrosion debris accordingly contaminate the pond water of Magnox reactors to significant levels. By contrast, the fuel of the steam-generating heavy water

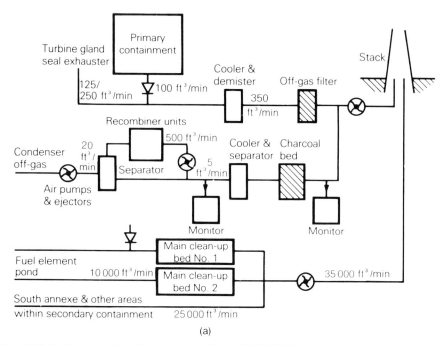

Fig. 6.8 (a)  Clean-up plant for gaseous effluent (SGHWR)

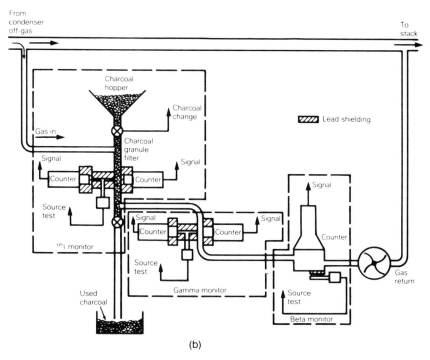

Fig. 6.8 (b)  Typical monitor in condenser off-gas (SGHWR)

reactor is specifically designed to operate in a hot-water-steam environment at a pressure of 65 bar, so that when irradiated fuel is subsequently stored in a cold water pond virtually no corrosion of the cladding material takes place. As the fuel material is uranium dioxide, which is quite passive in water, even failed fuel pins with exposed fuel material do not undergo any significant attack in the storage pond. Water purity in the SGHWR ponds at Winfrith has therefore been maintained by relatively simple, well-established treatment processes and application of good standards of housekeeping. A clean-up plant provides for filtration and ion exchange, and the possibility of organic growths resulting from exposure to indoor and underwater lighting is limited by careful control of the supply of dissolved nutrient salts.

Corrosion products from the material of the primary coolant and feed circuits deposited as scale on fuel cladding during operation tend to flake off from irradiated fuel and settle on the pond floor, from which they can be successfully removed by underwater suction devices. The levels of fission product concentrations in the relatively great bulk of water in the pond system are substantially below the International Commission on Radiological Protection (ICRP) standards for normal drinking-water supplies.

### 5.4 The heavy water moderator

Substantial quantities of tritium are formed in the operation of the SGHWR as a result of neutron capture by the deuterium and there are currently around 11 curies per litre of $D_2O$ as a consequence. The present tritium level therefore gives rise to some 440 000 curies in the total inventory of heavy water. However, as has already been stated, the operation of the prototype continues to endorse the design tenet that a heavy water system can be trouble-free provided it operates at essentially atmospheric pressure and at low temperature, since these conditions considerably facilitate the satisfactory containment of this active material. The system has been shown to be of very high integrity and leakage of tritiated water extremely small—the amounts discharged in liquid wastes predominantly from analytical samples having averaged only 50 curies per month, while tritiated water vapour discharged to atmosphere has amounted to less than 30 curies per month. Problems of personnel exposure to tritium have been minimal, though radiological control procedures are applied during any maintenance work on the moderator circuit and ancillary equipment.

### 5.5 The pressure tubes

An important advantage that was designed into the steam-generating heavy water reactor system is that major pressure components of the reactor coolant circuit can be replaced with minimum down time. Although defects are most unlikely and none have occurred, this feature of 'rebuildability' was convincingly demonstrated in the second year of operation when, without removing fuel from adjacent channels, six experimental zirconium–niobium alloy pressure tubes were replaced by standard units which have Zircaloy-2 tubes. The whole programme proceeded smoothly, the exchange of the last channel assembly and reinstatement of the primary pressure circuit occupying only five days.

Pressure-tube integrity can be endorsed by simple inspection techniques during each refuelling operation. Devices have been developed for visual, acoustic, and measurement surveys of the entire internal surface of a channel assembly. These can rapidly monitor the full length of a channel for incipient material defects, scan for surface damage such as might result from fretting between a fuel assembly and the pressure tube, and determine the extent of creep deformation.

### 5.6 **Operation at power**

Few health physics problems have been encountered in normal circumstances when the reactor is operating without any defective fuel. Radiolysis and activation of the primary coolant occur as it passes through the core, producing large quantities of the radioactive isotopes of nitrogen and oxygen. Principal among these, in diminishing abundance, are nitrogen-16, nitrogen-13, and oxygen-19 with respective half-lives of 7 seconds, 10 minutes, and 29 seconds. Such nuclides are also produced in reactors cooled by carbon dioxide, but in these the radioactivity is entirely confined within the coolant circuit. In the SGHWR, because of the use of a direct cycle, the radioactive gases are carried over in the steam phase to the turbine. Consequently, there are high radiation levels near steam pipes and close to the turbine itself (up to 250 mR/h).

The nitrogen and oxygen activation products together with hydrogen and oxygen formed by radiolysis of primary coolant water are removed in the condenser off-gas and piped to the clean-up plant. This incorporates catalytic hydrogen–oxygen recombiners to eliminate the risk of explosion, a cooler, and a large charcoal delay bed through which the off-gases pass before release to atmosphere. Delay time within this system is sufficient to ensure complete decay of the short-lived nitrogen and oxygen activation products.

Protection of staff from other sources of external radiation is not a serious problem, since the bulk of the reactor plant is inside the primary containment which provides very adequate shielding. Various chemical clean-up processes associated with active circuits require some access during plant operation, and work on irradiated fuel in the storage pond can also involve small irradiation doses.

## 6 Final assessment of the UK programme

So far as can be judged from the viewpoint of today, the UKAEA constructed a very satisfactory prototype, as regards both the system itself and the longer-term development of the pressure-tube reactor as a generic type. One can undoubtedly criticize certain features of the Winfrith plant, nevertheless the construction, commissioning, and operating programme can only be termed a success and any strictures are more of detail than of principle. The claims made for the SGHWR system all seem to have been endorsed by the building of the prototype and by its operation. Improvements have also been demonstrated as a result of operational and maintenance experience, and there would seem to be no basic technical or economic reason which should, in principle, have prevented the reactor from taking its place in the electricity generating industry for which it had been specifically developed.

In spite of this, the decision taken by the Government in 1974 to adopt it as the chosen system for the third British nuclear power programme was overturned two years later by the same Government and further development was cancelled.

The background to this alteration in policy has been well reviewed by the former Chairman of the UKAEA, Sir John Hill. The annual increase in the demand for electricity fell sharply from previous estimated values. Factors included increased costs, efforts in energy conservation, the depressed state of the national economy, and competition from natural gas. As a result, the role of the SGHWR programme was changed. In 1974, it was thought that the initial construction of 4000 MW(e) at Sizewell and Torness—to come on-line in the first half of the 1980s—would be a modest but reasonable start to a growing programme. The fall in demand was such that even this moderate programme on the timescale envisaged could not be fully justified. Given these changed circumstances, it was hard to avoid the conclusion that the limited funds available for nuclear research and development and for the regeneration of the nuclear industry could be better employed.

Additionally, it was evident in formulating the reference design of the commercial version of the SGHWR that modifications were needed to take account of the different safety criteria which had grown up since the design phase of the prototype. In combination with the prospect of a smaller and stretched-out programme, these modifications were recognized as almost certainly leading to an unacceptably high cost of electricity generation for the early stations of the new type. In short, it was realized that the nation had embarked upon the launch of a 'new' reactor system at just the time when the UK's home market for reactors—the firm base for our nuclear construction industry—was too small to sustain a unique system. It was hard not to conclude that, as a result of a late start on its commercial exploitation phase in conjunction with a potentially slow installation rate, the SGHWR would be too late to compete with established systems abroad.

While, then, there were no fundamental technical problems thrown up by the commercial design of the SGHWR, it was judged to be difficult to identify any specific feature of it which gave it a clear cut and unambiguous advantage over other commercial established water reactor systems.

# 7 A brief survey of heavy water power reactors in other countries

## 7.1 **Introduction**

If one looks beyond the UK, it is evident that heavy water power reactors constitute a minority choice compared with an overwhelming, world-wide adoption of light water concepts based on American designs. This should not, however, be taken to imply that their adherents regard heavy water reactors as being in any way of an ephemeral nature; Canada has had an extremely long history of active participation in nuclear research and development, throughout which its major effort has been steadfastly and successfully devoted to the attainment of commercial heavy water reactors. The association of the Italians and Japanese with this type of reactor has also been of a reasonably long-term nature, though the amount of actual construction does not begin to compare with the Canadian achievement. West Germany made only a relatively brief entry into the field. They constructed a small pressure-vessel demonstration reactor (MZFR) which became operational late in 1965—all the other countries so far mentioned have been exclusively involved with pressure-tube designs—and then sold an extremely large (at the time) commercial version to Argentina, a country just entering the field of nuclear power generation. Thereafter, apart from a limited experience with a heavy water tube reactor cooled by gas, the KNN Niederaichbach HWGCR, German construction of thermal reactors has reverted to a concentration upon light water designs. France, India, Pakistan, Sweden, and Switzerland have also been interested in heavy water power reactors to a greater or lesser extent, but the European countries mentioned have now dropped them from their programmes. Canada remains the supreme exponent of the technology.

In this brief survey no attempt has been made to describe in detail the characteristics of the modern CANDU power reactor system nor to relate the full history of its very successful development as a large-scale electricity generation option. However, any chapter on heavy water reactors must refer the reader to the Canadian achievement.

## 7.2 **Canada**

The Canadian heavy water experimental reactor ZEEP was the first critical assembly outside the USA, so that Canada is clearly one of the pioneer nations in nuclear technology. Canada has always maintained an interestingly 'different' attitude to certain aspects of nuclear power compared with other early 'nuclear' nations. Thus, she has always insisted on very simple fuel cycle management schemes.

It is therefore not surprising that Canada has based most of her nuclear studies on reactor systems which are both moderated and cooled by heavy water. Most, but not all, since Canada has maintained a smaller but ultimately, in a technical sense, very successful interest in a structurally rather similar reactor design that uses organic liquid as moderator and coolant. Principally, however, interest has concentrated on this double use of heavy water in association with natural uranium fuel. This is in line with one of the main tenets of the Canadian nuclear policy: that there should be neither enrichment plant nor any reprocessing involved with a nuclear power programme. The Canadians have emphasized—particularly to emerging countries with little nuclear technology but indigenous uranium ore reserves—that a reactor fuelled with natural uranium eliminates the complexity of enrichment.

The Canadian nuclear power programme contains, then, a number of postulates which appear to put it at variance with other countries with nuclear programmes. On turning to the sort of reactor built for power generation by the Canadians, the same degree of variance from what British engineers would probably regard as the most favoured design solution is often evident.

Originally, around 1956/7, having employed vertical, tank-cum-calandria designs for their low-power research reactors, NRX and NRU, the Canadians took the decision to base their nuclear power demonstration (NPD) reactor on a pressure-vessel design. The system was generically designated CANDU (*Can*adian *D*euterium *U*ranium) and would have been the heavy water analogue of the PWR, using natural uranium fuel in oxide form clad in zirconium alloy. The pressurized heavy water would have attained a temperature approaching 300 °C by passing through the core without any bulk boiling in the coolant. This would then have been circulated through external pipework to steam generators producing saturated steam from light water feed. However, regular UK–Canada technical exchange meetings at that time were discussing with enthusiasm the application of pressure-tube designs to large power reactors, and the Canadians saw the potential for constructing CANDU in that way. The vessel version of CANDU was shortly afterwards cancelled and, some months later, it was replaced by a pressure-tube design. Surprisingly, however, given previous experience with vertical designs, the design concept for NPD-2 involved fuel located in horizontal channels. Now, whilst the fuel is in the form of short, unconnected multi-rod bundles, the way in which the individual assemblies rest on the bottom of the circular pressure tubes and are progressively pushed along the channels by the successive insertion of new bundles from one end of each tube is very reminiscent of the fuel arrangements in the earlier British graphite-moderated reactors GLEEP, BEPO, and Windscale 1 and 2. Dr. W. B. Lewis who was responsible for the design of the Canadian pressure-tube reactors had been very closely associated with the design and construction of these early British reactors, so this may be the reason why this distinctive arrangement was adopted.

Technically, there are important differences between the systems in this respect of fuelling. Thus, in the graphite-moderated reactors, a single uranium-metal fuel rod, clad in aluminium, slid along the bottom of a circular channel formed in substantial graphite blocks; in the Canadian reactor, a bundle of seven fuel pins with Zircaloy-2 cladding and spaced apart from each other by a *two-start, helical 'rip'* formed in the clad of the outer pins, are pushed along the bottom of a relatively thin pressure tube also made, in NPD-2, of Zircaloy-2. The limited burn-up possible with natural uranium necessitates refuelling being undertaken whilst the reactor is at power, and this requires relatively sophisticated charge–discharge machinery to be developed and installed. The coolant circuit arrangement that was chosen resulted in the direction of flow being opposite in adjacent channels, but only one remotely operated fuel-handling machine was provided to service each of the two reactor faces. Accordingly, the refuelling procedure requires both machines to be fully operational and to locate on a particular channel tube. They then remove the seal plugs which normally close each end of the tube against a coolant pressure, in NPD, of in excess of 69 bar. A new bundle of fuel is next inserted into the reactor by the machine at the inlet end of the channel, and this operation pushes out a *spent* assembly from the outlet end of the channel into the machine that is connected there. In NPD, there are 132 channels (producing an electrical generation of 22 MW), and each channel accommodates nine fuel assemblies, of which eight will be within the active zone of the core. After a refuelling operation is completed, and the machines have resealed the channel tube successfully, they are free to move to the next channel requiring their attention. This may have the coolant flow in the opposite direction so that the new fuel has to be inserted by the machine which previously received the spent assembly, with the role of the machine at the other end of the reactor also reversed.

So far as British design principles are concerned, this arrangement transgresses at least three cardinal rules. (1) heavy water is employed at substantial pressure so that the possibility of leakage of this extremely expensive commodity is markedly increased; (2) it is also used at relatively high temperature so that any leakage into the reactor vault inevitably causes a not inconsiderable fraction to flash into steam, and (3) it is necessary as a regular operational procedure to open up the main coolant circuit whilst at working pressure, so that coolant is potentially able to exist in two phases in a LOCA (loss of coolant accident) situation.

There are also subordinate disadvantages compared with SGHWR practice, such as the enhanced possibility of scuffing and pick-up resulting from similar metals being in pressure contact (rubbing-pairs) as the fuel moves along its channel tube or if individual bundles *rattle* or *chatter* under the influence of coolant flow. These apparently enhanced prospects for damage to the internal surfaces of the pressure tube are, unfortunately, not associated with the excellent opportunities for bore-monitoring present in the SGHWR concept every time a channel is completely void, as it is during refuelling. On the other hand, the progressive movement of small sections of fuel through a CANDU reactor achieves near optimum fuel utilization. The need to circulate the pressurized heavy water to steam generators means there is a considerable investment in this liquid, similarly at high pressure, outside the immediate confines of the reactor, together with many additional potential leakage sites in the tubes and tube fixments inside the steam generators.

AECL's initial operating experience with NPD, and the larger Douglas Point Plant capable of generating 208 MW(e), which followed it, revealed a number of problems. Pumps, valves, and other circuit components did not have the required reliability; seal plugs and flanges leaked so that heavy water leakage rates became unacceptably high, and leakage exceeded the capacity of the collection system. Heavy water became down-graded by mixing with light water leakage. Radiation fields increased about all circuit components as a result of the transport of corrosion products, making maintenance increasingly difficult, while the man-rem levels that were experienced became too high for commercial plant. However, AECL reacted vigorously—strengthening its research and development programme in 1970, and creating an entirely new division in its organization with the sole purpose of improving component reliability. Facilities were constructed so that full-scale components could undergo the most searching endurance testing, and development work was initiated with the necessary redesign to eliminate such weakness as were revealed. Changes in component design and plant layout were also undertaken to improve accessibility, reduce the total number of circuit components, and, as a result, reduce the man-rem burden incurred by maintenance personnel. This impressive work has been very successful and is continuing.

In broad terms, the NPD design concepts are still retained in current CANDU commercial reactors: there remain very substantial quantities of heavy water under high pressure being circulated to tube-and-shell steam generators, and there are still remotely controlled, charge–discharge machines regularly working upon the pressurized coolant circuits with the reactor on full power. There is a marked similarity between the standard CANDU and PWR steam generator units, but the Canadian experience is much better than that of the PWRs, possibly as a result of differences in detail design, quality control, water treatment, and choice of materials. AECL have mounted a substantial development programme to minimize the consequences of any steam generator failure. This includes the establishment of techniques to reduce radiation fields within the heat exchangers, procedures for tube plugging, and eddy current inspection methods.

Whilst not exactly associated with design principles, the rather obvious feature of CANDU-PHW, as a class of reactor, which conflicts with the ideas of other countries, is the orientation of the system: horizontal channels as opposed to vertical disposition of the fuel. So far as British techniques are concerned, the choice of a horizontal reactor would raise substantial technical problems in achieving output control by changing the level of the bulk moderator—the low-pressure heavy water—in the calandria, even if it did not, in fact, entirely preclude the use of this principle. Recourse would thus have to be made to using movable neutron absorbers to control the output of a CANDU. This tends to result in both the length and diameter of a core having to be changed whenever a design is undertaken for a new size of reactor, so that the idea of using a standard fuel element within a standard pressure tube, which would have characterized SGHWR designs over a wide range of outputs, cannot be applied to the CANDU system. Clearly, the different output reactors that have been built display a strong family resemblance, but the diameter and length of pressure tubes, for example, both increase as the reactor sizes are increased. This prevents replication of major components which, in UK thinking, was seen as leading to substantial economy if the reactors are destined for multiple production. It also leads to manufacturing problems. Thus, the techniques for rolling the joints combining the pressure tube with the end sections to make a complete channel assembly for a 600 MW CANDU reactor must be similar to, but not precisely the same as those employed for NPD, and so detailed modifications in procedures have to be devised to achieve comparable results with the larger, longer tubes. These differences are not merely academic: substantial problems were encountered in advancing from the Douglas Point (208 MW(e)) to the Pickering size of reactor (514 MW(e)). Thus in addition to the loss of potential savings in production that would have followed from complete standardization based upon dimensions which have been fully endorsed in prototype construction and operation, there is also the possibility of incurring operational problems because the consequences of rolling larger tubes etc. have not been fully appreciated.

The basis of this argument seems soundly made. Nevertheless, the AECL can point to a steadily increasing number of CANDU reactors built, building, or on order throughout Canada and the world, whereas the British SGHWR has been dropped from the UK's nuclear power programme and exists only as a prototype at Winfrith. The single-mindedness of the Canadian programme, supported by superb engineering, has succeeded in producing CANDU reactors which lead the world league table of availability.

A list of CANDU reactors in operation, under construction, or committed is given in Table 6.2. This does not include the reactors which are the subject of current negotiation between the Canadians and Romanians. It will be seen that all save Gentilly 1 are of the PHW type. The exception is designated BLW, and is a direct-cycle variant, using boiling light water as coolant in vertical channels. As such, it resembles the SGHWR, but the dimensions and materials were selected to permit operation with natural uranium. There are no plans to build more of this particular form of CANDU, but studies have been made for larger reactors of this general type using low-enrichment fuel. Figures 6.9(a) and 6.9(b) illustrate schematically the layout of the PHWR and BLWR variants of CANDU.

In summary, the Canadian achievement with CANDU has been impressive. The 'prototype' power plant, Douglas Point, encountered problems, but the first commercial station, Pickering 'A'—the first nuclear power

Table 6.2 CANDU reactors in operation, under construction or committed

| Name of reactor or power station | Location | Type | Electrical output, MW | Date of first power |
|---|---|---|---|---|
| NPD | Ontario | PHWR | 22 | 1962 |
| Douglas Point | Ontario | PHWR | 208 | 1967 |
| Pickering A | Ontario | PHWR | $4 \times 514$ | 1971–3 |
| Gentilly 1 | Quebec | BLWR | 250 | 1971 |
| Kanupp | Pakistan | PHWR | 125 | 1971 |
| Rajasthan 1 | India | PHWR | 207 | 1972 |
| Bruce A | Ontario | PHWR | $4 \times 740$ | 1976–8 |
| Gentilly 2 | Quebec | PHWR | 600 | 1978 |
| Rajasthan 2 | India | PHWR | 207 | 1978 |
| Madras 1 | India | PHWR | 220 | 1979 |
| Point Lepreau | New Brunswick | PHWR | 600 | 1980 |
| Cordoba | Argentina | PHWR | 600 | 1980 |
| Pickering B | Ontario | PHWR | $4 \times 514$ | 1981–3 |
| Madras 2 | India | PHWR | 220 | 1981 |
| Wolsung 1 | Korea | PHWR | 600 | 1981 |
| Narora 1 | India | PHWR | 220 | 1982 |
| Narora 2 | India | PHWR | 220 | 1983 |
| Bruce B | Ontario | PHWR | $4 \times 750$ | 1983–6 |
| Darlington | Ontario | PHWR | $4 \times 800$ | 1986–8 |

station expressly for electricity-generation to incorporate four units of major size—came into service on schedule and has demonstrated high average load factors. This, in spite of difficulties with cracks in pressure tubes which required a large number of channels to be replaced in two of the reactors. The second and larger four-unit station, Bruce 'A', has now come into operation, and this includes some major design variations from Pickering 'A'.

Much of the success with this closely-knit programme can be attributed to resolute dedication to a system which has quite specific features that are fostered by the absence of enrichment and reprocessing plants. It is also the result of a thorough integration of all forces—Ontario Hydro, AECL Power Projects, and AECL R&D—into one team and to its close relationship with the safety authority—AECB.

## 7.3 Italy

Compared with the very substantial attainments of the CANDU programme, the study of heavy water power reactors in Italy has resulted in very little actual reactor construction. The Italians have, however, long appreciated the extent to which they will ultimately have to rely upon nuclear power because of their lack of indigenous fuel resources and took very early steps to acquire experience in this field. The Italians aimed for a versatile approach, and this nation was the very first to have full-size power plants in operation incorporating reactors cooled by gas, pressurized water, and boiling water,

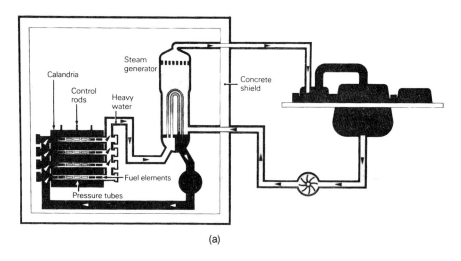

(a)

Fig. 6.9 (a) Diagrammatic arrangements of CANDU–PHWR and CANDU–BLWR: (a) CANDU–PHWR

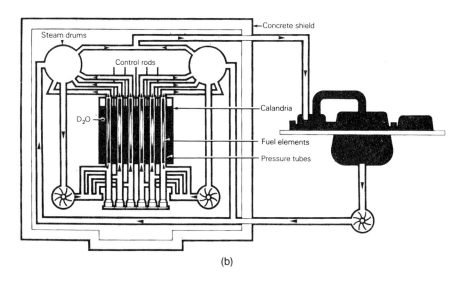

(b)

Fig. 6.9 (b) CANDU–BLWR

so that the characteristics of these systems could be evaluated in representative service conditions. These plants were bought from the UK and the USA, but prominent in their own nuclear research and development, the Italians have maintained a sustained interest in a reactor concept given the name Cirene. This has taken various forms but has always employed heavy water as the bulk moderator. In many ways, the project has followed a very similar course to the UK's SGHWR, an important common feature being that the studies have been based upon the ultimate size of reactor that will be required as opposed to some prototype design. Slow but steady progress has been made with not inconsiderable use of experiments to ensure the validity of data used in the parametric surveys aimed at producing a reactor specifically suited to Italy's needs later this century. Incidentally, this looking towards the future is common with a number of countries who appear to see the application of heavy water power reactors as a long-term exercise, frequently after a first phase of nuclear power derived from American-type light water reactors.

For some time, the results of these studies have pointed to Cirene more closely resembling the SGHWR than it does CANDU, in that it would incorporate vertical pressure tubes and employ boiling light water as its coolant. The fuel would, however, be natural $UO_2$ arranged in short multi-rod bundles so that the reactor physics differs from the SGHWR, involving operating with a significantly positive coolant void coefficient. This has had to be taken into account when designing the control and safety systems, and quite different solutions are proposed compared with SGHWR practice. Thus, the load control of the Cirene reactor also ensures the spatial stability of the core and consists of a number of special vertical tubes distributed throughout the lattice. The amount of reactivity absorbed in each of these tubes can be varied by changing the void fraction of a two-phase mixture of helium and light water containing boron 'poison', which is circulated throughout the particular tube. The control of the void fraction inside a reactivity control tube is achieved by means of an electro-pneumatic valve located in the line from the feed of poisoned light water. Actuation of the valve results in variation of the liquid flow rate and the driving signal is independent for each reactivity channel.

Whilst the design of Cirene (Fig. 6.10(a)) has varied in detail as the various studies were completed, it is estimated that the basic concept can be fully demonstrated in a prototype producing some 35 MW net electrical generation from a thermal output of 120 MW. It is expected that the prototype will go critical in 1983 at Latina.


## 7.4 Japan

There are close similarities between the Italian and Japanese approaches to nuclear power. Japan also lacks adequate indigenous fossil fuel resources and was soon importing various types of power reactor to gain practical experience with plant operating under actual generating network conditions.

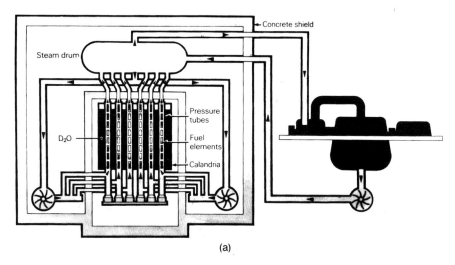

Fig. 6.10 (a)  Diagrammatic arrangements of CIRENE and FUGEN: (a) CIRENE

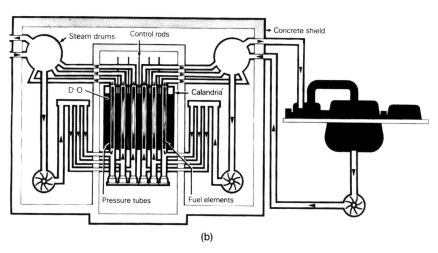

Fig. 6.10 (b)  FUGEN

As in Italy, individual Japanese power companies installed British and American reactors and were then able to choose between gas-cooled and water-cooled systems for their particular requirements. It would seem they have selected the PWR as the vehicle for the first phase of their nuclear power programme, and their manufacturing industry is also now committed to this system. However, for the longer term, nuclear research and development on a national basis has been conducted on a co-ordinated programme including both thermal and fast reactors. The advanced thermal reactor (ATR) system is known as Fugen—one of the traditional supporters of the Buddha—and is a vertical pressure-tube reactor employing enriched uranium dioxide fuel, boiling light water coolant, and heavy water moderation (Fig. 6.10(b)). It therefore resembles SGHWR, and irradiation of Fugen fuel has been carried out in the Winfrith reactor. There has also been substantial liaison between the UK and Japan over the years both in the experimental and theoretical fields, and a number of the SGHWR computer codes have been adapted for the ATR assessments.

The construction of a substantial prototype designed to produce 165 MW of electricity was begun at Tsuruga City on Honshu, in late 1970, and this became critical in March 1978. All functional tests were then undertaken including a period of operation at full power, so that the licence to work the plant was issued by the Japanese regulatory body in March 1979.

## 7.5 Germany

The German contribution to the development of $D_2O$-moderated reactors was made primarily by the firm of Siemens and related to systems fuelled by natural uranium. In the 1960s, this company designed and built two prototype reactors of quite different type. The first was designated MZFR and was a PHWR of pressure vessel construction. This had a net electrical output of 51 MW and became operational at Karlsruhe in December 1965. The second was a pressure-tube reactor employing $CO_2$ as the coolant. It was broadly similar in form to the GCHWR which had been studied by the UKAEA at the beginning of its development of the pressure-tube system. The German prototype was built at Niederaichbach and designed for an electrical output of 100 MW. This was commissioned in 1972.

The MZFR incorporated components which do not differ to any great extent from those of the light water PWR. The core geometry is appropriate to the use of heavy water as both coolant and moderator, while additional equipment was provided to allow refuelling with the reactor at load. Plant was also supplied to reprocess any heavy water leakage. Construction occupied exactly five years, and although some problems were encountered with the main coolant circulating pumps, the fuel handling machine, and the seal plugs which closed the access ports on the pressure vessel cover, all the contractual stipulations were fulfilled, so that the plant could be handed over in November 1966.

Siemens applied the experience gained with MZFR to produce designs for natural-uranium-fuelled PHWRs of central power station sizes, and it was such a plant that the company tendered in 1967 for the 320 MW(e) unit sought by Argentina for installation at Atucha. Against competition from the SGHWR and CANDU, the German firm gained the order, with the aid of an international trade deal. MZFR details were modified as a result of co-operation between Siemens and the Swedish ASEA and the French SOCIA companies—for example, the fuelling machine was modelled on that developed for the Swedish reactor at Marviken. The Atucha reactor became critical early in 1974 and achieved full power later in the same year. Since then, operation has been generally satisfactory for a pioneer application but, commercially, Siemens has become part of the German consortium, Kraftwerk Union, KWU, which has subsequently adopted a policy of constructing light water reactors.

This change in commercial policy also curtailed the development of the Niederaichbach concept. This, however, suffered a series of technical problems so that it did not become critical until December 1972, some two years behind schedule, after a long period of commissioning. It was shut down in 1974 after its performance had fallen appreciably short of expectations and when the new KWU organization had judged that it was evident its design could not be economically extended to full-scale commercial application.

## 7.6 Sweden

The Swedish contribution to this branch of reactor technology has been concentrated upon the use of heavy water as both moderator and coolant in reactors of the pressure-vessel type.

Following work with early research reactors, the Swedish utilities decided to construct a 65 MW(t) PHWR at Agesta to gain general experience for larger power reactors. The plant, which was designated R-3 and given the sobriquet Adam, was of interest since its output was split between electrical generation (10 MW) and district heating. It became critical in July 1963 and behaved very well in a technical sense. However, it always operated at a loss financially, which increasingly concerned the sponsors of the project, so that it was ultimately decided to cease operation in July 1974 after the reactor had served for some time as a training centre for operating personnel.

The next stage in the Swedish programme was aimed at simplifying the plant by introducing a direct-cycle system, i.e. the elimination of external steam generators by allowing the heavy water coolant to boil within the core. A BHWR was therefore designed which required a complete reappraisal of control philosophy and also the development of a $D_2O$ steam turbine. The design selected for construction was intended to produce 132 MW(e), but the concept was extrapolated in economical assessments to much larger sizes, the cores to be accommodated in concrete reactor pressure vessels.

Construction started at Marviken at the end of 1963, the reactor being designated R-4 and named Eve. During the early construction phase, the sponsors began to be seriously concerned about the cost of the project, and the idea was mooted of introducing nuclear superheat with the object of enhancing the electrical generation by some 60 MW. However, this scheme had to be abandoned because of the technical difficulties that were foreseen in coupling, in a nuclear sense, the boiling and superheating zones of the core in a pressure-vessel reactor. Problems were experienced in establishing a reliable control system and, whilst construction was nominally completed by November 1968 and the plant was then subjected to a six-month test period, it was decided to abandon the project before the reactor was fuelled and filled with heavy water.

This marked the termination of heavy water reactor technology in Sweden and 185 tonnes of $D_2O$ were sold to AECL for its CANDU programme. It was estimated that it would be too expensive to convert the Marviken reactor to a light water system and, for a time, it was considered it should be completed as a conventional oil-fired power station. This idea was eventually dropped because of the possibility of atmospheric pollution, and it was decided to use the plant for a series of comprehensive nuclear safety tests. To date, some 16 coolant blow-down tests have been performed from full pressure, providing a substantial endorsement of the concrete reactor containment, which is of the pressure-suppression type. These have more recently been followed by LOCA tests, which are claimed to have been very successful in demonstrating how a core structure behaves under such severe accident conditions.

## 7.7 France

The first French power reactors were graphite-moderated, gas-cooled systems not unlike the British Magnox plant. Consideration was next given to a GCHWR which retained carbon dioxide as coolant, embodied pressure-tube construction, and employed unpressurized heavy water as moderator. A prototype designated EL-4 and capable of producing 70 MW(e) was built at Brennilis near Quimper, construction starting in the summer of 1962 and criticality being achieved in December 1966. Beryllium was originally intended for the canning material though the initial charge was eventually clad in stainless steel; at a later date, it was intended to improve the neutron balance by using a zirconium–copper alloy which had been successfully irradiated in experimental reactors. Power began to be shipped to the national grid in July 1967, and operation appeared to be very satisfactory until February 1968. Cracks then developed however in tube welds in 5 of the 16 steam generators, which were of different design to those incorporated in the gas-cooled, graphite-moderated plant, and it was necessary for the plant to be closed down. A further new design of steam generator was produced and tested, and the plant returned to criticality with the new components in April 1971. Operation continued satisfactorily until 1975, when the plant was damaged by two explosions caused by bombs claimed to have been planted by Breton Separatists. By this time French construction of PWRs was well established, and so it was decided not to return EL-4 to working order.

## 7.8 Switzerland

Swiss interest in heavy water systems culminated in the construction of a 30 MW thermal GCHWR pressure-tube reactor at Lucens, building starting in August 1962 and criticality being obtained in December 1966. A net electrical output of 7.6 MW was generated, and this passed to the national grid in January 1968. All proceeded satisfactorily until the 21st January 1969, when at least one pressure tube ruptured causing an immediate loss of the $CO_2$ coolant. Several fuel elements melted in the LOCA, and the calandria also suffered damage to the extent that 23 $m^3$ of heavy water were spilt into the reactor vault. The effects of the incident were confined to the primary containment, so there was no injury or exposure of operating staff and no excessive releases to the environment.

The authorities decided that the project should be abandoned and the plant was made available for decontamination and dismantling experiments. All fuel and active materials were recovered and the whole reactor vault was successfully cleared of plant and equipment by 1973.

# 8 Research, test, and experimental reactors

Out of the 110 countries who are members of the IAEA, 17 have constructed research, test, and experimental reactors which can use heavy water as moderator. These total 58 in all. The first to go critical was CP-3 at Palos Park in Illinois, USA in 1944. This tank-type reactor, originally fuelled with natural uranium and producing an output of 300 kW and a maximum thermal flux of $10^{12}$ n cm$^{-2}$ s$^{-1}$, was shut down in 1950 to be rebuilt as CP-3. In its new form, it regained criticality later in 1950 with uranium fuel enriched to 2 per cent. It then had an output of 275 kW but a substantially increased maximum thermal flux of $3.4 \times 10^{12}$ n cm$^{-2}$ s$^{-1}$.

The first critical assembly outside the USA also employed heavy water as moderator. This was ZEEP (zero-energy-experimental-pile) built by Atomic Energy of Canada Ltd. (AECL)—the newly founded Crown Company—at its Chalk River Establishment in Ontario. This was also a tank reactor. It went critical in September 1945 and served with a variety of fuels until it was shut down in 1970. Originally, ZEEP was used for experiments leading to the design of the first large Canadian reactor, NRX. ZEEP cost $200 000 and needed an operating staff of 4–5 professionals aided by 3–4 technicians. Its ouput was nominally 100 W and it produced a maximum thermal flux of $10^8$ n cm$^{-2}$ s$^{-1}$.

Because the building of NRX (Canadian National Research experimental) represented an important early step in reactor technology that was completely outside any military programme involving nuclear energy, it is appropriate to mention that it is a tank-type reactor fuelled with metallic natural uranium, or enriched uranium, or plutonium alloyed with aluminium, clad in aluminium. The fuel is arranged in a hexagonal lattice and is moderated by heavy water. The reflector consists of graphite blocks placed outside the aluminium core tank.

The nominal heat output is 40 MW and the coolant is light water. Separation is therefore necessary between moderator and coolant and this is achieved by forming the reactor tank as a calandria. A single fuel element is then loaded into each tube which is connected by pipework to headers in the coolant circuit. The fuel is arranged vertically as in ZEEP, an interesting engineering variation from the very first graphite-moderated reactors, which accommodated their fuel in horizontal holes formed in the blocks that made up their moderating stack. NRX was designed by Defence Industries Ltd. in consultation with a mixed team of Canadian, British, and US scientists. This was of course a time of close integration between the British and Canadian nuclear research programmes, and the first director of AECL was Dr. (later Sir) John Cockcroft, subsequently Director of Harwell. NRX was built by Fraser Brace Co. Ltd. and its cost is estimated to have been $10M including the building to house it. It required a total of 26 professional and 45 support staff to operate it on a three-shift basis. Criticality was first achieved in July 1947 and full power developed in May 1948. The maximum thermal neutron flux is $6.4 \times 10^{13}$ n cm$^{-2}$ s$^{-1}$, and the purpose of NRX has been to permit neutron physics measurements to be made, to produce isotopes, and to test fuel elements and materials in reactor environments.

To put these early heavy water reactors into their proper historical perspective, it should be remembered that the construction of the Chicago Pile, CP-1, by the Metallurgical Laboratory of the University of Chicago for the United States Atomic Energy Commission (USAEC) began in October 1942. The unique function of this was to demonstrate experimentally the first self-sustaining, controlled nuclear chain reaction. The pile, as reactors tended then to be called (CP-1 did look like a 'pile' of bricks), was graphite moderated and its natural uranium metal fuel bars, or *slugs*, were cooled by atmospheric air. It became critical at 3.25 p.m. on 2 December 1942 and was dismantled without any ceremony in the following March to provide the basis for CP-2. Heavy water reactors, starting with CP-3, were therefore early upon the new nuclear scene. A study of Table 6.3 shows how widely they have been used. Mostly they are of simple, tank construction so that they can generally accept a wide range of fuel lattice arrangements provided a suitable base plate is available. They have been used to obtain basic physics data or more specific reactor physics information, such as lattice experiments, to further studies of some particular power reactor concept. Substantial use has been made of this type of heavy water reactor to produce isotopes and

Table 6.3 Heavy water research reactors built by member countries of IAEA in order of first achieving criticality

| Year of first criticality | Reactor name or designation | Country | Location | Type: enrichment: moderator/coolant | Output | Maximum thermal flux ($n\,cm^{-2}\,s^{-1}$) | Date of shutdown |
|---|---|---|---|---|---|---|---|
| 1944 | CP-3 | USA | Palos Park | tank: 7% U: $D_2O$ and $H_2O$ | 300 kW | $1 \times 10^{12}$ | 1950 |
| 1945 | ZEEP | Canada | Chalk River | tank: variable fuel:$D_2O$ | 100 W | $1 \times 10^{8}$ | 1970 |
| 1947 | NRX | Canada | Chalk River | tank: natural U: $D_2O/H_2O$ | 33 MW | $6.4 \times 10^{13}$ | |
| 1948 | EL-1 (Zoe) | France | Fontenay-aux-Roses | tank: natural U: $D_2O$ | 150 kW | $1 \times 10^{12}$ | |
| 1949 | TR | USSR | Moscow | tank: 2% U: $D_2O$ | 2.5 MW | $2.5 \times 10^{13}$ | |
| 1950 | CP-3' | USA | Palos Park | tank: 2% U: $D_2O$ | 275 kW | $3.4 \times 10^{12}$ | 1963 |
| 1951 | JEEP-1 | Norway | Kjeller | tank: natural U: $D_2O$ | 450 kW | $1.5 \times 10^{12}$ | 1973 |
| 1952 | EL-2 | France | Saclay | tank: natural U: $D_2O/CO_2$ | 2 MW | $7.7 \times 10^{12}$ | 1965 |
| 1953 | PDP | USA | Aiken | tank: variable fuel: $D_2O$ | 1 kW | $1 \times 10^{8}$ | |
| 1954 | CP-5 | USA | Argonne | tank: 90% U: $D_2O$ | 5 MW | $1 \times 10^{14}$ | |
| (1962) | Dimple | UK | Harwell Winfrith | tank: U or Pu: variable | 100 W | $3 \times 10^{8}$ | 1960 |
| 1956 | Aquilon | France | Saclay | tank: natural U: $D_2O$ | 1–100 W | $1 \times 10^{8}$ | |
| | DIDO | UK | Harwell | tank: 93% U: $D_2O$ | 22.5 MW | $2 \times 10^{14}$ | 1967 |
| 1957 | EL-3 | France | Saclay | tank: 1.35% U: $D_2O$ | 20 MW | $1 \times 10^{14}$ | |
| | NRU | Canada | Chalk River | tank: natural U: $D_2O/H_2O$ | 110 MW | $2.5 \times 10^{14}$ | |
| | PLUTO | UK | Harwell | tank: 93% U: $D_2O$ | 22.5 MW | $2 \times 10^{14}$ | |
| 1958 | HRE-2 | USA | Oak Ridge | aqueous homogeneous: 93% U: $D_2O$ | 5.2 MW | $1.1 \times 10^{14}$ | 1961 |
| | DMTR | UK | Dounreay | tank: 93% U: $D_2O$ | 25 MW | $2 \times 10^{14}$ | 1969 |
| | Hifar | Australia | Lucas Heights | tank: 93% U: $D_2O$ | 10 MW | $1.7 \times 10^{14}$ | |
| | MITR | USA | Cambridge | tank: 90% U: $D_2O$ | 5 MW | $2.4 \times 10^{13}$ | |
| | PRR | USA | Pawling | tank: 92.5% U: $D_2O$ | 5 W | $2 \times 10^{8}$ | 1971 |
| | PSE | USA | Aiken | tank: natural U: $D_2O$ | negl. | $1 \times 10^{6}$ | |
| | R-B | Yugoslavia | Vinca | tank: natural U: $D_2O$ | negl. | — | |
| 1959 | HBWR | Norway | Halden | tank: 1.5% U: $D_2O$ | 25 MW | $4.5 \times 10^{13}$ | |
| | Ispra-1 | Italy | Ispra | tank: 20% U: $D_2O$ | 5 MW | $8.9 \times 10^{13}$ | |
| | R-O | Sweden | Studsvik | tank: natural U: $D_2O$ | 50 W | $1 \times 10^{8}$ | |
| | R-A | Yugoslavia | Vinca | tank: 2% U: $D_2O$ | 10 MW | $6.5 \times 10^{13}$ | 1973 |

Table 6.3 (*cont.*):

| Year of first criticality | Reactor name or designation | Country | Location | Type: enrichment: moderator/coolant | Output | Maximum thermal flux ($n\,cm^{-2}\,s^{-1}$) | Date of shutdown |
|---|---|---|---|---|---|---|---|
| 1960 | Cirus | India | Trombay | tank: natural U: $D_2O/H_2O$ | 40 MW | $6.3 \times 10^{13}$ | |
| | Diorit | Switzerland | Würenlingen | tank: natural U: $D_2O$ | 20 MW | $3.5 \times 10^{13}$ | |
| | DR-3 | Denmark | Risø | tank: 93% U: $D_2O$ | 10 MW | $1.5 \times 10^{14}$ | |
| | JRR-2 | Japan | Tokai | tank: 90% U: $D_2O$ | 10 MW | $1.8 \times 10^{14}$ | |
| | PRTR | USA | Richland | pressure tubes: Pu and U: $D_2O$ | 70 MW | $1 \times 10^{14}$ | 1969 |
| | SPERT-2 | USA | Idaho Falls | tank: 93% U: $H_2O$ or $D_2O$ | transient | $1 \times 10^{14}$ | 1965 |
| | ZED-2 | Canada | Chalk River | tank: variable fuel: $D_2O$ | 150 W | $1 \times 10^{8}$ | |
| 1961 | AHCF | Japan | Tokai | aqueous homogeneous: 20% U: $D_2O$ | 50 W | $1 \times 10^{9}$ | |
| | FR-2 | West Germany | Karlsruhe | tank: natural U: $D_2O$ | 44 MW | $1 \times 10^{14}$ | |
| 1962 | Nora | Norway | Kjeller | tank: natural U: $D_2O$ and $H_2O$ | 100 W | $5 \times 10^{8}$ | |
| | Zerlina | India | Trombay | tank: variable fuel: $D_2O$ | 400 W | variable | |
| | DAPHNE later Delphi | UK | Harwell | tank: 80% U: $D_2O$ | 100 W | $1.3 \times 10^{9}$ | |
| | FRJ-2 (DIDO) | West Germany | Jülich | tank: 80% U: $D_2O$ | 23 MW | $1.7 \times 10^{14}$ | |
| 1963 | HWCTR | USA | Aiken | tank: 93% U: $D_2O$ | 61 MW | $1.2 \times 10^{14}$ | |
| | JRR-3 | Japan | Tokai | tank: natural U: $D_2O$ | 10 MW | $2.2 \times 10^{13}$ | |
| | PRCF | USA | Richland | tank: Pu, 2.35% U: $H_2O$ or $D_2O$ | 10 kW | variable | |

| Year | Name | Country | Location | Description | Power | Flux |
|---|---|---|---|---|---|---|
| 1964 | GTRR | USA | Atlanta | tank: 93% U: $D_2O$ | 5 MW | $2 \times 10^{13}$ |
| | JUNO | UK | Winfrith | tank: variable fuel: $D_2O$ or $H_2O$ | 100W | $3 \times 10^{8}$ |
| | R-1 | Sweden | Stockholm | tank: natural U: $D_2O$ | 1 MW | $1 \times 10^{12}$ |
| 1965 | Venus | Belgium | Mol | tank: 7% U: $D_2O/H_2O$ | 500 W | $1 \times 10^{9}$ |
| | ALRR | USA | Ames | tank: 94% U: $D_2O$ | 5 MW | $1 \times 10^{14}$ |
| | ECO | Italy | Ispra | tank: natural U: $D_2O$/organic | 1 kW | $1.3 \times 10^{9}$ |
| | EOLE | France | Cadarache | tank: variable enrichment: $D_2O$ | 10 MW | variable |
| | HFBR | USA | Upton | tank: 90% U: $D_2O$ | 40 MW | $7 \times 10^{14}$ |
| | WR-1 | Canada | Pinawa | tank: 2.4% U: $D_2O$/organic | 60 MW | $9.3 \times 10^{13}$ |
| 1966 | JEEP-2 | Norway | Kjeller | tank: 2.5% U: $D_2O$ | 2 MW | $4 \times 10^{13}$ |
| 1967 | ESSOR | Italy | Ispra | tank: natural and 90% U: $D_2O$ | 40 MW | $2 \times 10^{14}$ |
| | NBSR | USA | Gaithersburg | tank: 90% U: $D_2O$ | 10 MW | $1 \times 10^{14}$ |
| | Pelinduna Zero | South Africa | Pelindaba | tank: 2% U: $D_2O$ | negl. | $2 \times 10^{6}$ |
| 1969 | BR-3/VN | Belgium | Mol | vessel: 7% U: $D_2O$ and $H_2O$ | 40.9 MW | $7.3 \times 10^{13}$ |
| 1971 | High flux reactor | France | Grenoble | tank: 93% U: $D_2O$ | 60 MW | $1.5 \times 10^{15}$ |

another very important role they have also discharged is to study the behaviour of materials and sub-assemblies under conditions of irradiation. It will be seen from Table 6.3 that the bulk of these reactors continue to be in commission or available for operation: of 58 'builds', there have been 15 shutdowns, but two of these were to allow rebuilding in an improved form. Construction of such reactors was at its peak in the period 1955–65. The last noted, the high-flux reactor at the Institut Laue-Langevin, Grenoble, went critical for the first time as recently as 1971. It was built as a tank-type heavy water reactor to provide maximum flexibility, and has the capability of attaining a maximum thermal neutron flux as high as $1.5 \times 10^{15}\,\mathrm{n\,cm^{-2}\,s^{-1}}$. This high-intensity source of thermal neutrons offers great potential for research in nuclear physics of the condensed state, where it can be used to investigate the properties of amorphous substances and liquids. Such facilities are extremely expensive, and so the reactor was originally funded in equal shares by France and West Germany. Britain has now joined the project through the Rutherford Institute sponsored by the Science Research Council.

The British contribution to these associated classes of heavy water reactors has been substantial. Basically, there have been two main designs: a low-power reactor for the study of the physics of reactor systems moderated by light or heavy water (or a mixture of these), or by an organic liquid, and a much more powerful design for the irradiation of reactor materials, neutron diffraction experiments, and large engineering loops.

Two very similar low-power reactors were designed and built by the UKAEA: DIMPLE and JUNO. DIMPLE (Deuterium-moderated pile of low energy) was originally constructed at Harwell in 1954 and used there until late 1960. It was then dismantled and the components moved by road to the then new Atomic Energy Establishment at Winfrith. It was rebuilt there and recommissioned in June, 1962—hence the entries in Table 6.3. JUNO was built at Winfrith to replace the Authority's NERO reactor and it uses much of the instrumentation, control circuitry, and biological shield of the earlier plant. The reactor was commissioned in April 1964.

Both DIMPLE and JUNO were of the tank type with a maximum thermal power of 100 W and peak neutron flux of $3 \times 10^8\,\mathrm{n\,cm^{-2}\,s^{-1}}$, the principal difference being that the main dump tank of DIMPLE is located in a separate pit some distance from the reactor. Shielding is provided by an array of concrete blocks having a thickness of 1.2 m. Containment and monitoring equipment have been provided so that plutonium-bearing fuel can be used. Within the reactor tank, a wide variety of core structures can be assembled within a limit of 2.6 m diameter by 3 m high, depending upon the base plate and support arrangement that is employed. This permits a range of differing core configurations: large pressure-tube systems; small, close-packed systems, typical of light water reactor lattices; or a mixed critical system consisting of a central fast reactor region driven by a peripheral thermal reactor zone. The moderating liquid can be heated to 90°C.

Two alternative types of emergency shutdown equipment have been provided—safety rods and a rapid dump of the moderating fluid into a cavity below the core structure but within the reactor tank. The latter method of fast shutdown is particularly appropriate for the investigation of close-packed systems where the provision of channels to allow the insertion of safety rods would produce an unacceptable perturbation of the core parameters. Control can be effected either by variation of moderator level or by movement of coarse and fine control rods. These can be of variable composition and number, and the difference between fine and coarse control is achieved by different gear speeds.

Experimental equipment includes automatic flux-scanning gear with punched card data presentation for use in measurements of critical size and power distribution. Pulse source equipment is available to measure the kinetic characteristics of the system under study and to provide alternative means of measuring changes in reactivity. DIMPLE was also used to test solid-state devices instead of relays in all the interlock and safety circuits. Dynamic logic methods are employed in which the 'safe' state of any control input is indicated by continuous switching of the appropriate circuits. Automatic corrective actions or alarms are initiated when the switching ceases.

DIMPLE continues to be available for use; JUNO was dismantled in 1978, though the components have been retained.

The more powerful British heavy water research reactors are generically of the type originally known as *materials-testing reactors*. Their achievement is somewhat wider than DIMPLE and JUNO if only because they are more numerous; as a basic design, they number six of which, however, only three were built in the UK. The remainder were exported to Australia, Denmark, and Western Germany, forming the first appreciable construction 'run' of essentially similar plant in the nuclear era. The design was formulated by the Authority with manufacture and construction undertaken, at home and abroad, by Messrs Head Wrightson Processes. When it is remembered that the building programme started at Harwell in 1955 and was completed in 1962 at Jülich, Germany, it can be readily appreciated that it represented a very important early step in introducing high-integrity nuclear technology and techniques into the fields of commerce and plant manufacture. The extremely satisfactory performance attained from all the reactors involved, and their ability to accept subsequent modification and extension, is an impressive endorsement of the quality of the initial design and workmanship.

The pioneer reactor of the series was DIDO—the name being derived from the chemical composition of heavy water, DDO. The second was PLUTO (power loop testing reactor). This was built alongside DIDO at AERE Harwell, the reactors being commissioned on 7 November 1956 (DIDO) and 28 October 1957 (PLUTO). A third reactor was built in the UK at Dounreay. This was originally to be called PLUTO-2 but was always known as DMTR (Dounreay materials testing reactor). It was in use from 1958 until 1969 when it was closed down as part of a reorganization programme. The other DIDO-class reactors were HIFAR (high flux Australian reactor) commissioned at Lucas Heights in January 1958, DR-3 commissioned at Risø as the third Danish reactor in August 1959, and finally FRJ-2, commissioned at Jülich, Germany, in November 1962.

The design is a tank-type reactor, the tank being of aluminium and of cylindrical shape with a diameter of 2 m. It can accommodate a core which is a vertical cylinder having a height of 610 mm and a diameter varying between 840 mm and 875 mm. The thermal output from this class of reactor can be as high as 25 MW (achieved in DMTR), having risen to this value from about 8.5 MW as originally conceived. The design originally provided for a central flux of approximatley $1 \times 10^{14}$ n cm$^{-2}$ s$^{-1}$, but the higher power now possible corresponds to a maximum thermal flux of $2.3 \times 10^{14}$ n cm$^{-2}$ s$^{-1}$ with a maximum fission flux of $1.7 \times 10^{14}$ n cm$^{-2}$ s$^{-1}$. Heavy water at negligible pressure is used both as moderator and coolant, the bulk outlet temperature not exceeding 65 °C as the coolant leaves the core. The heavy water is circulated by 'canned' pumps to heat exchangers below the biological shield under the reactor tank. In these exchangers the heat generated in the core is transferred to light water which is then cooled in forced-draught cooling towers located outside the reactor containment. These reactors have a heavy water reflector zone within their tanks surrounding the core region and a graphite reflector arranged outside the aluminium reactor tank. Around the graphite reflector is a light water-cooled, lead thermal shield, 102 mm thick, the whole structure being positioned inside a steel tank lined with boral sheeting that is 65 mm thick. This outer steel tank is located within a barytes concrete biological shield, the top of the reactor being provided with a water shield tank arranged beneath a thick steel top plate in a ring mounting. The outside of the DIDO concrete biological shield is finished in the form of a 10-sided prism, whereas subsequent reactors had the outside of their shields built in the simpler form of a rectangular prism. Thus, the thickness of the barytes concrete at DIDO is 1.52 m and not less than 1.2 m at the other reactors.

There have been substantial development of the fuel elements since the reactors were first designed. The core of DIDO consists of 25 elements arranged in rows of 4,6,5,6,4. The other reactors can accommodate 26 fuel elements in rows of 4,6,6,6,4. In either case, a near rectangular lattice is achieved. Each element consists of assemblies of plates of enriched uranium-aluminium sandwiched between two thin aluminium sheets, edge-welded and then rolled up to a slightly cambered section. Originally, each plate was assembled with 10 others, and suitable edge spacers, to form an approximately 'square' box-sectioned element. This was then fitted with inlet and outlet nozzles to form a duct through which the coolant flowed, up between the individual plates. Such configurations were very common among materials-testing reactors of the time. More recently, in the British reactors, the fuel plates, still rolled to a curved section, have been arranged as involutes between two concentric aluminium tubes to form an annular assembly. The inner tube, in the centre of each assembly, provides a 52 mm diameter *hole* into which experimental irradiation assemblies can be inserted.

In addition to test facilities provided within the newer designs of fuel elements, horizontal and vertical holes are arranged in the heavy water reflector region, and further holes are incorporated in the graphite reflector outside the aluminium reactor tank.

An indication of the scope of these reactors may be gained from a brief summary of the work that has been undertaken in them. DIDO has been used mainly for the irradiation of samples of reactor materials, neutron diffraction experiments, and engineering loops. The behaviour of graphite under irradiation has received detailed study, and a high-pressure water loop has been installed serving several facilities connected with the development of pressurized-water reactors. The reactor is linked by compressed air lines to several out-lying buildings at Harwell, from which samples can be pneumatically fed into the core region and, similarly, extracted from the core and delivered back into the building. This is called the 'rabbit' facility. At present, some fifty experiments are undergoing continuous irradiation. PLUTO, as its name implies, was built primarily to accommodate large engineering rigs and loops and incorporates a number of large-diameter holes for this purpose. One loop developed and installed in 1961 was specifically for testing prototype fuel for the DRAGON high-temperature, gas-cooled reactor at Winfrith. Automated neutron beam and associated diffraction equipment is also in use. Otherwise there is a continuous programme of irradiation of reactor materials involving more than thirty experiments. DIDO and PLUTO both operate on a cycle of 28 days (of 24 hours), each cycle includes a 3–4 days shutdown. During its operational life-time, DMTR was used for thermal neutron irradiation experiments in support of all the major projects then part of the UKAEA's reactor programme. In particular, two very large loops demonstrate the diversity of work that can be done in this type of research reactor. Thus, the graphite loop supported the first two generations of the British gas-cooled reactor family by circulating carbon dioxide at varying temperatures and high pressure through a matrix of reactor-grade graphite. The phenomenon investigated was the mass transfer of graphite from the hottest section of the loop to the cooler portions. This migration of the moderator was expected to cause problems during the lifetime of both Magnox and advanced gas-cooled reactors (AGRs), which are based upon a 'permanent' graphite moderating stack in their core structure. Pumps with gas-bearings were developed to circulate the gas, and it was demonstrated that the mass transfer could be minimized by dosing the $CO_2$ with methane, the optimum amount of additive depending upon the working conditions in the reactor. The boiling water loop supported the development of the steam-generating heavy water reactor (SGHWR) by investigating the corrosion of zirconium alloys in the actual steam conditions experienced in this type of reactor. This programme was assisted by a substantial chemistry group working at Dounreay. DMTR was also used for a series of fuel development experiments, which included work for the sodium-cooled, fast reactor and also for the high-temperature gas-cooled reactor (HTGCR) programme, which was associated with fuel formed from compacts of small-diameter spheres of uranium dioxide coated with silicon carbide. A large neutron radiography development programme was also carried through in DMTR using a neutron beam from the core.

The work on DIDO, PLUTO, and DMTR was greatly assisted by DAPHNE (DIDO's and PLUTO's handmaiden for nuclear experiments), which first went critical in 1962 and was shut down in 1972. This was a zero-energy reactor sharing, however, most of the construction details of the DIDO class of reactors and used to measure the reactivity, fluxes, and other nuclear data of cores to be installed in DIDO and PLUTO. This use as an auxiliary or mock-up of the two highly utilized reactors at Harwell gave the reactor its name.

The reactor was built adjacent to the swimming-pool reactor LIDO at AERE, Harwell, and could be loaded with fuel elements, rigs, and beam holes to simulate DIDO, PLUTO or other reactors of this class. Core configurations with either 25 or 26 fuel elements could be accommodated together with the appropriate control and shut-off devices. By measuring the reactivity of a particular core loading and the flux distributions through rigs, the behaviour of the high-flux reactors could be predicted and optimized with great precision and the least waste of their operating time. DAPHNE could also be used to measure safety parameters, such as prompt neutron lifetime, to compare calculations with measurement, and to test improved designs of fuel elements, control mechanisms, and rigs. The reactor did not operate at powers above 100 W, and so there was no requirement for a cooling system. The investment of heavy water was therefore only some 8 tonnes compared with 10 tonnes in the actual DIDO class. Operation at this very low power also meant that the radioactivity of the fuel was kept to a minimum, so that dismantling the reactor to make changes to the core was a simple undertaking.

When DAPHNE was shut down, it was to permit the reactor to be used to support the design of a British high flux-research reactor, being briefly renamed DELPHI in this application. The scheme to build a separate UK system was then abandoned, however, on the grounds of cost and, as has already been mentioned, Britain now shares use of the high-flux reactor at the Institut Laue-Langevin, Grenoble, with France and Germany. DAPHNE/DELPHI has not been dismantled but remains out of use at Harwell.

To conclude this review of heavy water research reactors, it should be observed that it is unlikely that countries with established nuclear power programmes will need to construct any further reactors of this type. There may be a need for small reactors in the larger universities, and ultimately some replacements may be wanted; however, there are no sufficiently close analogues for us to predict what future generations of nuclear technologists may require. Some developing countries may find it desirable to build their own research reactors but, in the main, the principal work for which the bulk of research reactors were created has been largely completed. Comparing the systems that have emerged, there is little doubt that the tank type, using a liquid moderator, was an extremely attractive choice, given the variety of lattices that could be built in them. The use of heavy water introduced the possibility of using natural uranium fuel, but reference to Table 6.3 shows that many reactors of this type were fuelled with quite highly enriched material to generate the high neutron fluxes which some experiments required. It is also evident from Table 6.3 that some countries constructed heavy water research reactors because they had already identified themselves with the possible use of $D_2O$ as moderator and, perhaps also as coolant, for a subsequent power reactor programme.

From today's viewpoint, the Canadian development path might be considered as being strongly associated with heavy water from the beginning. No such alignment should be read into the British use of heavy water research reactors. Whilst Britain's use of these systems has been as extensive as any, the prime reason for their adoption was their great flexibility. This is seen from their employment to support the various stages of the British gas-cooled reactor programme, in their use to advance light water reactor technology, and in their continuing association with the UK's fast reactor development. Our heavy water research reactors represent a substantial national investment: their installation and operation has rewarded the country with a wealth of knowledge and information across the full width of the several reactor technologies in which we are concerned.

## Suggestions for further reading

From the extremely large library of literature that is available, the following two sources might be regarded as starting points for further reading.

INSTITUTION OF MECHANICAL ENGINEERS: *The SGHWR Symposium, May 1967.* I Mech. E. London (1967).
THE BRITISH NUCLEAR ENERGY SOCIETY. *Proc. Conf. on Steam Generating and other Heavy Water Reactors 14–16 May, 1968.* BNES, London (1968).

# 7

# Novel reactor concepts

J. SMITH

In the early days when nuclear power development was in its formative stage, not surprisingly a wide range of ideas came forward. In these the proponents sought desirable performance characteristics through completely new concepts, or tried to exploit technological experience already being accumulated in new ways which were believed to have special attractions. An important feature of a system is its coolant, which is a major determining factor in the engineering and other development aspects. This had an important influence on the pattern of new ideas, and in this article the choice of coolant is taken as a framework for the descriptions.

The chapter begins by considering the ways in which gas-cooling has been studied for concepts other than the advanced gas-cooled reactor and the high temperature gas-cooled reactor. The outstanding example is the gas-cooled fast reactor, which has a number of variants, of which the most unusual is the dissociating-gas scheme at present under investigation in the USSR. In the field of liquid cooling, investigations into the use of organic fluids as alternatives to water as coolants are briefly described.

The concluding section of the article covers in outline the field of fluid-fuel reactors, in which the reactor appears rather as a chemical plant producing heat, the fuel in the fluid form having both potential heat-transfer and safety advantages as well as possibilities of 'internal' reprocessing and recycling of fuel. The difficulties of the technologies involved in the various schemes are described, the molten-salt system receiving the main consideration as being the one most extensively pursued.

## Contents

# 1 Introduction

The main reactor types which are now either in commercial use or are approaching that stage are usually characterized by the type of *coolant* they employ—gas and water for thermal rectors, and liquid metal (sodium) for fast reactors. Because of the major influence of the coolant and its technology on the development programme for a reactor, it is not surprising that there has over the years been a significant effort devoted to examining whether 'accepted' coolants could be exploited in a different way. Nor is it surprising that studies of coolants of a completely different nature have been made, where these have been thought to confer some special advantage. In these cases, the aim has usually been to utilize so far as possible a fuel technology already being developed.

Both of these main groups envisaged the use of reactor core assemblies in the form of solid fuel within a can or clad. However, a radically different approach was taken by the proponents of fluid-fuel reactors. They reasoned that if the heat was generated within the coolant itself, the heat-transfer limitations within the reactor core, which often are the restricting features of solid fuel designs, would be avoided. They also hoped to gain further benefits. Firstly, a fluid fuel would avoid the costs and problems of fabricating fuel to high precision, often under active conditions; secondly, burn-up damage limitations on fuel would be eliminated, thereby minimizing the amount of fuel recycling; and thirdly, a self-contained system could be envisaged with its own reprocessing plant built in to the reactor complex.

In this chapter, a selection of types from the two major classes will be described. It is not intended to be comprehensive but rather to illustrate the salient features, and the successes and problems.

# 2 The gas-cooled fast reactor

Industrial experience with gases ranges over a wide selection of substances and operating conditions, and it was only to be expected that there would be attempts to use this technological base not only for thermal reactors, but also for fast reactors. The reactor experience accumulating from the early UK decision to adopt gas-cooled thermal reactors was a further inducement. The fuel ratings desirable in a fast reactor are considerably higher than those adequate for a thermal reactor, so a gas with its recognized heat-transfer limitations has to be used at very high mass flow rates. If the power for circulating the coolant is not to be unacceptably high, the gas must be at high pressure. The effects of circulation failure or of a leak depressurizing the system are aspects that immediately come to mind as having obvious safety implications.

By the mid-1960s there were two basic types of gas-cooled fast reactor (GCFR) under consideration: the helium-cooled system which was intended to build on the technological experience gained from the thermal HTR programme, and the steam-cooled system.

## 2.1  Steam-cooled fast reactors

Although steam is a superheated vapour rather than a gas, in general character it fits better in the latter classification than any other. Water clearly cannot be used as a fast reactor coolant because of its strong neutron-moderating properties, but it was thought that steam, with its lower density, could perhaps be accommodated without serious detriment to nuclear performance. Engineers are familiar with steam and there was the possibility of a direct cycle to a turbine which would keep down plant costs. At the time of the design studies there were experimental reactor plants in the US which were BWRs with superheaters, and it was hoped to make use of this experience and extend it to fast reactor applications. Studies were made in Europe and in America over a number of years. In several instances the ideas were extended to the use of supercritical-pressure 'steam' in an endeavour to achieve the high fuel ratings considered desirable. It did not in the end, however, prove to be practicable to reconcile adequately the conflict between the desired operating conditions and the fault conditions arising where a depressurization accident was assumed to take place. Reactivity changes due to the drop in steam density, disruptive forces if rapid pressure changes occurred, and the problems of maintaining an adequate coolant flow following the accident, all combined unfavourably. Furthermore, the thermal reactor experiments in the USA and the fuel development programme were encountering difficulties of fuel and can corrosion in superheated steam and in dealing with the consequences of fuel can failure. Consequently, around 1970, by common consent, the steam-cooled fast reactor scheme was abandoned. It is to be noted nevertheless that the USSR graphite-moderated water-cooled reactors have a super-heating section in the fuel channels, as discussed briefly in chapter 4.

## 2.2  Helium-cooled fast reactors

In contrast, the helium-cooled system continued to generate interest, and it is still under examination both in Europe and in the USA as an alternative (or 'back-up') to the sodium-cooled fast reactor. With a *neutron transparent* coolant such as helium, a good nuclear performance with high breeding gain is achievable without having complex fuel and fertile material arrangements in the reactor core or postulating the development of fuel materials such as uranium/plutonium carbides. Changes in coolant density have negligible nuclear effects and so do not cause power transients. Failure of the pressure circuit might, however, result in problems of safe removal of the residual heat when the system is depressurized. The resolution of the problem of depressurization clearly was basic to the further development of ideas on GCFRs. Whereas the sodium fast reactor system (LMFBR) can claim to have a large heat capacity in its sodium circuit as well as a significant natural circulation capability in the event of failure of the forced circulation system, gases do not have the heat capacity of liquids, and although natural circulation can be an effective means of removing decay heat when the system remains pressurized, should pressure be lost the capability is very limited. This means that the reactor proposals must rely very heavily on what are often called *engineered safeguards*, that is, careful provision of back-up cooling systems with supplementary power supplies to ensure maintenance of adequate coolant flow in all circumstances. Even these, however, fail to engender the confidence necessary if major coolant pressure envelope failures have to be considered, and it was not until the early 1960s, when in the context of the gas-cooled thermal reactor programme the prestressed concrete reactor vessel (PCRV) was evolved, that an acceptable solution became available.

The redundancy which can be built into a PCRV is considered to reduce the prospect of gross circuit failure to a negligible level, and the problem becomes one of ensuring that the inevitable *closures* or access points to the PCRV (for boilers, circulators, fuel handling, etc.) are in a form which will restrict any failure to a small opening. Only a slow rate of depressurization needs then to be considered, and there is then more time in which to introduce the necessary back-up cooling facilities.

As well as aiming to exploit the engineering technology acquired from the gas-cooled thermal reactors, the currently popular version of the GCFR endeavours to utilize fuel technology gained from the liquid-metal-cooled fast reactor programme. The possible effects of the high external coolant pressure of gas compared with sodium on fuel can behaviour (e.g. creep collapse failure) are eliminated by a pressure-equalization system which ensures that there is only a small differential pressure between the main coolant stream and the interior of the fuel pins. The individual pins are attached at their fixed end to a fuel assembly base plate which contains galleries connecting each pin to an extract system (See Fig. 7.1 for an illustration of the principle). By suitable selection of the position of the common interconnection point between main coolant and the fuel pin extract system, the desired differential is maintained over the required range of flow conditions, and in addition, the fission-product gases emanating from the fuel are prevented from building up pressure within the pin. Each pin has an absorption bed built into it to provide a trap and delay feature,

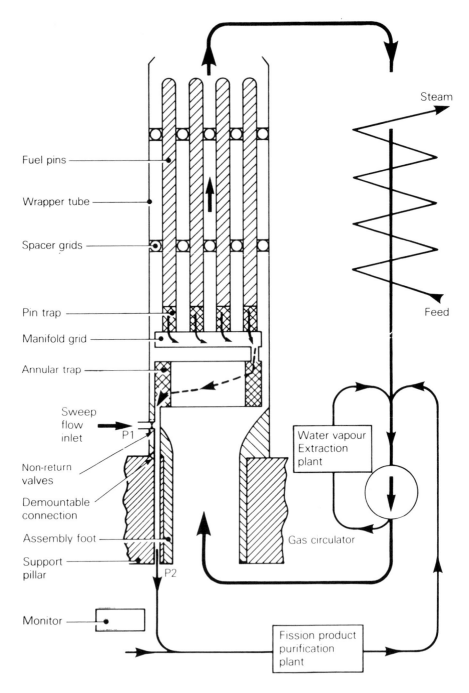

**Fig. 7.1 Gas-cooled fast reactor—balanced pressure fuel pin principle**

and the fission products which pass beyond that point are dealt with in the clean-up circuit. The proponents argue that by adopting this scheme LMFBR fuel technology, combined with other materials experience, can be extensively used—so much so that the fuel arrangement can, by a 'synthesis' technique, be endorsed to a sufficient degree to justify going to a demonstration unit of substantial size (300–600 MW(e)) as the next move. From the point of view of timescale of introduction, this is important to the GCFR as otherwise it could lag very seriously behind the LMFBR.

Single-pin experiments have been conducted in the US sufficiently long for high burn-ups to be reached in tests in EBR II (13 per cent in one case). These tests have shown the effectiveness of the fission-product trap within the pin and encouraged hope that blockages of the balancing system will not occur. In addition to the US tests, a major test of a 12-pin assembly is taking place under joint German/Belgian sponsorship in the BR-2 experimental reactor which is expected by 1981/2 to yield further important information, having already reached a burn-up of 28 000 MW d/t.

Should the results be favourable, perhaps the main uncertainty would then be the radiation damage effects in the fuel assembly. The high-energy neutron doses arising in fast reactors can cause significant *swelling* of structural materials and the extent of swelling is very sensitive to temperature. A GCFR has a very *hard* neutron spectrum, which gives it good breeding properties, but results in high damage doses. There are temperature differentials in a GCFR fuel assembly of suffecient size for differential swelling effects to be a potential source of trouble. Even though a neutron-transparent coolant such as helium gives the designer useful freedom for allowances on fuel pin spacing and mountings, the constraints are still sufficient for this interaction effect between components of the fuel assembly to remain as one of the major issues.

Choice of materials to minimize swelling effects is clearly important, but in compensation, the GCFR's potential for selection is made easier by the absence of the graphite, which in thermal gas-cooled systems is the dominating factor in determining coolant composition. Conditions can be selected which give a relatively benign condition, and which should also ensure that even with 'leaky' fuel pins there is no serious attack on the oxide pellets.

Over the last ten years two major studies have been made of the helium-cooled GCFR using balanced-pressure fuel, in Europe and in the US. Both use the *pod* concept of PCRV used in UK advanced gas-cooled thermal reactors of the Heysham–Hartlepool type (AGRs), and proposed for the various HTR schemes. Both also, in order to attain the fuel ratings considered necessary for a fast reactor, have to invoke coolant pressures significantly greater than those on which there is thermal reactor experience (about 10 MPa compared with 4 MPa in AGRs). The version investigated by the Gas Breeder Reactor Association (GBRA) uses electrically-driven circulators—following UK experience—and an upward flow of coolant in the reactor core, which is intended to exploit natural circulation potential as fully as possible. The designs in the US (carried out by the HTR proponents, General Atomic) not surprisingly followed their HTR scheme of steam-driven circulators and had a downward flow core, with fuel hanging freely from an above-core structure. This latter feature was adopted because it was considered that in the event of a coolant failure extreme enough for the fuel to melt, it would be able to fall away freely to a post-failure heat removal and dilution bed provided in the bottom of the reactor cavity of the PCRV. This, however, militated against use of natural circulation, and following a re-evaluation of the US programme, it has been decided to go to an upward-flow core arrangement, and to adopt electrically-driven circulators. This makes it possible to carry out full-flow commissioning tests on the coolant circuit in its complete form as well as on the components. Thermal reactor experience has shown only too clearly the importance of such tests, for example, for vibration checking.

The general layout of the latest GBRA scheme, shown in Fig. 7.2, typifies the upward flow version, and a full description is given in reference (1). The US design is reported in reference (2). As with HTRs, there have been studies of a GCFR directly coupled with a gas turbine. The schemes have been subjected to first-stage safety analyses, from which it is claimed that the differential pressures in the system would not lead to disruptive and dangerous effects in the event of internal duct failures; but if, as seems to be favoured, the alternator is mounted externally to the pressure vessel, the reliability and integrity of the seal on the turbine–alternator shaft would be important in establishing safety.

Figure 7.3 shows a simplified flow diagram for the steam-generating version, and it will be seen that engineered safeguards to maintain coolant flow and residual heat removal are extensive. Although natural convection will remove decay heat with the system still pressurized, there are special provisions for supply of electricity to auxiliary motors on the main circulators by generators continuously operating, which can provide the necessary low power from the residual heat generation, while there is a further emergency decay-heat removal system which can also be invoked. These provisions are made to cover depressurization and loss of circulator power accidents.

There are primary and secondary shut-off control rod systems, in which the absorbers are actuated by two different types of mechanism to give diversity. The secondary units are of the more unconventional type. They have in their outlet end a *fuse* which, on detection of unduly high outlet temperatures, will melt and release the absorber material, thus providing an *independent* shutdown method. The proponents of the GCFR argue that by using a gas, a material that is in its 'highest' phase, they avoid effects due to phase changes on reactivity and on heat removal and flow conditions. They argue furthermore that since the effects of coolant density on reactivity are small (which incidentally also rules out the danger of prompt criticality accidents), they are able to exploit the substantial temperature increases permissible as a monitoring feature. Other calculations of the effect of water

1 Reactor core and blanket     6 Lower cavity

2 Steam generator unit     7 Helium and fission gas treatment plant

3 Circulator unit     8 Shielding

4 Emergency cooling loop     9 Pre-stressed concrete reactor vessel

5 Refuelling pantograph

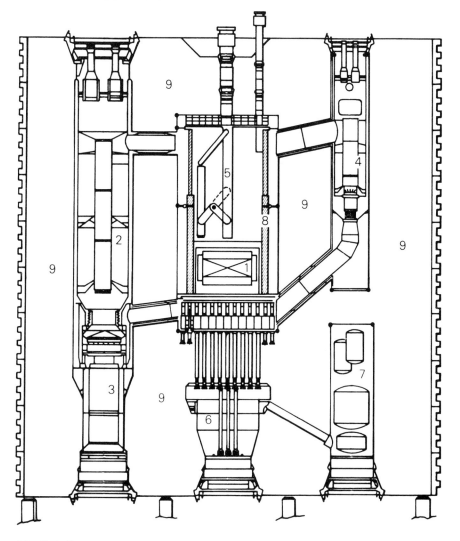

Fig. 7.2 Gas-cooled fast reactor—nuclear steam supply system

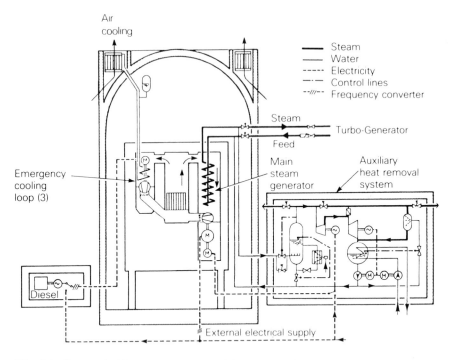

Fig. 7.3 Gas-cooled fast reactor—general heat removal system

ingress into the reactor core due to steam generator leaks into the coolant have shown that the change in reactivity over the likely *rapid-event* range is small. This avoidance of rapid reactivity variations eases considerably the duties required of the engineered safeguards, particularly when combined with the slow rate of depressurization conferred by the PCRV. If the further studies in progress can give confidence that faults in one fuel assembly cannot propagate across the core, it may well be possible to avoid completely the eventuality of core meltdown. Even so, it is considered prudent in current schemes to include the investigation of meltdown accidents and to provide a secondary containment.

A further argument advanced in favour of the GCFR is the forecast capital cost. Because there is no need for an intermediate circuit, the general plant arrangement can be very similar to that of a gas-cooled thermal reactor. Some items can in fact be eliminated (for example, the graphite stack), and the core cavity in the PCRV can consequently be smaller. These potential gains may be reduced by the need for a *core-catcher* and a secondary containment, but it is contended by those who have made detailed analyses that the net effect is a capital cost very similar to that of a gas-cooled thermal reactor. This, according to recent studies, is about 20 per cent above that of an LWR. Given the realization of the fuel cycle cost savings of a fast reactor relative to a thermal reactor as uranium prices rise, this modest capital cost penalty would soon be offset, and the GCFR would then be in direct economic competition with LWRs.

Table 7.1 gives typical parameters of a current GCFR scheme. It will be seen that it has a high breeding gain, but that, because of the heat removal limitations of gas even when highly pressurized, the fuel inventory optimized against an overall cost criterion is greater than that of an LMFBR. The net result, nevertheless, is still to give the GCFR an attractive doubling time. The higher fuel inventory has its direct effect on the rate at which reactors could be introduced in the earlier stages of a programme where the fuel is supplied from an initial thermal-reactor-generated plutonium stock.

Table 7.1 Parameters for 1200 MW(e) GRB4

| | | |
|---|---|---|
| Temperature (core outlet) | °C | 560 |
| Temperature (core inlet) | °C | 260 |
| Coolant pressure (core inlet) | MPa | 9.0 |
| Core pressure drop | MPa | 0.24 |
| Pumping power | MW(e) | 124 |
| Net efficiency | % | 35 |
| Peak linear rating | $kWm^{-1}$ | 40 |
| Mid-cycle fissile enrichment | % | 13.2 |
| Peak burn-up | MWd/t | 100 000 |
| Peak fast neutron fluence | $10^{-27}$ $m^{-2}$ | 2.5 |
| $(E > 0.1 \text{ MeV})$ | | |
| Refuelling interval (0.75 LF) | y | 1 |
| Core fuel in-pile time (0.75 LF) | y | 3 |
| Burn-up reactivity | % | 0.6 |
| Start-up fissile core inventory | kg/MW(e) | 3.92 |
| Breeding ratio | — | 1.40 |
| System doubling time | y | 11.8 |
| Net fissile Pu production | $kg \, MW(e)^{-1} \, y^{-1}$ | 0.287 |

## 2.3 Alternative forms of GCFR

The parameters chosen for the helium-cooled GCFR were selected at a time when rapid rates of expansion in electrical power requirements across the world, particularly the developing countries, were forecast, and the desire was to be competitive with the LMFBR in terms of potential installation rate. Low fuel inventory was a key factor. However, the substantial reduction in forecast demand which has taken place in the last few years has led several organizations to consider whether some of the requirements could be relaxed, and if so, whether a version of the GCFR more closely akin to current AGR thermal reactor technology would provide a valuable initial stage, even if the helium version already described were to be the one ultimately required. Although little has yet been published on this work, it is clear that a concept with quite attractive breeding capabilities is possible. However, restriction of the pumping power and gas pressures to AGR levels, as well as using $CO_2$ as the coolant, means that the fuel inventory is at the high end of the fast reactor scale. This would restrict the rate at which such units could be introduced into the generating system. Proponents, however, argue that feeding these reactors with enriched uranium as well as plutonium will not only overcome the uranium difficulty, but that doing this would, in the long run, achieve valuable savings in total demand, because these reactors would displace thermal systems which are less efficient in fuel utilization. The fuel for such a system would be sealed pins, again in an aim to use existing technology as far as possible, but there would be a need to demonstrate an adequate solution to the conflicting effects of the external pressure on the fuel clad during reactor operation, and the danger of clad failure due to internal fission gas pressure during a depressurization accident.

## 2.4  Extensions of the GCFR

These considerations are linked, of course, to the combinations of thermal and fast reactors based on the Pu–$^{235}$U–$^{238}$U fuel cycle, which have been studied with the aim of not only minimizing cumulative uranium demand, but also the peak annual demand for uranium. It is, however, also necessary to consider the contribution which thorium supplies could make, and General Atomic in the US, in particular, has examined in detail in a number of studies the so-called 'symbiosis' of fast–thermal systems, the latter using $^{233}$U generated from thorium in fast reactors. The basic physics characteristics of the $^{233}$U–Th cycle under fast reactor conditions do not challenge those of the Pu–$^{238}$U cycle, so there is not the incentive to employ it in the core of a fast reactor, but generation of $^{233}$U in the Th-blanket of a fast reactor, the core of which operates on the Pu–U cycle, is an alternative which has always attracted interest. The systems studies indicate that one suitably designed GCFR with a thorium radial blanket could supply the $^{233}$U necessary to maintain the fuelling of several HTRs operating on the $^{233}$U–Th cycle. In the example taken, the plutonium generated in the GCFR axial blanket would be sufficient just to maintain the fuel requirements of the parent reactor. The ratio of power obtained from HTRs to that in the GCFR is principally a function of the equivalent breeding ratio of the fast reactor and the HTR conversion ratio chosen. Figure 7.4 taken from reference (3) shows a flow sheet for three HTRs coupled to one GCFR. To reach this ratio of thermal reactor power to fast reactor power calls for designs of both reactors to have nuclear performances at the high end of the scale. It also clearly involves the interrelation of two processing and fuel fabrication plants. Furthermore, close integration of the operating programmes of the two reactor types is needed if the requisite fuel flow rates are to be maintained. This will be particularly so in the early stages of a programme of reactors until the averaging effects of a large number of units can smooth out variations.

The good nuclear qualities of the GCFR have led to two other favourable features being pointed out by its proponents. It is possible to design a GCFR core which, within itself, generates more plutonium than it burns. This means that there is no need to refuel the core for reactivity reasons until very

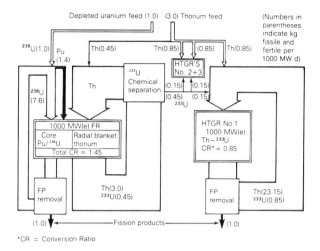

(Numbers in
parentheses
indicate kg
fissile and
fertile per
1000 MW d)

Fig. 7.4 Gas-cooled fast reactor—fast reactor/thermal reactor combination

high burn-ups have been achieved: $2-3 \times 10^5$ MW d/t is possible before the reactivity worth of individual fuel assemblies becomes too low. To do this requires very large cores with a high total fuel inventory, and this would penalize their introduction. But even if fuel element damage aspects prevent the very high burn-ups being reached, a fuel cycle can be envisaged which has interesting and attractive possibilities. Fuel withdrawn from the core will have a higher plutonium content than is needed for fresh feed fuel. In the reprocessing therefore, it requires to be diluted rather than—as is the case in all other fast reactor concepts—have fissile material added. This can be done by adding natural or depleted uranium, or the material from the blankets, and there is no need at any stage to have plutonium as a separate material. The surplus 'diluted' material could go towards fuelling a further reactor.

The other claim for the GCFR is that it can be designed to *incinerate* the higher actinide waste products. It has sufficient spare neutrons to deal with the fission products which are difficult to separate from the actinides in the reprocessing cycle, and in its hard spectrum the actinides can be made to serve as a fissile fuel. When power demands were levelling out, and at a time of plutonium plenty, it could be possible either to modify GCFR core designs in power reactors for this function, or alternatively to operate special *burner* reactors.

## 2.5 Other gaseous coolant schemes

So far, this discussion of gas-cooled fast reactors has kept to the use of 'conventional' gases such as helium and $CO_2$. Attempts have been made in various ways to improve both the heat-transfer and the heat-transport properties of gases. The use of solid particles to 'load' the gas was at one time a popular approach. The search was for a combination in which dust neither deposited in undesirable places nor caused damage to the circuit materials by erosion or corrosion. No specific proposals ever emerged from this work.

The Russians have explored extensively a very different approach. They have examined the possibilities of using a gas in which chemical dissociation and reassociation occurs in the temperature range of interest. The heat involved in these reactions can be regarded as the equivalent of a latent heat, and the heat-transport properties are significantly enhanced. For their investigations they have chosen the reactions $N_2O_4 \rightleftharpoons 2NO_2 \rightleftharpoons 2NO + O_2$, the dimerization on the left of the equation being considered a useful contribution to the overall effectiveness, even though it occurs at the lower end of the practicable temperature range. These nitrogen oxides can be used as the working fluid of the system, and the Russian schemes envisage a direct cycle. The thermal characteristics of the oxides are such that the condensing cycle which gives the best efficiency can nevertheless be arranged so that all heat transfer in the reactor is in the gaseous phase. Avoiding two-phase conditions eliminates problems due to change of cooling regime or reactivity fluctuations due to big coolant density changes. Table 7.2 gives the main parameters of two alternative cases. The important thermodynamic point which had to be established was whether the reactions proceeded sufficiently rapidly for the various equilibria to be established at the appropriate points in the cycle. A test loop supported the predictions that this would be so, and it is not now considered to be a problem by the proponents.

Not only does the cycle gain in being direct by reducing the number of components in the system, but turbines are about five times smaller than steam turbines of equivalent power output and the good heat-transfer properties keep down the size of the heat exchange units outside the reactor. However, in order to obtain the desired efficiency, the cycles would in some cases be using top pressures as high as 17.0 MPa. The safety studies have tried to make as much use as possible of the reserves of liquid which a condensing cycle contains in order to maintain coolant flow both in the event of loss of circulating power and of circuit fracture. Leaks are said to be obvious, even when very small, because of the distinctive colour of the coolant.

The Russians have devoted considerable effort to materials investigations, and claim good results without having to incorporate exotic alloys. The 'gas'

Table 7.2 Characteristics of thermodynamic cycles of $N_2O_4$ nuclear power plant

|  | Case 1 | Case 2 |
|---|---|---|
| Reactor thermal power, MW | 3253 | 2400 |
| Electrical output, MW | 1030 | 1100 |
| Gas temperature at reactor inlet, °C | 200 | 370 |
| Gas temperature at reactor outlet, °C | 450 | 570 |
| Gas pressure at reactor inlet, MPa | 17.1 | 13.7 |
| Cycle efficiency, % | 32.3 | 47.4 |
| Net station efficiency, % | 30.7 | 44 |
| Both are condensing cycles (condenser pressure in Case 1 is 0.21 MPa) | | |

is subject to radiolysis, some of the products being incondensible gases which must be extracted if they are not to interfere with heat transfer in the condenser. Rigid exclusion of water vapour is of major importance to avoid acid formation, and the experiments indicate that certain corrosion products which form precipitate from the gas phase and must be filtered out. The fuel concept to which particular attention has been given is of the matrix type, $UO_2/PuO_2$ particles being embedded in a metal matrix, which is then clad in stainless steel. This gives double protection against fission-product leaks, a feature of considerable advantage in a direct-cycle system. Although this results in significant neutron losses in structural material, the good fuel conductivity, together with the good coolant properties, permits high core power densities and fuel ratings. The operating temperatures are low enough for uranium metal to be considered for the blankets. Very high core volumetric ratings are claimed—up to 100 MW m$^{-3}$—with core fuel inventories for a 1500 MW(e) system of 1600—1900 kg of Pu. The breeding ratio quoted is 1.5 to 1.6, so the doubling time is as low as five years.

The Russian programme has been in progress since the 1960s. Reference (4) gives a detailed account of work to 1969, and the most recent presentation was to the IAEA Conference at Salzburg in 1977.[5,6] It is not known whether any decisions have been made about a test reactor.

# 3 Liquid coolants

The only hydrogen-containing liquids other than water that could act both as a coolant and moderator and which have been put to practical test have been certain organic materials. Although there was some interest in them for large power units, one of the significant factors influencing their initial consideration was the search for a marine reactor. Any alternative to a water system had to match it for size and weight, which made hydrogen moderation an essential. At the time that the interest was shown the relative economics of oil firing—with cheap oil—and nuclear power were the vital criteria. When compared with water, the improved efficiency and lower operating pressure which these compounds offered were in their favour.

The organic materials selected were the polyphenyls, in particular the terphenyls, which occur as three isomers and are a by-product of the manufacture of biphenyl by pyrolysis of benzene. The commercial mixture mostly considered (for economic reasons) was Santowax R. Organic materials, because of their structure, are very susceptible to breakdown as a result of radiation. The terphenyls decompose to hydrogen, to some light fractions, and, of particular importance, also to heavy ones, but it seemed from the early work that the rate of decomposition was not excessive. Furthermore, the presence of the products reduced the rate of decomposition as the level built up, the heavier products remaining soluble in the original material to a considerable degree. Although it was necessary, therefore, to provide a coolant treatment plant, and to provide a steady supply of make-up coolant, the requirements did not in the initial investigations seem unacceptable. The chief concern, not resolved in the earlier stages, was fouling of heat-transfer surfaces by deposition of decomposition products. The physical properties of the terphenyls allowed the use of temperatures up to about 400 °C for the maximum heated surface, with only modest pressurization of the circuit. The heat-transfer qualities were significantly poorer than water, but the higher temperatures differences permissible were an offset against this handicap. Plate arrangements of fuel with high surface area were one design solution explored.

A wide range of standard structural materials for the circuits showed excellent compatibility, and the principal problem was that of finding a material compatible with the hydrogen present in the coolant which also had sufficiently low neutron-absorption properties to be usable for fuel canning and core structure. The early work tended to rule out zirconium because of hydriding effects, and this search eventually led to a proposal to use SAP (sintered aluminium product) which, despite its limited ductility, appeared to be a possibility. The chemistry and materials aspects of organic coolants are set out in reference (7).

Interest, in fact, grew to the level where a reactor experiment (OMRE— organic moderated reactor experiment) was built and operated in the US at Idaho Falls for several years. A small power demonstration reactor (PIQUA) was also built. It became increasingly clear, however, that the economic argument did not provide sufficient incentive. The cost of replacement coolant and the rather expensive form the fuel had to take turned out to be significant factors, and work in the US drew to a close about the mid-1960s.

However, in the meantime, the Canadians had developed an alternative approach, which they followed up at the Whiteshell Establishment of Atomic Energy of Canada Ltd. The prime Canadian interest has always been in reactors which can be fuelled with natural uranium, and, as one of their exploratory investigations, they were interested in ascertaining what might be used in place of heavy water ($D_2O$) as a coolant, while still retaining natural uranium as the reactor fuel. Their concept was a pressure-tube reactor with a terphenyl cooling, and heavy water moderation in a standard calandria layout. This had the additional advantage that only a small volume of coolant was effectively under neutron radiation (a volume not much greater than that in the pressure tubes). Consequently, the decomposition, rate, and therefore the processing and replacement costs for the organic coolant, were very much reduced compared with those of a pressure-vessel type. If however the calandria design was to be able to use natural uranium fuelling, a canning and pressure-tube material with the required low neutron-absorption properties had to be established. Experiments were conducted in and out of radiation fields, beginning with SAP as the selected material, but as the work progressed it was found that careful control of impurities and 'treatment' of the terphenyl by certain additions avoided fouling, and that zirconium alloys could be used at surface temperatures up to 500 °C. A test reactor has operated successfully at Whiteshell. This work was reported to the 1964 Geneva Conference.[7] Attractive though the ultimate version seemed to be to its supporters, and even though in Italy there was also sufficient interest for a similar scheme called ORGEL to be prepared, there never have been proposals for moving towards a power reactor.

# 4 Fluid-fuel reactors

## 4.1 The general philosophy

Fluid-fuel reactors represent a very different and radical approach. Not only are there possibilities of the general benefits which have been referred to in the introduction to this chapter, such as elimination of core heat-transfer problems and fuel fabrication, and unlimited burn-up, but also there are some more specific potential gains.

For example, expansion due to a rise in temperature in the reactor core reduces not only fluid density, but also the amount of fissile material in the core—thus reducing reactivity. The system offered the prospect therefore of being self-regulating, and the reactor experiments that were operated showed that the classical control rod absorber system was not necessary. Nor was it considered necessary to have a shutdown system. The combination of self-regulation and the ability to dump fuel to a previously prepared storage system, with a geometry designed to render criticality impossible and which also has built-in independent heat removal, was considered the better arrangement from a safety point of view. However, although using a self-heated fluid may perhaps overcome heat-transfer problems within the reactor core, the heat generated still has to be transferred through a heat-exchange system to the working fluid or a secondary coolant. Again, although the inventory of fuel in the core itself can be small, it has to be remembered that the fluid fills the whole of the primary circuit, and if care is not taken to keep its volume to a minimum, fuel inventories can become excessive. Furthermore, fission products will be released in the fuel solution, and the materials of the primary circuit must be compatible with the resulting mixture. It may be necessary to process the fuel solution for this reason, in addition to any requirements imposed by neutron economy or other fuel cycle aspects. Finally, the whole of the circuit will become highly active owing to fission products and also as a result of some activation due to delayed neutrons.

Formidable though these problems look, the potential attractions of low fuel-cycle costs were sufficient to engender considerable activity, particularly during the 1950s, extending in one instance to the mid-1970s. The principal work was in the US, and a major incentive was provided by the interest in the use of thorium as fertile material in thermal reactors. By the time the limiting features and criteria had been properly understood and applied, only the three systems described in §§4.2–4.4 survived to any significant extent.

In a thermal reactor the $^{233}$U–Th cycle has little margin before it ceases to breed, and all the schemes have, of necessity, aimed for the utmost neutron economy. Provision has to be made for maintaining fission products at a low level by continuous processing where possible—xenon is particularly important in this respect. In some cases too the protactinium (Pa) product intermediate between thorium and $^{233}$U (which has a 27 day half-life) has to be removed, as neutron capture in Pa not only prevents $^{233}$U being formed, but loses a further neutron from the cycle. This effect is proportional to neutron flux level, and as fluid-fuel reactors aim to have very high in-core fuel ratings, it is important to minimize the extent to which protactinium is present in these high-flux regions. This was one of the reasons why all the early concepts envisaged a two-zone reactor with fuel only in the core solution, which was fed from a thorium-containing blanket via a suitable processing system (a layout comparable with that in a fast reactor).

## 4.2  The liquid-metal thermal reactor

A system based on the use of a fuel solution of $^{233}$U in bismuth was proposed at the first Geneva Conference of 1955. Not only has bismuth a low neutron absorption, it is one of the few metals which will dissolve uranium to the concentration desirable from a physics point of view. The reactor proposed was again of the two-zone type. The moderator was graphite and this was also the material proposed for the boundary between core and blanket. There is no solution of thorium in bismuth which would give the fertile material concentration which the neutronics needed, and so a slurry of thorium bismuthide in bismuth was first postulated. Clearly, with a slurry of this kind where there is some solubility of the thorium component, careful control of conditions is highly important, and indeed at a later stage attention was turned to slurries of thoria ($THO_2$) in bismuth. Mass transfer of constituents of circuit materials from hot to cooler regions is a general problem with bismuth. Stainless steels cannot be used because the phenomenon is particularly serious with nickel—and even with Cr–Mo steel the bismuth has to have an oxygen getter (Mg) and an inhibitor (Zr) added to it. Bismuth is, of necessity, present in large quantities in the core and under irradiation it yields a polonium isotope with $\alpha$-and $\gamma$-activities comparable to plutonium. These, and other technological problems, together with a limit of 1.05 on the breeding ratio even with favourable assumptions, led to work on the system being discontinued in the late 1950s.

## 4.3  The homogeneous aqueous reactor

One system which did however reach the reactor experiment stage at the Oak Ridge National Laboratory in the US was the homogeneous aqueous reactor (HAR). This also was a two-zone reactor which used $D_2O$ as the solution base and moderator to give good neutron economy. One of its principal problems was the need to operate it at about 14.0 MPa to achieve reasonable efficiency, and any leaks that might arise from such a high-pressure active circuit could clearly cause difficulties. Viewed from the present day there would be many questions to be asked about the large primary circuit with its big inventory of *free* fission products, and the consequent containment requirements. Furthermore, if rapid depressurization were needed, blowing down (and cooling) the fluids would present considerable difficulty. The power reactor had as core fluid a solution of uranyl sulphate ($UO_2SO_4$) in heavy water ($D_2O$). Once again, the blanket fluid had to be slurry, this time of thoria ($THO_2$) in $D_2O$, because there was no solution of thorium which gave the concentration needed for the requisite nuclear performance. The 5 MW (thermal) reactor experiment which was run from 1957 to 1961 had a Zircaloy-2 core vessel, the outer (strength) vessel being of carbon steel lined with stainless steel. From an operational point of view, the maintenance of pressure balance between the core and blanket circuits is essential if the core vessel (which has to be thin-walled for reasons of neutron economy) is not to be damaged. The reactor was known as HRE-2, its main parameters are given in Table 7.3.

Table 7.3 Design parameters of HRE-2 ($D_2O$ blanket)

|  | Core | Blanket |
|---|---|---|
| Power, heat, kW | 5000 | 220 |
| Pressure, MPa | 13.8 | 13.8 |
| Vessel: |  |  |
|   inside diameter, m | 0.81 | 1.52 |
|   thickness, mm | 9.52 | 111.8 |
|   material | Zircaloy-2 | Stainless steel clad carbon steel |
|   volume, m$^3$ | 0.290 | 1.55 |
| Specific power, MW m$^{-3}$ | 17 | 0.14 |
| Solution | $UO_2SO_4$–$D_2O$ | $D_2O$ |
| Uranium concentration, g $^{235}$U/kg $D_2O$ | 9.6 | 0 |
| Circulation rate, m$^3$ s$^{-1}$ | 0.0252 at 256° C | 0.0145 at 278° C |
| Inlet temperature | 256° C | 278° C |
| Outlet temperature | 300° C | 282° C |
| Volume of gas generated, m$^3$ s$^{-1}$ at s.t.p. | 0.0272 | 0.00037 |

HRE-2 encountered many problems. There were stress-corrosion troubles during commissioning, and operation at power had only taken place for a short time when the core vessel was breached. This was eventually attributed to poor flow distribution in the vessel, which allowed some regions of the core solution to overheat. At just below 300 °C an aqueous solution of $UO_2SO_4$ separates into two phases, one of which contains a high concentration of $UO_2SO_4$. This could form in local regions of poor circulation causing the temperature to rise progressively. Plating-out could occur, and the resulting high local heat generation would damage the vessel. This is, in fact, what happened in HRE-2, and is a clear illustration of one of the difficulties of a fluid-fuel system—keeping track of the fissile material and ensuring no 'hide-out' or local concentration in taking place.

Although the processing of the thoria from an HAR blanket could follow conventional solvent extraction lines, treatment of the core solution for fission-product removal was not easy. The high-energy deposition from the fission fragments, all of which is released in the solution, caused considerable radiolytic decomposition of the $D_2O$. While this helped in sparging of the rare gas fission products, it resulted in a need for a recombination plant of substantial size. High efficiency was also needed if there were not to be undue losses of $D_2O$. Corrosion products and some of the fission products are basically insoluble in water and the hydraulic cyclones proposed for extracting the particulates did not turn out to be as efficient as had been hoped. It was not really surprising therefore that, although the core vessel was repaired after the breach in 1958 (a considerable feat in itself and illustrative of what can be accomplished by remote techniques), the reactor only continued in rather limited operation as a single-zone system until 1961, when the programme was terminated.

In parallel with this there had always been interest in the possibilities of a single-zone reactor with a slurry which combined fuel and fertile material. There was an early US proposal for a power reactor based on this principle, which never materialized. Nevertheless, studies of the chemical and materials aspects associated with these slurries continued, and a subcritical assembly was built in Holland in the early 1960s. It was intended that this would be followed by a low-power test reactor, but interest gradually declined and it now seems that no work is being done.

### 4.4  Molten-salt thermal reactors

The fluid-fuel reactor which enjoyed the most extended run of investigatory life was the molten-salt system. This began with an aircraft reactor experiment, which operated successfully in 1954 in a 'proof-of-principle' short-term test at a power level of 2.5 MW(t) and at temperatures up to 860 °C. The fuel was a solution of $UF_4$ in other fluorides with Inconel-clad beryllia as the moderator. Those who took part in this programme realized from the early stages that, potentially, the molten-salt system was a candidate as a power reactor, and a group at ORNL began work in 1957. A further reactor experiment (MSRE) was built and operated successfully at a power level of 7.3 MW(t) until the end of 1969. Once again, the original concept was of a two-zone system. As ideas progressed, however, and particularly as new processing methods which gave promise of rapid extraction of protactinium were developed, the simpler single-zone system which avoids the core–blanket interface vessel became the preferred design. Table 7.4 gives a selection of the relevant parameters.

Figure 7.5, taken from reference (9), is a simplified flow diagram of this

Table 7.4 Characteristics of a 1000 MW(e) molten-salt breeder reactor

| | |
|---|---|
| Reactor thermal power, MW | 2250 |
| Net electrical output, MW | 1000 |
| Overall plant efficiency, % | 44 |
| Fuel salt inlet/outlet temperatures, °C/°C | 566/704 |
| Coolant salt inlet/outlet temperatures, °C/°C | 454/621 |
| Steam conditions | 24.3 MPa. 538 °C |
| Core height/diameter, m/m | 4.0/4.3 |
| Salt volume fraction in core, % | 13 |
| Salt volume fraction in under-moderated regions, % | 37 and 100 |
| Average core power density, MW m$^{-3}$ | 39 |
| Estimated core graphite life, years | 4 |
| Total salt volume in primary system, m$^3$ | 48.7 |
| Thorium inventory, kg | 68 000 |
| Fissile inventory, reactor plus processing plant, kg | 1470 |
| Breeding ratio | 1.06 |
| Exponential doubling time at 80% LF, years | 21 |

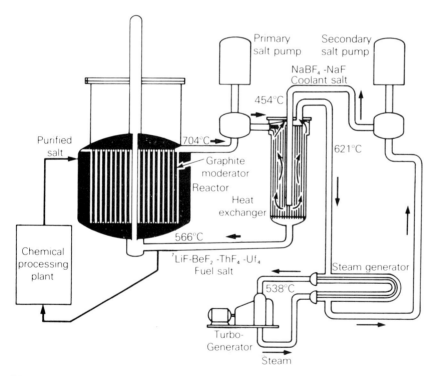

Fig. 7.5 Molten salt breeder reactor—flow diagram

version. A separate moderator has to be provided and it consists of bars of graphite, through which the salt circulates. These bars must be removable and replaceable as they suffer considerable radiation damage. By varying the graphite/fuel-salt ratio across the core, the outer region can be made to function as a form of blanket—that is, most of the neutrons are captured in thorium. The fuel salt is $^7$LiF–BeF$_2$–ThF$_4$–UF$_4$ (note the need for isotope-separated lithium for neutron-economy reasons), which has to be pre-heated to 500 °C to melt. The heat produced is transferred in external heat exchangers to a secondary coolant of sodium fluoroborate. This secondary salt was envisaged as generating supercritical steam with a top temperature of 538 °C. Helium is injected into the fuel salt to assist centrifugal separation of the fission-product gases. A small side-stream of fuel salt is passed through a processing plant which has the important duty of removing protactinium as rapidly as possible, so keeping the protactinium level in the circuit low; the processing plant also has the duty of removing the salt-soluble fission products. This plant represents a major engineering problem in itself as the conditions are arduous and the choice of materials compatible with the fluoride salts and the liquid-metal extraction solutions which have to be used is very limited. Molybdenum may be the only possibility.

Figure 7.6, also taken from reference (9), is an excellent illustration of the basis on which a thermal reactor using the $^{233}$U–Th fuel cycle has to work if it

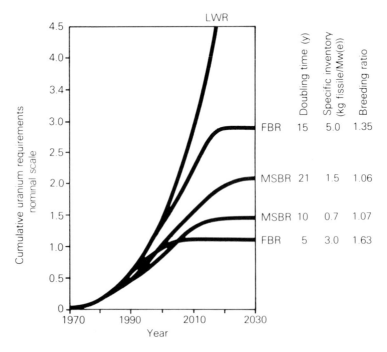

Fig. 7.6 Molten salt breeder reactor—effect of breeder characteristics on uranium requirements

is to breed successfully. The breeding ratio is very low at 1.07, but he low fuel concentrations needed in the solutions mean that inventory is low also. The net performance is very good, but clearly with such a limited breeding ratio the designer is working to close margins. Fission products must be kept to low levels, neutron loses must be kept to a minimum, and protactinium must be kept out of high-flux regions.

It was decided to shut down MSRE after 1969, not because it was no longer operable, but because it was considered it had provided all the information it usefully could. It is of interest that it operated initially on $^{235}$U, then on $^{233}$U, and finally on Pu. There has always been some controversy as to how a $^{233}$U–Th system should be started initially, and the Pu experiment was intended to demonstrate the feasibility of operating with this as the fissile material. Successful though this was in itself, the question of how to operate a fuel cycle is still left unanswered. The Pu can provide neutrons to generate $^{233}$U from the thorium in the salt, which is extracted for use in other reactors, but the chemistry is such that it seems uneconomic to attempt to extract Pu from the salt. It is therefore burnt out as far as possible, and the residue, together with the charge of salt, rejected. This does not seem, therefore, to be a clear-cut solution to the problem of finding an alternative to highly enriched $^{235}$U as starting material.

It is of interest that substantial dismantling of MSRE was achieved very satisfactorily, results being obtained on fission-product deposition and take-up. Although this was only on small-scale plant, the experience from an active-handling point of view was encouraging.

Following shutdown of MSRE, the Oak Ridge programme concentrated on a number of important items requiring development if the system was to be of commercial interest. Materials problems had been encountered. In the graphite, Xe and iodine were being taken up to the detriment of the nuclear performance—a non-wetting graphite was needed. The circuit material specially developed for the system by ORNL (Hastelloy N) was deteriorating in its properties as a result of intergranular attack caused, it was believed, by the fission product tellurium. It was also susceptible to damage from helium formation in a neutron flux. Tritium formed from the primary salt was presenting difficulties as it diffused (as does hydrogen) through the primary and secondary heat exchanger tube walls and led to activity in the steam circuit. Barrier films on the tube surfaces were being tested, the normal oxide film in some cases being effective. In addition, further work in the processing field was clearly needed. The research and development programme continued for a number of years, but as time passed, the scale gradually decreased, until in the mid-1970s the support funding was terminated. Apart from some limited work in France, there is no activity in the fluid-fuel thermal reactor field.

## 4.5  Fluid-fuel fast reactors

In parallel with these thermal reactor activities there was naturally interest in whether a fluid-fuel fast reactor could be developed. Safety was considered to be a particular point in its favour because of the ability to dump the fuel to ever-safe tanks with a prepared cooling system. The solid-fuel fast reactor's melt-down accident was therefore eliminated. Against this, from a perform-ance point of view, was the difficulty that fast reactors require a much higher fissile proportion in their fuel than do thermal systems, so that if external hold-up in the primary circuits was high, inventories could be high with consequent detriment to the doubling time. Among the early contenders was a scheme envisaging the use of $UO_2/PuO_2$ particles in sodium. There were difficulties in avoiding flocculation—additives had to be used—and the general problem of the stability of the slurry under reactor operating con-ditions seemed a formidable one. Interest in this gradually declined, while parallel work at Los Alamos, New Mexico, pursued investigations of molten plutonium metallic alloys contained in cans (or a calandria) cooled by liquid metal. Some of the fluid-fuel advantages were retained, and the disadvan-tages associated with circulating the liquid fuel through an external circuit were eliminated. The alloys considered were eutectics of plutonium with iron, cobalt, and nickel, the chief interest being in the Pu–Fe eutectic with a melting point of 410 °C. These eutectics are highly corrosive, and finding a suitable container material was a major problem. Refractory metals were the only ones possible, and alloys of tantalum with tungstem and yttrium were tried. A 1 MW low-power test unit, the Los Alamos Molten Plutonium Reactor Experiment (LAMPRE), ran for a time around 1960.

As interest declined in these difficult concepts, there was something of a hiatus in the fluid-fuel fast reactor field, until continuing progress on the molten-salt thermal reactor programme led to consideration of whether some form of molten salt might be used. The fluoride salts used in the thermal system were not suitable for nuclear reasons, and interest turned towards the use of molten chlorides. Solutions were available for both fuel and blanket requirements; for example, 60% NaCl–37% $UCl_3$–3% $PuCl_3$ (molar) for the fuel, and 60% NACl–40% $UCl_3$ for the blanket, but the high melting-point of 577 °C puts a premium on an efficient, economic and highly reliable preheating system. The two-zone system was generally considered more likely to achieve the desired nuclear performance. In order to minimize fuel inventory, consideration was at one stage given to a direct-cooling arrangement, in which a curtain of lead drops was delivered to the outer region of the core vessel to act both as a heat-removal medium and as a means of maintaining the required circulation in the core. The difficulties of ensuring proper circulation, and of separating lead and salt, particularly in preventing the latter being carried through the heat-exchanger circuits with consequent 'high-out' possibilities and activation effects, led to this concept being abandoned, and attention being turned to the externally-cooled version (the same as in the thermal reactors). Keeping the circuit volume to an absolute minimum is clearly vital, but the best doubling times that could be achieved were at the higher end of the range (35–40 years) unless separated chlorine ($^{37}$Cl) could be used to improve the nuclear economy by enhancing breeding ratio.

Choice of secondary coolant is important in the externally cooled version, good heat-transfer and heat-transport properties keeping down the sizes of the heat exchangers. Although sodium was suggested at one stage, there were doubts concerning its usefulness because of its reaction with the salts, which could make any leakage a problem. Attention turned to lead, which is compatible with both the salt and with the steam, but which involves development of a further new technology, and restricts the choice of materials. There is also a mis-match in the required salt temperatures (which need to be high to avoid freezing), and in the temperature range suitable for the lead and for the steam generators. This necessitated substantial re-circulation in the lead circuit, thus adding to cost and complication.

The next stage was to turn to gas-cooling as giving greater flexibility. In order to reach the required heat-removal levels, the gas obviously needed to be at high pressure, and for safety reasons, therefore, both the primary and secondary circuits (apart from small auxiliary sections) had to be confined in a pre-stressed concrete vessel. The temperatures are such that a gas-turbine cycle gives good efficiency, and the rather improbable-sounding combination gives a compact layout with useful supplementary features; for example, the primary circuit can always be kept under compression by choice of gas pressure, thus minimizing the chance of failure and the effect of leaks.

As with all the salt systems, there are materials problems, particularly at the high end of the temperature range, and molybdenum or one of its alloys, e.g. TZM (titanium–zirconium–molybdenum), seems to be the only likely possibility as the circuit material. The chemical stability of the salt mixtures seems reasonable, though exclusion of oxygen and nitrogen is important. Sulphur from $^{35}$Cl and some fission products are potential precipitating species. Processing could be carried out by aqueous solvent extraction, at some cost in external hold-up, and causing difficulties if separated chlorine is to be recovered. High-temperature processing has the potential benefits of being close-coupled, of reducing inventory, and of conserving $^{37}$Cl if this is used. It appears feasible but, as with the thermal salt systems, there are very difficult materials and remote-handling operational problems to be solved.

The alternative which has been outlined by Taube[10] is an internally cooled scheme with several zones, in which the heat from the fuel salt is removed by fertile salt circulating in the secondary side of a heat-exchanger system extending through the core region. The fertile salt also forms the blanket. The low fuel inventory means that the potential performance would be high if the problem could be solved of finding a suitable material for the heat-exchange system that will withstand the irradiation conditions and not detract unduly from the nuclear performance. Taube has also pointed out that the high flux conditions needed to give the capability for incineration of actinides and destruction of harmful fission products, e.g. $^{90}$Sr and $^{137}$Cs, could be met by a suitable combination of fast and thermal molten-salt reactors.

## 5  The current position

In this chapter, some of the more unusual proposals for nuclear power reactors have been examined. The aim has been to illustrate the key aspects which led their proponents to formulate them, to see how far the original concepts were realized, and to show what has been the outcome of the investigations into the various systems proposed. Embarking on and demonstrating a new technology on an engineering scale is a daunting project, which quite apart from any technical difficulties can be a major obstacle to commercialization. It now seems that the versions most likely to survive will be those which synthesize a new concept by adapting technologies already being developed.

## Acknowledgement

The author would like to thank the Gas Breeder Reactor Association, the British Nuclear Energy Society, and Pergamon Press Ltd., for permission to reprint illustrations. He would also like to thank P. Haubenrich (author of *Molten Salt Reactor*) for similar permission.

# References

1. GAS BREEDER REACTOR ASSOCIATION. *GBR4 design description.* GBRA (1975).
2. *Gas-cooled fast breeder reactor engineering and design: special issue. Ncl. Engng. and Des.*, **40**(1) (1977).
3. FORTESCUE, P. *Ann. Nucl. Sci. and Engng.*, **1**, 21 (1974).
4. KRASIN, K. A. (ed.). *Dissociating gases as heat transfer media and working fluids in power installations.* Amerind Publishing, Private Ltd., New Delhi (1975).
5. KRASIN, A. K., NESTERENKO, V. B., TVERKOUKIN, B. E., ZELENSKÝ, P. Ph., NAUMOV, V. A., GOL'TSEV, V. P., KOVALEV, S. D., KOLYKHAN, L. I. Basic performance problems and prospects for 1200–1500 MW gas-cooled fast reactors with the dissociating coolant. In *Proc. Conf. on Nuclear Power and its Fuel Cycle-Salzburg May 1977.* (pp. 569–83). IAEA–CN–36/332. International Atomic Energy Agency, Vienna (1977).
6. NESTERENKO, V. B., SHAROVAROV, G. A. KOVALEV, S. D., TRIBNIKOV, V. P. Physical and technical aspects of nuclear and chemical safety of atomic power stations with fast reactors cooled with $N_2O_4$. In *Proc. Int. Conf. on Nuclear Power and its Fuel Cycle—Salzburg May 1977.* pp. 725–37. IAEA–CN-36/322. IAEA (1977)
7. DAWSON, J. K. AND SOWDEN, R. B. *Chemical Aspects of Nuclear Reactors*, vol. 3, sect. 2, p. 38 ff. Butterworths, London (1963).
8. CAMPBELL, W. M. *et al.* Development of Organic Liquid Coolants. In *Proc 3rd UN Int. Conf. on the Peaceful Uses of Atomic Energy, May 1964, A/CONF. 28/P/15,* United Nations (1964).
9. HAUBENREICH, P. *Molten salt reactors—concepts and technology, J. Brit. Nucl. Energy Soc.*, **12**(2), 147 ff. (1973).
10. TAUBE, M. (ed.). *Fast reactors using molten chloride salts as fuel.* EIR Bericht, N.332 Würenlingen (1978).

# 8

# Prospects for fusion

P. A. DAVENPORT

The life span of normal stars is measured in billions of years, during which time they radiate huge amounts of energy. This energy is continuously replenished from exothermic nuclear transformations taking place in the stars' hot interiors and resulting in the process of nucleosynthesis by which heavier elements are formed from hydrogen, the dominant constituent of stars. As this explanation of the source of stellar energy became accepted, the possibility of achieving such thermonuclear reactions under terrestrial conditions presented itself: the surface of the earth is rich in suitable light elements with the potential of providing a virtually inexhaustible supply of energy.

It comes as no surprise that the conditions necessary for the realization of controlled thermonuclear reactions (or fusion reactions as they are now known, since the light elements are fused together) are rigorous and extreme, far beyond the realms of normal experience. The essence of the problem is to heat a small mass of hydrogen isotopes to stellar temperatures, sustaining it in virtually complete isolation until a net energy gain accrues. Necessary criteria for achieving this are that the product of the number density of the reactants, $n$, and the time they are kept isolated, $\tau$, should exceed a certain threshold value, given by $n\tau > 10^{14}$ cm$^{-3}$ s for the most favourable fusion reaction, and that the temperature of the reactants should be at least 100 million K.

Current research is pushing towards these criteria along two distinct paths: (1) the isolation of tenuous hot hydrogen plasma by strong magnetic fields (magnetic confinement) and (2) transient operation at densities many times that of solid hydrogen, achieved by laser induced implosion (inertial confinement).

Progress towards the plasma conditions needed for power production has been steady and rapid, especially in magnetic-confinement systems of the tokamak type. World-wide confidence in the successful demonstration of fusion power is such that in recent years much attention has been paid to conceptual fusion reactor designs, in which the diverse disciplines of plasma physics and electromechanical and nuclear engineering are brought to bear upon the problems likely to be encountered in the realization of a practical and economic fusion power station.

## Contents

# 1  Physics background

The nuclei of atoms are composed of protons and neutrons, collectively called nucleons: the number of protons defines the element and the number of neutrons its isotope. Nuclear energy is liberated when these constituent nucleons are rearranged in nuclear reactions to form product nuclei which are more tightly bound than the reactants. The resulting decrease in the total mass appears as energy according to Einstein's celebrated formula $E = mc^2$. This phenomenon is quantitatively presented in the well-known binding energy curve (Fig. 8.1), which shows the binding energy per nucleon of the

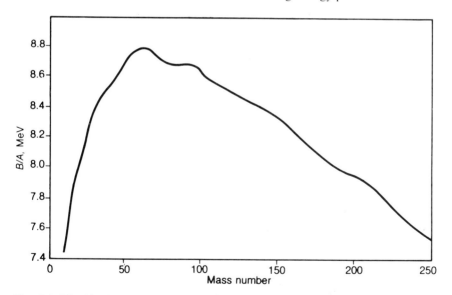

Fig. 8.1 The binding energy per nucleon (B/A) of the most stable isobar of atomic weight A

stable isotopes as a function of atomic weight. From it two things are immediately apparent. Firstly, the amount of energy liberated in nuclear reactions, being measured in MeV, is roughly a million times greater than that in chemical reactions, which is typically in the eV range. Secondly, the most tightly bound nuclei are seen to be those of medium atomic weight, between 50 and 60, so that such elements are the least useful for nuclear exploitation.

Thus nuclear energy can in principle be released to two distinct ways, by breaking up heavier nuclei or by combining those which are lighter; the first process is known as nuclear *fission*, the second as nuclear *fusion*. Fission is the process fundamental to the operation of nuclear power stations; self-sustaining fission processes are essentially man-made phenomena, with one remarkable known exception, the extinct Oklo reactors where spontaneous chain reactions in uranium occurred some $1.8 \times 10^9$ years ago.[1] Nucleosynthesis by fusion is the main energy source of all normal stars: it is truly the universal energy source.

Self-sustaining nuclear reactions are much more difficult to achieve than their chemical counterparts. Fusion reactions are inhibited because nuclei carry positive electric charge and are thus constrained by electrostatic repulsive forces (the *Coulomb* barrier) from approaching each other closely. For fusion to occur the nuclei approach each other with sufficient speed (see Fig. 8.2) for the Coulomb barrier to be penetrated, after which the short-range nuclear attractive forces come into play, allowing a nuclear reaction to take place.

In the central core of normal stars the temperature is high enough for the random velocities of the nuclei to be adequate for self-sustaining fusion reactions to proceed. Because the kinetic energy necessary to achieve penetration of the Coulomb barrier is thermally derived, the process is known as *thermonuclear fusion*. The central temperature of such stars is thought to be approximately $2 \times 10^7$ K, which corresponds to a mean thermal energy of 2 keV.

On the other hand, the Coulomb potential for even singly charged nuclei is about 1000 keV, much higher than the average thermal energy. Fortunately particles do not need to have energies above the Coulomb potential to penetrate the barrier, as classical mechanics would indicate. Particles of lower energy have a finite probability of tunnelling through the barrier, a phenomenon explicable by wave-mechanics, though the probability of their doing so falls sharply with decreasing energy. Thus the cross-section for a nuclear fusion reaction does not have a threshold corresponding to the Coulomb potential, but is appreciable at lower energies. These considerations have two immediate consequences: first, the main contribution to nuclear processes will come from particles at the high-energy end of the Maxwellian velocity distribution which appertains in stellar interiors (cf. Fig. 8.3), and second, only encounters between nuclei of low electric charge, and therefore low atomic number, will be significant in energy production.

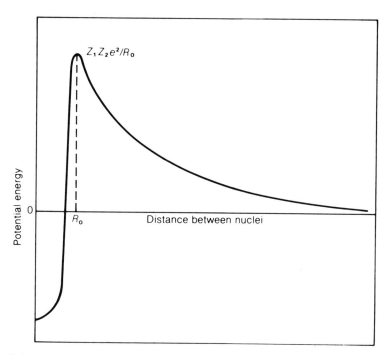

Fig. 8.2 Variation of Coulomb potential energy with distance between nuclei

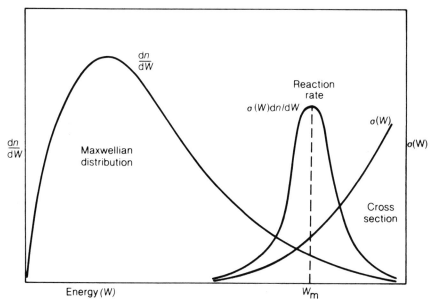

Fig. 8.3 The nuclear reaction rate $\sigma(W)$ d$n$/d$W$ showing the main contribution from the high-energy tail of the Maxwellian distribution at $W_m$

Hydrogen is by far the main constituent of normal stars, and the most important overall reaction is the conversion of protium (hydrogen with mass number $= 1$) to helium:

$$4H + 2e^- \rightarrow {}^4He + 2\nu + 26.7 \text{ MeV.}$$

The considerable energy release reflects the fact that $^4He$ is a particularly tightly bound nucleus. This many-body reaction could not of course take place directly; it is the net consequence of two separate chains of simpler reactions involving only two-body reactions and radioactive decay of short-lived product nuclei. These are respectively the proton–proton chain and the carbon–nitrogen–oxygen (CNO) cycle, which are detailed Tables 8.1 and 8.2

Table 8.1 The proton-proton reaction

| | |
|---|---|
| ${}^1H + {}^1H \rightarrow {}^2D + e^+ + \nu$ | $+ 1.44 \text{ MeV} (14 \times 10^9 \text{ y})$ |
| ${}^2D + {}^1H \rightarrow {}^3He + \gamma$ | $+ 5.49 \text{ MeV} (6 \text{ s})$ |
| ${}^3He + {}^3He \rightarrow {}^4He + {}^1H + {}^1H$ | $+ 12.85 \text{ MeV} (10^6 \text{ y})$ |

Table 8.2 The carbon cycle

| | |
|---|---|
| ${}^{12}C + {}^1H \rightarrow {}^{13}N + \gamma$ | $+ 1.95 \text{ MeV} (1.3 \times 10^7 \text{ y})$ |
| ${}^{13}N \rightarrow {}^{13}C + e^+ + \nu$ | $+ 2.22 \text{ MeV} (7 \text{ min})$ |
| ${}^{13}C + {}^1H \rightarrow {}^{14}N + \gamma$ | $+ 7.54 \text{ MeV} ((2.7 \times 10^6 \text{ y})$ |
| ${}^{14}N + {}^1H \rightarrow {}^{15}O + \gamma$ | $+ 7.35 \text{ MeV} ((3.2 \times 10^8 \text{ y})$ |
| ${}^{15}O \rightarrow {}^{15}N + e^+ + \nu$ | $+ 2.71 \text{ MeV} (82 \text{ s})$ |
| ${}^{15}N + {}^1H \rightarrow {}^{12}C + {}^4He$ | $+ 4.96 \text{ MeV} ((1.1 \times 10^5 \text{ y})$ |

The proton–proton reaction is dominant in the lower-temperature, fainter stars and the carbon cycle is dominant in the higher-temperature, brighter stars. The sun is near the dividing line between the two processes; most solar energy comes from the proton–proton reaction, with a small but significant contribution from the carbon cycle. A third nuclear process is thought to occur during the later stages of stellar evolution, when a star has exhausted the hydrogen in its core and relies on the conversion of helium to carbon to replenish the energy it radiates.

The times quoted in parentheses in Tables 8.1 and 8.2 are the mean reaction times, calculated for the conditions quoted: in other words they represent the life expectation of an individual nucleus in the centre of a typical star. It will be noted that these overall reaction chains are too slow by many orders of magnitude ever to be exploited as terrestrial energy sources. Fortunately, as will be seen later, there are other reaction chains between light nuclei, rare in stars but relatively plentiful and accessible on the earth's surface, which proceed sufficiently rapidly to make controlled thermonuclear fusion feasible.

## 2  Fusion reactions

The nuclear reactions which occur between the isotopes of light elements at energies corresponding to thermonuclear temperatures (1–100 keV) can be studied with comparative ease in the laboratory, by electrostatically accelerating charged nuclei of the required species and allowing them to impinge on suitable targets. Such studies began in the 1930s at Cambridge; by the middle 1950s most of the raw nuclear data necessary for assessing the feasibility of fusion power was readily available. It appeared that there were two overall reactions involving terrestrially abundant isotopes which had adequately high reaction rates:

$$3D \rightarrow {}^4He + p + n + 21.6\,MeV, \text{ and}$$
$$D + {}^6Li \rightarrow 2\,{}^4He + 22.4\,MeV.$$

As in the case of helium synthesis in stellar interiors, these equations are the net sum of chains of simpler processes. The first, the D–D reaction, involves four distinct thermonuclear processes:

$$D + D \rightarrow {}^3He + n + 3.3\,MeV$$
$$D + D \rightarrow T + p + 4.0\,MeV$$
$$D + T \rightarrow {}^4He + n + 17.6\,MeV$$
$$D + {}^3He \rightarrow {}^4He + p + 18.3\,MeV$$

The second, the D–T reaction, involves three processes, only the first of which is thermonuclear:

$$D + T \rightarrow {}^4He + n + 17.6\,MeV$$
$$n + {}^6Li \rightarrow {}^4He + T + 4.8\,MeV$$
$$n + {}^7Li \rightarrow {}^4He + T + n - 2.5\,MeV$$

The n–Li reactions are necessary to breed the required tritium, which is radioactive with a half-life of 12.35 years and hence is not naturally available.

The D–T reaction is the fastest thermonuclear process known, and thus most fusion reactor concepts are based on it. Besides it and the D–D reaction, the only other reaction offering slight promise of using naturally occurring isotopes as fusion fuel is a chain reaction based on the reaction between hydrogen and lithium,

$$p + {}^6Li \rightarrow {}^3He + \alpha + 4\,MeV$$

The energetic reaction products (marked *) can interact with fresh fuel thus:

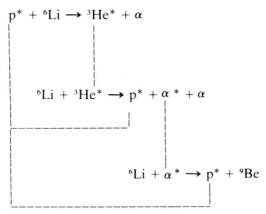

$$p^* + {}^6Li \rightarrow {}^3He^* + \alpha$$

$${}^6Li + {}^3He^* \rightarrow p^* + \alpha * + \alpha$$

$${}^6Li + \alpha * \rightarrow p^* + {}^9Be$$

provided they are not first slowed down by encounters with electrons. This necessitates operation at very high temperature, with consequent high radiation losses. Thus the possibility of exploiting the p–$^6$Li reaction in a practical system is doubtful.

There is even less prospect for that other attractive suggestion, the neutron-free proton–boron reaction

$$p + {}^{11}B \rightarrow 3\,\alpha + 8.7\,\text{MeV},$$

because recent calculations using refined cross-section data have shown the maximum nuclear energy yield to be several times less than unavoidable radiation losses, meaning that in any practical system an unacceptably high proportion of the power produced would need to be recirculated simply to keep the system running. These inevitable losses, known as *bremsstrahlung*, take the form of continuous radiation emitted by charged particles, mainly electrons, when they are deflected by the Coulomb fields of other charged particles. Bremsstrahlung from thermonuclear plasmas lies in the soft X-ray region of the spectrum and its energy distribution peaks at a wavelength inversely proportional to temperature; for a temperature of 10 keV this peak is at a wavelength of about 0.06 nanometres. The absorption length of such radiation in typical plasma conditions is about 1000 km, so that it will escape from terrestrial systems but be reabsorbed in stellar interiors.

Representative measured reaction-rate parameters, $<\sigma v>$, for the two most favourable reactions are:

|      | $T(\text{keV})$ | $<\sigma v>(\text{m}^3/\text{s})$ |
|------|------|------|
| D–T  | 20   | $4.3 \times 10^{-22}$ |
| D–D  | 40   | $1.6 \times 10^{-23}$, |

which lead to reaction times $(1/(n<\sigma v>)$, where $n$ is the particle density) ranging from seconds in envisaged self-sustaining fusion systems $(n \approx 10^{22})$ to much shorter times in the pulsed inertial devices $(n \approx 10^{32})$ which will be described later. The terrestrial sources and abundance of the three isotopes concerned are detailed in the appendix, from which it will be seen that they collectively represent a huge resource of energy potentially available to mankind. For all practical purposes, the problem of controlled nuclear fusion lies not in the magnitude of the fuel resources nor in extraction and manufacturing costs, but in developing practicable means of releasing the energy latent in these resources.

## 3 Basic problems

The main problems in attempting to adapt the principles of stellar energy to terrestrial applications arise from the vast diminution required in size and in timescale. In contrast, the temperature needs to be somewhat higher in order to speed up the reaction rate and make good the radiation losses inherent in terrestrial systems. But a modest increase of an order of magnitude or so will suffice, since the nuclear reaction probabilities increase extremly rapidly with rise of temperature in the range of concern. This higher operating temperature, coupled with a better choice of nuclear fuel, enables reaction times to be reduced from aeons to seconds or less.

The consequences of the extreme dimensional scaling are well illustrated by comparing a self-consistent set of conditions taken from a typical solar model with the corresponding quantities which current design studies predict for the nuclear island of a quasi-steady fusion power station.

| Quantity | Reacting core of star | Core of fusion reactor | Approx. ratio |
|---|---|---|---|
| Temp. (keV) | 1.5 | 10 to 20 | 10 |
| density (kg/m³) | $1.5 \times 10^5$ | $5 \times 10^{-7}$ | $3 \times 10^{-12}$ |
| pressure (MPa) | $4 \times 10^{10}$ | 0.4 | $1 \times 10^{-11}$ |
| scale length (m) | $1 \times 10^8$ | 1 to 10 | $2 \times 10^{-8}$ |
| power (MW) | $4 \times 10^{20}$ | 4000 | $1 \times 10^{-17}$ |
| power density (MW/m³) | $1 \times 10^{-4}$ | 10 | $1 \times 10^5$. |

In the centre of a star, gravitational forces hold the hot reacting material together throughout the lifetime of the star and support the extremely high densities and pressures necessary for the slow thermonuclear reactions to be utilized, but there is no way of producing such gravitational forces here on Earth. This circumstance underlies the greatest single problem facing fusion research, the problem of confining extremely hot matter, or plasma as it is called, with a high degree of thermal insulation for times long enough to create a subtantial energy surplus. Hence the achievement of plasma confinement has been the main focus of attention to date in controlled fusion research.

There is a secondary disadvantage arising from our inability to exploit gravitational confinement. The core of a normal star is thermally stable, since any decrease in thermonuclear reaction rate will be counterbalanced by the energy released by the ensuing gravitational contraction. Because of the sharp dependence of nuclear cross sections on temperature, any quasi-steady terrestrial fusion reaction will need to rely on some other temperature dependent property to achieve thermostatic operation. In confined plasmas of terrestrial dimensions, this could be provided by the inherent plasma radiation (bremsstrahlung) in the X-ray region of the spectrum, which increases with temperature and is not self-absorbed.

## 4  Towards a solution

Ideas of producing, in laboratory conditions, thermonuclear reactions between nuclei of light elements appear to have been current some years before the discovery of the fission process in heavy elements. In 1932 Gamow, still in the Soviet Union, was invited by Bukharin to head a controlled thermonuclear reactor (CTR) project, which aimed at producing reactions between lithium nuclei and protons by depositing the entire electric power of the Moscow industrial district in a thick copper wire, suitably impregnated with an Li–H mixture; this project seems to have fallen through.[2] In 1936 Oliphant reported unsuccessful attempts to observe D–D reactions in an electrical discharge tube. There was much speculation about CTR between physicists during the Second World War, but this has not yet been documented. The first subsequent experiments on fusion were begun independently in the late forties at three UK universities, Oxford, London, and Liverpool, and at laboratories in the USA and the USSR somewhat later.

Apart from the scientific challenge inherent in so difficult an enterprise, the motivation of those concerned was to demonstrate the feasibility of a new and everlasting source of terrestrial energy, fuelled by abundant and evenly distributed light elements whose extraction and preparation are already routine industrial processes, unattended by any special hazards. The complete fusion cycles do not of themselves yield radioactive wastes, and unlike fossil fuels, they do not contribute to the $CO_2$ content of the atmosphere; the only reaction product giving rise to long-term radioactivity would be the surplus neutron flux, whose effects would be confined to the reactor structure, and furthermore are capable of being alleviated by a wise choice of materials.

The optimistic enthusiasm for fusion in these early days was by no means confined to the scientists who took up the challenge. Security restrictions were imposed because of the apparent potential of fusion reactions in deuterium to provide abundant neutrons for producing nuclear material for military applications, and the three research programmes of the three groups in the UK, the USA, and the USSR were classified as secret by their respective governments. They thus worked in theory completely isolated from each other and from the physics community at large, but in practice spurred on by occasional leaks and rumours. The first public disclosure of a major effort on fusion came from a quite unexpected source, when I.V. Kurchatov, visiting the United Kingdom with Soviet leaders Bulganin and Kruschev in April 1956, volunteered to lecture at Harwell on thermonuclear research in the USSR. This dramatic event quickly led to a relaxation of security on their corresponding research programmes on fusion by the Western powers, allowing the publication in the open scientific literature of several papers which together pointed to comparable research effort in the UK and the USA; by the end of 1956 mutual collaboration between these countries had been established, but still on a classified basis.

The following year was momentous; the first British H-bomb test, in May, was successful; the most ambitious fusion experiment to date, ZETA (zero-energy thermonuclear assembly) was commissioned at Harwell; and the appetite of the UK national press was whetted by rumour and counter-rumour. In the last quarter of 1957 there appeared on average at least two news features per week on controlled fusion, despite its classified status. Mounting public pressure, from MPs, leader writers, and correspondents to the press, sometimes alleging US pressure to stifle a British breakthrough—there was an atmosphere of national mourning there following the success of *Sputnik I*—culminated in the prearranged simultaneous publication in *Nature* (25 January 1958) of seven papers on the controlled release of thermonuclear energy (two British, five American) heralded by an editorial 'Harnessing nuclear fusion'.

In the heady glare of attendant publicity, press conferences, and television interviews, cautious utterances by discreet and modest scientists became misconstrued by the media, leaving the general public in no doubt that 'the supply of energy to the human race is secured for all time' (*Sunday Times*). Only six months later, when careful experiment had shown that the nuclear reactions in ZETA were not truly thermonuclear, the media depicted the experiment as a fiasco, despite the objective and factual assurances of the former and current Directors of Harwell, Cockcroft and Schonland. Perhaps for the first time the general public became aware of the vicissitudes attending scientific endeavour; *The Times* concluded that, 'only during the past five months has the mixture of advance and disappointment common in research become public property'. In retrospect, ZETA is seen as the forebear of the *tokamak* and the *reversed-field pinch*, two of today's most promising magnetic fusion devices.

The complete global declassification of fusion research came in time for its first major exposition at the Second Geneva Conference on the Peaceful Uses of Atomic Energy in September 1958. It there became apparent, first, that the achievment of controlled thermonuclear fusion was proving to be far more difficult than had been earlier imagined, and second, that the United Kingdom contribution to the state-of-the-art was far from negligible.

From the smaller countries participating in the 1958 Geneva Conference, Sweden's contribution to fusion reactor studies was notable. In the ensuing years, fusion research has provided an outstanding example of international collaboration towards a scientific goal, whose attainment would be of lasting benefit to all.

## 5 Alternative approaches

To restate the problem of plasma confinement, it is that of producing and isolating high-temperature matter with adequate thermal insulation and particle containment, so that the nuclear energy generated can exceed the total energy needed to assemble and heat the plasma and to make good the energy lost from the plasma by conduction, convection, and radiation. A hot plasma formed from light elements consists of electrons and fully ionized atoms, or nuclei. Their interaction with condensed matter is strong, and unless the plasma is physically separate from its surroundings it will be rapidly cooled, thus quenching the thermonuclear reactions.

The degree of isolation required can be specified in terms of the *energy confinement time*, $\tau_e$, the total heat in the hot plasma divided by the total rate of energy loss. The minimum value of $\tau_e$ for net energy production from a plasma of density $n$ in thermodynamic equilibrium at a temperature $T$ can be expressed in terms of the inequality $n\,\tau_e > f(T)$ (known as the Lawson criterion, after the British physicist who first enunciated it).[3] The function $f(T)$ depends on the particular fusion reaction concerned; graphs of $f(T)$ versus $T$ for the most favourable terrestrial fusion reactions are shown in Fig. 8.4, which is calculated for a Maxwellian energy distribution of ions and electrons, both at the temperature $T$. It will be seen that for the D–T reaction, $f(T)$ has a broad minimum with a value of about $1.6 \times 10^{20}$ s/m³ at a temperature $T$ of 20 keV. Somewhat lower values result from more favourable assumptions concerning the energy distribution of the particles, for example by having the ions hotter than the electrons, but these carry attendant disadvantages.

How are hot plasmas to be contained? There is no way under terrestrial conditions to obtain the strong gravitational fields which hold stars together in space, and a plasma unrestrained by some long-range field of force will simply expand at a velocity corresponding to the random thermal velocity of its more massive constituents, the nuclei, with the electrons following so as to maintain charge neutrality. At the required temperatures this thermal velocity $V_{th}$ is about $10^6$ m/s, so that for a plasma of a radius of one metre, the inertial confinement time $\tau_i \approx R/V_{th}$ is only about one microsecond, much too short to satisfy Lawson's criterion at the plasma densities so far considered.

The approaches that have been directed towards the problem of plasma containment divide broadly into the two main categories of magnetic-

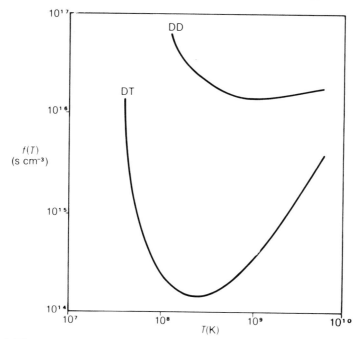

Fig. 8.4 The Lawson criterion; f (T) as a function of temperature for the most favourable fusion reaction, D-T, and D-D

confinement and inertial-confinement techniques. We shall begin by considering magnetic-confinement techniques, as chronologically the earlier of these two approaches.

Because plasma consists of charged particles, it is an electrical conductor and it will therefore interact with electric and magnetic fields. The direct confinement by external electric fields of a plasma whose density is high enough to be of thermonuclear interest is not possible because plasma in bulk is almost exactly neutral, to better than one part in a million. Thus the external electric forces on the oppositely charged nuclei and electrons just cancel out. Such an objection does not apply to the use of magnetic fields for confinement. In a magnetic field the charged particles no longer move outwards in straight lines, but are constrained to perform helical orbits in the direction of the magnetic field (Fig. 8.5); this applies to particles of both negative and positive charge. It is thus possible to devise suitable configurations of magnetic fields which will confine plasma for times long enough for it to react; these configurations are known as magnetic *bottles* or *traps*. Plasma is lost from them when the particle orbits are disturbed by collisions, field inhomogeneities, turbulence, and instabilities of various kinds, but by subtle and complex magnetic topology such plasma losses can be reduced to an acceptable level. Some idea of the quality of magnetic engineering necessary can be gained by reflecting that in one second, the sort of particle containment time required, ions will travel about 1000 km and electrons about 60 times as far.

The magnetic traps that have been devised since the earliest exploratory studies can be assigned to one of two main magnetic-field configurations, the *open line*, with cylindrical geometry, and the *closed line* with toroidal geometry. Both configurations have their fundamental advantages and drawbacks. Open-line systems, though topologically simpler, suffer from end-losses of plasma, absent in closed-line systems. Closed-line systems on the other hand cannot avoid radial field gradients which have adverse effects on particle orbits (Fig. 8.6); also the centre of the confined plasma is less accessible.

An example of an open-line system is the *mirror machine*, pioneered in the USA where it still has its devotees as a possible reactor system; it is shown in its simplest form in the diagram. The plasma is confined between a pair of magnetic 'mirrors', regions of increasing magnetic field strength, which have the effect of reflecting most of the charged particles which constitute the plasma. This is because each particle behaves as a current loop with an invariant magnetic moment $\mu = \frac{1}{2}mV_\perp^2/B$, where $V_\perp$ is its velocity transverse to the magnetic field $B$. Thus if the parallel velocity $V_{\parallel}$ is small enough, the particle will have insufficient total energy $\frac{1}{2}m(V_\perp^2 + V_{\parallel}^2)$ to be able to surmount the potential barrier $\mu B$ in a region of increasing $B$ and will fall back into the region of lower $B$ between the mirrors (see Fig. 8.7 and 8.8). Mirror confinement thus depends on the velocity ratio $V_{\parallel}/V_\perp$ of a particle being small enough; if it is sufficiently increased by Coulomb collisions with other particles or by electric field fluctuations resulting from turbulence, then it will escape through the first mirror it subsequently encounters. This can be visualized with the help of the velocity–space diagram in which the velocity components of the particles are plotted. The condition for mirror confinement is that the velocity vectors should lie outside the *loss-cone* corresponding to the critical value of $V_{\parallel}/$

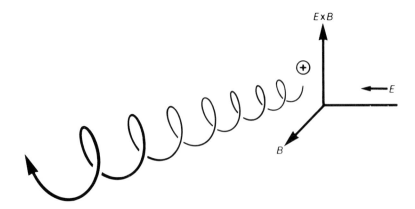

Fig. 8.5 The actual orbit of a gyrating particle in space

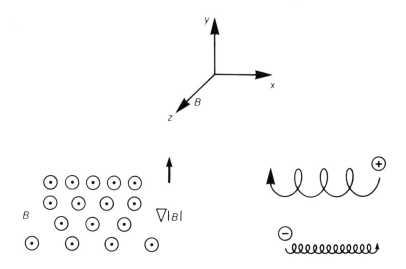

Fig. 8.6  The drift of a gyrating particle in a non-uniform magnetic field

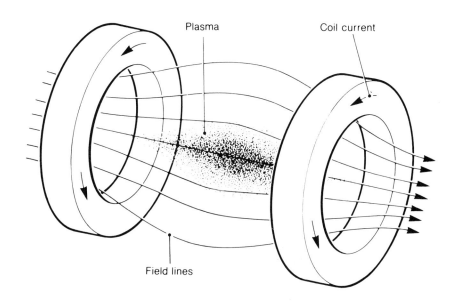

Fig. 8.7  Simple magnetic mirror trap

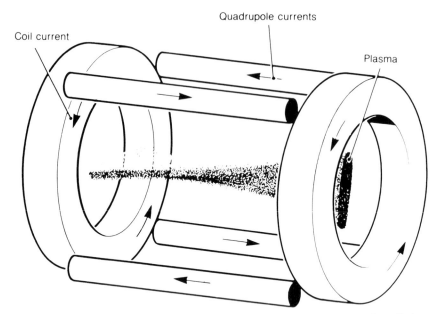

Fig. 8.8 Mirror trap with 'minimum B' coils added to form a magnetic well, i.e. a region from which |B| increases in every direction

$V_\perp$ ; particles scattered into this loss-cone quickly escape from the system (Fig. 8.9).

Mirror confinement occurs on a vast scale in the magnetosphere, the region, some tens of earth radii in diameter, permeated by the geomagnetic field. Here there are huge natural magnetic-mirror traps, shaped like doughnuts and lying several hundred miles above the surface of the earth. They were discovered in 1958 by J.A. Van Allen, who flew instruments in the US satellite *Explorer I*, and are named after him. These radiation belts contain large accumulations of charged particles, partly arising from the spontaneous decay of neutrons (see Fig. 8.10).

As earlier indicated, an alternative to open-line cylindrical geometries is the closed-line toroidal trap. The earliest attempts to construct closed-line traps exploited the *pinch effect*, the tendency of a current-carrying fluid conductor to be inwardly compressed by its self-magnetic field $B_\theta$ . This effect had been recognized about the turn of the nineteenth century and was subsequently studied in cylindrical geometry in gaseous discharge tubes. To make a closed-line trap the pinched current channel must be located inside a toroidal discharge tube. This can be achieved by changing the magnetic flux linking the tube, so that the plasma behaves as the single-turn secondary of a transformer (Fig. 8.12). Simple torodial pinches thus seem to be restricted to a.c. or pulsed operation.

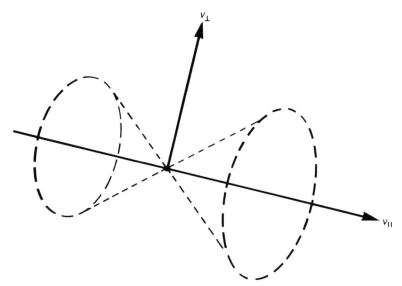

Fig. 8.9 The mirror loss cone in velocity space. Particles whose velocity vector lies within the cone pass through the mirror at their first encounter

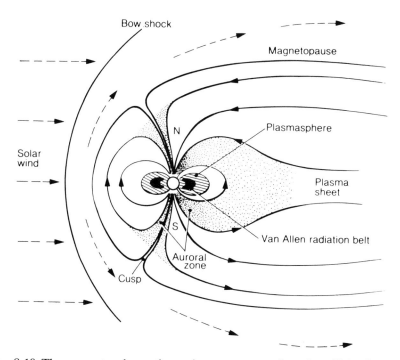

Fig. 8.10 The magnetosphere, the region some tens of earth radii in diameter, permeated by the earth's magnetic field. Its boundaries are largely shaped by the solar wind, a supersonic stream of charged particles from the sun. Near the earth charged particles are trapped in the Van Allen radiation belts

It rapidly emerged that they are also very unstable to *sausage* and *kink* instabilities, so-called from their general appearance when viewed on a microsecond timescale. Such deformations result in a crowding of magnetic lines, and hence an increase in magnetic pressure, at the plasma boundary with the tightest curvature, which accentuates the deformations and causes rapid growth (Fig. 8.11).The most violent of these instabilities can however be suppressed by the superposition of a toroidal magnetic field $B_\phi$, that is one with its lines-of-force in the direction of the circular axis of the torus. Such a field, combined with that of the plasma current, produces helical magnetic lines of force whose pitch varies with their minor radius $r$. This attribute, called magnetic shear, stabilizes the more dangerous modes of instability, because a perturbation aligned with the magnetic field at one radius will encounter field lines at an angle as it grows to a larger radius (see Fig. 8.13). The stabilizing effect of magnetic lines embedded in plasma can loosely be pictured as akin to the stiffening properties of fibre reinforcements in composite materials. Pursuing this analogy, a sheared field resembles a multi-ply material with progressively oriented plies.

Stability can be further improved by increasing the superimposed toroidal field so that $B_\phi / B_\theta$ exceeds the geometrical factor $R/r$, the *aspect ratio* of the toroidal plasma. This is a characteristic of the *tokamak* concept (the name is derived from the Russian words for 'toroidal magnetic chamber'). The quantity $q = rB_\phi/RB_\theta$ is a measure of the pitch or helicity of the resultant magnetic field lines; $q = 1$ corresponds to a line having just one rotation round the magnetic axis per circuit of the torus. $q$ is known as the *safety factor* because the higher the value of $q$, the higher the mode number $n$ of the first kink instability to which the plasma is susceptible; perturbations with $n < q$ tend not to grow in toroidal systems because the distance round the torus is less than one wavelength. Since $B_\theta$ is due to the axial current flowing in the plasma, it follows that the smaller the plasma current, the greater the stability against such gross deformations.

There is another class of toroidal closed-line traps, in which the confining magnetic fields are entirely produced by currents in rigid conductors, no current being deliberately induced in the plasma for this purpose. One of these is the *levitron* (Fig. 8.14), in which the $B_\theta$ field is provided by current flowing in a solid ring situated at the circular axis of the torus. In some experimental systems the ring is superconducting and so can be levitated indefinitely by electromagnetic forces; this avoids the use of material supports which would interact with the plasma. Their awkward topology makes levitron systems unpromising candidates as fusion reactors, but they have played a very important role in the understanding of plasma stability. Another is the *stellarator trap* (Fig. 8.15), in which the required combination of poloidal ($B_\theta$) and toroidal ($B_\phi$) fields is provided entirely by currents flowing in coils external to the toroidal tube containing the plasma. The poloidal field component is produced by helical multipole coils which carry oppositely directed currents in adjacent windings. The further one moves inwards from this coil structure, the more nearly the magnetic fields of the

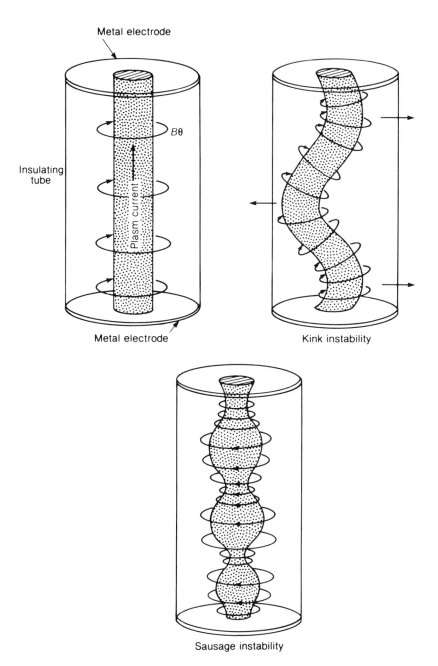

Fig. 8.11 Cylindrical pinch discharge and typical instabilities. Note the concentration of magnetic field lines where the curvature is tightest, causing the instabilities to grow

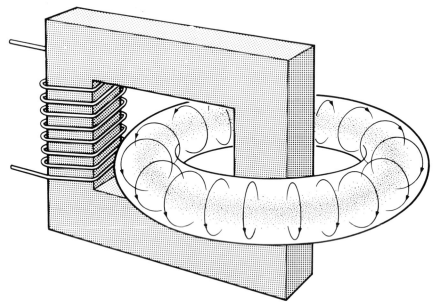

Fig. 8.12 A schematic drawing of a toroidal pinch system showing the iron-cored pulse transformer and the discharge-tube torus

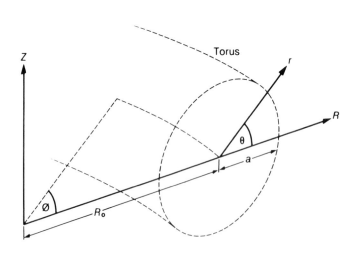

Toroidal geometry. Instabilities are described by the two integers $m$ and $n$, where

$$\zeta = \zeta(r) \exp i (m\theta + n\varnothing)$$

Fig. 8.13 Toroidal geometry

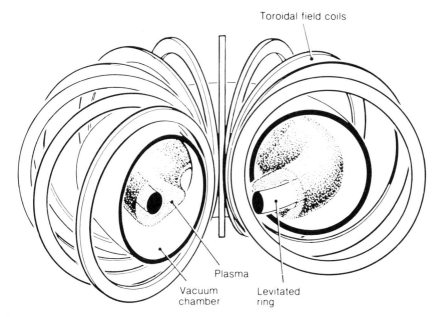

Fig. 8.14 Levitron geometry

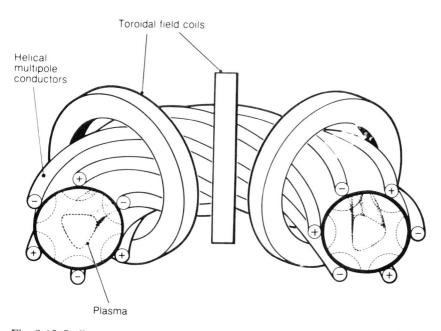

Fig. 8.15 Stellarator geometry

individual windings annul one another, so large currents are required to produce adequate field strength near the plasma surface. Hence stellarator traps are expensive in magnetic engineering; they do however offer the one certain prospect of continuously operating toroidal traps.

Having looked at magnetic-confinement systems, let us now return to consider the second of the two main approaches discussed earlier, inertial confinement of plasma. Since the inertial-confinement time $\tau_i$ is proportional to radius, and Lawson's criterion specifies a minimum value of the product $n\,\tau$, it follows that the minimum quantity $M$ of thermonuclear fuel which can react efficiently when inertially confined is proportional to the inverse square of the density, i.e. $M \propto n^{-2}$. It turns out that at normal solid DT densities the explosive yield of an efficient inertial thermonuclear device would be equivalent to several kilotonnes of TNT, which is much too large to power a conventional electricity generating station. But the recent concept of laser compression (Fig. 8.16), by which a small spherical target, uniformly irradiated, might be adiabatically compressed until its density is increased by a factor of as much as $10^4$ before it is heated to thermonuclear temperatures, has revived the practicability of inertial fusion devices, since this would decrease their yield per shot by $10^8$ to the manageable value of about 10 kg TNT, or 50 megajoules. Large-scale research on laser fusion is now being undertaken, particularly in USA and USSR, and parallel developments utilizing particle beams for compression and heating are currently being explored.

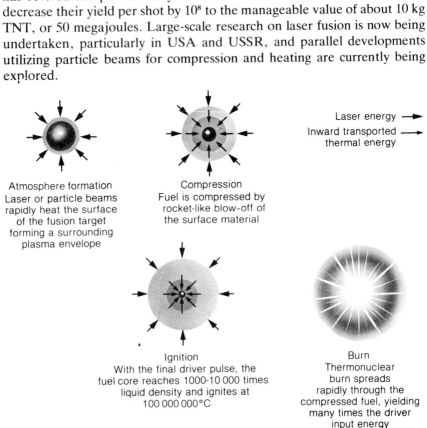

Laser energy →
Inward transported →
thermal energy

Atmosphere formation
Laser or particle beams
rapidly heat the surface
of the fusion target
forming a surrounding
plasma envelope

Compression
Fuel is compressed by
rocket-like blow-off of
the surface material

Ignition
With the final driver pulse, the
fuel core reaches 1000-10 000 times
liquid density and ignites at
100 000 000 °C

Burn
Thermonuclear
burn spreads
rapidly through the
compressed fuel, yielding
many times the driver
input energy

Fig. 8.16 Inertial confinement

# 6 Present status

It has already been shown that the criteria for producing a net energy gain from a thermonuclear plasma can be expressed as minimum values, requiring simultaneous achievement, of the ion temperature and of the product of the ion density and confinement time; for a DT plasma these were approximately $T = 20$ keV and $n\tau = 1.6 \times 10^{20}$ s/m³. Not surprisingly, it has become common practice to compare the success and the potential of major confinement experiments by displaying their operating regimes on a diagram of $T$ versus $n\tau$, as in Fig. 8.17; the target for a fusion reactor is the top right-hand corner.

Such a presentation of course needs some qualification; for instance, it does not bring out what is perhaps the most important aspect, the progress during the past quarter of a century from almost total ignorance, via guided empiricism, to the qualified understanding of today. Formerly results reflected the attributes of individual apparatus and laboratory skills; nowadays plasma temperatures of 1 keV can be sustained as a matter of routine for times approaching a second in many laboratories across the world. It is this ability to create and control high-temperature matter which constitutes the major advance in fusion science.

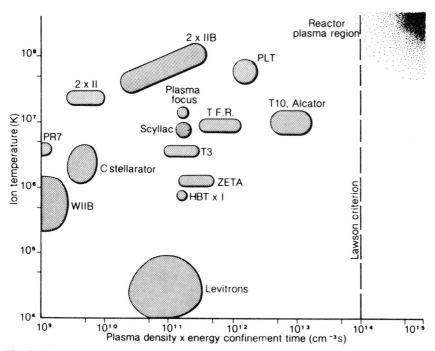

Fig. 8.17 Regimes in some confinement experiments: ion temperatures versus $n\tau_e$

It is perhaps worth mentioning at this point that the magnetic-confinement systems in which this progress has been demonstrated are not operated with a DT plasma, but with one of ordinary hydrogen (sometimes called protium) or deuterium. The reason lies in the radioactivity of tritium, which suffers beta decay with a half-life of some 12 years. It is hazardous because it easily enters biological systems, particularly in the form of the oxide HTO. Whilst the handling of tritiated materials on a small scale is a routine glove-box procedure, its presence in large-scale apparatus is a practical complication. It is moreover an unnecessary one, since at the present stage of research the dynamic behaviour of confined hydrogen plasma is little affected by its isotopic mix. The nuclear reaction rate in a deuterium plasma is down by a factor of approximately 50 compared with DT, but it is still quite adequate to provide data from which many of the properties of a DT plasma can be inferred with confidence.

This progress towards the achievement of Lawson's criteria has mainly been made with toroidal magnetic traps operating in the tokamak mode, that is with the safety factor $q > 1$ (see Fig. 8.18). The world fusion community was alerted to the superior performance of tokamak systems in 1969 by the publication of impressive results from the Soviet Union, confirmed by a joint Culham/Kurchatov team. Since then tokamaks have steadily proliferated, as will be seen from Table 8.3 which gives parameters of those

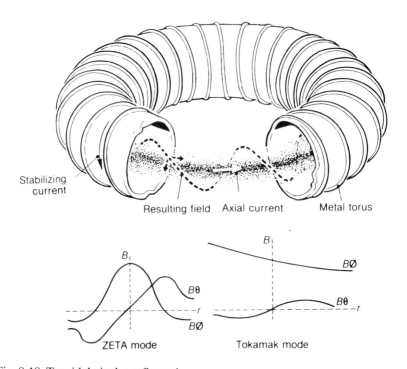

Fig. 8.18 Toroidal pinch configuration

currently operating and planned. Figure 8.19 illustrates the steady improvement in confinement quality and ion temperature in toroidal pinches, in more recent years operating in the tokamak mode. A major factor has been the improvement of experimental techniques, particularly with regard to plasma cleanliness and auxiliary heating. The latter is needed because the

Table 8.3 Large tokamaks under construction

| Plasma parameters | Europe JET (Culham) | Japan JT60 (Mukaiyama) | USA TFTR (Princeton) | USSR T15 (Moscow) |
|---|---|---|---|---|
| Major radius, $R$ (m) | 2.96 | 3.03 | 2.5 | 2.4 |
| Minor radius, $a$ (m) | 1.25 | 1.0 | 0.85 | 0.7 |
| Half-height, $b$ (m) | 2.10 | 1.0 | 0.85 | 0.7 |
| Torodial magnetic field on axis, $B_T$ (T) | 2.7 | 4.5 | 5.2 | 3.5 |
| Plasma current, $I_\phi$ (MA) | 3.8 | 2.7 | 2.5 | 1.4 |
| Electron density, $n$ (m$^{-3}$) | $10^{20}$ | | $10^{19}$ | $10^{20}$ |
| Ion temperature, $T_i$ (keV) | 10 | | 18 | 5 |
| Energy confinement time, $\tau_e$ (s) | 1 | | 1 | 0.3 |
| Pulse duration (s) | 20 | 5 | | 5.0 |
| Gases used | H,D,T | not T | H,D,T | not T |
| Operation date | end 1983 | 1983 | 1981/2 | end 1983 |

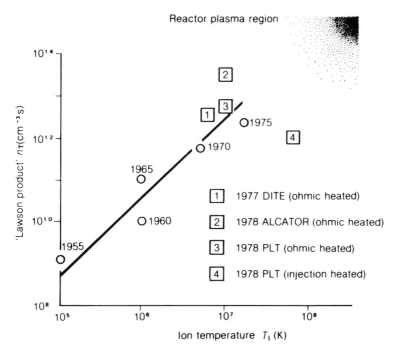

Fig. 8.19 Progress with toroidal systems towards reactor plasma conditions. The squares refer to toroidal pinches operating in the tokamak mode

electrical conductivity $\sigma$ of hot plasma increases with temperature according to the formula $\sigma \propto T^{3/2}$, so that the resistive heating due to the induced plasma current $I_\phi$ becomes steadily less effective as the temperature rises. But there is an upper limit to $I_\phi$, due to the onset of disruptive instabilities, and auxiliary heating is necessary to reach thermonuclear temperatures in tokamaks. It can be supplied by injecting into the plasma powerful beams of particles, uncharged (or *neutral*) atoms, which may be of the same species as those constituting the plasma. The particles must of necessity be neutral so that they are not deviated appreciably by magnetic fields and can penetrate the confining fields unhindered. The energetic neutrals then become ionized by collisions, and their directed energy becomes randomized in the plasma. To achieve heating, the injection energy must be well above the mean thermal energy of the background plasma. In practice this is assured since the energy necessary to penetrate well into the plasma is much greater than this.

The efficiency of neutral injection heating for tokamaks has been strikingly demonstrated on the Princeton Large Torus (PLT). Using injectors developing a total power of 2.4 MW, a central ion temperature of 75 million degrees and an electron temperature of at least 46 million degrees were achieved, both being the highest yet recorded in tokamak devices.

In tokamak systems the requirement that $q > 1$ means that the toroidal component of magnetic field $B_\phi$ is much stronger than the poloidal component $B$; as Fig. 8.18 shows, the value of $B_\phi$ changes litle across the plasma boundary. But in equilibrium the sum of the plasma pressure $p$ and the magnetic field pressure $B^2/2\mu_0$ is everywhere constant, and hence the ratio $\beta$ of the plasma pressure to the magnetic pressure is small; the highest value of $\beta$ so far achieved in a tokamak is about 2 per cent. This means that the applied magnetic field is being used inefficiently; low $\beta$-values constitute an economic penalty, and 10 per cent is the minimum value contemplated in most reactor concepts.

Toroidal pinch systems in which $B_\theta$ and $B_\phi$ are comparable can exhibit stable confinement provided the magnetic lines of force are twisted so as to provide the requisite shear. Such operation is sometimes called the ZETA mode, as it was first observed during the quiescent period of the discharge cycle in that machine. Figure 8.18 shows that $B_\phi$ changes sign at an intermediate value of the minor radius, and hence this confinement system has become known as the reversed-field pinch (RFP). Such self-stabilizing behaviour tended to be regarded as somewhat fortuitous until it was satisfactorily explained by recent theoretical work which predicts that under suitable conditions a toroidal plasma will relax naturally into such a stable configuration.[4] This has led to a resurgence of interest in RFP devices, which in addition to their advantage of a higher $\beta$, can operate at higher plasma current densities and offer the possibility of reaching thermonuclear temperatures by ohmic heating alone. Most of the experimental work supporting the potential of the RFP has been done with small-scale apparatus and confinement times have been predictably short, since they tend to scale as the square of the radius; there is now a temporary lull while larger RFP systems are being designed and constructed (see Fig. 8. 20).

It has been noted that the geometrically simple open-ended magnetic traps suffer from end losses. In the simple mirror these are such that the ratio of DT reaction rate to particle loss rate (which is independent of density) is insufficient at any temperature to meet the Lawson $n \tau$ criterion; the ratio $Q$ of the thermonuclear power output to the power needed to maintain the plasma is only marginally greater than unity in the most favourable circumstances. But such is the attraction of a steady-state system with topological simplicity that dogged attempts to overcome the loss problem continue. Some invoke combinations of mirror traps, in linear array or toroidally linked (see Figs 8.21 and 8.22). Other attempts rely on direct electrical conversion of the energy losses so efficiently that high recirculating power fractions could be tolerated.

Another approach to confinement in cylindrical geometry is to increase the length of the system many times, so that the relative importance of the end losses is reduced, and to heat the plasma by axial electron or laser beams. If these heated solenoid concepts, as they are called, are to achieve fusion reactor conditions they will possess some heroic parameters; lengths of half a kilometre and solenoidal magnetic fields of 150 kilogauss are envisaged.

Inertial-confinement studies are now almost exclusively concerned with the compression and heating of small pellets of thermonuclear fuel (D or a DT mix) to the domain of density and temperature where significant amounts

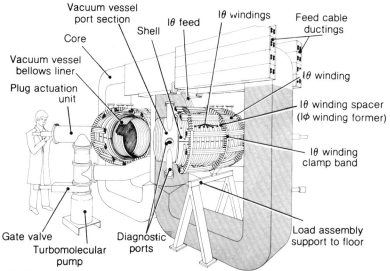

Fig. 8.20 Reversed field pinch experiment HBTX 1A (under construction at Culham)

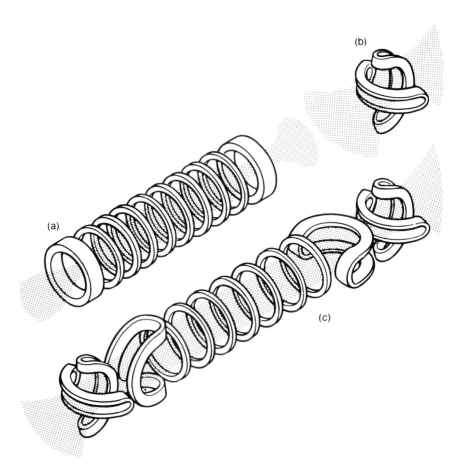

Fig. 8.21 Evolution of the Tandem Mirror
    a)  The simple solenoidal mirror
    b)  The minimum-B mirror
    c)  The tandem mirror, consisting of a solenoid plugged at both ends with
    minimum-B mirrors
    (Courtesy Lawrence Livermore National Laboratory)

of power are produced. Symmetric irradiation by laser beams causes ablation (vaporization) of the outside of the pellet, triggering an inwardly converging shock which compresses the central region and simultaneously heats it to reacting temperature, resulting in an inertially confined 'burn' followed by a rapid expansion which quenches it. The product of the density and radius of the compressed pellet, $\rho R$, is a direct measure of the efficiency of a compression device; for DT at a temperature of 10 keV, the Lawson criterion becomes $\rho R \approx 1\text{kg m}^{-2}$ and the fraction of the nuclear fuel burned is $\rho R/(60 + \rho R)$. Values of $\rho R$ approaching 30 are thus desirable for reactor purposes.

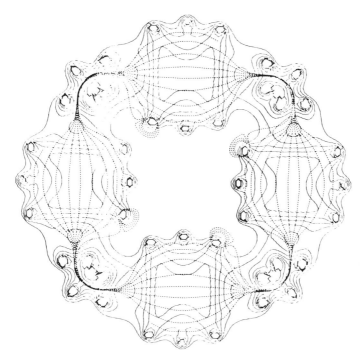

Fig. 8.22 A toroidally-linked mirror configuration, showing the field lines (dotted) and the magnetic contours. There are 8 mirrors and the contours are at 0.5 Tesla intervals from 1.0 to 3.5 Tesla

Rapid experimental progress has been made since the inception of laser compression in 1973, and theoretical predictions of its effectiveness have been quantitatively verified up to the current limitations of laser performance (see Fig. 8.23). The pellet targets used are glass microballoons, typically 100 $\mu$ m in diameter and containing 10 ng of DT, frozen on to the inner wall of the shell to eliminate back-pressure which would otherwise hinder compression. With the Shiva device at Livermore, California, implosions have compressed DT to over 50 times its normal liquid density, and a thermonuclear neutron yield of $2.7 \times 10^{10}$ has been obtained by irradiating special pellets.

The development of high-power lasers is crucial to the fruition of laser compression systems. Not only will they need to provide a precisely tailored pulse of adequate energy density; they will also need to perform with an efficiency to match the pellet power gain $Q$. Because lasers have efficiencies of only a few per cent, and furthermore are expensive, much current attention is being given in USA and USSR to the prospects of using pulsed particle beams to compress and heat fusion pellets. It is planned to exploit

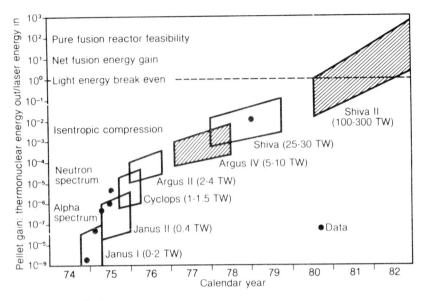

Fig. 8.23  Laser fusion progress

either electron beams, which can be produced efficiently and focused onto small targets by existing technology, or heavy ion beams (e.g. caesium ions of mass 133, singly charged) which have very short stopping lengths in the target material and thus allow more flexibility in target design.

In this section a review has been given of the status of current experimental approaches which seem to offer the best prospects of reaching the minimum conditions necessary to achieve a net power gain from thermonuclear reactions, a view reinforced by the scale of effort and expenditure accorded to them. But there are of course many other approaches being pursued today which for one reason or another enjoy less general popularity, but which nevertheless have made and continue to make solid contributions to our knowledge of the behaviour of hot plasma. In addition to stellarators and systems with buried conductors, which have been described earlier, these include shock-heated devices in cylindrical geometry (theta pinches and imploding liner systems); detailed accounts of them will be found in literature referenced in the bibliography.

High-temperature plasma exhibits collective behaviour because its elements exert forces on one another over large distances, and its behaviour is much more complicated than that of a dense, collision-dominated fluid. Because of this very complexity, progress in fusion research has tended to come mainly from experiment; nevertheless, the important contribution that theoretical physics has made to the understanding of plasma behaviour should not be overlooked. Scaling laws, stability criteria, equilibria in toroidal-confinement systems, and plasma wave phenomena are just a few subjects which have yielded to theoretical explanation. The increasing importance of computational physics in reconciling the mass of data from complex experiments with relevant (and often equally complex) theoretical models is also noteworthy.

## 7 Fusion reactors

In parallel with experiments on plasma confinement which have been pushing towards the reactor domain of plasma parameters with larger and better engineered devices, the last decade has seen a growing activity in the detailed study of fusion reactor systems. This demonstrates not only confidence that the necessary temperatures and confinement times will be achieved but also that understanding of high-temperature plasma behaviour has reached a stage where extrapolation to reactor conditions has adequate precision and validity. The ultimate object of these reactor studies is the engineering realization of a fusion-powered electricity generating station acceptable to the public, the manufacturing industry, and the utilities; this represents a long and protracted development programme, as was the case with fission reactors. The preliminary work so far carried out has defined in outline the main problems, pointed to the direction of their solution, and assessed the feasibility of fusion power.

Nearly all the work has been based on a nuclear island exploiting the DT reaction, because the physical conditions required are the least stringent and because the specific nuclear power output is the highest for a given plasma pressure. This has two immediate consequences: it imposes the need to breed tritium within the system, (see Fig. 8.24), and the necessity to cope with the flux of fast (14 MeV) neutrons produced by the DT reaction; these consequences are of course interdependent, since the neutrons are used to breed tritium in a surrounding blanket containing lithium, via the nuclear reactions quoted earlier. The possibility of exploiting the DD or the p+ $^{11}$B cycles for fusion power has attracted some consideration, since they use plentiful elements for fuel and the neutrons they produce are fewer in number and lower in energy, but systems studies show that neither is likely to yield an economically acceptable power even in those concepts, mentioned earlier, where the ion energy distribution is artificially elevated.

The conceptual reactor design study has proved to be a powerful approach in providing not only fundamental insight into key technological problems and their solutions, but also in identifying areas of plasma physics research needing more attention. Most recent designs for magnetic fusion reactors have been based on the tokamak system, and a convergence of solutions and a process of optimization are now emerging. Their main common design features are:

| | |
|---|---|
| thermal power | 2–5 GW |
| overall efficiency | 30–40 per cent |

| | |
|---|---|
| major radius, $R$ | 10–15 m |
| minor radius, $a$ | 2–5 m |
| toroidal field, $B$ | 3–8 T |
| plasma current, $I$ | 10–20 MA |
| current pulse duration | 100–1000 s. |

Implicit in these figures are the plausible but as yet unattained plasma parameter $\beta$ of 5–10 per cent, and the achievement of quasi-steady-state operation, requiring fuelling and exhaust systems to maintain a clean plasma for long burn times. The technological requirements are for a containment chamber whose walls are capable of withstanding the radiation and particle flux from the plasma, equivalent to a power loading of a few MW/m², without the need for frequent replacement, and for a combined blanket and shield region which performs the three functions of (1) tritium breeding, (2) heat removal (to turbines), and (3) radiation shielding of the magnetic field coils (which for efficiency and economy are superconducting). A generic design embodying these features is shown in Fig. 8.25, which also depicts the range of ambient temperature and neutron flux encountered.

Conceptual design studies have not so far uncovered any insuperable technological obstacle which may make the realization of fusion reactors impracticable. Not all aspects have been examined in great depth because it has been judged premature to deploy as much effort in this direction as in the more urgent area of plasma physics problems whilst uncertainties remain there.

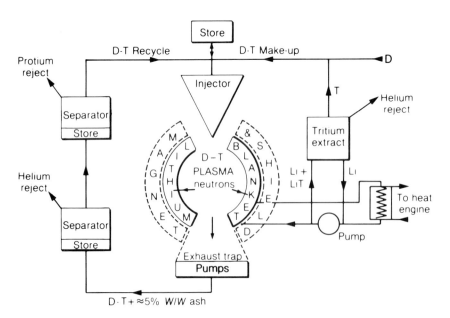

Fig. 8.24 Schematic fuel circuits of a fusion reactor

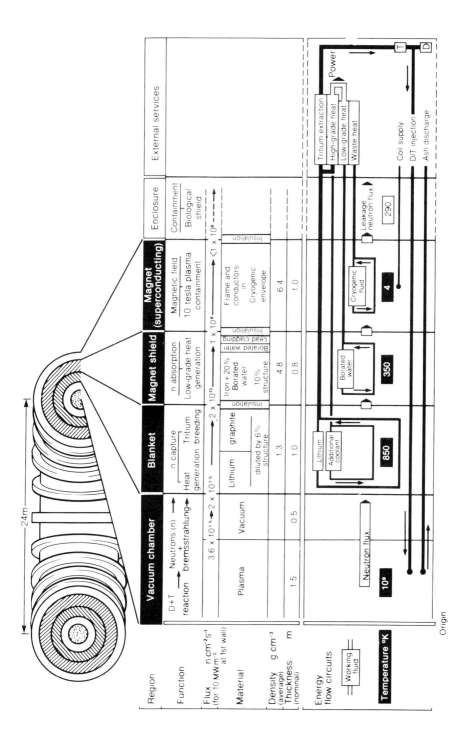

Fig. 8.25 Fusion reactor blanket design

Nevertheless, several features common to all magnetic systems have received thorough investigation. Prominent among these is the design of the blanket, whose primary function is to breed the tritium fuel which is not available naturally. The only route is via neutron-induced reactions in lithium, which may be used in a variety of forms: liquid metal (its melting point is 180 °C), molten salts (e.g. fluorides), or solid compounds (e.g. lithium aluminate). The essential feature is that the breeding ratio should exceed unity, more tritium being produced than is used up in the reactor. Neutronic calculations show that this can be achieved in a number of compact blanket designs, whose radial thickness can be as little as 500 mm. Heat transfer from the blanket can utilize the breeding material itself, if it is in liquid form, but helium cooling with a solid breeding material is preferred because the total tritium inventory in the blanket is reduced, and furthermore the electromagnetic losses associated with pumping conducting liquids through strong magnetic fields are avoided. Methods of extraction and purificaton of the bred tritium are at the conceptual stage. Although no serious technical difficulties have been uncovered, blanket reprocessing will require active chemical engineering plant of novel design and may present unforeseen problems. One major concern will be efficient scavenging so that routine tritium leakage is kept down to the currently permissible level of a few milligrammes per day.

Developments in the design and construction of the external field coils essential for magnetic systems give confidence in the practicability of engineering the large and complex structures called for in some designs. Superconducting materials are dictated because the power dissipated in conventional copper coils would be prohibitively large, comparable with the total electrical output of the reactor. Currently favoured materials are niobium–titanium and niobium–tin, maintained at superconducting temperatures by liquid helium coolant. The coils need to be protected from the residual neutron flux, to prevent heat deposition and material degradation, which will require a shield about one metre thick, and a massive support structure is required to withstand the electromagnetic forces between coils. It seems indeed that considerations of structural integrity rather than the intrinsic properties of superconductors set limits to the magnetic field strength available.

Much attention has also been paid, particularly by the UKAEA, to the problems of routine maintenance and repair of the blanket and first wall of the reactor. Economic considerations place a premium on high wall loading, with the result that those components nearest the plasma will have a much shorter service life than the rest of the reactor. Solutions to these problems have been embodied in the Culham Conceptual Tokamak Reactor (CCTR) designs, and involve the use of remotely operated machines to remove and replace blanket and shield components without dismantling the magnetic coils, which can thus remain at cryogenic temperatures.[5] The cooling (or warming) of such large superconducting magnets would occupy several days; with this delay obviated, the reactor down-time occasioned by a routine maintenance schedule is shortened to a few hours, a feature of great operational importance.

The scale and complexity of fusion reactor designs will be apparent from Fig. 8.26 to 8.31, which show a representative tokamak reactor, a magnetic-mirror concept embodying very large-scale direct conversion, and a laser fusion reactor design. Another facet of fusion reactor studies is the assessment of their environmental impact in the broadest sense; several studies have been completed, including one by the IAEA.[6] On the preliminary evidence available they conclude in particular that the potential total biological hazard presented by a fusion reactor is an order of magnitude less than that of a typical fission reactor. The decay heat due to the irradiation of the structure is small (0.5 per cent of the thermal output) and can be passively dispersed. The reaction zone contains only small quantities of fuel, which is replenished as it is consumed; there is thus no possibility of a dangerous nuclear excursion or runaway. The most serious hazard is the release of tritium into the biosphere during operation. At currently accepted radiation levels, the routine escape of tritium from a 2000 MW(e) reactor must be kept below 15 mg per day, but this does not appear to present unsurmountable design problems. It will be necessary during and after normal operation of

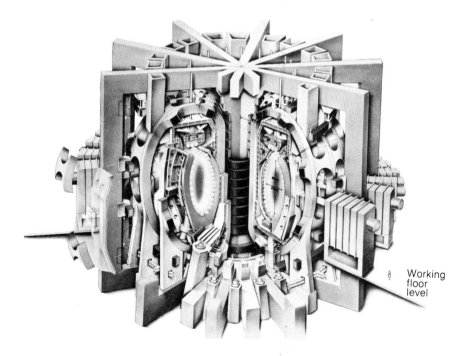

Working floor level

Fig. 8.26 Culham conceptual tokamak reactor Mk. II

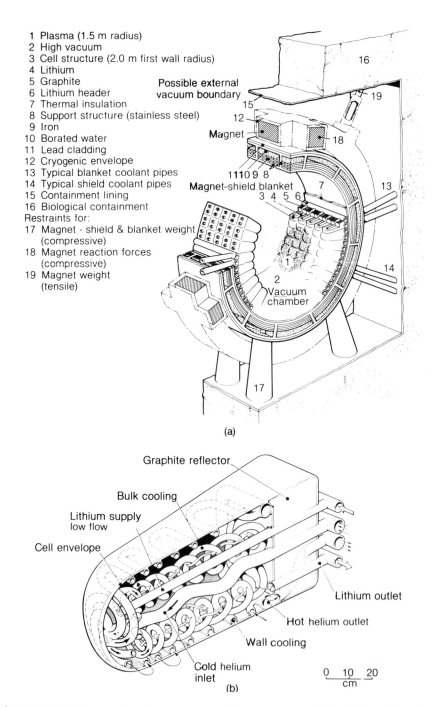

1 Plasma (1.5 m radius)
2 High vacuum
3 Cell structure (2.0 m first wall radius)
4 Lithium
5 Graphite
6 Lithium header
7 Thermal insulation
8 Support structure (stainless steel)
9 Iron
10 Borated water
11 Lead cladding
12 Cryogenic envelope
13 Typical blanket coolant pipes
14 Typical shield coolant pipes
15 Containment lining
16 Biological containment
Restraints for:
17 Magnet - shield & blanket weight
   (compressive)
18 Magnet reaction forces
   (compressive)
19 Magnet weight
   (tensile)

Possible external
vacuum boundary

Magnet

Magnet-shield blanket

11 10 9 8

3 4 5 6

1

2
Vacuum
chamber

(a)

Graphite reflector

Bulk cooling

Lithium supply
low flow

Cell envelope

Lithium outlet

Hot helium outlet

Wall cooling

Cold helium
inlet

0   10   20
     cm

(b)

Fig. 8.27 (a) Constructional arrangement of the components of the Culham tokamak
               reactor
          (b) A helium cooled blanket cell from the Culham reactor

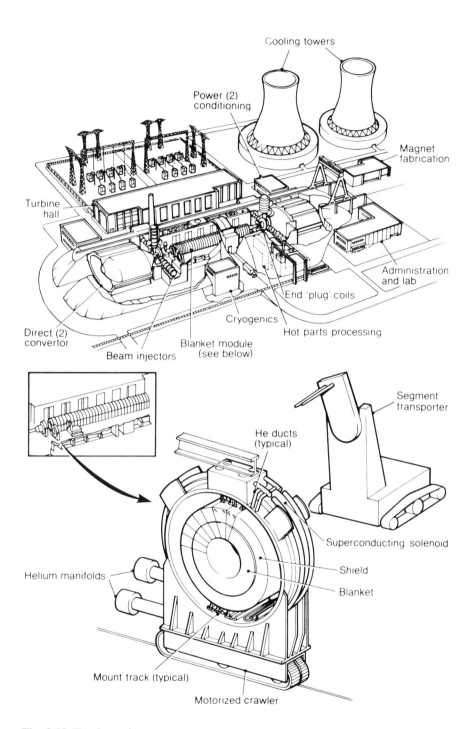

Cooling towers

Power (2) conditioning

Magnet fabrication

Turbine hall

Administration and lab

Direct (2) convertor

End 'plug' coils

Cryogenics

Hot parts processing

Beam injectors

Blanket module (see below)

Segment transporter

He ducts (typical)

Superconducting solenoid

Helium manifolds

Shield

Blanket

Mount track (typical)

Motorized crawler

Fig. 8.28  Tandem mirror reactor

Fig. 8.29  Blanket module for tandem mirror reactor (courtesy Lawrence Livermore National Laboratory)

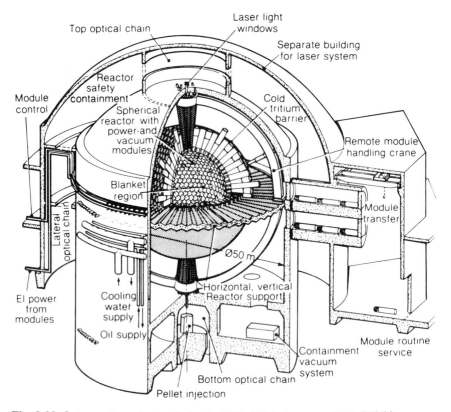

Fig. 8.30 *Saturn*, a laser-fusion design by KFA Jülich (courtesy KFA Jülich)

the reactor, to replace and to dispose of activated structural components. Although these are solid and demand no chemical reprocessing, and furthermore their activity can be controlled by proper design and appropriate choice of materials, they will constitute a form of radioactive waste which will have to be dealt with. Another pertinent conclusion is that the land despoliation consequent upon the large-scale operation of fusion reactors and arising from the acquisition of both fuel and constructional materials is no worse than that associated with fission reactors; nuclear power in general has great environmental advantages over fossil-fuelled power, because the vastly higher energy density of the fuel makes mining, transport, and waste disposal operations much easier to contain.

## 8   Other applications

So far consideration has been given to the thermonuclear plasma in its role as a primary energy source, the energy ouput appearing as the kinetic energy

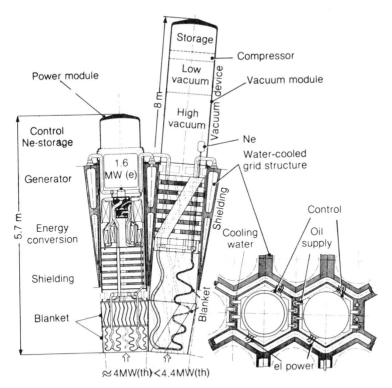

Fig. 8.31 *Saturn*, constructional details of the power and vacuum modules

of fast neutrons and charged particles, as electromagnetic radiation mainly in the ultra-violet and soft X-ray regions of the spectrum, and as the heat produced by exothermic tritium-breeding reactions in the blanket. In fusion power station concepts this aggregated energy is converted into electricity as efficiently as possible. But there are other potential applications of fusion in which these different manifestations of energy are exploited as separate entities; of particular industrial interest are the possibilities of nuclear and chemical fuel production, of radioactive waste reduction, and of the reclamation of raw materials from industrial wastes and civic refuse. But very few studies of such potential applications have been made and the subject must be regarded as highly speculative.

One exception is the hybrid concept, which envisages some of the neutrons produced from fusion reactions being used to breed fissile material from abundant naturally occurring thorium-232 or uranium-238. Extensive studies of this concept are being carried out in the USA and USSR. They are based on magnetically confined reacting plasmas and their attraction lies in using as energy amplifiers fusion systems which would otherwise fail to achieve the economic criteria necessary for a pure fusion reactor. The scheme hinges on the neutron-multiplying properties of thorium and uranium incorporated in tritium breeding blankets. The surplus neutrons create fissile material by the following reactions:

$$^{238}U + n \longrightarrow {}^{239}U \xrightarrow[\beta]{} {}^{239}Np \xrightarrow[\beta]{} {}^{239}Pu$$

$$^{232}Th + n \longrightarrow {}^{233}Th \xrightarrow[\beta]{} {}^{233}Pa \xrightarrow[\beta]{} {}^{233}U.$$

The rate of production of fissile material in hybrid blankets is estimated to be a few kilogrammes per year per MW of plasma power.

Another application, similar in principle but not as fully investigated, is the use of fusion neutrons to transmute the long-lived radioactive wastes from fission reactors into more benign isotopes. The effective decay rate of the more noxious products, $^{90}Sr$, $^{99}Tc$, $^{129}I$, and $^{137}Cs$, and the actinides, would be accelerated via (n, 2n) and (n, $\gamma$ ) reactions, reducing the activity of the wastes to the level of that of naturally occurring uranium ores in some tens of years.

# 9  Future plans

World fusion research today is dominated by four large and well-defined groups, the European Community, Japan, the Soviet Union, and the United States. As was remarked earlier, there is a high degree of international collaboration in plasma physics and fusion research, all major results being published in the open literature and freely discussed at international conferences. It is not surprising therefore to find some similarities in approach and outlook between the respective groups. One obvious example is that each is commited to a large tokamak experiment aimed at achieving plasma conditions approximating to the Lawson criteria for the DT reaction (but not using tritium in the first instance). They are respectively JET (Joint European Torus) being built at Culham, UK, TFTR (Tokamak Fusion Test Reactor) at Princeton, NJ, T15 at Moscow, and JT60, being prefabricated by Japanese industry and sited near the Japan Atomic Energy Research Institute (JAERI), All these tokamaks are expected to become operational in the period 1982–4. A provisional list of their design parameters and predicted performance is shown in Table 8.3, and a cutaway view of JET is shown in Fig. 8.32.

The JET project is a collaborative enterprise involving the member states of the European Community together with Sweden and Switzerland. The experiment is being constructed on a site adjacent to the UKAEA Culham Laboratory; at 1979 prices it will cost about £150 million and should be commisssioned early in 1983. By then the JET team will number about 320, of whom 120 will be professional scientists and engineers. The essential object of JET is to obtain and study a plasma in conditions and with

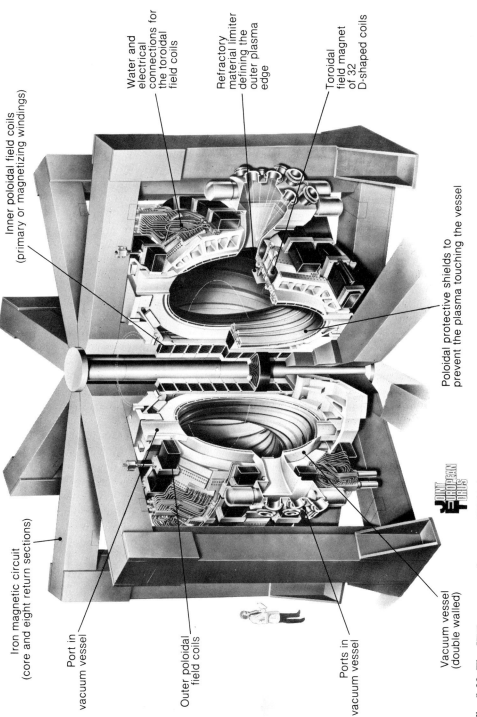

Inner poloidal field coils
(primary or magnetizing windings)

Water and
electrical
connections for
the toroidal
field coils

Refractory
material limiter
defining the
outer plasma
edge

Toroidal
field magnet
of 32
D-shaped coils

Iron magnetic circuit
(core and eight return sections)

Port in
vacuum vessel

Outer poloidal
field coils

Ports in
vacuum vessel

Vacuum vessel
(double walled)

Poloidal protective shields to
prevent the plasma touching the vessel

Fig. 8.32 The JET apparatus (courtesy JET joint undertaking)

dimensions which approach those needed in a fusion reactor. Its realization involves four main areas of work:

(1) the study of the scaling of plasma confinement properties,
(2) the study of plasma interaction with material container walls,
(3) the study of plasma heating methods, and
(4) operation with reacting D–T plasmas.

The three phases of the planned experimental programme, outlined in Table 8.4, retain a high degree of operational flexibility.

Notwithstanding their common devotion to tokamak experiments, all countries seem to have adopted the strategy of pursuing fusion studies on a broad front, for the next several years at least, apparently having decided that the time is not yet ripe to commit themselves to one single reactor concept. In their individual choices of alternative approaches, some noteworthy differences in emphasis are evident.

The United States has one of the largest programmes on laser-driven inertial confinement, and there is also substantial effort on alternative magnetic-confinement systems, particularly developments of the mirror machine principle. In parallel with these there is a large and growing commitment to the engineering and technological development necessary to build reactors.

In the Soviet Union more attention has been given to the basic physics of fusion research and less to reactor studies and technology. They have

Table 8.4 Outline of the JET Experimental Programme

| 1983 | 1985 | 1986 |
|---|---|---|
| Phase I: Exploratory Studies (using hydrogen plasma) | Phase II:[†] Improvement of Plasma Performance | Phase III: Fusion Studies (using deuterium–tritium plasma) |
| Establish a range of operating conditions | Enhance power supplies | Initiate fusion experiments using: |
| Scaling studies | Examine heating methods using internal structures | (a) non-Maxwellian beam-plasma methods, with compression |
| Work up to maximum plasma current for installed power supplies including use of D-shaped cross-section | Increase heating power | (b) thermonuclear self-heating if possible |
| Examine additional heating | Fit divertor if necessary | |
| Establish limits of operation | Try new limiter or wall materials | |
| Decide on future power supplies | | |
| Investigate impurity effects and | | |
| Decide whether a divertor is needed for impurity control | | |

† if results from Phase I are extremely favourable the Fusion Studies phase will be brought forward

another large inertial-confinement programme in addition to their emphasis on tokamaks.

The European programme is mainly concentrating on magnetic-confinement systems, the main alternative to the tokamak being the reversed-field pinch, on which large experiments in the UK and in Italy are planned. Inertial-confinement studies are maintained at low level. Fusion reactor studies and technology are an important part of the programme and will be co-ordinated in support of JET and future large experiments.

In the Far East, Japan has a comprehensive programme, much of it located in universities; it includes a rapidly expanding effort on inertially confined fusion, thought to be comparable to that in the USA. The People's Republic of China is now known to be actively engaged on plasma physics and fusion research, to an extent yet to be revealed with any precision. In Australia there is considerable university-based activity inclining to the more academic aspects of fusion research.

There is growing confidence that the large tokamak experiments presently under construction will achieve their aim of demonstrating the heating and containment scaling to near reactor conditions. The next step will involve a strong move towards solving engineering problems such as the performance of first-wall materials and maintenance of irradiated materials in the complex toroidal geometry of tokamaks. There are also fundamental problems that can be classified as physics still to be solved, notably the injection of plasma, control of impurity exhaust of power and particles, and the plasma performance when thermonuclear heat is being generated.

Investigation of all these problems will require a deuterium–tritium burning tokamak, producing thermonuclear power at high levels for long periods and this will be the next stage of CTR research. This stage will also help to clarify the overall economics of CTR, and the logistics of introducing thermonuclear reactors in the next century. The aim is to start the stage early in the 1990s. It should be emphasized that this stage is still very much experimental, seeking solutions to problems that can only be investigated when thermonuclear power is generated. The diversity of problems that should be investigated at any stage is of course a matter of judgement—for example, should the next stage test the tritium breeding cycle and the problems of large-scale tritium handling, or should the next stage test the economics and performance of the superconducting magnets necessary in an eventual reactor? One fact is certain: the overall scaling shows that from this stage onwards CTR tokamaks will be too expensive to be financed by individual nations and international co-operation will be needed. Such scientific co-operation has been very successful in the field of high-energy particle physics, but CTR is different in two important respects: the object is to test the device itself rather than provide a research tool using near-conventional technology, and little incentive is offered to those with mainly academic aspirations. The IAEA INTOR tokamak reactor study,[7] started late in 1978, could be the base from which such an international collaborative venture might grow.

## Acknowledgements

The author would like to thank the following people and organizations for Figures reproduced in this article: Lawrence Livermore Laboratory for Figure 21; Dr Ralph Moir of Lawrence Livermore Laboratory for Figures 28 and 29; KFA Jülich for Figures 30 and 31; and Dr B. Green (JET project) for Figure 32.

## Appendix: Terrestrial fusion fuel supplies

*Deuterium*, the isotope of hydrogen of mass 2, occurs abundantly, approximately one hydrogen atom in 6500 being of this heavier variety. The oceans of the world contain about 50 million million tons of deuterium which can be extracted easily and cheaply. Even if mankind's current total energy demands were to be met entirely from the fusion of deuterium, these terrestrial reserves would last longer than this planet is expected to be capable of sustaining life.

*Tritium*, the heaviest isotope of hydrogen, mass 3, does not occur naturally in significant quantities. It is radioactive with a half-life of about 12 years, so it cannot be stored indefinitely. The tritium burnt in fusion reactors will be bred from lithium during operation, but the first charge of a reactor will need to be manufactured from lithium using an external neutron source. Fortuitous sources of tritium, such as heavy water used as a fission reactor moderator, do not yield sufficient tritium for this purpose.

*Lithium*, is a relatively common metal, forming 65 parts per million (by weight) of the earth's crust; it is thus four times as plentiful as lead, for instance. The proven and inferred land resources of the western world are estimated to be six million tonnes, equivalent to about $6 \times 10^9$ megawatt-years of electrical energy and sufficient to provide the present US electricity consumption for at least two thousand years. Seawater is rich in lithium, containing 200 tonnes per cubic kilometre. This vast potential source of over $2 = 10^{11}$ tonnes has not yet been exploited because the current demand $(. \approx 5000$ tonnes p.a.) can be more economically provided from rich land resources.

# References

1. WEINBERG, A. Assessing the Oklo phenomenon. *Nature, Lond.*, **266**, 206 (1977).
2. GAMOW, G. *My world line: an informal autobiography.* Viking Press, New York (1970). Quoted in REINES, F. (ed.) *Cosmology, fusion and other matters* (George Gamow memorial volume). Adam Hilger, London (1972).
3. LAWSON, J. D. Some criteria for a power producing thermonuclear reactor. *Proc. Phys. Soc. B.*, **LXX**, 6 (1957).
4. TAYLOR, J. B. Relaxation of toroidal plasma and generation of reverse magnetic fields. *Phys. Rev. Lett.* **33**(19) 1139–1141 (1974).
5. MITCHELL, J. T. D. Blanket replacement in toroidal fusion reactors. *Proc. 3rd Topical Meeting on Technology of Controlled Nuclear Fusion, Santa Fe, New Mexico, May 1978.* Vol. II, pp. 954 ff. American Nuclear Society, Illinois (1978).
6. FLAKUS, F. N. Fusion power and the environment. *Atom. Energy Rev.*, **13**(3), 587–614 (1975).
7. INTOR GROUP International Tokamak Reactor Executive Summary of the IAEA Workshop, (1979). *Nucl. Fusion*, **20**(3), 349–388 (1980).

# Suggestions for further reading

### Stellar energy sources

SCHWARZSCHILD, M. *Structure and evolution of the stars*, Princeton University Press (1958).
CLAYTON, D. D. *Principles of stellar evolution and nucleo-synthesis*, McGraw Hill, New York (1968).

### Fusion energy conversion

POST, R. F. Nuclear fusion. *Annual Rev. Energy*, **1**, 213–255 (1976).
POST, R. F. AND RIBE, F. L. Fusion reactors as future energy sources. *Science*, **186**, 397–407 (1974).
MILEY, G. H. *Fusion energy conversion.* American Nuclear Society (1976).

### Status reviews of fusion research

PEASE, R. S. Experimental guide-lines in controlled thermonuclear research. *Physica*, **82B** + **C**, no. 1, 1–18 (1976).
PEASE, R. S. *The potential of controlled nuclear fusion. Contemp. Phys.* **18**(2), 113–135 (1977).
VILEKHOV, E. P. AND KINTNER, E. E. The current state and prospects for development of controlled thermonuclear fusion, *Trans. Amer. Nucl. Soc.*, **25**, pp. 193–200 (1977).
INTERNATIONAL FUSION RESEARCH COUNCIL [IFRC]. Status report on controlled thermonuclear fusion. *Nucl. Fus.* **18**(1), 137–149 (1978).
PEASE, R. S. Progress and problems in controlled nuclear fusion, (lecture given at a BNES meeting in London on 9 March 1978). *Nucl. Energy*, **17**(4), 271–281 (1978).

## Magnetic confinement

FURTH, H. P. Tokamak Research. *Nucl. Fus.* **15**(3), 487–553 (1975).

MOIR, R. W. *Preliminary design study of the tandem mirror reactor.* Lawrence Livermore Laboratory Report *UCRL–52302* (1977).

BODIN, H. A. B. AND KEEN, B. E. Experimental studies of plasma confinement in toroidal systems. *Rep. Progr. Phys.*, **40**(12) 1415–1565 (1977).

GIBSON, A. The JET project *Atom*, No. 254, 326–338, (1977).

THE JET PROJECT. CEC REPORT *EUR–JET–R7*, pp. 51, iv. JET Joint Undertaking, Abingdon (1978).


## Inertial confinement

LESSING, L. A. Lasers blast a shortcut to the ultimate energy solution. *Fortune*, 220–23, 322, 326, 328, 330 (May 1974).

EMMETT, J. L., NUCKOLLS, N., AND WOOD, L. Fusion power by laser implosion. *Scient. Am.* **230**(6), 24–37 (1974).

BOYER, K. Status of laser fusion research. *IEEE Trans. Nucl. Sci.* **22**(1), 38–44 (1975).

BOOTH, L. A., FREIDWALD, D. A., FRANK, T. G. AND FINCH, F. T. Prospects of generating power with laser-driven fusion .*Proc. IEEE*, **64**(10), 1460–1482 (1976).

ARNOLD, R. C. Heavy-ion beam inertial confinement fusion, *Nature, Lond.*, **276**(2683), 19–23 (1978).

YONAS, G. Fusion power with particle beams. *Scient. Am.* **239**(5), 40–51 (1978).

MOTZ, H. *The physics of laser fusion.* Academic Press, London (1979).


## Fusion reactor systems

RIBE, F. L. Fusion reactor systems. *Rev. Mod. Phys.* **47**(1) (1975).

KAMMASH, T. *Fusion reactor physics: principles and technology.* Ann Arbor Science Publishers Inc., Ann Arbor, Michigan (1975).

STEINER, D. Fusion reactor technology. *Proc. IEEE*, **63**(11), 1568–1608 (1975).

DAVIS, J. W. AND KULCINSKI, G. L. Major features of DT tokamak fusion reactor systems, *Nucl. Fus.* **16**(2) 355–373 (1976).

JASSBY, D. L. Neutral-beam-driven tokamak fusion reactors. *Nucl. Fus.* **17**(2), 309–365 (1977).

STACEY, W. M. AND ABDOU, M. A. Tokamak fusion power reactors. *Nucl. Technol.*, **37**, 29–39 (1978).

PARKINS, W. E. Engineering limitations of fusion power plants. *Science.* **199**(31) 1403–1408 (1978).

FURTH, H. P. Progress towards a tokamak fusion reactor. *Scient. Am.*, **241**(2), 39–49 (1979).

# Glossary

*Absolute filter*: an efficient filter for removing particulate matter from gases.

*Absorbed dose*: the energy deposited by ionizing radiation per unit mass of material (such as tissue).

*Absorber*: material which reduces radiation by removing energy from it; the magnitude of the reduction depends on the type of radiation, the type of material, its density, and its thickness. A sheet or body of such material.

*Absorption process*: a process where a component in a gaseous or liquid mixture is removed by contacting the gaseous or liquid mixture with a liquid or solid absorber. Silica gel is a common solid absorber.

*Accelerator*: a machine for producing high-energy charged particles by electrically accelerating them to very high speeds. Types include betatron, cyclotron, synchrotron, Cockcroft-Walton, Van de Graaff, tandem generator, and linac (linear accelerator).

*Accident transient*: the variation with time of the neutron population, the power, and the temperature of a reactor following some postulated accident.

*Actinides*: actinium and the elements following it in the Periodic Table; the most important are actinium, thorium, uranium, neptunium, plutonium, americium, and curium. Many of them are long-lived alpha-emitters.

*Activation*: the process of inducing radioactivity by irradiation, usually with neutrons, charged particles, or photons.

*Activation cross-section*: effective cross-sectional area of target nucleus undergoing bombardment by neutrons, etc. Measured in barns.

*Active*: often used to mean radioactive.

*Active area*: part of a laboratory where radioactivity may be present, and where exposure to individuals is constantly controlled.

*Activity*: the number of nuclear disintegrations occurring per unit of time in a quantity of a radioactive substance. Activity is measured in curies. Often used loosely to mean radioactivity.

*Activity, specific*: the activity per gram of material.

*Additive compounds*: compounds formed by additive reactions, in which a double bond is converted into a single bond by the addition of two more atoms or radicals.

*Adiabatic*: without loss or gain of heat.

*Adsorption*: the retention of dissolved substances on the surface of a substance (adsorbent).

*Advanced gas-cooled reactor (AGR)*: the successor to the Magnox reactors in the UK nuclear power programmes. It uses slightly enriched oxide fuel canned in stainless steel with graphite moderator and carbon dioxide coolant.

*Aerosol*: a colloidal system such as a mist or fog, in which the dispersion medium is a gas.

*Aggressive salts*: these are chemicals like sodium and magnesium hydroxides or chlorides which are very corrosive if they exist in high concentrations at high temperature (200–300 °C).

*Alara*: as low as reasonably achievable.

*Alpha particle*: a charged particle having a charge of 2 and a mass of 4 atomic mass units. It is emitted in the decay of many heavy nuclei and is identical with the nucleus of a helium atom, consisting of two protons and two neutrons.

*Alpha (radiation)*: helium nucleus emitted by some radioactive substances, e.g. plutonium-239.

*Americium*: artificially made transuranic element.

*Anharmonic*: said of any oscillation system in which the restoring force is non-linear with displacement, so that the motion is not simple harmonic.

*Anhydrous*: a term applied to oxides, salts, etc. to emphasize that they do not contain water of crystallization or water of combination.

*Anisotropic*: said of crystalline material for which physical properties depend upon direction relative to crystal axes. These properties normally include elasticity, conductivity, permittivity, permeability, etc.

*Annihilation*: spontaneous conversion of a particle and corresponding antiparticle into radiation, e.g. positron and electron, which yield two gamma-ray photons each of 0.511 MeV.

*Annihilation radiation*: the radiation produced by the annihilation of a particle with its corresponding antiparticle.

*Antibody*: a body or substance invoked by the stimulus provided by the introduction of an antigen, which reacts specifically with the antigen in some demonstrable way.

*Antigen*: material which sensitizes tissues in an animal body by contact with them and then reacts in some way with tissues of the sensitized subject *in vivo*, or with his serum *in vitro*.

*Aqueous phase*: the 'watery' solution in a solvent-extraction process.

*Argillaceous rocks*: sedimentary rocks having a very small mineral grain size, as in clay.

*Aspect ratio (torus)*: the ratio $R/r$ of the major to minor radii.

*Atom*: the smallest particle of an element, which has the chemical properties of that element. An atom consists of a comparatively massive central nucleus of protons and neutrons carrying a positive electric charge, around which electrons move in orbits at relatively great distances away.

*Atomic*: strictly, relating to the behaviour and properties of entire atoms—nuclei and orbital electrons; it is more usually a synonym for *nuclear*, it is as in 'atomic energy'.

*Atomic absorption spectrometry*: a method of physical analysis. A small aliquot of the sample to be analysed is introduced into a flame, the heat from which excites the outer electrons of the atoms causing the emission of light characteristic of the constituent elements. This light preferentially absorbs light of identical wavelengths when this, emitted from a standard comparison lamp, is passed through the flame. This provides a means of identifying elements of a given material as well as an estimate of the amount of each present from the extent of absorption.

*Atomic displacement cross-section*: a measure of the probability of a neutron displacing an atom from its normal position in the crystal lattice structure of a material. The probability is expressed as a target area or cross-section.

*Atomic mass unit (amu)*: one-twelfth of the mass of an atom of carbon-12. Approximately the mass of an isolated proton or neutron.

*Atomic weight*: the average mass of the atoms of an element at its natural isotopic abundance, relative to that of other atoms, taking carbon-12 as the basis. Roughly equal to the number of protons and neutrons in its nucleus.

*Atomic number (Z)*: the number of protons in an atomic nucleus. Nuclei with the same atomic number but different mass numbers are isotopes of the same chemical element.

*Attenuation*: reduction in intensity of radiation in passing through matter.

*Autoclave*: a vessel, constructed of thick-walled steel for carrying out chemical reactions under pressure and at high temperatures.

*Average effective dose-equivalent*: the measure of the risk from exposure to radiation, which takes account of the different sensitivity of various organs of the body and allows for the effects of different types of radiation.

*Azimuthal power instability*: eccentric neutron behaviour which results in uneven nuclear conditions in the reactor.

*Backfitting*: making changes to plants already designed or built.

*Background*: (1) in discussing radiation levels and effects, it refers to the general level of natural and man-made radiations against which a particular added radiation component has to be considered; (2) in discussing radiation measurement techniques, it may also include spurious readings due to the *noise* characteristics of the instrument and its power supplies, and to the presence of local radioactive contamination, etc.

*Bare sphere critical mass*: the mass of pure fissile material which if formed into spherical shape, with no outer layer of neutron reflecting material, will just sustain a chain fission reaction.

*Barn*: a unit of area ($10^{-24}$ cm$^2$) used for expressing nuclear cross-sections.

*Barytes*: barium sulphate, a common mineral in association with lead ores, occurring also as nodules in limestone and locally as a cement of sandstones.

*Base load*: in electricity generation, the minimum steady power demand on the system over a period.

*Batch (process)*: a process not operated continuously.

*Bearing pads*: pads attached to the outer faces of fuel rod wrappers which contact with similar pads on neighbouring elements either initially (restrained core) or as a result of distortion (free-standing core).

*Bearing resonances*: low-speed synchronous whirling frequencies, which are determined by the inertia of the rotor and the support stiffness of the bearings.

*Bearing systems*: the supports to hold a rotating shaft in its correct position.

*Becquerel (Bq)*: the new unit of activity in the SI System; it is equivalent to 1 disintegration per second or roughly $2.7 \times 10^{-11}$ Ci.

*Belt grinding*: an abrasive belt process for removal of a thin surface layer from a tube outer surface.

*Benchmark*: a name of American origin to describe a well-defined problem or experiment which then provides a reference standard for inter-comparison of various methods of solution or prediction.

*Beta particle*: an electron or positron emitted from a nucleus in certain types of radioactive disintegration (beta-decay).

*Beta quenching*: rapid cooling of uranium from the $\beta$-phase region.

*Beta radiation*: nuclear radiation consisting of $\beta$-particles.

*Bifurcate*: twice forked, forked.

*Binder/binderless routes*: pelleting methods employing or not employing a binder to assist the powder compaction.

*Binding energy*: the energy theoretically needed to separate a nucleus into its constituent protons and neutrons; it gives a measure of the stability of the nucleus.

*Biological shielding*: heavy concrete shielding erected around certain sections of plant containing radioactive materials in order to protect the operators from nuclear radiations.

*Biosphere*: that part of the earth and the atmosphere surrounding it, which is able to support life.

*Blanket*: *fertile* material (usually depleted uranium) arranged round a fast reactor core to capture neutrons and create more fissile material (usually plutonium); in a fusion reactor the blanket may be of lithium to capture neutrons and create more tritium.

*Bled-steam feed-heating train*: a series of heat exchangers in which steam is *bled* or extracted from the main expansion path through a steam turbine and is used to raise the temperature of feed water being returned to a steam generator from the condenser at the exhaust of the turbine.

*Blowdown*: rejection of liquid from a vessel under pressure to reduce dissolved solids.

*BNFL*: British Nuclear Fuels Limited.

*Boiling-water reactor (BWR)*: a light water reactor in which the water is allowed to boil into steam which drives the turbines directly.

*Boltzmann equation*: the fundamental particle conservation diffusion equation based on the description of individual collisions, and expressing the fact that the time rate of change of the density of particles in the medium is equal to the rate of production less the rate of leakage and the rate of absorption.

*Boral sheeting*: a composite formed of boron carbide crystals in aluminium with a cladding of commercially pure aluminium.

*Bore grinding*: a grid abrasion process to remove a thin surface layer from a tube bore.

*Boron*: element important in reactors, because of large cross-section (absorption) for neutrons; thus, boron steel is used for control rods. The isotope $^{10}$B on absorbing neutrons breaks into two charged particles $^7$Li and $^4$He, which are easily detected, and is therefore most useful for detecting and measuring neutrons.

*Boron counter*: an ionization chamber or proportional counter for detecting thermal neutrons by their interaction with boron-10 nuclei.

*Brachy-therapy*: treatment of tumours by radiation from sources placed in or near to the tumour.

*Branching*: alternative modes of radioactive decay which may be followed by a particular nuclide.

*Brazing*: the process of joining two pieces of metal by fusing a layer of brass or spelter between the adjoining surfaces.

*Breeder*: short for fast breeder reactor.

*Breeding*: the process of converting a fertile isotope, e.g. $^{238}$U into a fissile isotope, e.g. $^{239}$Pu. Fast reactors can be designed to produce more fissile atoms by breeding that are lost by fission. The process is also referred to as conversion.

*Bremsstrahlung*: X-rays produced when rapidly moving charged particles, e.g. electrons, interact with matter (from German 'braking radiation').

*Broad-group library*: a set of nuclear cross-sections tabulated as average values over a few (about 40) relatively broad energy groups.

*Buffer tank*: a vessel usually charged with a gas connected to a system containing a liquid, allowing the liquid to be expelled from the system and the out-flow brought to rest in a controlled manner by the cushioning effect of the gas in the tank as it is compressed above the liquid.

*Bundle*: see *fuel assembly*.

*Burn-up*: (1) in nuclear fuel, the amount of fissile material burned up as a percentage of the total fissile material originally present in the fuel; (2) of fuel element performance, the amount of heat released from a given amount of fuel, expressed in megawatt-days per tonne.

*Burst*: a fuel cladding defect which allows fission products to escape into the coolant; it need not be more than a very small crack or pin-hole.

*Busbar*: an electric conductor of large current capacity connecting a number of circuits.

*Butex*: name given to dibutyl ether of diethylene glycol, an organic liquid used in solvent-extraction processes.

*Calandria*: a closed tank penetrated by pipes so that liquids in each do not mix.

*Calorimetry*: the measurement of thermal constants, such as specific heat, latent heat, or calorific value; such measurements usually necessitate the determination of a quantity of heat, by observing the rise of temperature it produces in a known quantity of water or other liquid.

*Campaign*: the period from plant start-up to plant shutdown in a nuclear fuel reprocessing operation—usually a few months long.

*Can*: the container (usually of metal) in which nuclear fuel is sealed to prevent contact with the coolant or escape of radioactive fission products etc., sometimes also to add structural strength and to improve heat transfer. It may be made of Magnox, Zircaloy or stainless steel and may carry fins to increase the rate of heat tranfer.

*CANDU*: a type of thermal nuclear power reactor developed in Canada and widely used there; it uses natural (unenriched) uranium oxide fuel canned in Zircaloy, and heavy water as moderator and coolant.

*Canyon concept*: a reprocessing plant layout favoured in the USA, where the plant is constructed partly below ground in concrete-lined vaults or canyons.

*Capture*: the process in which a particle (e.g. a neutron) collides with a nucleus and is absorbed in it.

*Carbon-dating*: a means of dating by measuring the proportion of radioactive carbon. Atmospheric carbon dioxide contains a constant proportion of radioactive $^{14}C$, formed by cosmic radiation. Living organisms absorb this isotope in the same proportion. After death it decays with a half-life $5.57 \times 10^3$ years. The proportion of $^{12}C$ to the residual $^{14}C$ indicates the period elapsed since death.

*Carbon-14*: see *carbon-dating*.

*Carcinogenesis*: the production and development of cancer.

*Carcinoma*: a disorderly growth of epithelial cells which invade adjacent tissue and spread via lymphatics and blood vessles to other parts of the body.

*Carrier-free*: a carrier-free preparation of an isotope consists of atoms of that isotope alone.

*Cartridge*: a unit of nuclear fuel in a single can.

*Cascade*: (1) in nuclear fuel processing etc., a progressive sequence of operations in which the process material flows from one stage of the plant to the next, for example, in solvent-extraction processes (in gaseous diffusion or gas centrifuge separation processes there may be several hundred almost identical stages—flow may be in both directions); (2) the emission of gamma rays by a radioactive nucleus in sequence separated by a very short time interval.

*Cascade (ideal)*: a cascade in which the flow is graded along it to avoid mixing losses, yielding maximum separative power.

*Cascade (jumped)*: a cascade in which the product or waste from a stage is not necessarily connected to the stage above or stage below, respectively. If the cut is small, the waste may be connected to a point several stages below in the cascade.

*Cascade (squared off)*: a cascade built in sections whose outlines follow those of an ideal cascade—each section of a squared-off cascade is a square cascade.

*Catalytic hydrogenation*: chemical reactions in which molecular hydrogen is added to the organic compound in the presence of a secondary element or compound, the catalyst. The catalyst effectively takes no part in the reaction but is essential for its completion. Typical hydrogenation reactions are the hydrogenation of olefins to paraffins, aromatics to napthenes, and the reduction of aldehydes and ketones to alcohols. Typical catalysts are Ni, Pd, $V_2O_5$, etc.

*Cation exchange*: the process by which suitable solid agents such as zeolites, artificial resins or clays can remove cations (i.e. positively charged atoms or molecules) from solution by exchanging them with another cation. Commonly used in water softening whereby calcium and magnesium ions are replaced by sodium ions; the removed ions are held by solid cation exchanger. In the context of radioactive waste disposal the process would be typically the removal ot strontium and plutonium ions by replacement by sodium or potassium ions from a mineral.

*Cave*: a heavily shielded compartment in which highly radioctive materials can be safely kept, handled or examined by remote manipulation; sometimes called a *hot cell*.

*Centrifuge*: apparatus rotating at very high speed, designed to separate solids from liquids, or liquids from other liquids dispersed therein.

*Ceramic*: hard pottery-like materials having high resistance to heat, e.g. oxides and carbides of metal; it is used for nuclear fuels operating at high temperatures.

*Cermet*: an intimate mixture of metallic and ceramic particles which combine some of the desirable qualities of both, e.g. for reactor fuels.

*Chain reaction*: a process which, once started, provides the conditions for its own continuance. In the chain reaction of nuclear fission, neutrons cause nuclear fission in uranium or plutonium, producing more neutrons, which cause further fissions, and so on.

*Channelling*: the escape of radiation through flaws in the moderator or shielding of a reactor, etc. leading to high levels of radiation in the regions affected.

*Charcoal delay bed*: beds of charcoal to which gases can be admitted providing hold-up or delay by absorption and desorption. In a nuclear plant such delays are used to allow the decay of activity.

*Charge-face*: of a nuclear reactor, that face of the biological shield through which the fuel is inserted.

*Charge/discharge machine*: a mechanical device for inserting or removing fuel in a nuclear reactor without allowing the escape of radiation and, in some reactors, without shutting the reactor down.

*Charged particles*: nuclear or atomic particles which have a net positive or negative electric charge; they include electrons (and beta particles), positrons, protons, deuterons, alpha particles, and positive or negative ions of any of the chemical elements, but not neutrons.

*Chemisorption*: irreversible adsorption in which the adsorbed substance is held on the surface by chemical forces.

*Chopping*: process of cutting nuclear fuel into small lengths.

*Cisternography*: visualisation of body spaces.

*Cladding*: the protective layer, usually of metal, covering the fuel in a nuclear fuel element.

*Clean critical assembly*: a reactor in its initial stage before irradiation has caused changes in the fuel composition; the material composition is usually very well known.

*Closed cycle, closed-cycle cooling*: a completely enclosed path, e.g. a Magnox coolant circuit.

*Cluster*: see *fuel assembly*.

*Coastdown*: the process of slowing down of a pump or turbine once the drive mechanism has stopped or been disengaged.

*Coated fuel particle*: a compact of nuclear fuel coated with a refractory material which restricts release of fission products.

*Codecontamination*: the decontamination of U and Pu together from fission products.

*Cold criticality*: the establishment of a low-power, nuclear fission chain reaction under conditions of essentially zero heat generation.

*Cold drawing*: a continuous cold metal working process in which a tube is pulled through a die, reducing its wall thickness and/or diameter.

*Cold pilgering-type operation*: an intermittent metal-working process to reduce tube wall thickness and/or diameter.

*Cold work*: the plastic deformation of a material at temperatures far below the melting-point, carried out usually in order to increase its strength.

*Collimator*: a device to confine radiation to a narrow beam by preferentially shielding against radiation in other directions.

*Column*: vertical, cylindrical apparatus for carrying out a chemical operation, e.g. solvent extraction, absorption, etc.

*Commissioning*: running a machine etc. (e.g. a nuclear reactor) up to power and checking that it complies with the specifications before the supplier hands it over to the customer.

*Committed dose-equivalent*: the total integrated dose-equivalent over 50 years to a given organ or tissue from a single intake of radioactive material into the body.

*Common-mode failure*: the failure of two or more supposedly independent parts of a system, e.g. a reactor, from a common external cause or from interaction between the two parts.

*Compound nucleus*: a highly excited nucleus, of short lifetime, formed as an intermediate stage in a nuclear reaction, e.g. $^{236}$U prior to fission.

*Concentration limits*: a technique of criticality control using concentration limits as the control.

*Concentration pulse*: a term coined to describe an enhancement, which exists for a particular period only, in the concentration of one of the constituents of a general medium. In the case of a radioactivity concentration pulse, a graphical plot of specific activity against time demonstrates a profile rising to a maximum and, thereafter, falling towards the usual background level.

*Condenser off-gas*: incondensable gases isolated in the steam condenser at the exhaust of a turbine that have to be drawn off if the condensing process is to continue efficiently.

*Conditioning*: the addition of chemicals to a solution in order to adjust the chemical composition; it is usually carried out in a 'conditioner'.

*Confinement time*: in nuclear fusion research, the average life-time of a particle in a plasma containment system.

*Constant-volume feeder*: see *CVF*.

*Contactor*: (reprocessing) generic term for solvent-extraction apparatus.

*Containment*: physical boundaries constructed to confine radioactive material from a reactor or plant used in reprocessing.

*Contamination*: the presence of unwanted radioactive matter, deposited on solid surfaces, or introduced into solids, liquids or gases.

*Continuous operation*: a method of operation of a process or plant in an unbroken sequence (see *batch*).

*Continuous refuelling*: replacing fuel channels one at a time at the required interval rather than in a batch at longer intervals. Refuelling with the reactor on-load is a particular case.

*Control rod*: a rod of neutron-absorbing material (e.g. cadmium, boron, hafnium) moved in or out of the reactor core to control the reactivity of the reactor.

*Control rod worth*: the reactivity change resulting from the complete insertion of a fully withdrawn control rod into a critical reactor under specified conditions.

*Conversion*: a term used for breeding (see separate entry) when the main fissile element consumed is different from the main fissile element bred. Thus the term applies to reactors fuelled with $^{235}$U/$^{238}$U as opposed to $^{239}$Pu/$^{238}$U.

*Conversion factor*: the ratio of the number of fissile nuclei produced by conversion to the number of fissile nuclei used up as fuel.

*Converter reactor*: a reactor in which the conversion process takes place, but in which breeding, with a net gain of fissile material, does not.

*Coolant*: the gas, water or liquid metal circulated through a reactor core to carry the heat generated in it by fission (and radioactive decay) to boilers or heat exchangers.

*Cooling, radioactive*: progressive diminution of radioactivity, especially of nuclear fuel after removal from a reactor. This is accompanied by a diminution of heat output, so the word may be used in either sense.

*Cooling pond*: a water-filled tank in which used fuel elements are placed while cooling (in the radioactive and the thermal senses) is allowed to proceed; the water provides both radiation shielding (conveniently transparent) and means of removing the heat of radioactive decay.

*Core-catcher*: commonly used to describe the device designed to retain the products of the melted reactor core after a postulated accident. It may be within the reactor vessel (internal) or below the vessel (external).

*Core follow*: a mathematical technique for constraining the theoretical prediction of the behaviour of a reactor to follow the measured behaviour during operation.

*Core, homogenous*: core materials so distributed that the neutron characteristics of the reactor are homogenous.

*Core power*: the rate of production or use of energy in the core.

*Core, reactor*: the central region of a nuclear reactor containing the fuel elements, where the chain reaction of nuclear fission proceeds.

*Coriolis coupling*: the coupling between vibrations and rotations in a molecule.

*Coulomb barrier*: the potential barrier between charged particles due to mutual electrostatic repulsion.

*Counter*: an instrument for counting pulses of radiation, or the electric pulses that these cause, and displaying or recording them in digital form; also used loosely for any form of radiation detection or measuring instrument.

*Counter-current*: opposing flows, as for example, where the organic phase carrying U + Pu flows in one direction, while the aqueous phase containaing fission products flows in the other direction.

*Counting rate*: the rate at which radioactive events, for example, the emission of beta particles, are registered by the measuring device; it is usually expressed in counts per minute, counts per second etc. The counting rate is less than the total radioactive disintegration rate by a factor which expresses the overall counting efficiency of the particular measuring device for the radiation in question.

*Coupled control system*: a form of power station control which is inherently load following; in the nuclear application, the system is sometimes referred to as 'core follows turbine'.

*Coupled hydrodynamic–neutronic instability*: in a BWR thermohydraulic instability is complicated by a feedback through the link between the amount of steam in the core (voidage) and the power generated in the fuel. This feedback effect can be dominant when the time constant of a hydraulic oscillation is close to the same magnitude as the time constant of the fuel element. Strong nuclear-coupled thermohydrodynamic instabilities occured in early experiments at Idaho where a metallic fuel with a low time constant was operated in a low-pressure boiling-water flow.

*Creep (radiation)*: the time-dependent, non-reversible dimensional change in a material subject to both a mechanical load and a neutron flux. Radiation creep is caused by the stress-induced preferential segregation of the atoms displaced by irradiation damage. It is observed at temperatures too low for thermomechanical creep to occur. At higher temperatures, e.g. in a reactor core, both types of creep may occur simultaneously.

*Creep (thermal or thermomechanical)*: the low, time-dependent, non-reversible dimensional change in a material when subject to a mechanical load less than that required to produce plastic deformation. Creep is very temperature-dependent, significant only at temperatures above about 0.4 of the melting-point on the absolute scale.

*Critical, criticality*: a nuclear reactor or other assembly of fissile material is said to have 'gone critical' when its chain reaction has just become self-sustaining.

*Critical mass*: the amount of fissile material needed to maintain a nuclear chain reaction.

*Critical material*: the material in which the concentration of radioactivity resulting from a given discharge is highest, when expressed as a fraction of the appropriate derived working limit.

*Critical pathway*: the pathway by which most radioactivity reaches the critical material.

*Critical population group*: the group of persons whose radiation doses, resulting from a given practice, are highest.

*Criticality incident or excursion*: inadvertent accumulation of fissile material into a critical assembly, leading to criticality and the sudden and dangerous emission of neutrons, gamma rays and heat.

*cross-section, nuclear*: the target area presented by a nucleus to an approaching particle relating to a specified nuclear interaction, e.g. capture, elastic scattering, fission. The cross-section varies with the type of nucleus, the type of energy of the incident particle and the specified interaction. Cross-sections are measured in barns, and give a measure of the probability of the particular reaction.

*Cryostat*: low-temperature thermostat.

*Curie (Ci)*: the unit of radioactivity, being the quantity of radioactive material in which $3.7 \times 10^{10}$ nuclei disintegrate every second. Originally it was the activity of 1 gram of radium-226. The curie has now been superseded in the SI system by the becquerel (Bq), equal to 1 disintegration per second.

*Curium*: artificially made transuranic element.

*Cut*: the fraction of the feed to a separation stage which emerges in the product stream.

*CVF (constant-volume feeder)*: a rotating device with 'buckets' on the ends of radial arms, which scoops up liquid and delivers it at a rate proportional to the rotational speed.

*Cycle*: in solvent extraction, used to denote one complete sequence of extraction, scrubbing and stripping.

*Cyclotron*: an accelerator in which charged particles follow a spiral path in a magnetic field and are accelerated by an oscillating electric field.


*Dating, radioisotope*: determination of the age of an archaeological or geological specimen by measuring its content of a radioactive isotope in relation to that of its precursor or decay product, or of its stable isotope; applied particularly to radio-carbon dating of archaeological specimens. (See *carbon dating*).

*Daughter product*: the nuclide immediately resulting from the radioactive decay of a parent or precursor nuclide. If it is radioactive, it will in due course become a parent itself.

*D–D*: symbol for reaction between two nuclei of deuterium atoms.

*Decade*: any ratio of 10:1

*Decay chain*: a series of radionuclides each of which disintegrates into the next, until a stable nuclide is reached.

*Decay constant (decay, law of radioactive)*: the probability per unit time that a nucleus will decay spontaneously. If the number of nuclei is $N$, the rate of decay $dN/dt$ and the decay constant $\lambda$, then the law of radioactive decay is: $dN/dt = -\lambda N$.

*Decay heat*: the heat produced by radioactive decay, especially of the fission products in irradiated fuel elements. This continues to be produced even after the reactor is shut down.

*Decay product*: synonym for *daughter product*.

*Decay, radioactive*: the disintegration of a nucleus through emission of radioactivity. The decrease of activity due to such disintegration.

*Decommissioning*: the permanent retirement from service of a nuclear facility and the subsequent work required to bring it to a safe and stable condition.

*Decommissioning, stage of*: the sequence of stable stages of partial or total decommissioning.

*Decontamination*: the process of removing radioactivity or any other unwanted impurity.

*Decontamination factor*: the ratio of the proportion of contaminent to product before treatment to the proportion after treatment.

*Delayed neutrons*: neutrons resulting from fission but emitted a measurable time after fission has taken place. They play an essential part in nuclear reactor control.

*Depleted uranium*: uranium with less than the natural content (0.71 per cent) of $^{235}$U, e.g. the residue from an isotope enrichment plant or from a nuclear fuel reprocessing plant.

*Derived working limit (DWL)*: a limiting value for the amount of radioactive material which may be present continuously in a given situation without risk that the basic international dose limitations will be exceeded.

*Detector, radiation*: a device for detecting and counting individual radiation pulses or for measuring radiation intensity. The variety of radiations and the kinds of measurements that need to be made require many forms of detector, which include: Geiger-Müller counters, proportional counters, solid and liquid scintillation counters, fission and ionization chambers, and semi-conductor detectors.

*Deuterium*: the hydrogen isotope of mass 2, 'heavy hydrogen'. (See *heavy water*.)

*Deuteron*: the nucleus of a deuterium atom, comprising one neutron and one proton.

*DFR*: the Dounreay fast reactor.

*DFR fuel*: fuel used in DFR, an alloy of enriched uranium and molybdenum.

*Diagrid*: the structure supporting the core, blanket and radial shield rods, which also distributes the coolant flow amongst these items.

*DIDO*: nuclear reactor situated at Harwell.

*Die-filling*: filling of a die (container) with powder prior to compressing into a pellet.

*Diffractometer*: an instrument used in the examination of the atomic structure of matter by the diffraction of X-rays, electrons, or neutrons.

*Diffusion*: in general, the random movements of particles through matter. Specifically used for: (1) diffusion of a gas through a porous membrane, notably in the enrichment of uranium by the diffusion of uranium hexafluoride gas; (2) the movement of fission neutrons through a moderator.

*Diffusion plant*: a plant for the enrichment of uranium in the $^{235}$U isotope by gaseous diffusion of uranium hexafluoride through a porous membrane.

*Direct cycle*: where the turbine is driven by coolant directly received from the reactor, i.e. one primary circuit.

*Disequilibrium*: the converse of the stable condition described under *secular equilibrium* in the particular case where a transient preferential separation of one or more of the members in the isotopic decay chain upsets a previously established equilibrium state which then takes a time, dependent upon the longest half-life members disturbed in the chain, to re-establish.

*Dishing*: a shallow spherical or truncated conical depression in one or both end faces of a UO$_2$ fuel pellet.

*Disintegration*: any transformation of a nucleus, either spontaneous or by interaction with radiation, in which particles or photons are emitted. It is used in particular to mean radioactive decay.

*Disposal*: the removal from man's environment of unwanted or dangerous material, notably nuclear waste, to a place of safety, without the intention of retrieving it later.

*Distillation*: purification of a liquid by boiling it and condensing and collecting the vapours.

*Distribution coefficient*: the ratio of the total concentration of a substance in the organic phase (regardless of its chemical form) to its total analytical concentration in the aqueous phase.

*Divergence*: a nuclear chain reaction is said to be divergent when the rate of production of neutrons exceeds the rate at which they are lost, so that the fission reaction increases in intensity or spreads through a larger volume of material.

*DNA*: deoxyribonucleic acid (see *nucleic acid*).

*Doppler broadening; Doppler coefficient; Doppler constant; Doppler effect*: when the velocity of an atom (derived from its thermal energy) is comparable to that of interacting neutrons, the proportion of neutrons affected by resonances in the neutron cross-sections changes with temperature. This is because the resonances are effectively broadened as the velocity of the atom increases with temperature (Doppler broadening). The effect is important for a few isotopes present in quantity in a typical LMFBR, for example $^{238}U$. As temperature increases, capture in $^{238}U$ increases and reactivity decreases; this is called the Doppler effect and is a very useful safety feature. The Doppler temperature coefficient is $dR/dT$ where $R$ is reactivity and $T$ temperature. Because this coefficient is found to be approximately inversely proportional to the absolute temperature $T$, a Doppler constant $D$ is also defined, where $D = T\,dR/dT$.

*Dose commitment*: future radiation doses inevitably to be received by a person or group, e.g. from a radioactive material already incorporated in the body.

*Dose-equivalent*: the absorbed dose multiplied by a quality factor to measure the biological effectiveness of radiation irrespective of its type in rems or sieverts.

*Dose-equivalent (effective)*: the *dose-equivalent* to the whole body having the same risk of causing biological harm as an exposure of part of the body.

*Dose-radiation*: generally, the quantity of radiation energy absorbed by a body.

*Dose-rate*: the dose absorbed in unit time, e.g. rems per year.

*Dosemeter, Dosimeter, Dose-rate meter*: an instrument which measures radiation doses or dose-rates

*Doubling time*: (1) in a divergent reactor, the time taken for the neutron flux density, and therefore the power, to double; (2) in a breeder reactor, the time taken to produce new fuel equivalent to a full replacement charge in addition to the fuel consumed during this time.

*Down time*: the period during which a reactor is shut down for routine maintenance.

*Dragon*: a high-temperature gas-cooled reactor experiment operated at Winfrith, UK, by an OECD project team from 1964 to 1975.

*Dry cooling tower*: a cooling system which uses the atmosphere as a heat sink by a combination of a jet condenser, closed water circuit, heat exchangers cooled by air, and a cooling tower.

*Dry well*: the region around the reactor vessel which is kept 'dry' during normal operation, but through which steam and water would discharge in the event of a loss of coolant accident.

*Dryout margin*: the factor relating a heat flux employed in a boiling system with the critical heat flux which would cause the heating surface to be blanketed by the vapour phase thereby raising the temperature of this surface.

*D–T*: symbol for reaction between nuclei of deuterium and tritium.

*Ductility*: the maximum dimensional change per unit length of a mechanically loaded material, as measured just before the point of failure, excluding any region of gross deformation.

*EBR*: experimental breeder reactor.

*Eddy diffusion*: the mixing of isotopes due to turbulent motion.

*Effective dose-equivalent*: the dose-equivalent to individual organs multiplied to give the 'effective whole-body dose-equivalent'.

*Effluent*: a waste stream from a chemical process, usually gaseous or liquid.

*Egg boxes*: a constructional feature of fast reactor fuel sub-assemblies.

*Eigenfunction*: the solution of an equation compatible with the boundary conditions associated with possible values of a parameter of the equation (the eigenvalue).

*Elastic scattering*: the outcome of collisions between particles, in which the total kinetic energy of the system is unchanged, but the directions of motion of the particles are altered (i.e. the particles simply bounce off one another with no net energy loss).

*Electromagnetic radiation*: radiation having the nature of electromagnetic waves. In the nuclear context it includes gamma rays and X-rays.

*Electromagnetic separation*: the separation of ions of different masses by deflection in a magnetic field.

*Electron*: one of the stable elementary particles of which all matter consists. It carries a single unit of negative electric charge equal to $1.6 \times 10^{-19}$ coulombs and has a mass of $9 \times 10^{-31}$ kilograms.

*Electronegative residuals*: a term that refers to the particular impurities that might form electronegative ions left in the gas after attempts have been made to remove them. Successful operation of a proportional counter depends upon the gas it contains being as pure as possible especially in regard to certain molecular species (e.g. $Cl_2$, $O_2$, $NH_3$, $H_2O$, etc.) which readily form electronegative ions. Formation of electronegative ions limits, or may even completely prevent, the operation of the counter in its intended mode.

*Electronvolt*: a unit in which energy is measured in the study of nuclear particles and their interactions. It is equal to the change in energy of an electron crossing a potential difference of 1 volt. Abbreviation eV and multiples keV ($10^3$ eV) and MeV ($10^6$ eV).

*Electrophoresis*: motion of colloidal particles under an electric field in a fluid, positive groups to the cathode and negative groups to the anode.

*Element, chemical*: a simple substance which cannot (by normal chemical means) be broken down into simpler components. All its atoms have the same number of protons in their nuclei and therefore occupy the same place in the Periodic Table.

*Eluant*: liquid which is added to an ion-exchange column and passes through it carrying the desired product which it has removed from the column.

*Emanometer*: radon monitor.

*Energy, atomic*: popular, though not strictly accurate, synonym for nuclear energy.

*Energy containment time*: the total energy of a confined plasma divided by the rate of energy loss from it.

*Energy fluence*: the energy intensity integrated over time of a short pulse of radiation.

*Energy loss discrimination*: a technique by which charged particles that may be identical in certain respects, e.g. in mass and momentum-to-charge ratio, are separately distinguished by means of the different rates at which they lose energy in passing through a thin detector (thin meaning the path length within the counter, over which the energy loss is measured, is small compared with the total range of the particle).

*Energy, nuclear*: the energy released when the particles constituting the nuclei of atoms undergo rearrangement, especially through neutron-induced fission in uranium or plutonium.

*Engineered storage*: storage of spent fuel or high-activity wastes in facilities specially constructed to ensure safe keeping until such time as processing or disposal is undertaken.

*Enrichment*: the process of increasing the abundance of fissionable atoms in natural uranium (which contains 0.7 per cent of fissle isotope $^{235}U$). This is usually done in either the *centrifuge* process (where isotopes are separated by centrifugal force) or the *diffusion* process (where a series of screens retards the heavier isotopes).

*Environmental pathway*: the route by which a radionuclide in the environment can reach man, e.g. by progressive biological concentration in foodstuffs.

*EURATOM*: European Atomic Energy Community.

*Event tree*: a diagram which, starting from some initiating event, identifies the possible courses of an accident by a series of branches expressing the respone 'YES' or 'NO' to the question 'does this engineered safety feature work?'

*Eversafe*: a description given to plant whose dimensions are limited so that a critical quantity of plutonium (or highly enriched uranium) cannot be accommodated.

*Excitation*: the addition of energy to a system, e.g. a nucleus, transferring it from its 'ground' state to an 'excited' state.

*Excursion*: a rapid increase of reactor power above the set levels of operation. This increase may be deliberately caused for experimental purposes or it may be accidental.

*Exothermic*: accompanied by the evolution of heat.

*Expansion coefficient*: the fractional expansion (i.e. the expansion of unit length, area, or volume) per degree rise of temperature.

*Exponential assembly*: an experimental sub-critical assembly into which thermal neutrons are introduced at one face. The neutron flux density in the assembly decreases exponentially with distance from this face. Used in studies of reactor physics, etc.

*Extraction (in solvent extraction)*: the transfer of a dissolved substance from an aqueous phase to an organic phase.

*Extraction column*: solvent-extraction apparatus where the contacting is done in a vertical column with aqueous inlet at top and organic inlet at bottom.

*Fall-out*: deposition of radioactive dust etc. from the atmosphere, resulting from the explosion of nuclear weapons or from accidental release.

*Fast neutrons*: neutons travelling with a speed close to that with which they were ejected from the fissioning nucleus, typically about 20 000 km s$^{-1}$.

*Fast reactor*: a nuclear reactor in which most of the fissions in the chain reaction are caused by fast neutrons, travelling with a speed close to which they were ejected from the fissioning nucleus. It contains no moderator, and is capable of generating more fissile material than it consumes.

*Fault conditions*: any condition in the reactor or plant which is a departure from the designed normal operating condition and which could lead directly or indirectly to an automatic shutdown, to damage or to an accident.

*Fault tree*: a diagram representing possible initiating events and sequences of successive failures that could lead to an accident.

*Fecundity*: capacity of a species to undergo multiplication.

*Feed*: a solution introduced into an extraction system.

*Feed train*: the series of components which exist between the power station condenser and the steam generator through which the feedwater must pass. Typically these components will be condensate booster pumps, de-aerators, feed pumps, and feed heaters.

*Feedwater*: the water, previously treated to remove air and impurities, which is supplied to a boiler for evaporation.

*Fertile*: material such as uranium-238 and thorium-232 which can be transformed by neutron absorption into fissile $^{239}$Pu or $^{233}$U, respectively.

*Fine-group library*: a set of nuclear cross-sections tabulated as average values over many energy groups, usually about 2000.

*Fine-structure experiment*: one in which the detailed variation of the neutron reaction rates from the moderating material, through the coolant, the structural materials, and into the fuel is measured.

*FINGAL*: fixation in glass of active liquid.

*Fissile*: a material readily capable of undergoing fission when struck by a neutron, notably $^{235}$U, $^{233}$U, $^{239}$Pu, and $^{241}$Pu.

*Fission-counter*: a detector consisting of a tube lined with fissile material or filled with a fissile gas which detects neutrons by the ionization produced in it by the fission products.

*Fission, fast*: fission induced by fast neutrons.

*Fission gas bubble swelling*: a swelling of $UO_2$ resulting from the accumulation of fission gases into bubbles, mainly on grain boundaries.

*Fission neutrons*: neutrons produced at the time of fission.

*Fission, nuclear*: the splitting of a heavy nucleus usually into two nearly equal fast-moving fragments, accompanied by fast neutrons and gamma rays. Fission may be either spontaneous or induced by the absorption of a particle or a high-energy photon.

*Fission product(s)*: the nuclide(s) formed when a fissile material undergoes nuclear fission.

*Fission spectrum*: (1) the energy distribution of prompt neutrons in fission of a specified nuclide (2) the energy distribution of prompt gamma radiation arising from the fission; (3) the range and abundance of nuclides formed from the fission.

*Fission, spontaneous*: a mode of radioactive decay in which a heavy nucleus undergoes fission without being excited by any external cause. It occurs in $^{238}U$ as a mode of decay having a very long half-life.

*Fission, thermal*: fission induced by thermal neutrons.

*Fission track dating*: a physical method of dating applicable to glassy materials such as obsidian. Such materials frequently contain the isotope $^{238}U$, which in undergoing spontaneous fission, forms two, energetic, heavy nuclei, which lose energy in travelling through the crystal lattice and cause visible damage along their paths (fission tracks). On formation, at high temperatures, the material is fully annealed, meaning all previous crystal lattice damage is reformed and no tracks are present. Thus, from a physical count of the number per unit area in a section of a sample and knowledge of the specific U concentration producing tracks, it is possible to estimate the elapsed time since the material's formation.

*Fission yield*: the number of nuclei of a particular mass resulting from 100 fissions (strictly the chain fission yield).

*Fissium*: collective term to describe all the fission products formed by nuclear fission of a fissile material.

*Flask*: a heavily shielded container used to store or transport radioactive material, especially used nuclear fuel.

*Flame photometry*: as for *atomic absorption* a method of physical analysis, but in this case the light emitted from the samples introduced into a flame is directly examined. The presence of a particular element is demonstrated by the identification of that element's characteristic light output seen as discrete lines when viewed through a suitable spectroscope, and the intensity depending upon the quantity present.

*Flashing-off*: the process whereby steam is formed from hot water by a reduction in the pressure of the system.

*Flocculation*: separation of radioactive waste products from water by coagulation of an insoluble precipitate.

*Floc process*: a process where material is precipitated as a mass of fine particles.

*Flow path redundancy*: a way of providing a series of alternative parallel routes for a water flow, so that if one route becomes blocked or fails, other parallel routes are still available.

*Flowsheet*: a schematic plan giving details of all the steps in the process including quantities of chemicals required.

*Fluidized bed*: if a fluid is passed upward through a bed of solids with a velocity high enough for the particles to separate from one another and become freely supported in the fluid, the bed is said to be fluidized.

*Fluoroscopy*: examination of objects by observing their X-ray shadow shown on a fluorescent screen.

*Flux density*: for a given point (especially in the core of a nuclear reactor) the number of neutrons or other particles incident per second on an imaginary sphere centred at that point, divided by the cross-sectional area (1 cm²) of that sphere. It is identical with the product of the population density of the particles and their average speed

*Forced-draught cooling towers*: a system used for cooling fluids, where the coolant— generally atmospheric air—is drawn or forced through a 'rain' of the fluid to be cooled under the action of motor-driven fans.

*Form factor*: the ratio of the effective value of an alternating quantity to its average value over a half-period.

*Fossil fuel*: fuel derived from fossilized organic matter; includes coal, crude oil, and natural gas.

*Fractional electrolysis*: concentration or separation of isotopes by application of electrolysis.

*Fracture machanics*: a method of analysis which can establish the response of a loaded structure to flaws or cracks postulated to be present at various locations in the material of the structure.

*Free electron–hole pairs*: the particle pairs formed in solid crystalline material when neutral impurity atoms contained in the material are dissociated by radiation or an electric field, and an electron is physically separated from the remaining positively charged atom (the whole).

*Free energy*: the capacity of a system to perform work, a change in free energy being measured by the maximum work obtainable from a given process.

*Freon*: a halogenated hydrocarbon. Used in refrigeration and as aerosol propellants.

*Frequency following*: the adjustment of the load of an electricity generator by attempting to restore the instantaneous frequency of the supply to a nominal value.

*Fretting*: a uniform, sometimes very rapid, removal of metal from one or both of two contacting surfaces between which there is relative movement or periodic impacting; sometimes called fretting corrosion.

*Froth flotation*: a method of separating particles of different densities by the use of frothing agents to vary the density of the liquid in which they are suspended.

*Fuel assembly*: a group of nuclear fuel elements forming a single unit for purposes of charging and discharging a reactor. The term includes bundles, clusters, stringers, etc.

*Fuel channel*: a channel in a reactor core designed to contain one or more fuel assemblies.

*Fuel cycle*: the sequence of steps involved in supplying and using fuel for nuclear power generation. The main steps are mining and milling, extraction, purification, enrichment (where required), fuel fabrication, irradiation (burning) in the reactor, cooling, reprocessing, recycling, and waste management and disposal.

*Fuel cycle equilibrium*: when the isotopic compositions of feed fuel, output level, and wastes become constant in a system where fuel is returned to the reactors.

*Fuel element*: a unit of nuclear fuel which may consist of a single cartridge, or a cluster of thinner cartridges (pins).

*Fuel inventory*: the total amount of nuclear fuel invested in a reactor, a group of reactors, or an entire fuel cycle.

*Fuel pin threaded end plug*: a modified fuel pin in which the end plug which seals the hollow Zircaloy-2 tube is extended somewhat, and the external surface is threaded to allow it to screw into the lower and upper tie plates and thus act as a structural member.

*Fusion, nuclear*: a reaction between two light nuclei resulting in the production of a nucleus heavier than either, usually with release of excess energy.

*Gamma ray*: very short wavelength electromagnetic radiation, emitted during many types of nuclear reaction.

*Gamma-ray spectrometry*: an analytical technique whereby radionuclides are identified and measured by determining the energies and intensities of the gamma rays they emit during radioactive decay.

*Gangue*: the portion of an ore which contains no metal.

*Gas counter*: Geiger counter into which radioactive gases can be introduced.

*Gas counting*: counting of radioactive materials in gaseous form. The natural radioactive gases (radon isotopes) and carbon dioxide ($^{14}C$) are common examples.

*Gaseous diffusion*: name given to the practical separation process based on the principle of molecular diffusion.

*Gaseous wastes*: generic term to denote gaseous fission products (e.g. iodine, krypton, etc.) or gaseous chemical wastes (e.g. steam, oxides of nitrogen, etc.).

*Gas–graphite reactors*: gas-cooled, graphite-moderated reactors such as Magnox, AGR, and high-temperature gas-cooled reactors.

*Gas lift*: technique of lifting liquor from one level to a higher level by entraining liquor in gas bubbles under pressure in a narrow tube.

*Geiger-Müller counter*: a simple and well-established form of radiation detector which produces electrical pulses at a rate related to the intensity of the radiation. Commonly called a 'Geiger counter'.

*Gel precipitation (of fuel)*: a process for converting liquid metal nitrate into solid mixed-oxide spheres using a gelling agent. The plutonium is co-precipitated with uranium from the nitrate to produce the required enrichment or composition for fast reactor fuel. The process involves fairly simple fluid-handling procedures and can produce fuel with a very low dust content.

*Geometric limitation*: a method of criticality control which prevents neutron multiplication by appropriate design of the container of fissile material.

*Germanium*: a metalloid element; it occurs in a few minerals, including coal, and has exceptional properties as a semiconductor.

*Glassification*: see *vitrification*.

*Glove box*: a form of protection often used when working with alpha-emitting radioactive materials. Gloves fixed to ports in the walls of a transparent box allow manipulation of work within the box without the risk of inhalation or contact.

*Granulocytes*: a group of blood cells of the leucocyte division.

*Graphite*: a black crystalline form of carbon used as a moderator and/or reflector of neutrons in many nuclear reactors.

*Gray (Gy)*: the unit of absorbed radiation dose (replacing the rad under the SI system).

*Guide thimbles*: another name for support tubes in PWR fuel elements.


*Half-life*: the characteristic time taken for the activity of a particular radioactive substance to decay to half of its original value—that is, for half the atoms present to disintegrate. Half-lives vary from less than a millionth of a second to thousands of millions of years, depending on the stability of the nuclide concerned.

*Half-life, biological*: the time required for the amount of a particular substance in the body to be halved by biological processes.

*Halides*: fluorides, chlorides, bromides, iodides, and astatides.

*Handed*: arranged as a mirror image of an adjacent section.

*Hanger bar*: the portion of a suspension section (q.v.) of a fuel assembly by which the fuel and scatter plug are hung from the seal plug.

*Hard*: of radiation, having a relatively high penetrating power, i.e. energy.

*HARVEST*: highly active residues vitrification engineering study.

*Head end*: that part of the reprocessing scheme before solvent extraction, i.e. fuel receipt, fuel breakdown, fuel dissolution, liquor clarification, and conditioning.

*Health physics*: the study of persons exposed to radiation from radioactive materials.

*Heat exchanger*: device for transferring heat from one body of fluid to another.

*Heavy water (deuterium oxide, $D_2O$)*: water in which the hydrogen is replaced by 'heavy hydrogen' or deuterium. Because of the very low neutron absorption cross-section of deuterium, heavy water makes an excellent moderator and is used in, e.g. CANDU and SGHWR nuclear reactor. It is present in ordinary water at one part in about 5000.

*Helical multipole coils*: helical coils in which adjacent conductors carry opposing currents to produce a multipole magnetic field.

*Heterogeneous core*: one in which fertile or blanket sub-assemblies are loaded within the boundaries of the highly enriched core zone.

*Hex*: a colloquialism for uranium hexafluoride ($UF_6$), the gas used in isotope enrichment plants.

*High-level*: of radioactive wastes, those that require continuous cooling to dissipate the heat of radioactive decay.

*Homogeneous diffusion reactor model*: a mathematical description of a reactor in which the reactor is represented as a homogeneous medium of average material composition and the diffusion approximation of neutron transport is assumed to apply.

*Honeycomb grid*: description of a constructional feature of fast reactor fuel sub-assemblies

*Hopping mechanism*: molecules adsorbed on a surface can move by hopping from one adsorption site to another.

*Hot*: jargon for highly radioactive.

*Hot cell*: see *cave*.

*Hot spot*: (1) the point of highest temperature in a reactor fuel or its cladding; (2) a restricted area of comparatively intense radiation or radioactive contamination.

*Hulls*: small lengths of fuel pin cladding left after dissolution of the fuel.

*Hydrocarbon*: a compound of hydrogen and carbon.

*IAEA*: International Atomic Energy Agency

*Immiscible*: that cannot be mixed.

*Immunology*: the science dealing with the various phenomena of immunity, induced sensitivity, and allergy.

*Imploding linear systems*: fusion devices in which cylindrical plasmas are created by the implosion of material lining the reactor vessel.

*In coincidence*: two nuclear events occurring within a fixed (normally extremely short) space of time are said to be in coincidence. The fixed time is called the resolving time.

*Indirect cycle*: where the turbine is driven by steam produced from the heat of the reactor coolant, i.e. a primary and a secondary circuit.

*Inertial confinement*: short-term plasma confinement arising from inertial resistance to outward forces.

*Individual risk*: the probability of harm to an individual.

*Induced activity*: activity of the radionuclides produced within materials by neutron irradiation.

*Inelastic collision*: a collision in which kinetic energy is not conserved. With neutrons, a collision with a nucleus in which part of the initial energy is released as a gamma ray. A neutron is emitted from the nucleus which is of lower energy than the incident neutron.

*Inelastic scattering*: the outcome of collisions between particles, in which some energy is absorbed or emitted, i.e. they do not simply bounce off one another.

*Inert gas*: helium, neon, krypton, xenon, and radon are the so-called inert gases.

*Infarct*: death of tissue resulting from the arrest or sudden insufficiency of circulation in the artery supplying the part.

*INFCE*: International Nuclear Fuel Cycle Evaluation.

*Infra-red radiation*: invisible electromagnetic radiation with wave-lengths between those of visible light and those of radio waves, i.e. from approximately 0.8 m to 1 mm.

*Integro-differential equation*: a mathematical equation involving terms in which quantities are required to be integrated, and terms in which quantities require differentiation.

*Intermediate-level waste*: all those wastes not included in the categories 'high level' and 'low level'.

*Invariant magnetic moment*: the magnetic moment associated with a gyrating charged particle is an invariant of motion.

*In vitro*: literally 'in glass', term used to describe the experimental reproduction of biological processes in isolation from the living organism.

*In vivo*: term used to describe biological processes occurring within the living organism.

*Ion*: an electrically charged atom or group of atoms.

*Ion exchange*: interchange between ions of charge. The process is used to remove species from aqueous solution by exchange with other species held on an insoluble solid compound.

*Ion pair*: a positively charged ion together with the electron removed from it by ionizing radiation.

*Ionization*: the process of creating ions by dislodging or adding orbital electrons.

*Ionization chamber*: a device for measuring the intensity of ionizing radiation. The radiation ionizes the gas in the chamber and the rate at which ions are collected (on oppositely-charged electrodes) is measured as an electric current.

*Ionization continuum*: the energy region above the threshold for ionization of an atom.

*Ionization radiation*: radiation which removes orbital electrons from atoms, thus creating ion pairs. Alpha and beta particles are more densely ionizing than gamma rays or X-rays of equivalent energy. Neutrons do not cause ionization directly.

*Irradiate*: to expose to irradiation, particularly to penetrating forms such as gamma rays and neutrons. Often used to mean the 'burning' of fuel in a nuclear reactor. Also used as a measure of the extent of this burning, expressed in megawatt-days per tonne.

*Irradiation swelling*: changes in the density and volume of materials due to neutron irradiation.

*Isomers (nuclear)*: nuclides having the same number of neutrons and of protons, but having different internal energy levels.

*Isomorphism*: the name given to the phenomenon whereby two or more minerals crystallize in the same class of the same system of symmetry and develop very similar forms.

*Isothermal temperature coefficient*: the rate of change in reactivity with temperature from one uniform temperature over the whole reactor to another uniform temperature.

*Isotopes*: forms of the same element having different atomic weights.

*Isotopic abundance*: in a specimen of an element, the percentage of atoms having a particular mass number (i.e. of a specified isotope).

*Isotropic*: said of a medium, the physical properties of which, e.g. magnetic suscepti-bility or elastic constants, do not vary with direction.

*Jet nozzle process*: process whereby isotope separation is obtained by the fast flow of uranium hexafluoride in a curved duct.

*Joule*: a unit of energy; 1 kW h = 3.6 million joules.

*K-capture*: a radioactive transformation whereby a nucleus captures one of its orbital electrons. Usually accompanied by emission of electromagnetic radiation.

*Laminar flow*: flow in which adjacent layers do not mix, except at molecular scale.

*Laser compression*: compression of matter induced by impinging laser beams

*Laser fusion*: fusion achieved by spherically symmetrical laser compression (q.v.)

*Lattice*: (1) the regular pattern of fuel arrangement within a reactor core; (2) the arrangement of atoms in the structure of a crystal.

*Lattice constants*: the simple factors, which characterize the neutron physics of the lattice, such as the fast fission factor, the resonance escape probability, the fuel absorption probability, and the number of neutrons produced per absorption in the fuel.

*Leaching*: the dissolution of a substance from a solid containing it, e.g. uranium from ore or fission products from vitrified radioactive waste.

*Lead-time*: the expected time required from the placing of an order for a plant to the commercial operation of the plant.

*Leakage*: the net loss of particles (e.g. neutrons, gamma rays) from a region or across a boundary of a region.

*Leaning post*: a structural item in the fast reactor core against which a group of six adjacent sub-assemblies are located or are sprung to provide radial restraint. Control rods are usually located and moved inside leaning posts. Leaning posts may be used also to contain instruments, experiments, and non-standard sub-assemblies.

*Levitron*: in fusion research, a toroidal magnetic trap formed by levitating a current-carrying ring in the plasma chamber.

*Licensee*: the holder of a licence issued by the regulatory body to perform specific activities related to the siting, construction, commissioning, operation, and decommissioning of a nuclear plant.

*Light water*: ordinary water, as distinct from heavy water.

*Light water reactor (LWR)*: a reactor using ordinary water as both moderator and coolant. The term embraces boiling-water reactors and pressurized-water reactors.

*Limiter*: an aperture defining the boundary of a plasma.

*Liquid scintillation counting*: a method for measuring the rate of decay of a radioactive isotope, frequently a β-emitter, in which the material containing the isotope is converted to a liquid, which is then mixed with another liquid called a scintillant. The scintillant contains molecules of a solute (fluors), which fluoresce when excited by a transfer of energy, from, for example, an emitted β-particle. The resultant emission of light quanta (flashes of light), the number of which is proportional to the total energy of the initiating β-particle, are registered as a simple pulse by a closely placed photomultiplier tube (or tubes). The number of pulses counted in unit time (the counting rate) is thus related to the disintegration rate (activity) of the sample being measured.

*LMFBR*: liquid-metal-(cooled) fast breeder reactor.

*load factor*: the ratio of the average load during a year (or any other selected period) to the maximum load occurring in the same period. It is also used for the ratio of total output of a generating unit in a period to its designed or reference capacity.

*Load throw-off*: the rapid rejection of the load on an electric generator.

*LOCA*: loss of coolant accident. The conditions which might arise when the coolant level in the primary vessel, or in the secondary circuit, falls.

*Loop-type reactor*: one in which the primary coolant is piped outside the main vessel to external heat exchangers (c.f. *pool-type* reactors).

*Loss-cone*: in a *magnetic trap* (q.v.) the region of velocity space occupied by escaping particles.

*Low-level waste*: generally, those wastes which because of their low radioactive content do not require shielding during normal handling and transport. In the UK, it is usually interpreted as those solid, liquid or gaseous wastes that can be disposed of safely by dispersal into the environment.

*LWR (light water reactor) box*: the term used to describe the square-cross-section sheath which encloses the fuel pins comprising the fuel element of an LWR.

*LWR fuel*: fuel used in LWRs, usually slightly enriched $UO_2$.

*Magnetic confinement*: in fusion research, the use of shaped magnetic fields to confine a plasma.

*Magnetic field*: effect produced in the region around a conductor carrying an electric current. It exerts a force on any moving electric charge, causing charged particles to travel in helical paths about magnetic field lines.

*Magnetic field line*: often called 'line of force', an imaginary line showing the direction of the magnetic field. The density of these lines is often used to denote field strength.

*Magnetic mirror*: when a particle gyrating round a line of force moves from weaker to a much stronger field, it can be reflected back. This arrangement is called a magnetic mirror.

*Mangetohydrodynamics (MHD)*: the study of the motion of an electrically conducting fluid in the presence of a magnetic field.

*Magnox*: (1) an alloy of magnesium containing small amounts of aluminium and beryllium developed for cladding natural uranium metal fuel used in the Calder Hall reactors and the power stations of the first British nuclear power programme (the Magnox reactors): it absorbs few neutrons and does not react with the carbon dioxide gas used as reactor coolant (hence the name—'*magnesium, no oxidation*'); (2) the generic name given to the type of gas-cooled graphite-moderated reactor using Magnox-clad fuel, on which Britain's first nuclear power programme was based.

*Mandrel*: a rod inserted into a pellet die prior to filling with powder to form a central hole in the pressed pellet.

*Mass balance area (MBA)*: section of plant or process area that can be isolated in order to determine the quantity of fissile material present.

*Mass defect*: the difference between the mass of a nucleus and the sum of the masses of its constituent nucleons.

*Mass–energy equivalence*: confirmed deduction from relativity theory, such that $E = mc^2$, where $E$ = energy, $m$ = mass, and $c$ = velocity of light.

*Mass limitation*: technique for criticality control where the total mass of fissile material present is limited.

*Mass number*: the total number of neutrons and protons in an atomic nucleus.

*Mass spectrometer*: an analytical instrument in which accelerated positive ions of a material are separated electromagnetically according to their charge-to-mass ratios. Different species can be identified and accurate measurements made of their relative concentrations.

*Mathematical modelling*: the representation of a real physical process by a series of mathematical equations, whose solution thus aids the understanding of the process.

*Maximum permissible level/body burden/concentration*: the maximum permitted value for such quantities as radiation dose-rate, quantity or concentration of a radionuclide, as determined by health physics considerations. Usually based on the recommendations of the International Commission on Radiological Protection.

*Meson*: one of a series of unstable particles with masses intermediate between those of electrons and nucleons, and with positive, negative or zero charge.

*Metastasis*: a secondary tumour.

*MeV*: million electronvolts—a measure of energy.

*Micrometre*: one millionth of a metre.

*Milling capacity*: the quantity of uranium ore which can be handled by the fabricating plant. The throughput of uranium ore in a uranium processing mill will be restricted by the milling capacity.

*Missile shield*: a steel or concrete structure placed over the upper head of the reactor vessel to protect the containment building from missiles such as ejected control rods.

*Mixed mean temperature, mixed outlet temperature*: the most useful mean value of the temperature which, when multiplied by the mass flow of fluid and its specific heat, gives the transport of heat along a passage.

*Mixed oxide fuel*: fast reactor fuel consisting of intimately mixed dioxides of plutonium and uranium (which may be depleted). In a fast reactor the plutonium undergoes fission, while the uranium acts as a fertile material for breeding.

*Mixer–breeder*: the upper axial breeder zone of the fast reactor, which forms part of the core sub-assemblies and is a separate assembly of large-diameter wire-wrapped pins.

*Mixer–settler*: a solvent-extraction plant unit comprising two inter-connected tanks, in one of which two immiscible fluids of different densities are stirred together, then allowed to separate out in the other. The plant will comprise several such units arranged in a cascade with the two liquids flowing in opposite directions, one carrying the product and the other the impurities.

*Moderating ratio*: a figure of merit which accounts for both a moderator's ability to slow down neutrons and its propensity to capture them by absorption.

*Moderator*: the material in a reactor used to reduce the energy, and hence the speed, of fast neutrons to thermal levels, so far as possible without capturing them. They are then much more likely to cause fission in $^{235}$U nuclei. The most important moderators are graphite, water, and heavy water.

*Moderator coefficient*: the rate at which the reactivity of the system increases with temperature.

*Mole*: the amount of substance that contains as many entities (atoms, molecules, ions, electrons, etc.) as there are atoms in 12 g of $^{12}$C. It replaces in SI the older terms gram-atom, gram-molecule, etc. and for any chemical compound will correspond to a mass equal to the relative molecular mass in gram.

*Molecular diffusion*: the process by which molecules pass through very fine pores. The lighter molecules move faster and travel more easily down the holes.

*Monazite sands*: deposit from which thorium is mainly extracted.

*Monte Carlo method*: statistical procedure when mathematical operations are performed on random numbers.

*MOX*: mixed oxide fuel.

*MTR*: materials testing reactor (e.g. DIDO, PLUTO).

*MTR fuel*: enriched uranium–aluminium alloy clad in aluminium metal.

*Multipass steam superheater*: an arrangement for raising the temperature of steam above the boiling temperature associated with its pressure, wherein the steam flows successively through separate paths arranged through the heat source—for example—the core of a nuclear reactor.

*Multiphoton dissociation*: the process in which a molecule is dissociated by the absorption of many photons of the same energy.

*Multiplication factor*: the ratio of the rate of production of neutrons by fission in a nuclear reactor or assembly to the rate of their loss, symbolized by the letter $k$ 1. When the reactor is operating at a steady level, $k = 1$, when divergent, $k > 1$, when shut down, $k < 1$.

*Multi-start helical finning*: a series of separate helical fins formed on the outer surface of a tube.

*Muon*: subatomic particle with rest mass equivalent to 106 MeV. It has unit negative charge, a half-life of about $2\,\mu s$, and decays into electron, neutrino, and antineutrino. It participates only in weak interactions without conservation of parity.

*Myocardium*: the middle layer of the heart, consisting of cardiac muscle.

*Neutrino*: a particle having no mass or charge which is emitted in radioactive beta decay along with an electron. Although of great interest as an elementary particle, it is of little concern in nuclear technology.

*Neutron*: an elementary particle with mass of 1 atomic mass unit (approximately $1.67 \times 10^{-27}$ kg), approximately the same as that of the proton. Together with protons, neutrons form the nuclei of all atoms. Being neutral, a neutron can approach a nucleus without being deflected by the positive electric field, so it can take part in many types of nuclear interaction. In isolation neutrons are radioactive, decaying with a half-life of about 12 minutes by beta emission into a proton.

*Neutron absorber*: substance with the property of absorbing neutrons, e.g. boron, gadolinium, etc.

*Neutron absorption*: a nuclear interaction in which the incident neutron joins up with the target nucleus.

*Neutron economy*: for good neutron economy losses of neutrons must be kept to a minimum. In a reactor, some neutrons are unable to take part in the chain reaction because they are captured in the nuclei of the reactor material or they leak out of the system.

*Neutron-induced voidage*: the presence in a material of voids, which are small aggregates of vacancies stabilized by gas atoms; these give macroscopic increases in the volume of non-fissile components subjected to irradiation by neutrons.

*Neutron poisons*: see *neutron absorber*: absorbers dissolved in solutions of fissile materials or incorporated into plant equipment which holds or processes fissile materials are soluble or fixed poisons, respectively.

*Neutron scatter plug*: an arrangement in a fuel assembly which prevents neutrons streaming out from a reactor core along the coolant pipework by causing such neutrons to be deflected back into the core.

*Neutron spectrum*: the energy distribution of a neutron population.

*Neutron yield*: the average number of fission neutrons emitted per fission.

*NII*: Nuclear Installations Inspectorate.

*Noble gases*: another name for the inert gases, comprising the elements helium, neon, argon, krypton, xenon, and radon-222. Their outer (valence) electron orbits are complete, thus rendering them inert to most chemical reactions.

*Nomogram*: chart or diagram of scaled lines or curves for the facilitation of calculations.

*Non-invasive*: not interfering directly with bodily tissues.

*Nuclear (energy etc.)*: resident in, derived from or relating to atomic nuclei (rather than to entire atoms—see *atomic*).

*Nucleic acids*: the non-protein constituents of nucleoproteins. Nucleic acids play a central role in protein synthesis and in the transmission of hereditary characteristics.

*Octahedral symmetric anharmonicity*: a term in the potential energy of a polyatomic molecule which is octahedrally symmetric.

*Off-gas*: gaseous effluents from a process vessel, a chemical process, or a nuclear reactor.

*OK*: odourless kerosene, a hydrocarbon mixture used as a diluent for the extractant in a solvent-extraction process.

*Oklo*: the uranium mine in Gabon, West Africa, where the first evidence of a natural nuclear reactor was discovered. Thousands of millions of years ago the isotopic abundance of $^{235}U$ was much higher than it is today. Local concentrations at Oklo were high enough, when moderated by incoming water, to form critical assemblies in which fission took place over periods of millions of years.

*Once-through fuel cycle*: where the fuel is used only once, and is not reprocessed for reuse.

*Once-through type boiler*: high-pressure boiler using superheated (dry) steam.

*One-group cross-section*: a single, average cross-section value for the whole range of neutron energies of interest.

*Orbital*: of electrons, revolving around the atomic nucleus at considerable distances (on the atomic scale) from it. The number of possible orbits is limited, and when an electron changes orbit, energy is given off (as light or X-rays), or absorbed.

*Osteosarcoma*: a malignant tumour derived from osteoblasts, composed of bone and sarcoma cells.

*Oxide fuel*: nuclear fuel manufactured from the oxide of the fissile material. Oxides will withstand much higher temperatures and are much less chemically reactive than metals.

*Parameter*: one of the measurable characteristics or limits of a given design, system or operation. For example, in a nuclear reactor the pressure vessel dimensions, coolant temperature limits, neutron doubling time, radiation levels, power output, etc.

*Parent*: the immediate precursor of a *daughter product* in radioactive decay processes.

*Particles, elementary*: particles which are held to be simple; in nuclear energy these are neutrons, protons, electrons, positrons, photons. In the study of high-energy physics there are a great many other 'elementary' particles such as mesons, pions, quarks, neutrinos, etc.

*Partition coefficient*: the ratio of the total concentration of a substance in the organic phase to its total concentration in the aqueous phase.

*Passage grave*: one of the main categories of megalithic or chamber tomb.

*Pebble bed (core)*: a loose bulk of pebble-shaped fuel compacts made from graphite and uranium dioxide powder.

*Perfusion*: use of radioisotopes to measure the blood flow per unit volume of the organ under investigation.

*Periodic Table, chart*: a chart of the chemical elements laid out in order of their atomic number so as to bring out the relationship between atomic structure and chemical etc. properties.

*PFR*: prototype fast reactor, at Dounreay, Scotland.

*PFR fuel*: a mixture of uranium and plutonium dioxide used as fuel in the PFR.

*Photochemical*: the chemical effects of radiation, chiefly that due to visible and ultraviolet light.

*Photodissociation*: dissociation produced by the absorption of radiant energy.

*Photomultiplier*: a device in which electrons striking its first stage, owing to absorbtion of a photon of light, are multiplied in number so as to produce an easily detectable electrical pulse at the last stage.

*Phytoplankton*: planktonic plants.

*Pickling*: a chemical process for cleaning metallic surfaces by the dissolving of a thin layer of metal.

*Pile*: the former name for a nuclear reactor, particularly of the graphite-moderated type.

*Pilot plant*: small-scale plant used by chemical engineers to study behaviour of larger plants which they are designing.

*Pin, fuel*: a very slender fuel can, as used, for instance, in fast reactors.

*Pinch effect*: the constriction of an electric discharge due to the action of its own magnetic field.

*Pitchblende*: a particularly rich ore of uranium, historically important in the discoveries of radioactivity by Becquerel and the Curies. Radium was first discovered in this mineral.

*Planchet*: plain disc of metal.

*Plankton*: small plants and animals living in the surface waters of the sea.

*Plasma*: an electrically neutral gas of free ions and electrons (i.e. electrons that have been stripped from the original atoms).

*Plasma temperature*: temperature expressed in degrees K (the thermodynamic-temperature) or in electronvolts (the kinetic temperature); 1 keV = 10 000 K.

*Plate crevice (tube to tube)*: the very narrow gap which is formed when a tube is located in the drilled hole in the thick tube plate.

*Plateout*: a general term used to encompass all processes by which material is removed from suspension to form a coating (or plating) on exposed surfaces.

*Plenum (fuel pin)*: a space provided at one or both ends of a fuel pin for the fission gases released by the fuel during operation.

*PLUTO*: materials testing reactor at Harwell.

*Plutonium (Pu)*: the important fissile isotope $^{239}$Pu produced by neutron capture in all reactors containing $^{238}$U. Higher isotopes, plutonium 240, 241, 242, and 243, are produced in lesser quantities by further captures.

*Poison*: see *neutron poison*.

*Poison, burnable*: a neutron absorber deliberately introduced into a reactor system to reduce initial reactivity, but becoming progressively less effective as burn-up progresses. It thus helps to counteract the fall in reactivity as the fuel is burned up.

*Poisson distribution*: statistical distribution characterized by a small probability of a specific event occurring during observations over a continuous interval (e.g. of time or distance).

*Poloidal field*: the magnetic field generated by an electric current flowing in a ring.

*Polonium*: radioactive element, symbol Po. Important as an alpha-ray source relatively . free from gamma emission.

*Polyatomic*: a molecule containing many atoms.

*Polymerization*: the combination of several molecules to form a more complex molecule having the same empirical formula as the simpler ones. It is often a reversible process.

*Pool-type reactor*: one in which the primary coolant circuits (i.e. including the intermediate heat exchangers and primary pumps) are within the primary vessel (cf. loop-type reactor).

*Positron*: the anti-particle of the electron, having a positive charge instead of the more usual negative charge. It is the only anti-particle of significance in the context of nuclear power.

*Positron annihilation*: a positron, or positive electron, can annihilate with a negative electron to produce electromagnetic radiation consisting of two gamma rays of 0.511 MeV energy emitted (if the positron and electron are at rest) in exactly opposite directions (180 ° to each other). If the electron is in a solid, the distribution of angle between the gamma rays provides information on the distribution of electron velocities in the solid, and hence about the physical structure of the material.

*Posting*: the name given to the transfer of radioactive or dangerous materials such as plutonium from store to sealed glove box or plant or from plant to plant etc. in such a manner that the materials are never exposed to personnel.

*Power density*: rate of production of energy per unit volume (of a reactor core).

*Power ramping*: an increase in reactor power from a pre-existing level to a higher level. The term is generally applied to a fairly rapid increase after a prolonged period at the lower level.

*Precursor*: in a radioactive decay chain any nuclide which has preceded another.

*Pre-equlibration conditioning*: preliminary treatment of an aqueous phase in order to convert the material to be solvent extracted into the most suitable chemical form.

*Pressure-suppression containment*: a form of reactor primary containment employing *pressure-suppression ponds* to reduce the pressure inside the containment, following an accident in which there was an escape of a condensable reactor coolant.

*Pressure-suppression ponds*: a large volume of water used to reduce the pressure in, say, the primary containment of a reactor cooled by boiling or pressurized water after a release of coolant, by condensing the steam fraction and the vapour flashed-off from the escaping liquid.

*Pressure-tube reactor*: a class of reactor in which the fuel elements are contained in a large number of separate tubes, through which the coolant water flows, rather than in a single pressure vessel. Examples include Britain's SGHWR, and Canada's CANDU.

*Pressure-vessel*: a reactor containment vessel, usually made from thick steel or prestressed concrete, capable of withstanding high internal pressures. It is used in gas-cooled reactors and light water reactors.

*Pressurized-water reactor (PWR)*: a light water reactor in which ordinary water is used as moderator and coolant. The water is prevented from boiling by being kept under pressure and is circulated through a boiler in which steam is raised in a separate circuit for the turbo-alternators.

*Primary circuit*: the coolant circuit which removes heat directly from the core.

*Primary separation plant*: that part of a nuclear fuel reprocessing plant where the bulk of the fission product decontamination occurs, and plutonium and uranium are separated from each other.

*Prismatic core*: vertical columns of graphite prisms.

*Proliferation*: in the nuclear policy context, an increase in the size of nuclear weapons arsenals worldwide; this may involve the escalation of nuclear weapons capability to countries which currently are not nuclear weapons states.

*Prompt*: of neutrons or gamma rays, emitted immediately upon fission, or other interaction.

*Prompt critical*: the state of achieving criticality in a reactor by means of the prompt neutrons alone and therefore without the control effected through the delayed neutrons.

*Proportional counter*: a detector for ionizing radiation which uses the proportional region in a discharge tube characteristic, where the gas amplification in the tube exceeds unity but the output pulse remains proportional to initial ionization.

*Protium*: lightest isotope of hydrogen of mass unity ($^1$H) most prevalent naturally. The other isotopes are deuterium ($^2$H) and tritium ($^3$H).

*Proton*: the nucleus of hydrogen atoms of mass number 1 and a part of all other nuclides. The number of protons in a nucleus of any element is the atomic number, $Z$, of that element.

*Pulsed inertial devices*: fusion systems relying on *inertial confinement* (q.v.)

*Purex*: generic name for solvent-extraction processes using TBP as the extractant.

*Pyrite*: sulphide of iron crystallizing in the cubic system. It is also known as iron pyrite.

*Pyrometallurgical process*: process using heat to refine or purify metals.

*Quality assurance (QA)*: a systematic plan of inspection necessary to provide adequate confidence that a product will perform satisfactorily in service.

*Quality control*: a statistically based procedure of operational checks and tests for the production of a uniform product within specified limits in accordance with design requirements.

*Quality factor*: of radiation, a factor used to express the biological effectiveness of different kinds of radiation.

*Quartz fibre electrometer*: an instrument which measures radiation exposures via the force between two charged quartz fibres; this charge is reduced as ionization occurs. It can be conveniently sized for monitoring personal doses, with an immediate visual display.

*Rad*: a unit of absorbed radiation dose, equivalent to $10^{-2}$ J/kg. The unit is being replaced by the SI unit, the gray (Gy), equal to 100 rads.

*Radiation*: electromagnetic waves, especially (in the context of nuclear energy) X-rays and gamma rays, or streams of fast-moving particles (electrons, alpha particles, neutrons, protons), i.e. all the ways in which an atom gives off energy.

*Radiation area*: an area to which access is controlled because of a local radiation hazard.

*Radiation damage*: undesired effects in a material arising from disturbance of the atomic lattice 'or from ionization caused by radiation. It is often deliberately incurred in the course of experimental work, especially on reactor materials.

*Radiation dose*: the quantity of radiation received by a substance.

*Radiation risk*: the risk to health from exposure to radiation.

*Radiation source*: a device which emits radiation, such as a quantity of radioactive material encapsulated as a sealed source, or a machine, e.g. for generating X-rays for medical purposes.

*Radioactive source*: any quantity of radioactive material intended for use as a source of radiation.

*Radioactive waste*: all materials arising from reactor operations which are deemed to be of no further value, but which, due to induced activity or contamination, or a combination of both, have a radionuclide content which exceeds a prescribed level.

*Radioactivity*: the property possessed by some atomic nuclei of disintegrating spontaneously, with loss of energy through emission of a charged particle and/or gamma radiation.

*Radioactivity, induced*: radioactivity that has been induced in an otherwise inactive material, usually by irradiation with neutrons.

*Radioactivity, natural*: the radioactivity of naturally occurring materials (e.g. uranium, thorium, radium, potassium-40).

*Radiobiology*: branch of science involving study of effect of radiation and radioactive materials on living matter.

*Radiocarbon dating*: (see *carbon dating*).

*Radiochemistry*: that part of chemistry which deals with radioactive materials, including the production of radionuclides etc. by processing irradiated or naturally occurring radioactive materials. The use of radioactivity in the investigation of chemical problems.

*Radiography*: a method of visually examining the interior of a specimen for defects etc. by passing a beam of penetrating radiation through it so that 'shadows' are cast by the denser or thicker parts. These can be examined on a fluorescent screen or a cathode ray tube, at the time, or recorded on photographic film. Medical diagnostic X-rays and industrial gamma-ray tests are the best known examples. A more recent development uses neutrons.

*Radioiodines*: radioactive isotopes of the element iodine.

*Radioisotope*: short for radioactive isotope.

*Radiolysis*: the chemical decomposition of material by radiation.

*Radionuclide*: radioactive nuclide.

*Radiotherapy*: treatment of disease by the use of ionizing radiation.

*Radiotoxicity*: a measure of the harmfulness of a radioactive substance to the body or to a specified organ following its uptake by a given process.

*Radon*: a zero-valent, radioactive element, the heaviest of the noble gases.

*Raffinate*: the waste stream remaining after the extraction of valuable materials from solution particularly in the reprocessing of fuel.

*Rare-earth elements*: a group of metallic elements possessing closely similar chemical properties. These are extracted from monazite, and separated by repeated fractional crystallization, liquid extraction, or ion exchange.

*Rare-earth fission products*: fission products which are rare earths.

*Raster*: a pattern of scanning lines arranged to provide complete coverage of an area.

*Rating (linear, mass, of fuel)*: the rate at which heat is generated in fuel. Mass rating is expressed in watts per gram of fuel; linear rating of a fuel pin is usually expressed in watts per unit length of pin.

*Reactant*: a substance taking part in a chemical reaction.

*Reactivity*: a measure of the ability of an assembly of fuel to maintain a neutron chain reaction. It is equal to the proportional change in neutron population between one generation and the next. In terms of $k_{eff}$, reactivity $= 1 - (1/k_{eff})$.

*Reactivity worths*: the reactivity change caused by a particular addition or removal of a material or sub-assembly to or from a reactor. The worth of an individual isotope is often expressed as the ratio of its reactivity at the core centre to that of a $^{239}$Pu atom.

*Reactor chemical*: a vessel, or part of a plant, in which a chemical reaction is maintained and controlled, usually as part of a production process.

*Reactor vessel*: the container of a reactor in which the fuel, moderator (if any) coolant and control rods are situated.

*Recuperator*: an arrangement whereby hot fluid leaving a circuit, heats the incoming fluid.

*Recycling*: the reuse of the fissionable material in irradiated nuclear fuel after it has been recovered by reprocessing.

*Reducing agent*: a substance that will remove oxygen from or add hydrogen or electrons to a second substance, itself being oxidized in the process. Reducing agents are often used in the separation of Pu from U.

*Reflector*: an extra layer of moderator or other material outside the reactor core designed to scatter back ('reflect') into the reactor some of the neutrons which would otherwise escape.

*Refractory elements*: metals with a high melting point such as vanadium, niobium, tantalum, molybdenum, titanium, and zirconium.

*Regulatory body*: a national authority or a system of authorities designated by national government, assisted by technical and other advisory bodies, and having the legal authority for issuing of licenses.

*Rem (roentgen-equivalent man)*: the unit of effective radiation dose absorbed by tissue, being the product of the dose in rads and the quality factor. The rem is being replaced by the SI unit, the sievert (Sv), equal to 100 rem.

*Reprocessing*: the processing of nuclear fuel after its use in a reactor to remove fission products etc. and to recover fissile and fertile materials for further use.

*Reserves*: of uranium, resources which can, with reasonable certainty, be recovered at a cost below a specified limit using currently proven technology.

*Residence time*: dwell time in a given section of a process.

*Resin, useful lifetime*: the period during which the resins arranged in a water purification plant of the ion-exchange type continue to be effective in service.

*Resolved resonance region*: region where the peaks in cross-section can be separately recognized.

*Resonance integral*: the integral cross-section value over all the energies within the resonance region, weighted by the reciprocal of the energy.

*Resonance (reaction rate)*: the high, often narrow, peak in a reaction rate curve as a function of energy which is observed when the incident particle excites a specific energy level in a compound nucleus.

*Resources*: of uranium, 'reasonably assured resources' refer to ore known to exist and to be recoverable within a given production cost. 'Estimated additional resources' refer to ore surmised to occur around known deposits or in known uranium-bearing districts.

*Reversed field pinch*: a toroidal magnetic trap in which the toroidal field changes sign at an intermediate minor radius.

*Reverse osmosis*: a process essentially akin to filtration, involving the use of extremely high-quality membranes for the removal of dissolved salts from solution.

*Rig*: an experimental device, especially one designed to enable work to be carried out in, or in close association with, a nuclear reactor or chemical plant.

*Roentgen*: a unit of exposure to radiation based on the capacity to cause ionization. It is equal to $2.58 \times 10^{-4}$ coulombs per kilogram in air. Generally an exposure of 1 roentgen will result in an absorbed dose in tissue of about 1 rad.

*Roll and shock*: motion imparted to a ship as a result of interaction with waves.

*Rose Bengal*: a red dye which has the property of accumulating in the liver.

*Rotameter*: instrument for measuring the flow of liquids or gases.

*Rotating nut and translating screw device*: a mechanical device consisting of a vertical threaded shaft together with a series of freely rotating roller nuts canted to match the lead angle of the threads on the shaft. As the shaft rotates the nuts turn within the threads of the shaft translating vertical motion to it, much as a turning nut would cause a bolt to raise or lower in a slot which prevents the bolt from turning.

*Rotor precession*: this is low-frequency orbiting motion superimposed on the high rotation frequency.

*Safety rod*: One of a set of additional reactor control rods used specifically for emergency shutdown and for keeping the reactor in a safe condition during maintenance etc.

*Salting out*: improving the extraction of a substance (in solvent extraction) by the addition of particular substances to the aqueous phase.

*Saturated fluid*: a liquid at its boiling-point corresponding to the imposed pressure.

*Saturated steam*: steam at the same temperature as the water from which it was formed, as distinct from steam subsequently heated.

*Saturation temperature*: the temperature at which the liquid and vapour phases are in equilibrium (at some given pressure). When the pressure is 1 atmosphere, the saturation temperature is called the boiling-point. Above the saturation temperature, the liquid phase cannot exist stably.

*Scattering*: general term for irregular reflection or dispersal of waves or particles. Particle scattering is termed elastic when no energy is surrendered during the scattering process—otherwise it is inelastic.

*Scintillation counter*: a radiation detector in which the radiations cause individual flashes of light in a solid (or liquid) 'scintillator' material.Their intensity is related to the energy of the radiation. The flashes are amplified and measured electrically and displayed or recorded digitally as individual 'counts'. (see *gamma-ray spectrometry*).

*Scoop*: a pipe which is used for extracting the product or waste from a centrifuge. It can also act as a braking mechanism and to stimulate friction flow in a machine.

*Scram*: the evacuation of a building or area in which it would be dangerous for the operators to remain; in US usage, emergency shutdown of a nuclear reactor or other potentially dangerous plant.

*Scrub*: the process of removing impurities from the separated organic phase containing the main extractable substances by treatment with fresh aqueous phase.

*Sealed-face production lines*: a design of production line for active fabrication in which all equipment is sited behind a single operating face, as distinct from the use of free-standing glove boxes.

*Sealed source*: a radiation source totally enclosed in a protective capsule or other container so that no radioactive material can leak from it.

*Seal plug*: specifically the pressure closure at the end of the channel of a tube reactor through which access may be gained for refuelling.

*Secular equilibrium*: the term given to the particular stable situation which exists in respect of the relative abundances of the isotopic members of a radioactive decay chain, e.g. A decays to B decays to C ... D etc., after a long decay period (long in terms of the decay constant of the longest half-life isotope at the head of the chain). Secular equilibrium is established when the quantity present of any given isotope within the chain is exactly balanced by, on the one hand, the rate of its production caused by the decay of the isotope one up in the chain, and, on the other hand, its own rate of decay to the isotope next down in the chain. At this point, the disintegration rates of all the members of the chain are equal.

*Semi-conductor detector*: detector for nuclear particles which makes use of a semi-conductor; this is a material whose resistivity is between that of insulators and conductors and can be changed by the application of an electric field.

*Separation factor*: this factor measures the difference in mole fraction of the desired isotope after passage through a separating element.

*Separation nozzle process*: in this process isotope separation is achieved by expanding gas through a nozzle.

*Separative work*: a measure, for costing purposes, of the work done in enriching a material such as uranium from the initial concentrations to the final desired enrichment.

*Serum protein*: protein in the blood serum; this is the fluid portion of the blood obtained after removal of the fibrin and blood cells.

*SGHWR*: steam-generating heavy water reactor.

*Shear*: a property of a twisted magnetic field whereby the amount of twist varies with depth; it is used in some plasma confinement experiments to reduce instabilities.

*Shearing*: cutting nuclear fuel into small lengths.

*Shield, biological*: a mass of absorbing material which reduces the level of ionizing radiation, e.g. from a reactor core, to an acceptably low level. It is usually made from high-density concrete, lead, or water.

*Shield, thermal*: in a reactor, a shield of thick metal plates placed inside the biological shield to protect it from damage by overheating.

*Shim*: a term used to describe a device which permits a small adjustment to be made. It may be used in relation to the small adjustment in reactivity control permitted by the use of soluble poison boric acid.

*Shock heating*: a method of heating plasma by a sudden increase in magnetic field.

*Shroud tube*: a vertical tube within the above-core structure, in which a control rod moves and is supported. Shroud tubes align with the control rod guide tubes; so that during operation a control rod may be partly within a shroud tube, partly within a guide tube, and partly within a leaning post.

*Shutdown, emergency*: the rapid shutdown of a reactor to remedy or forestall a dangerous condition.

*Shutdown power*: in a shutdown reactor, the continuing power output due to heat produced by fission product decay.

*Shutdown, reactor*: stopping the chain reaction of fission by inserting all control rods, making the reactor sub-critical (i.e. *k* less than 1).

*Silanes*: a term given to the silicon hydrides.

*Single reheat*: passing steam back to a superheater after it has been partially expanded in a turbine.

*Sintering*: a process of densifying and binding granules or particles. It is based on the increase in the contact area between granules resulting from the enhanced diffusion of molecules across the surfaces of granules when a material (such as mixed fuel oxide) is heated.

*Site licence*: a licence issued by the Nuclear Installations Inspectorate for the operation of a nuclear site. In the UK, under the 1965 Nuclear Installations Act, all nuclear sites (except those operated by government departments or the UKAEA—for which, however, similar conditions apply) must be so licensed.

*Skimmer*: this is used in the jet nozzle process to divide the flow into product and waste streams.

*Slab geometry*: a model of a reactor in which the fuel and moderating materials are represented as adjacent slabs. The slabs may be finite or of infinite extent.

*Slumping (fuel)*: the movement of molten fuel under gravity.

*Smear test*: a method of estimating the loose, i.e. easily removed, radioactive contamination upon a surface. It is made by wiping the surface and monitoring the swab.

*Societal risk*: the probabiity of harm to numbers of people.

*Sodium void coefficient*: the reactivity change resulting from the loss of all coolant, or from the loss of coolant in a specified region of the reactor.

*Sodium voiding*: the removal of sodium from a specific location in the reactor core, or from the whole core, or from a section of the core.

*Soft radiation*: radiation having little penetrating power.

*Sol*: a colloidal solution, i.e. a suspension of solid particles of dimensions in the approximate range $10^{-4}$–$10^{-6}$ mm.

*Solenoid*: a cylindrical coil of wire in which an electric current through the wire sets up a magnetic field along the axis of the coil.

*Sol-gel*: a process in which the aqueous sol (colloidal dispersion) is converted into gel spheres by partial dehydration.

*Solid waste*: radioactively contaminated waste in solid form.

*Solvent*: the organic phase which may consist of a mixture of extractants and/or diluent and/or modifier.

*Solvent degradation*: deterioration of the solvent brought about by high irradiation levels.

*Solvent recycle*: the return of the (organic) solvent to the solvent-extraction process for reuse, usually after purification.

*Somatic effects*: the effects of radiation on the body of the person or animal exposed (as opposed to genetic effects).

*Source term*: the mathematical expression describing the source of neutrons of a particular energy at a point in space.

*Spacer grids*: skeletal structures which hold the fuel pins in the required spatial positions at intervals along the length of a fuel element.

*Spatial modes*: the components of the neutron flux distribution that can occur in a reactor analogous to the fundamental and higher harmonic standing waves in a vibrating string, or on a vibrating surface.

*Spatial transient experiments*: measurements on the variation of properties, such as the neutron population or the reactor temperature with time and position, following some initiating disturbance.

*Sparger (sparge tube)*: a device consisting of a tube with holes in it through which water can be sprayed.

*Specific energy loss*: energy loss per unit path length.

*Specific heat capacity*: the quantity of heat which unit mass of a substance requires to raise its temperature by one degree.

*Spectrometer*: instrument used for measurements of wavelength or energy distribution in a heterogeneous beam of radiation.

*Spent fuel*: nuclear fuel which has reached the end of its useful life in a reactor.

*Spider connecter*: the device which links the control rod drive shaft to each individual absorber rod in the fuel assembly (see Fig. 4.8).

*Spiked seating*: fuel support allowing space for gases to circulate.

*Spike (sub-assembly)*: the lowest part of a sub-assembly; it fits tightly into the coolant supply plenum (diagrid), and so maintains both the geometry of the coolant flow and the positional rigidity of the sub-assembly.

*Spoil*: waste.

*Spot prices, spot market*: although the majority of uranium procurement is carried out under long-term contracts, some small quantities of uranium are bought for immediate delivery on the 'spot market' at prices which tend to be considerably higher.

*Sputter*: the term 'sputtering' is used to denote the process whereby atoms, or clusters of atoms, charged or uncharged, are released from an electrode (generally a metal), held at a negative potential, under the impact of bombarding ions.

*Square coupons*: a term to describe thin square slabs of reactor material. When assembled in appropriate numbers, a desired average reactor composition is obtained.

*Stage*: a concept in solvent extraction where complete equilibrium between phases is attained. In mixer–settlers the concept of a stage is often synonymous with the physical unit of one mixer and one settler.

*Standpipe*: an open vertical pipe connected to a pipeline, to ensure that the pressure head at that point cannot exceed the length of the standpipe.

*Steam driers*: devices used to remove the last traces of water from steam. They usually consist of a series of passages in which the steam is made to change direction abruptly. Water droplets having inertia tend to travel straight on, impinge on the sides of the passage and the water is collected in vertical gullies or drains.

*Steam-end pedestal*: the structure containing the bearing and shaft-sealing glands of a turbine, specifically at the end of the cylinder.

*Steam quality at TSV*: the measure of the temperature and pressure of steam at the turbine stop valve. From a knowledge of the temperature and pressure of the steam, the amount of superheat can be calculated.

*Stellarator*: in fusion research, a toroidal magnetic trap with the magnetic fields generated entirely by conductors placed around the torus.

*Step function*: one which makes an instantaneous change in value from one constant value to another.

*Stochastic, non-stochastic effects (of radiation)*: an effect is said to be stochastic if its probability is a function of the irradiation dose without threshold. If the severity of the effect is a function of the irradiation dose, with or without a threshold, then the effect is non-stochastic. Somatic effects can be stochastic or non-stochastic, while hereditary effects are usually regarded as stochastic.

*Stoichiometry*: exact proportion of elements to make pure chemical compounds.

*Storage*: emplacement in a facility, either engineered or natural, with the intention of taking further action at a later time, and in such a way and location that such action is expected to be feasible. The action may involve retrieval, treatment *in situ* or a declaration that further action is no longer needed, and that storage has thus become disposal.

*Stratosphere*: the earth's atmosphere above the troposphere.

*Stress corrosion*: crack propagation which depends on the combined action of chemical corrosion and tensile stress.

*Stringer*: see *fuel assembly*.

*Stripping foil*: a stripper, which is an essential feature of a tandem accelerator, in the form of a thin foil. Ions are accelerated towards the stripper which is maintained at the potential of the high-voltage terminal. In passing through it, molecular ions are totally disintegrated and atomic ions are stripped of the charge by which they were first accelerated and, in changing sign, are then freshly accelerated away from it. Sign changes can be either from +ve to −ve or, as in the case of carbon ion produced from a caesium *sputter* ion source (q.v.), from −ve to +ve.

*Strip solution*: the aqueous phase used for removing a particular solute from a loaded solvent or extract.

*Sub-assembly, fuel, blanket*: the basic, removable, unit of the fast reactor core. Each fuel sub-assembly has a steel, usually hexagonal, wrapper enclosing a bundle of fuel pins, usually with an inlet filter and a gag to regulate the coolant flow which removes heat from the fuel pin bundle. A blanket sub-assembly similarly contains blanket pins. The sub-assembly represents the least mass of fuel handled by the operator for movements to and from the core.

*Superconductor*: a material which exhibits zero resistance (at low temperature) to electric current flow.

*Supercritical*: exceeding the necessary conditions for the attainment of criticality.

*Superheat*: (1) the condition in which the temperature of a liquid is above the saturation temperature corresponding to the pressure in the liquid, where the degree of supherheat, $T$, is the difference between the superheated liquid temperature and the saturation temperature; (2) the temperature rise given to steam passing through a superheater.

*Superheat channel liner*: a thin metal shield arranged inside the pressure tube (of a reactor) in which the coolant is being superheated, allowing secondary coolant to be introduced so that the pressure tube can be maintained at a temperature associated with the boiling process.

*Surge tank*: an appropriately elevated or pressurized tank connected to, say, the suction or inlet main of a group of feedwater pumps. It is able either to accept surplus condensate or to supply extra condensate when there is demand for a change in the system flow-rate following a change in load before an equilibrium condition is re-established. In such an arrangement, the surge tank is said to be 'riding' on the feed pump suction.

*Suspension section*: an arrangement for 'hanging' a fuel assembly inside a reactor core.

*Swirler*: a device for mixing the coolant to enable fully representative recordings of outlet temperatures and possible fission product release to be obtained. In the case of PFR fuel this duty is performed by the mixer–breeder.

*Swirl vane separators*: devices to separate steam from water which involve imparting a centrifugal motion to the mixture, so that the water is flung out to the walls of the separator to be collected, whilst the steam passes up the centre of the device.

*Synchrotron*: an accelerator in which charged particles follow a circular path in a magnetic field and are accelerated by synchronized electric impulses.

*Tailings*: mine or mill wastes consisting of crushed uranium ore from which the uranium has been extracted chemically.

*Tails*: the depleted uranium produced at an enrichment plant, typically containing only 0.25 per cent of $^{235}U$.

*TBP*: tri-*n*-butyl phosphate—the solvent used as the extractant in the Purex process.

*Teletherapy*: treatment of tumours by radiations from outside the body.

*Theoretical or ideal stage*: a stage where equilibrium is not affected by chemical or physical influences.

*Thermal diffusion*: process in which a temperature gradient in a mixture of fluids tends to establish a concentration gradient. It has been used for isotope separation.

*Thermal neutrons*: neutrons in thermodynamic equilibrium (i.e. moving with the same mean kinetic energy) with their surroundings. At room temperature their mean energy is about 0.025 eV and their speed about 2.2 km/s.

*Thermal reactor*: a nuclear reactor in which the fission chain reaction is propogated mainly by thermal neutrons, and which therefore contains a moderator.

*Thermite-type reaction*: a strongly exothermic reaction between a metal and a metal oxide in which the oxide is reduced to the metal.

*Thermocouple*: a device for measuring temperature using the electrical potential produced by the thermoelectric effect between two different conductors.

*Thermoluminescent dating*: a further physical method of dating associated with the accumulated effect of radiation ($\alpha$-particles, $\beta$-particles and $\gamma$-rays) emitted in the decay of the radioactive constituents in the material concerned (e.g. pottery). In crystalline material (e.g. quartz) a fraction of the energy associated with this radiation is permanently stored in the crystal lattice of the minerals in the material (as electrons trapped at regions of imperfection). Subsequent heating of the material allows this energy to be released in the form of light (thermo-uminescence), the quantity being related to the product of the time since the last heating (firing in the case of pottery) and the quantity of radioactivity present. Thus, from an assessment of the total radioactive content an estimate of the elapsed time can be determined.

*Thermonuclear reaction*: a nuclear fusion reaction brought about by very high temperatures.

*Thermosetting resins*: solid (plastic) compositions in which a chemical reaction takes place while they are being moulded under heat and pressure. The product is resistant to further applications of heat (up to charring point), e.g. phenol formaldehyde, urea formaldehyde resins.

*Theta pinches*: cylindrical plasmas constricted by external currents flowing in the $\theta$-direction to produce solenoid magnetic fields.

*Thorium*: a naturally radioactive metal, the mineral sources of which are widely spread over the earth's surface. The main deposit from which thorium is industrially extracted is monazite sands.

*Thorium cycle*: a nuclear fuel cycle in which fertile thorium-232 is converted to fissile uranium-233.

*THORP*: the Windscale thermal oxide reprocessing plant.

*Threshold energy characteristic*: in the context of a nuclear reaction, the dependence upon incident neutron energy of an event, such as the probability of nuclear fission in some nuclei, which only takes place above a certain energy—the threshold energy of the incident neutron.

*Time-of-flight technique*: a way of measuring the neutron spectrum in which the energy or speed of neutrons is determined by the time taken by the neutrons to traverse a known distance.

*Tokamak*: in fusion research, a toroidal magnetic trap whose poloidal field is generated by the current in the plasma. Tokamak is an acronym of the Russian words meaning toroidal magnetic chamber.

*Toroidal*: having the shape of a torus.

*Toroidal magnetic field*: magnetic field generated by current flowing in a toroidal solenoid.

*Torus*: a tube bent round in a ring with the ends joined up together to give the shape of a motor-tyre tube or an American doughnut.

*Toxic*: of poison.

*Tracer*: a small amount of easily-detectable material fixed to some substance whose movement one wishes to follow. If this material is radioactive it can easily be detected by devices such as $\gamma$-cameras or other nuclear radiation detectors.

*Transient over-power*: an accident in which the reactor power exceeds the normal safe upper limit, but full coolant flow continues throughout.

*Transient, reactor*: a change in power level and/or a temperature which may be accidental or deliberate, causing other reactor or plant parameters to change from their steady-state values.

*Transition, nuclear*: a change in the configuration of a nucleus, usually either a disintegration or a change in internal energy level, accompanied by emission of radiation.

*Tri-normal*: a solution of reagent containing three gram-equivalents per litre.

*Transuranic elements*: the elements of atomic number 93 and higher which have heavier and more complex nuclei than uranium. They can be made by a number of nuclear reactions, including prolonged neutron bombardment of uranium.

*Trip*: rapid automatic shutdown of a reactor caused when one of the operational characteristics of the system deviates beyond a present limit. Often applied to spurious shutdowns caused, for example, by an instrument malfunction.

*Tritium*: the isotope of hydrogen having an atomic mass number of 3.

*Tritium unit*: a proportion of tritium in hydrogen of one part in $10^{18}$.

*Troposphere*: atmosphere up to region where temperature ceases to decrease with height.

*Tube-and-shell units*: a form of heat exchanger construction comprising a tube 'nest' supported within a casing referred to as the 'shell'. One fluid flows through the tubes, while the other passes over their outer surfaces, during which it is contained within the shell.

*Tube-sheet*: the thick cylindrical plate which is drilled to take the very large number, typically 5000, of tubes which go to make up a nuclear steam generator.

*Tunable dye laser*: a dye laser, consisting of an organic dye dissolved in a solvent, which fluoresces over a broad band of frequencies, enabling the output to be tuned.

*Turnings*: swarf produced by machining a piece of metal.

*Two-phase coolant*: a coolant which can exist in significant quantities in both the liquid phase and the gaseous phase.

*Two-start, helical 'rip'*: a helical, longitudinal fin (here, specifically, there would be two such fins at any section) produced in the outer surface of a fuel clad. Used in conjunction with fuel elements having 'plain' cladding to space one element from another along their whole length within a fuel bundle assembly.

*Unsealed source*: radioactive material that is not encapsulated or otherwise sealed, and which forms a source of radiation. For example, radioactive material in use as a *tracer*.

*Uranium*: the heaviest element found in appreciable quantities in the earth's crust. Natural uranium contains 0.0055 per cent $^{234}U$, 0.71 per cent $^{235}U$, (by weight) the remainder being $^{238}U$. Other isotopes of uranium are produced by irradiation in reactors.

*Vacuum chamber*: in fusion experiments, the vessel inside which the plasma is confined, so-called because it is first highly evacuated to remove impurities before the plasma is introduced. In actual confinement, the plasma is compressed by the magnetic field into a much smaller volume than the vacuum chamber, thus isolating it from the walls.

*Valence*: chemical unit of combining power.

*Value function*: a function representing the value of a quantity of separated material which depends on the number of separating elements required to produce it.

*Venturi tube*: a wasp-waisted tube used to measure the flow of fluids.

*Vibro-compaction*: mechanical vibration and compaction.

*Vicor glass tube*: a proprietory glass tubing (available in the USA).

*Vipak*: a vibratory compaction process.

*Vitrification*: the incorporation of radioactive waste oxides into glass (also known as glassification).

*Void coefficient*: the partial derivative of reactivity with respect to a void (i.e. the removal of the material) at a specified location within a reactor. It is equal to the reactivity coefficient of the material removed.

*WAGR*: Windscale advanced gas-cooled reactor.

*Warm twisting operation*: twisting at temperature of typially, 350–450 K.

*Waste disposal*: the consignment of radioactive wastes to areas or facilities from which there is no intent to retrieve them.

*Waste storage*: the consignment of radioactive wastes to areas or facilities from which there is an intent and capability to retrieve the waste for further treatment or disposal.

*Waste transport*: the movement of radioactive waste from one location to another.

*Water reactors*: nuclear reactors in which water (including heavy water) is the moderator and/or the coolant.

*Weldment*: a welded assembly.

*Wet (saturated) steam*: steam containing particles of unevaporated water.

*Whole-body monitors*: an assembly of large scintillation detectors, heavily shielded against background radiation, used to identify and measure the total gamma radiation emitted by the human body.

*WIMS*: the name of a family of computer codes used widely to predict the properties of thermal reactors. It is an acronym of Winfrith Improved Multigroup Scheme.

*Xenon effect*: the rapid but temporary *poisoning* of a reactor by the build-up of xenon-135 from the radioactive decay of the fission product iodine-135. Xenon-135 is a strong absorber of neutrons and until it (and its parent) has largely decayed away reactor start-up can be difficult.

*X-rays*: electromagnetic radiations having wavelengths much shorter (i.e. energy much higher) than those of visible light. X-rays with clearly defined energies are produced by atomic orbital electron transitions. X-rays produced by the interaction of high-energy electrons with matter have a continuous energy spectrum, and it is these that are generally used in medical X-ray machines (see *Bremsstrahlung*).

*Yellow-cake*: concentrated crude uranium oxide, the form in which most uranium is shipped from the mining areas to the fuel manufacturers.

*Zeolites*: one of a number of minerals consisting mainly of hydrous silicates of calcium, sodium, and aluminium, able to act as cation exchangers.

*Zero-power reactors*: reactors not requiring high power or large coolant flow, used to study reactor physics of various designs without the build-up of significant quantities of fission products.

*Zircaloy*: an alloy of zirconium and aluminium used for fuel cladding in water reactors.

*Zooplankton*: planktonic animals.

# INDEX